Texts in Theoretical Computer Science
An EATCS Series

More information about this series at http://www.springer.com/series/3214

Roberto Gorrieri • Cristian Versari

Introduction to Concurrency Theory

Transition Systems and CCS

Springer

Roberto Gorrieri
Dipto. di Informatica: Scienza e Ingegneria
Università di Bologna
Bologna, Italy

Cristian Versari
Lab. d'Inform. Fondamentale de Lille CNRS UMR 8022
Univ. des Sciences et Technol. de Lille
Villeneuve d'Ascq, France

Series Editors
Monika Henzinger
Faculty of Science
Universität Wien
Wien, Austria

Juraj Hromkovič
ETH Zentrum
Department of Computer Science
Swiss Federal Institute of Technology
Zürich, Switzerland

Mogens Nielsen
Dept. of Computer Science
Aarhus Universitet
Aarhus, Denmark

Grzegorz Rozenberg
Leiden Centre of Advanced Computer Science
Leiden University
Leiden, The Netherlands

Arto Salomaa
Turku Centre of Computer Science
Turku, Finland

ISSN 1862-4499
Texts in Theoretical Computer Science. An EATCS Series
ISBN 978-3-319-36638-8 ISBN 978-3-319-21491-7 (eBook)
DOI 10.1007/978-3-319-21491-7

Springer Cham Heidelberg New York Dordrecht London

Softcover reprint of the hardcover 1st edition 2015

Printed on acid-free paper

Springer International Publishing AG Switzerland is part of Springer Science+Business Media (www.springer.com)

To the memory of Nadia Busi

Foreword

In recent decades a large community of scholars has intensively investigated the theoretical bases of concurrency, driven both by scientific curiosity and by the fact that concurrent features appear in many different fields, such as biology, economics, medicine, and the social sciences. Many mathematical models have been developed for formalizing the behavior of concurrent systems, i.e., systems composed of agents that not only compute in isolation, but also by exchanging information with each other. Particularly relevant is therefore the representation of the interactions, or process, that may occur in these concurrent systems, which are hosted in an open environment.

Roberto Gorrieri and Cristian Versari have been leading scientists in the field, and have great experience in teaching the topics covered. The reader of this book is gently lead through the fascinating area of concurrency from very basic results to increasingly complex issues, always with great clarity and analytical rigor. A large number of examples and exercises help in uncovering and understanding the many subtleties of the presentation.

This monograph is therefore an excellent textbook for introducing undergraduate and graduate students, as well as people developing or using concurrent systems, to the theories of concurrency. Some aspects of concurrent systems are not considered here, those typical to more advanced models, like mobility of agents. After a course based on this book, however, a student will have all the knowledge and the techniques to face this intriguing and challenging topic, and many others.

The starting point of the intellectual journey proposed by the authors is the semantic structure, namely labelled transition systems, which provides us with the means and the tools to express processes, to compose them, to prove properties they enjoy, typically equality of syntactically different systems that behave the same. The rest of the book relies on the *Calculus of Communicating Systems* proposed by Milner. Tailored versions of this calculus are used to study various notions of equality between systems, and to investigate in detail the expressive power of the models considered.

Warm thanks are due to the authors of this book, for their successful work in bringing together the fundamentals of concurrency theory in an accurate, uniform and enjoyable volume.

Pierpaolo Degano

Acknowledgements

Roberto Gorrieri writes:

"It is difficult to acknowledge all the persons that in some way have influenced this book. I should mention, first of all, Ugo Montanari, who introduced me to the world of concurrency theory. Then, Pierpaolo Degano who, besides teaching me how to become a researcher, contributed to increase my interest in the field, by introducing me to Robin Milner during the TAPSOFT conference in Pisa in 1987. The work of Robin Milner has been the main source of inspiration for this book. Milner, by means of his seminal book on CCS, published in 1980 [Mil80], as well as the more recent textbooks [Mil89, Mil99], explained the basic ideas so clearly that any new author in the field would have trouble finding a new, attractive twist on the presentation of that theory.

Many other scholars have influenced the work reported in this textbook. First, I should mention Rob van Glabbeek, whose research on comparative semantics [vGl01, vGl93] has influenced the work reported in Chapter 2. Then, Davide Sangiorgi, whose research on the bisimulation proof technique [San12] offered hints in some parts of the book. Then Luca Aceto and co-workers, whose textbook [AILS07] has provided some good examples to work with. Finally, Jos Baeten and co-workers, whose book [BBR10] on variations of process algebra has been a source of inspiration for some technical development in Section 5.4.

I am also indebted to my Ph.D. students: Nadia Busi first, who unexpectedly passed away in 2007, but also Marco Bernardo, Riccardo Focardi, Gianluigi Zavattaro, Mario Bravetti, Alessandro Aldini, Roberto Lucchi, Claudio Guidi and Cristian Versari (of course). Working with them has been a very stimulating experience.

The list of persons who have read part of this book and gave substantial advice includes: Luca Aceto, Alessandro Aldini, Jos Baeten, Pierpaolo Degano, Maurizio Gabbrielli, Rob van Glabbeek, Catuscia Palamidessi, Davide Sangiorgi and the students of the master's course *Modelli e Sistemi Concorrenti* I taught in 2009–2013 at the University of Bologna. Other persons gave advice on some technical aspects: Eugene Lucks, Carla Piazza, Francesco Ranzato. They are all warmly thanked.

I would also like to thank Grzegorz Rozenberg for his encouragement to write this book, and Ronan Nugent for his assistance in the publishing process.

Last, but not least, I have to thank all the women of my family for their support: my mother Irma, my elder sisters Roberta and Daniela, my wife Nicoletta and, finally, my daughters Lucia, Emma and Caterina for not disturbing me too often with their homework."

Cristian Versari writes:

"I would like to thank first of all Roberto Gorrieri for involving me in this work. Then, Nadia Busi who guided me, together with Roberto, during my PhD studies. I would also like to thank Gianluigi Zavattaro for his invaluable support during the writing of the book, and Guillaume Madeleine for his accurate feedback and substantial advice.

My final thanks go to my parents and my beloved Francesca who supported me ceaselessly."

Contents

Chapter 1
Introduction

Abstract This introductory chapter outlines the main motivations for the study of concurrency theory and the differences with respect to the theory of sequential computation. It also reports the structure of the book and how to use it. Finally, some background material is briefly surveyed.

1.1 Motivation

Computer systems, implemented in hardware or software or as a combination of both, are supposed to offer certain well-specified services, so that their users can safely rely on them. However, often a computer system is not equipped with a proof that the specified service or property is guaranteed. In order to do so, one has first to define an abstract semantic model of the system (the *specification*), that can be used to study whether it satisfies the requested property. If so, then one has to use such a specification as the reference model to build the actual executable *implementation*, and possibly prove that the implementation is compliant with the specification. We call this production methodology the *specification-verification-implementation* methodology.

This kind of production methodology is largely used in more traditional and well-established engineering disciplines, such as in construction engineering, where a model of a construction, e.g., a bridge, is always designed, studied and proved correct, before being constructed. By contrast, in computer science and engineering this approach has been used extensively only recently, after some astonishing incidents in the 1990s, such as Intel's Pentium II bug in the floating point division unit in 1994. It is still common practice today to go directly to implementation: too often the specification-verification phases are missing (or are only very sketchy and informal), and correctness of the implementation is checked by testing a posteriori; however, as Dijkstra [Dij69] observed: "testing can be used to show the presence of bugs, but never to show their absence". Therefore, if the formal guarantee of correctness is a necessary requirement of the system, a formal specification must be provided and used as a basis to prove the correctness of the design first, and then, possibly, also of the implementation.

An important reason why the specification-verification-implementation methodology is not so widespread in computer engineering is the current limitations of the theoretical tools that can be used in support. On the one hand, semantic theories for modeling computer systems are often not easy, or are even mathematically difficult, so that an engineer would certainly not spend time on it, unless the payoff is very rewarding. In some cases, indeed, the effort is worthwhile: nowadays there is an increasing number of success stories, mainly related to hardware verification. On the other hand, there are intrinsic mathematical limitations to verification that are rooted in classic undecidable problems of computability theory, such as the *halting problem* (see Section 1.3.5). Therefore, in some cases, we are forced to live with partially unverified systems.

This book aims at offering a simple, introductory theory of concurrent, reactive systems that is mathematically well-defined, rich enough to offer mathematical tools for verification and expressive enough to model nontrivial, sometimes even complex systems. It is based on the semantic model of *labeled transition systems* [Kel76] and on the language CCS, proposed by Robin Milner [Mil80, Mil89, Mil99]. The main verification technique is based on *equivalence-checking*, where an abstract model of a system, described as a CCS process, is compared with a more detailed implementation of it, expressed in the same language. We will see that this technique is useful in some remarkable cases.

Of course, this simple theory does not cover all the possible aspects of the behavior of real-life systems; for instance, we are not dealing with real-time or mobility issues; nonetheless, extensions of this theory to include such additional features are possible, already well-investigated and can be profitably studied in more specific books, such as the second part of [AILS07] for real time or [SW01] for mobility.

The following subsections provide a historical perspective on the problem of the semantics of concurrency, which has led to the ideas that are at the base of the theory presented in this textbook.

1.1.1 Sequentiality, Nondeterminism and Concurrency

Classical programming languages, such as Pascal [JW+91], are sometimes denoted as *sequential*, to express their distinguished feature that any of their programs runs in isolation, without any interference by other programs that can run concurrently. Non-termination is considered a bad feature of such programs, as the goal of a sequential program is to compute a result; moreover, in case of termination, the result is unique because the computation is *deterministic*: at any time instant, the next computation step is uniquely determined. Therefore, the semantics of a sequential program is rather intuitive: it is roughly a (partial) function from the input values (or initial values of the program variables) to the output values (or final values of the program variables), if any; the operational behavior of a sequential program (i.e., *how* it computes) can be safely abstracted to a function (*what* it computes), with no details about the intermediate states of the computation. Hence, functions are the correct *semantic model* for sequential programs and the motto *programs-as-functions* well characterizes sequential programming.

Two sequential programs are *equivalent* if their semantics is the exact same function, independently of their operational behavior. For instance, let **:** = denote the assignment operator and **;** the sequential composition operator; then, the following two program fragments are equivalent

$x := 1$ we call p, and $x := 0\ ;\ x := x+1$ we call q

as they both compute the same function f: whatever initial value is attributed to x, at the end of the computation x holds value 1. A bit more formally, function f is a function that maps an association of the form (x,n), where x is the unique program variable and n is its initial value, to an association $(x,1)$, as 1 is its final value. More generally, a *store s* is a function from program variables to values, and the semantics of a sequential program is a function from stores (specifying the initial values of the program variables) to stores (specifying the final values of the variables). Therefore, formally, function f can be defined as $f(s) = s[1/x]$, meaning that given any initial store s, the final store is s where the association for x is updated to $(x,1)$.

This semantic equivalence is also a *congruence*, i.e., it is preserved by the operator of the language. For instance, since p and q above are equivalent, we have that $p;p$ and $p;q$ are equivalent (and they are both equivalent to p); in general, for any program r, we are sure that both $r;p$ and $r;q$ are equivalent, as well as $p;r$ and $q;r$.

The semantics of sequential programs is defined in a *compositional* way, meaning that for each *syntactic operator* of the language there is a corresponding *semantic operator* over functions. For instance, consider the compound program $r;t$, where the execution of program r is followed by the execution of program t, according to the syntactic operator of sequential composition. Then, if we assume that the semantics of r is function f_r and the semantics of t is function f_t, then the semantics of $r;t$ is obtained by combining f_r and f_t by means of the semantic operator of functional composition: $f_r \circ f_t$. For instance, considering the program fragments p and q above, the semantics of $p;q$ is function $f \circ f = f$, where f is the function $f(s) = s[1/x]$.

By extending a sequential programming language with an operator of parallel composition, which we denote by $- \parallel -$, one has the possibility to define programs composed of sequential threads that can execute concurrently on a shared memory. For instance, the program fragment $p \parallel q$ is now expressible:

$$x := 1\ \parallel\ (x := 0\ ;\ x := x+1)$$

where p and q are the two sequential programs defined above. We may wonder if the semantics of a parallel program is a function from initial values of the program variables (or initial stores) to final values of the program variables (or final stores) also in this enriched setting.

First observation: the result of the parallel program $p \parallel q$ is not unique, as the final value of x can be 1 or 2, depending on the actual execution ordering of the elementary assignments. In particular, 2 is the final value for x when the assignment of p is executed in between the two assignments of q. As we cannot make any assumption on the relative execution speed of p and q, we are to accept any possible intertwined ordering, so that both final values for x are to be considered admissible. Hence, the computation is *nondeterministic*. Nonetheless, it is possible to associate a function to $p \parallel q$, but a more complex function that associates to any possible initial store s (with arbitrary value associated to x) the two possible final stores $s[1/x]$ and $s[2/x]$; in general, such a function goes from initial stores to *sets* of final stores.

Second observation: program equivalence based on the identity of the computed functions is not a *congruence* for parallel composition, i.e., it is not preserved by the operator of parallel composition. As a matter of fact, we have noted that p and q are semantically equivalent; however, $p \parallel p$ and $p \parallel q$ are not equivalent, as they compute different functions: on the one hand, for $p \parallel p$ the final value of x can only be 1, no matter the ordering of the assignments they perform; on the other hand, for $p \parallel q$ the final value of x can be 1 or 2, depending on the actual ordering of execution.

Third observation: a consequence of the observation above is that no *compositional semantics* is definable over functions. As a matter of fact, a compositional semantics for parallel composition is definable only if a semantic operator $- \otimes -$ exists over functions, corresponding to the syntactic operator $- \parallel -$ on programs. Therefore, since the semantics of p and q is the same function f, the semantics of $p \parallel q$ should be $f \otimes f$, and the semantics of $p \parallel p$ should be $f \otimes f$ as well. However, we have already noted that $p \parallel p$ and $p \parallel q$ compute different functions and so $f \otimes f$ is not definable: the correct semantics for $p \parallel p$ and $p \parallel q$ cannot be computed when abstracting from the intermediate states of the computation, as the function-based semantics does by associating to p and q the same function f. Summing up, a compositional semantics for the parallel operator cannot be defined over functions.

Fourth observation: often a concurrent program is not meant to compute a result, but rather to offer a service, possibly forever; for instance, an operating system is a concurrent program that is assumed not to terminate. Therefore, a function is not an appropriate semantic tool for expressing the behavior of concurrent programs, as all the non-terminating programs would be equated, independently of the different services they offer.

In conclusion, the semantics of concurrent programs cannot be defined satisfactorily in terms of functions from initial values of the program variables to final values of the program variables.

1.1.2 Interaction, Communication and Process Algebra

If the semantics of a concurrent program is not a function, what is it? From the example above, we have understood that a concurrent program offers a much richer behavior:

- *Interaction* among different entities is possible and we should be concerned about *when and how* the program *interacts* with its *environment*, i.e., the outside world. For instance, program p and q above are not equivalent in the way they interact with the memory: contrary to p, program q interacts twice with the memory and the memory intermediate state (in between the two interactions) offers to the environment (i.e., to other programs interacting with the memory) the possibility to interact with x holding value 0; such an interaction capability of the memory is not possible when executing program p.
- *Nondeterminism* is often an inevitable effect of different relative speed of execution of independent threads. For instance, program $p \parallel q$ above is nondeterministic.
- *Non-termination* is often a desirable property of a concurrent system, as its duty is to offer a service that, in principle, should be available forever.

Therefore, in general, a concurrent system (or *reactive system*) is to be seen as a system that may react to stimuli from its environment and, in turn, influence its environment by providing feedback. A simple example of a reactive system is any coffee vending machine, where the environment is a customer interacting with it by inserting coins, selecting the kind of coffee, and, finally, receiving a cup of coffee from the machine. Other more complex examples of reactive systems include operating systems, communication protocols, software embedded in mobile phones, control systems for transportation (such as flight or railway control systems), and so on.

The key idea of interaction is crucially based on the assumption that all the involved entities should be considered *active*, i.e., they can compute autonomously and interact by *message-passing* (i.e., by synchronizing or communicating values) with the other entities. According to this intuition, data structures, as well as the memory variables, are not *passive* entities used by programs, rather they are to be considered as autonomous, active interacting entities, willing to communicate the value they store to any program requiring it, as well as willing to rewrite the value they store according to a write request by some interacting program.

Hence, the basic building blocks for the model of a reactive system should be the atomic, indivisible activities, i.e., the *actions*, the reactive system performs either in isolation or by interacting with the environment. The execution of an action determines a change in the current *state* of the system. Therefore, the suitable semantic model for a reactive system is a sort of state-transition automaton, called *labeled transition system*. (See Section 2.1 for an introductory example of how a coffee vending machines can be modeled by means of labeled transition systems.)

A labeled transition system modeling a real system can be enormously large, so that a manageable description of it is often mandatory. A *specification language* can be a good solution to this problem. Such a language:

- should be simple enough to be equipped with a clear, well-defined semantics in terms of labeled transition systems, so that it can provide linguistic support for describing succinctly such models, possibly in a compositional manner;
- should be expressive enough to be able to represent a large class of labeled transition systems, including at least all those with finitely many states and transitions;
- should be executable, so that the specification can be analyzed before being implemented (early prototyping);
- should provide support for a compositional analysis of the model.

Process algebras — such as CCS [Mil89], ACP [BK84a, BW90], CSP [Hoa85, Ros98], and Lotos [BoBr87, BLV95], just to mention a few — emerged about thirty years ago as good specification languages for reactive systems (see [Bae05] for a historical overview). They are composed of a minimal set of linguistic operators, equipped with simple and intuitive semantics, with the desire to single out a very basic formalism for concurrency. To partially explain the great variety of the different proposals, we should be aware that the basic operators of different process algebras are chosen according to different intuitions on the basic mechanism of computation. For instance, only considering the communication mechanism, we can recognize at least the following different features:

- *synchronous* vs *asynchronous*: the former when the send action and the receive one are performed by the interacting partners at the same time; the latter when the send action is decoupled from the receive one.
- *point-to-point* vs *multi-party*: the former when the involved partners are only two, the latter when several partners interact at the same time.

The possible four combinations may give rise to different process algebras with different expressive powers. For instance, we will see in Chapter 6 that synchronous point-to-point communication of CCS (also called *handshake* communication) and synchronous multi-party communication of CSP are not equally expressive.

The interesting features that a process algebra may possess can be dramatically diverse. They may include, besides many different forms of communication, also the aspects related to:

- forms of sequentialization (e.g., operators of *action prefixing* for CCS and of *sequential composition* for ACP);
- scoping of names (e.g., operators of *restriction* for CCS and of *hiding* for CSP),
- mobility (e.g., in the π-calculus [MPW92] channel names are communicable values, allowing for dynamic reconfiguration of the system),
- priority among actions (e.g., [CLN01, VBG09]),
- security (e.g., [RSG+, FG01, FGM02]),
- real time (e.g., [Yi91, NS94, HR95]),
- performance evaluation (e.g., [Hil96, BG98, H02]),

and so on. In some of the above cases, in order to equip the process algebra under investigation with a satisfactory semantics, the model of labeled transition systems needs to be enriched to include further information, e.g., about time. For an advanced overview on different aspects of process algebra we refer you to the *Handbook of Process Algebra* [BPS01].

The process algebra presented in this textbook is CCS, proposed by Robin Milner [Mil80, Mil89, Mil99]. It has been chosen mainly for its deep simplicity, elegance of its algebraic theory and good expressive power, and also because it has been extended smoothly to include other features, such as *mobility* with the π-calculus [MPW92, Mil99, SW01], *security* with SPA [FG01] and Crypto-SPA [FGM02], *real time* with TCCS [Yi91], and so on; however, such extensions are not discussed here.

1.2 Why This Book?

This section illustrates the main distinctive features of this book with respect to other books on process algebra, such as [Mil89, AILS07, BBR10, San12]. The intended reader of this section is an instructor (a person who already knows a lot about this theory), who may wish to know the pros (and cons) of this book.

The main motivation for this book is the adoption of a different methodological approach: this book first presents and discusses the semantic model, i.e., *labeled transition systems*, together with a variety of sensible behavioral equivalences over them; then it proposes suitable linguistic means to define objects of the semantic model, i.e., it proposes a process algebra able to express all the labeled transition systems of interest. As a matter of fact, a distinctive feature of this textbook is the presence of some *representability theorems*; for instance, one shows that all labeled transition systems with finitely many states and transitions can be represented, up to isomorphism, by processes of a subcalculus of CCS called finite-state CCS.

Another distinctive feature of this book is the discussion about the relative expressive power of subcalculi of CCS (and also of other process algebras in Chapter 5) and their precise relationship w.r.t. the well-known classification of formal languages (regular, context-free and context-dependent languages), as described in the Chomsky hierarchy (Section 1.3.3). Indeed, in many parts of the book emphasis is put on the similarities and differences w.r.t. the classical theory of automata and formal languages (see Sections 1.3.2, 1.3.3 and 1.3.4 for a short overview). Moreover, a student-level, detailed treatment of Turing-completeness within CCS is provided.

Expressiveness is also the key aspect in the study of encodability of additional operators in CCS. For instance, hiding and sequential composition are proved derivable in CCS. The proof of these results can be seen as a correctness proof of a compiler from a source language (typically CCS enriched with some additional operator) to a target language (typically CCS).

Expressiveness limitations of CCS are also investigated: even if CCS is Turing-complete, it cannot solve all the problems one may wish to solve in concurrency theory. We will see that a deterministic, symmetric, fully distributed solution to the well-known dining philosophers problem [Dij71] cannot be provided in CCS, because of its limited synchronization discipline: binary, point-to-point (or handshake) communication. An extension to overcome this inability is presented in Chapter 6, where Multi-CCS is introduced by extending CCS with atomic behavior and multi-party synchronization.

1.2.1 Structure of the Book

Chapter 1 contains a brief introduction to concurrency theory, as well as a description of the structure of the book and some background material.

Chapter 2 introduces the semantic model we use throughout the textbook: labeled transition systems (LTSs for short). They are equipped with a suitable set of different behavioral equivalences, ranging from isomorphism to trace equivalence, the latter being very similar to the classic *language equivalence* over finite automata (see Section 1.3.4).

Chapter 3 presents the *Calculus of Communicating Systems*, CCS for short, in particular its syntax (which slightly differs from [Mil89]) and its operational semantics in terms of LTSs. A large part of this chapter is devoted to studying various subcalculi of CCS, in order to investigate their relative expressive power and algorithmic properties, and to offer a large collection of case studies of increasing complexity, ranging from basic examples of vending machines, to more complex examples of counters, stacks and queues. Of particular interest is *regular* CCS, as it corresponds to finite-state LTSs, as well as *finitary* CCS, whose programs are finitely representable. This latter calculus is the CCS subcalculus that is mainly used throughout the book; it is shown to be Turing-complete (see Section 1.3.4 for a definition of Turing-completeness), even if it cannot represent all the possible LTSs; moreover, all the behavioral equivalences studied in Chapter 2 turn out to be undecidable for finitary CCS, while they are all decidable for regular CCS. Finally, this chapter introduces a richer variant of CCS, called *value-passing* CCS, which explicitly allows for the communication of values; this variant is proved to be as expressive as CCS.

Chapter 4 discusses the algebraic properties of various behavioral equivalences: as CCS is built around a set of operators, such as parallel composition, it is natural to study which properties of such operators hold w.r.t. a given behavioral equivalence. For instance, parallel composition is associative and commutative w.r.t. all the equivalences. This chapter also investigates whether the equivalences of Chapter 2 are actually congruences w.r.t. the CCS operators. It turns out that most of the equivalences are indeed respected by the CCS operators. Finally we discuss the problem of *axiomatizing* such behavioral congruences, i.e., of finding a suitable set of axioms that characterize syntactically the behavioral congruence under investigation.

Chapter 5 shows that some useful operators, proposed in other process algebras, are actually derivable in CCS, so that CCS turns out to be a reasonably expressive language. A large part of this chapter is devoted to studying the ACP sequential composition operator, whose semantics needs a proper extension of the LTS model. An encoding of an extension of CCS, enriched with sequential composition, into CCS is proposed and proved correct.

However, even if Turing-complete, CCS is not able to model some additional useful behavior, such as the atomic execution of sequences of actions as well as multi-party synchronization. To this aim, Chapter 6 introduces an extension to CCS, called Multi-CCS, that is able to model these behavioral aspects. Some classical concurrency control problems, such as the *concurrent readers and writers* [CHP71], can be now solved satisfactorily in Multi-CCS, while they are not solvable in CCS.

1.2.2 How to Use It

Note for the instructor: The book is the result of several years of teaching a master's course in Concurrency Theory at the University of Bologna. The intended audience is composed of advanced undergraduate (or graduate) students.

The core of the book is composed of Chapters 2, 3 and 4, which can be a good basis for a semester course, possibly complemented with a lab with verification tools, such as the Concurrency Workbench [CWB]. Chapter 4 could be taught before the second part of Chapter 3 (about CCS subcalculi and case studies), in case the instructor wishes to fully develop the theory of CCS before discussing applications. Chapter 5 is useful if the instructor thinks it is a good idea to expose the students to the problem of encoding one language into another; indeed, the chapter offers some examples of this sort, from very basic to more advanced. Chapter 6 is useful if the instructor wishes to discuss limitations of the CCS language and a possible extension to overcome some of them. Technically, this chapter is more involved, as it presents more advanced techniques, such as working with a structural congruence; moreover, it introduces the reader to non-interleaving semantics, such as *step* semantics.

1.3 Background

This textbook is intended for an audience of students who have been already exposed to some introductory courses in mathematics and theoretical computer science, in particular, basic courses on discrete mathematics, formal languages and automata theory, as well as computability. In this case, this section can be skipped.

Nonetheless, this book can also be read by those who have very little knowledge of these topics, as we have tried to be as self-contained as possible. As a matter of fact, this section introduces a few notions — some very basic, some more advanced — that are referred to in the text. Of course, their presentation is very succinct, with little explanation and few examples, and in no way is it intended to replace a thorough exposition that can be found in well-known textbooks completely devoted to these topics, such as [HMU01, Sip06, Koz97, FB94].

1.3.1 Sets, Relations and Functions

We assume the reader is familiar with the notion of set. Set $\mathbb{N}$ denotes the set of natural numbers, $\mathbb{N} = \{0, 1, 2, \ldots\}$. Set $\mathbb{R}$ is the set of real numbers. Given a set A, we write $x \in A$ to mean that x is an *element* of A, and $x \notin A$ to mean that x is not an element of A. Given two sets A and B:

- their *union* is $A \cup B = \{x \mid x \in A \text{ or } x \in B\}$;
- their *intersection* is $A \cap B = \{x \mid x \in A \text{ and } x \in B\}$;
- $A - B = \{x \mid x \in A \text{ and } x \notin B\}$ is the set difference of A and B;
- $\overline{A} = \mathscr{U} - A$ is the complement of A w.r.t. the *universe* set $\mathscr{U}$, containing all the elements of interest;
- we write $A \subseteq B$ to mean that A is a *subset* of B, i.e., that each element of A is also an element of B; if this is not the case, we write $A \not\subseteq B$, i.e., there exists at least one element in A which is not an element of B. Moreover, $A \subset B$ if $A \subseteq B$, but $A \neq B$.

Note that for any set A, A is a subset of A and also the *empty set* $\emptyset$, the set with no elements, is a subset of A. The set of all the subsets of A, called the *powerset* of A, is denoted by $\wp(A) = \{B \mid B \subseteq A\}$.

Given two sets A and B, the *Cartesian product* $A \times B$ is the set $\{(x,y) \mid x \in A \text{ and } y \in B\}$. An element (x,y) is called a *pair*. More generally, when the Cartesian product is among many sets, e.g., $A_1 \times A_2 \times \ldots \times A_k$, an element $(x_1, x_2, \ldots, x_k)$ is called a *tuple*.

Given $B = A_1 \times A_2 \times \ldots \times A_k$, a *relation* R is a subset of B, i.e., a set of tuples. A *binary relation* on a set A is a subset of $A \times A$. For instance, relation $S \subseteq \mathbb{N} \times \mathbb{N}$, defined as $S = \{(n,m) \mid \exists k \in \mathbb{N} \text{ such that } n \times k = m\}$, relates n and m if n is a divisor of m. Given a binary relation R, its *inverse* R^{-1} is the relation $\{(y,x) \mid (x,y) \in R\}$. For instance, $S^{-1} = \{(m,n) \mid \exists k \in \mathbb{N} \text{ such that } n \times k = m\}$, i.e., m is a multiple of n. The *identity relation* $\mathscr{I}$ on A is the relation $\{(x,x) \mid x \in A\}$.

Given two binary relations, R and S, on a set A, we define the *relational composition* $R \circ S$ as the set $\{(x,z) \mid \exists y \in A \text{ such that } (x,y) \in R \text{ and } (y,z) \in S\}$.[1]

A binary relation $R \subseteq A \times A$ is *reflexive* if $(x,x) \in R$ for all $x \in A$, i.e., if $\mathscr{I} \subseteq R$. Relation R is *symmetric* if whenever $(x,y) \in R$ then also $(y,x) \in R$, i.e., if $R^{-1} \subseteq R$.

[1] Our notation for the relational composition $R \circ S$ is not standard, as it is more customary to write the two arguments in reverse order: $S \circ R$. However, for the aims of our book, we prefer to adopt this order of arguments: $R \circ S$.

Relation R is *transitive* if whenever $(x,y) \in R$ and $(y,z) \in R$ then also $(x,z) \in R$, i.e., if $R \circ R \subseteq R$.

A relation R that is reflexive, symmetric and transitive is called an *equivalence relation*. Let $[a]_R$ denote the equivalence class of a w.r.t. the equivalence relation R, i.e., $[a]_R = \{b \in A \mid (a,b) \in R\}$; the set of its equivalence classes $\{[a]_R \mid a \in A\}$ determines a *partition* of A, i.e., $A = \bigcup_{a \in A} [a]_R$ and $[a]_R \cap [b]_R = \emptyset$ if $(a,b) \notin R$, while $[a]_R \cap [b]_R = [a]_R$ if $(a,b) \in R$.

A binary relation R on A is *antisymmetric* if for all $x,y \in A$, if $(x,y) \in R$ and $(y,x) \in R$ then $x = y$. R is a *partial order* if R is reflexive, antisymmetric and transitive. For instance, relation $S = \{(n,m) \mid \exists k \in \mathbb{N} \text{ such that } n \times k = m\}$ is a partial order: if $(n,m) \in S$ and $(m,n) \in S$, then there exist k_1 and k_2 such that $n \times k_1 = m$ and $m \times k_2 = n$; thus, $(m \times k_2) \times k_1 = m$, which is possible only if both k_1 and k_2 are 1, and so $n = m$, hence S is antisymmetric; moreover, S is trivially reflexive ($(n,n) \in S$ by choosing $k = 1$) and transitive (if $(n,m) \in S$, i.e., $n \times k_1 = m$, and $(m,p) \in S$, i.e., $m \times k_2 = p$, then $(n,p) \in S$ because $n \times k_1 \times k_2 = m$).

A relation R is a *preorder* if it is reflexive and transitive. For instance, relation T on a set B of persons, defined as $\{(x,y) \mid x \text{ is not taller than } y\}$, is a preorder; note that T is not a partial order: if x is as tall as y, then $(x,y) \in T$ and $(y,x) \in T$, but x and y are two different persons, i.e., $x \neq y$; also T is not an equivalence relation: if y is taller than x, then $(x,y) \in T$ but $(y,x) \notin T$, hence T is not symmetric.

Given a binary relation R on a set A, the *reflexive closure* of R is the relation R' such that: (i) $R \subseteq R'$, (ii) $\mathscr{I} \subseteq R'$, and (iii) R' is the least relation satisfying (i) and (ii) above. This can be formalized by saying that R' is the least relation satisfying the following inference rules:

$$\frac{(x,y) \in R}{(x,y) \in R'} \qquad \frac{x \in A}{(x,x) \in R'}$$

Such relation R' is simply $R \cup \mathscr{I}$. Note that any relation R'' such that $R' \subset R''$ satisfies the two rules, but it is not the least one.

Given a binary relation R on a set A, the *symmetric closure* of R is the relation R' such that: (i) $R \subseteq R'$, (ii) $R'^{-1} \subseteq R'$, and (iii) R' is the least relation satisfying (i) and (ii) above. Such a relation R' is simply $R \cup R^{-1}$.

Given a binary relation R on a set A, the *transitive closure* of R is the relation, denoted by R^+, such that: (i) $R \subseteq R^+$, (ii) $R^+ \circ R^+ \subseteq R^+$, and (iii) R^+ is the least relation satisfying (i) and (ii) above. This can be formalized alternatively by saying that R^+ is the least relation satisfying the following inference rules:

$$\frac{(x,y) \in R}{(x,y) \in R^+} \qquad \frac{(x,y) \in R^+ \quad (y,z) \in R^+}{(x,z) \in R^+}$$

For instance, if $A = \{x,y,z,t\}$ and $R = \{(x,y),(y,z),(z,t),(y,x)\}$, then $R^+ = R \cup \{(x,z),(x,x),(y,t),(y,y),(x,t)\}$. Note that relation $R' = R^+ \cup \{(z,z)\}$ satisfies the two rules, but it is not the least one.

Given a binary relation R on a set A, the *reflexive and transitive closure* of R, denoted by R^*, is $R^+ \cup \mathscr{I}$. This can be equivalently formalized as the least relation satisfying the following rules:

$$\frac{(x,y) \in R}{(x,y) \in R^*} \qquad \frac{x \in A}{(x,x) \in R^*} \qquad \frac{(x,y) \in R^* \quad (y,z) \in R^*}{(x,z) \in R^*}$$

A binary relation $R \subseteq A \times B$ is a *function* if for all $x \in A$ there exists *exactly one* $y \in B$ such that $(x,y) \in R$; in such a case, we use notation $R : A \to B$ where set A is called the *domain* and set B the *codomain* of R. We usually use letters $f, g, \ldots$ (or mnemonic names) to denote functions and we write $f(a) = b$ to express that $(a,b) \in f$; in such a case b is called the *image* of a under f. As an example, $double : \mathbb{N} \to \mathbb{N}$, defined as $double(n) = 2 \times n$, is a function associating to each number n in the domain, the even number $2 \times n$ of the codomain. Relation $R \subseteq A \times B$ is a *partial function* if for all $x \in A$ there exists *at most one* $y \in B$ such that $(x,y) \in R$. In such a case, we use notation $R : A \multimap B$.

As functions are relations, we can define their composition as we did for relations. Given two functions $f : A \to B$ and $g : B \to C$, their *composition* $f \circ g : A \to C$ is the function that associates to each $a \in A$ the value $g(f(a)) \in C$. The definition of functional composition scales also to partial functions in the obvious way.

A function $f : A \to B$ is *injective* (or *one-to-one*) if for all $x, y \in A$, $f(x) = f(y)$ implies $x = y$; equivalently, f is injective if $\forall x, y \in A$, $x \neq y$ implies $f(x) \neq f(y)$. For instance, function *double* above is injective. Function f is *surjective* (or *onto*) if for all $b \in B$ there exists an $a \in A$ such that $f(a) = b$, i.e., each element of B is in relation with at least one element of A. For instance, function *double* above is not surjective, because odd numbers are not the image of any number. Function f is *bijective* if it is both injective and surjective. For a bijective function $f : A \to B$, we can define its *inverse* $f^{-1} : B \to A$ as follows: $f^{-1}(b) = a$ if and only if $f(a) = b$.

A set A is *finite* if there exists $n \in \mathbb{N}$ and a bijective function $f : A \to \{1, 2, \ldots, n\}$. In such a case $|A| = n$ denotes the *cardinality* of A, i.e., the number of elements in A. A set A is *denumerable* if there exists a bijective function $f : \mathbb{N} \to A$. For instance, the set of even numbers $P = \{n \mid n = 2 \times k, k \in \mathbb{N}\}$ is denumerable: the required bijective function is function $double(n) = 2 \times n$. A set A is *countable* if it is either finite or denumerable. A set A is *uncountable* if there is no bijective function $f : \mathbb{N} \to A$. For instance, set $\mathbb{R}$ is uncountable; this can be proved by using Cantor's *diagonalization method* (see, e.g., [Sip06] for a detailed account of this method).

A *multiset* (or *bag*) M over a set A is an unordered, possibly infinite, list of elements of A, where no element of A can occur infinitely many times. It can be represented formally as a function $M : A \to \mathbb{N}$ such that $M(x)$ is the number of instances of element $x \in A$ in M. Given two multisets M_1 and M_2 over the set A, we write $M_1 \subseteq M_2$ if $M_1(x) \leq M_2(x)$ for all $x \in A$. A multiset M over A is *finite* if $M(x) > 0$ for only finitely many $x \in A$. Of course, if A is finite, then any multiset over A is a finite multiset. The set of all finite multisets over a set A is denoted by $\mathscr{M}_{fin}(A)$.

1.3.2 Alphabets, Strings, Languages and Regular Expressions

An *alphabet* A is a finite, nonempty set, e.g., $A = \{a,b\}$. An element a of A is called a *symbol*. A *string* (or *word*, or *trace*) over A is any finite length sequence of symbols of A, e.g., aaa or $abbba$. The *empty string* ε is the string composed of no symbols. We use w,x,y,z, possibly indexed, to represent arbitrary strings.

The *length* of w, denoted as $|w|$, is the number of occurrences of symbols in w; e.g., $|aaa| = 3$, $|\varepsilon| = 0$ and $|abbba| = 5$. The power of a symbol, say a^n, denotes a string composed of n occurrences of a. This can be defined inductively as follows:

$$a^0 = \varepsilon \quad \text{and} \quad a^{n+1} = aa^n.$$

Set A^* is the set of all the strings over alphabet A. For instance, if $A = \{a\}$, then $A^* = \{\varepsilon, a, aa, aaa, aaaa, \ldots\} = \{a^n \mid n \in \mathbb{N}\}$. Given any alphabet A, set A^* is countable, as it is possible to define a bijective function $f : \mathbb{N} \to A^*$. Intuitively, we can list all of its strings by first enumerating all the strings of length 0 (only ε), then those of length 1 (all the symbols of A, which we can assume to be ordered in some alphabetical order), followed by those of length 2 (which are ordered in the lexicographical order induced by the alphabetical order), and so on. We denote by A^+ the set $A^* \setminus \{\varepsilon\}$, i.e., the set of all the nonempty strings over A.

The *concatenation* of strings x and y is the string xy obtained by juxtaposition of the two. This operation is associative and the empty string is its neutral element:

$$(xy)z = x(yz) \qquad x\varepsilon = x = \varepsilon x.$$

Moreover, $|xy| = |x| + |y|$ and $a^n a^m = a^{n+m}$. The power x^n of a string x is the juxtaposition of n copies of x: $x^0 = \varepsilon$ and $x^{n+1} = xx^n$.

If $a \in A$ and $x \in A^*$, we write $\sharp(a,x)$ for the number of occurrences of a in x; for instance, $\sharp(a,abbba) = 2$, $\sharp(b,abbba) = 3$ and $\sharp(c,abbba) = 0$.

A *prefix* of x is any initial substring of x; formally, y is a prefix of x if there exists z such that $yz = x$. A *sufffix* of x is any final substring of x; formally, y is a suffix of x if there exists z such that $zy = x$.

A *language* L is any subset of A^*, $L \subseteq A^*$. For instance, if $A = \{a\}$, any of the following subsets of $A^* = \{a^n \mid n \in \mathbb{N}\}$ is a language: $\{a, aaaa\}$, A, A^*, $\{a^n \mid n = 2 \times k, k \in \mathbb{N}\}$, $\emptyset$, $\{\varepsilon\}$. Note that $\emptyset$ is the empty language, i.e., no string belongs to $\emptyset$, while $\{\varepsilon\}$ is a one-string language.

A language L is *prefix closed* if whenever $xy \in L$, then $x \in L$, for all $x, y \in A^*$. Hence, if L is prefix-closed, then $\varepsilon \in L$ or $L = \emptyset$.

Languages can be concatenated: $L_1 \cdot L_2 = \{xy \mid x \in L_1 \text{ and } y \in L_2\}$. For instance, if $L_1 = \{a, aaa\}$ and $L_2 = \{b, bb\}$, then $L_1 \cdot L_2 = \{ab, abb, aaab, aaabb\}$; moreover, if $L_3 = \{a^n \mid n = 2 \times k, k \in \mathbb{N}\}$, then $L_1 \cdot L_3 = \{a^n \mid n = 2 \times k + 1, k \in \mathbb{N}\}$.

We can define the power L^n of a language L as follows: $L^0 = \{\varepsilon\}$ and $L^{n+1} = L \cdot L^n$. The *iterate* (or Kleene star) L^* and the *positive iterate* L^+ of a language L are defined as follows:

$$L^* = \bigcup_{n \geq 0} L^n \qquad L^+ = \bigcup_{n \geq 1} L^n$$

Note that since A is a language, the set A^* of all strings over A is such that $A^* = \bigcup_{n\geq 0} A^n$, as required by the definition above. Note also that $\emptyset^* = \{\varepsilon\}$ because $\emptyset^0 = \{\varepsilon\}$.

Regular expressions over an alphabet A, ranged over by e (possibly indexed), are defined by means of the following syntax in Backus-Naur Form (BNF):

$$e ::= \mathbf{0} \mid \mathbf{1} \mid a \mid e+e \mid e\cdot e \mid e^* \mid (e)$$

where a is any symbol in A. The syntax above is ambiguous: for instance, $a+b\cdot c$ can be interpreted as $(a+b)\cdot c$ or $a+(b\cdot c)$. To solve this problem, we assume that the operators have a different binding strength: the iterate postfix operator * binds tighter than the binary infix concatenation operator $\cdot$, in turn binding tighter than the binary infix alternative operator $+$. With this convention, $a+b\cdot c$ represents $a+(b\cdot c)$. Moreover, the concatenation operator is often omitted, so that $e_1 \cdot e_2$ is simply denoted as e_1e_2. Examples of regular expressions are: ab^*, $a+\mathbf{0}$, $(a+b)^*$ and $a(a+b)c^*$. Regular expressions are used to denote languages as follows:

$$\begin{array}{lll}
\mathscr{L}[\mathbf{0}] = \emptyset & \mathscr{L}[e_1+e_2] = \mathscr{L}[e_1] \cup \mathscr{L}[e_2] & \mathscr{L}[(e)] = \mathscr{L}[e] \\
\mathscr{L}[\mathbf{1}] = \{\varepsilon\} & \mathscr{L}[e_1\cdot e_2] = \mathscr{L}[e_1]\cdot\mathscr{L}[e_2] & \mathscr{L}[e^*] = (\mathscr{L}[e])^* \\
\mathscr{L}[a] = \{a\} & &
\end{array}$$

For instance, regular expression $a\cdot b^*$ denotes the language $\mathscr{L}[a\cdot b^*] = \mathscr{L}[a]\cdot\mathscr{L}[b^*] = \{a\}\cdot(\mathscr{L}[b])^* = \{a\}\cdot(\{b\})^* = \{a\}\cdot\{b^n \mid n\in\mathbb{N}\} = \{ab^n \mid n\in\mathbb{N}\}$.

A language L is *regular* if there exists a regular expressions e such that $L = \mathscr{L}[e]$. For instance, $L = \{a^n \mid n = 2\times k+1, k\in\mathbb{N}\}$ is regular because $L = \mathscr{L}[a(aa)^*]$.

1.3.3 Grammars and the Chomsky Hierarchy

Languages can be generated by means of *grammars*. A *general* grammar G is a tuple (N,T,S,P), where N is a finite set of *nonterminals* (ranged over by capital letters $A,B,C,\ldots$), T is a finite set of *terminals* (the symbols of the alphabet), $S\in N$ is the initial nonterminal and P is a finite set of *productions* of the form $\gamma\rightarrow\delta$, with $\gamma,\delta\in(T\cup N)^*$.

The language generated by a general grammar G, denoted by $L(G)$, is given by the set of all strings (or words) $w\in T^*$ derivable by rewriting from S: formally, $L(G) = \{w\in T^* \mid S \longrightarrow^* w\}$, where $S\longrightarrow^*\gamma$ is the minimal relation induced by the following axiom and inference rule:

$$\frac{}{S\longrightarrow^* S} \qquad \frac{S\longrightarrow^*\alpha\gamma\beta \quad \gamma\rightarrow\delta\in P}{S\longrightarrow^*\alpha\delta\beta} \quad \text{where } \alpha,\beta\in(T\cup N)^*$$

The form of the productions can be restricted in some way, thus yielding a classification of grammars, called the *Chomsky hierarchy*.

A grammar G is *right-linear* if all of its productions are of the form $B \to a$ or $B \to bC$ or $B \to \varepsilon$, where $B, C \in N$ and $a, b \in T$. For instance, grammar $G_1 = (\{S,A\}, \{a,b\}, S, \{S \to aA, A \to bA, A \to \varepsilon\})$ is right-linear and its generated language $L(G_1) = \{ab^n \mid n \in \mathbb{N}\} = \mathscr{L}[ab^*]$. It can be proved that the class of languages generated by right-linear grammars coincides with the class of regular languages, i.e., the class of languages denoted by regular expressions.

The intersection of two regular languages is a regular language. For instance, $\mathscr{L}[ab^*] \cap \mathscr{L}[a^*b] = \{ab^n \mid n \in \mathbb{N}\} \cap \{a^n b \mid n \in \mathbb{N}\} = \{ab\} = \mathscr{L}[ab]$, which is a finite language, hence regular.

A grammar G is *context-free* if all of its productions are of the form $B \to \gamma$, with $B \in N$ and $\gamma \in (T \cup N)^*$. A typical example of a context-free grammar is $G_2 = (\{S\}, \{a,b\}, S, \{S \to aSb, S \to \varepsilon\})$, which generates the language $L(G_2) = \{a^n b^n \mid n \in \mathbb{N}\}$. A language L is context-free if there exists a context-free grammar G such that $L(G) = L$. The class of context-free languages includes the class of regular languages because a right-linear grammar is also a context-free grammar. Such inclusion is strict because there are context-free languages that are not regular, e.g., language $L(G_2)$.

The intersection of a context-free language with a regular language is a context-free language. For instance, $L(G_2) \cap \mathscr{L}[(aa)^*(bb)^*] = \{a^{2n} b^{2n} \mid n \in \mathbb{N}\}$, which is context-free.

A context-free grammar G is in *Greibach normal form* if all of its productions are of the form $B \to a\beta$, where $a \in T$ and $\beta \in N^*$. In case ε belongs to the language to be generated by grammar G, it is admitted an ε-production $S \to \varepsilon$ for the initial nonterminal S, provided that S never occurs on the right-hand-side of any production. It can be proved that for any context-free grammar G, there exists a context-free grammar G' in Greibach normal form such that $L(G) = L(G')$.

A grammar G is *context-dependent* (or *monotone*) if all of its productions are of the form $\gamma \to \delta$, with $\gamma, \delta \in (T \cup N)^+$ and $|\gamma| \leq |\delta|$. As above, an ε-production $S \to \varepsilon$ for the initial nonterminal S is allowed, provided that S never occurs on the right-hand side of any production. A typical example of a context-dependent grammar is $G_3 = (\{S,B\}, \{a,b,c\}, S, \{S \to aSBc, S \to abc, cB \to Bc, bB \to bb\})$, which generates the language $L(G_3) = \{a^n b^n c^n \mid n \geq 1\}$. A language L is context-dependent if there exists a context-dependent grammar G such that $L(G) = L$. The class of context-dependent languages includes the class of context-free languages because a context-free grammar in Greibach normal form is also a context-dependent grammar. Such inclusion is strict because there are context-dependent languages that are not context-free, e.g., language $L(G_3)$.

1.3.4 Finite Automata and Turing Machines

A *nondeterministic finite automaton* (NFA, for short) M is a tuple (Q,A,δ,F,q_0) such that Q is a finite set of *states*, A is a finite alphabet of *input symbols*, $q_0 \in Q$ is the *initial state*, $F \subseteq Q$ is the set of *final states*, and δ is the *transition function* of type $\delta : Q \times (A \cup \{\varepsilon\}) \to \wp(Q)$.[2]

A configuration is a pair (q,w) with $q \in Q$ and $w \in A^*$. Configurations can evolve according to the following rules:

$$\frac{}{(q,w) \longrightarrow^* (q,w)} \qquad \frac{(q,w) \longrightarrow^* (q',\sigma w') \qquad q'' \in \delta(q',\sigma), \sigma \in A \cup \{\varepsilon\}}{(q,w) \longrightarrow^* (q'',w')}$$

An NFA $M = (Q,A,\delta,F,q_0)$ recognizes (or accepts) a string $w \in A^*$ if there exists a final state $q \in F$ such that $(q_0,w) \longrightarrow^* (q,\varepsilon)$, i.e., there is a path starting from the initial state q_0 in the automaton that, by reading w, leads to a final state. The automaton M recognizes the language $L[M] = \{w \in A^* \mid \exists q \in F.(q_0,w) \longrightarrow^* (q,\varepsilon)\}$.

A *deterministic finite automaton* (DFA, for short) M is an NFA (Q,A,δ,F,q_0) such that $\delta(q,\varepsilon) = \emptyset$ and $|\delta(q,a)| = 1$ for all $q \in Q$ and $a \in A$. In other words, δ has type $\delta : Q \times A \to Q$. Therefore, DFAs are a subclass of NFAs. However, from an expressiveness point of view, they are equivalent: given an NFA M, we can construct a DFA M' such that $L[M] = L[M']$ (by means of the *Rabin-Scott subset construction* [RS59]).

The class of languages recognized by finite automata coincides with the class of regular languages.

Informally, a (deterministic) *Turing machine* is composed of a finite control, which can be in any of a finite number of states, and a tape, of unbounded length, which is divided into cells; each cell can hold one symbol from a given alphabet. A tape head, which is positioned at one single tape cell, can read the content of that cell or write onto it. Initially, the input is written on the tape, while all the other cells hold a special symbol called *blank*, and the tape head is positioned at the leftmost symbol of the input. At each stage of the computation, the Turing machine reads the symbol in the cell pointed to by the tape head and, depending also on the current state, it writes a symbol in that cell, transits to the next state and moves the tape head — left or right — to the next cell. If the Turing machine eventually stops in a final state, then the input is accepted. If it stops in a non-final state, then the input is rejected. But it may also never end its computation for a given input. Hence, a Turing machine M computes a partial binary function $g_M : A^* \multimap \{0,1\}$, where $g_M(w) = 1$ if w is accepted by M, $g_M(w) = 0$ if M stops on w in a non-final state, but $g_M(w)$ may be undefined when M never ends, i.e., it is unable to accept/reject w.

Turing machines can be adapted to compute partial binary functions on any countable set B, notably $\mathbb{N}$. Turing machines can also be adapted to compute par-

[2] Function δ can be equivalently defined as a subset of $Q \times (A \cup \{\varepsilon\}) \times Q$, i.e., as a set of triples of the form (q,a,q') for $q,q' \in Q$ and $a \in A \cup \{\varepsilon\}$.

tial functions from $\mathbb{N}$ to $\mathbb{N}$. A function $f : \mathbb{N} \multimap\!\!\to \mathbb{N}$ is *Turing-computable* if there exists a Turing machine that computes it. A formalism is *Turing-complete* if it can compute all the Turing computable functions. A few examples of Turing complete formalisms are the *lambda calculus* [B84], *Counter Machines* (see Section 3.5.1), as well as any programming language that includes **while**, **if-then-else**, assignments and sequential composition [BJ66]. Hence, function f is Turing-computable if there exists an algorithm that computes it.

1.3.5 Decidable and Semi-decidable Sets and Problems

Given a set $B \subseteq \mathbb{N}$, its *characteristic function* $f_B : \mathbb{N} \to \{0,1\}$ and its *semi-characteristic function* $g_B : \mathbb{N} \multimap\!\!\to \{1\}$ are defined as follows:

$$f_B(x) = \begin{cases} 1 & \text{if } x \in B, \\ 0 & \text{if } x \notin B \end{cases} \qquad g_B(x) = \begin{cases} 1 & \text{if } x \in B, \\ \text{undefined} & \text{if } x \notin B \end{cases}$$

Set B is *decidable* (or *recursive*) if its characteristic function f_B is Turing-computable, i.e., there exists an algorithm that can compute such a function. A few examples of decidable sets are: $\emptyset$ ($f_\emptyset$ is the constant function $f_\emptyset(x) = 0$ for all $x \in \mathbb{N}$), $\mathbb{N}$ ($f_\mathbb{N}$ is the constant function 1), any finite subset of $\mathbb{N}$, $P = \{n \mid n = 2 \times k, k \in \mathbb{N}\}$. Set B is *undecidable* if its characteristic function f_B is not Turing-computable.

Set B is *semi-decidable* (or *recursively enumerable*) if its semi-characteristic function g_B is Turing-computable, i.e., there exists an algorithm that can compute such a function. Of course, any decidable set B is also semi-decidable, because the algorithm computing the characteristic function f_B can be easily adapted to compute the semi-characteristic function g_B. Examples of semi-decidable sets that are not decidable are less easy to find. Suppose $Z_0, Z_1, Z_2 \ldots$ is an enumeration of Turing machines. Then, $B = \{x \mid Z_x \text{ with input } x \text{ terminates}\}$ is semi-decidable as its semi-characteristic function g_B is Turing-computable. It can be proved that B is undecidable, as its characteristic function f_B is not Turing-computable: this is related to the so-called *halting problem*, discussed below. An example of a non-semi-decidable set is $C = \{x \mid Z_x \text{ with input } x \text{ does not terminate}\}$.

B is *effectively decidable* if it is decidable and the algorithm, computing the characteristic function f_B, can be explicitly exhibited. An example of a decidable but not effectively decidable set can be constructed as follows. Take $A = \{0\}$ and the non-semi-decidable set C described above; then consider $A \cap C$, which can be either A itself, in case $0 \in C$, or the empty set, in case $0 \notin C$. Of course, both A and $\emptyset$ are effectively decidable with their associated algorithms, so for sure there exists an algorithm that computes $f_{A \cap C}$ (i.e., $A \cap C$ is decidable), but we are unable to decide which of the two algorithms is the right one, because we are unable to test $0 \in C$, hence $A \cap C$ is not effectively decidable.

A language $L \subseteq A^*$ is a countable set, hence we can define characteristic and semi-characteristic functions for it. It can be proved that context-dependent languages are all decidable (or recursive), but there exist decidable languages that are not context-dependent. The class of languages generated by general grammars coincides with the class of semi-decidable (or recursively enumerable) languages.

A problem can be usually seen as a particular function, and sometimes such a problem is decidable (or *solvable*) if and only if its corresponding function is computable. For instance, the famous *halting problem* for Turing machines can be formulated as the following function *halt*. Given an enumeration of Turing machines, $Z_0, Z_1, Z_2, \ldots$, function $halt(x,y)$ — where x is an index of a Turing Machine and y is an input — is defined as

$$halt(x,y) = \begin{cases} 1 & \text{if } Z_x(y) \text{ terminates} \\ 0 & \text{otherwise.} \end{cases}$$

Solving the halting problem for Turing machines means being able to compute function *halt*. Unfortunately, function *halt* is not computable and so the halting problem is unsolvable. If *halt* were computable, then we could compute also function $K(x) = halt(x,x)$. If function K were computable, we could even compute function G, defined as

$$G(x) = \begin{cases} 1 & \text{if } K(x) = 0 \\ \text{undefined} & \text{if } K(x) = 1. \end{cases}$$

But now since G is computable, there should exist a Turing machine that computes G; suppose Z_j is this Turing machine. Then, we get a contradiction when computing $G(j)$: either $G(j) = 1$, which is possible only if $K(j) = 0$ and so $Z_j(j)$ must diverge, contradicting the fact that $G(j)$ returns 1; or $G(j)$ is undefined, which is possible only if $K(j) = 1$ and so $Z_j(j)$ converges (also a contradiction). As the only assumption we make in our reasoning is that function *halt* is computable, we can conclude that *halt* is not computable, and so the halting problem is undecidable.

Another well-known problem is the *membership problem*: given an enumeration $G_0, G_1, G_2, \ldots$, of a class of grammars over an alphabet A and given an enumeration $w_0, w_1, w_2, \ldots$, of strings in A^*, we want to decide if $w_i \in L(G_j)$ for all $i, j \in \mathbb{N}$. The problem has an obvious associated function *Mem*, defined as

$$Mem(x,y) = \begin{cases} 1 & \text{if } w_x \in L(G_y) \\ 0 & \text{otherwise.} \end{cases}$$

The computability of function *Mem* depends on the considered class of grammars: it is computable for context-dependent grammars, while it is not computable for general grammars. Therefore, the membership problem is solvable for context-dependent grammars while it is not for general grammars.

There are also problems whose associated functions are always computable. For instance, given an enumeration of programs $p_0, p_1, p_2, \ldots$, for a given programming language, and given a decidable predicate P over such programs, the *existential P problem* is solvable if there exists an index i such that $P(p_i)$ is true. Formally, given the computable function f_P, defined as

$$f_P(x) = \begin{cases} 1 & \text{if } P(p_x) \text{ is true} \\ 0 & \text{otherwise,} \end{cases}$$

the function $exist_P$ associated to the existential P problem is defined as

$$exist_P(x) = \begin{cases} 1 & \text{if } \exists i \text{ such that } f_P(i) = 1 \\ 0 & \text{otherwise.} \end{cases}$$

Function $exist_P$ is computable for any programming language because it is either the constant function 1 or the constant function 0.[3] However, we say that the existential P problem is solvable for a given formalism only if $exist_P$ is the constant function 1.

[3] To be precise, such a function is computable, but may be not *effectively* computable, i.e., an algorithm does exist for sure, but we may be unable to exhibit it explicitly.

Chapter 2
Transition Systems and Behavioral Equivalences

Abstract Transition systems are introduced as a suitable semantic model of reactive systems. Some notions of behavioral equivalence are discussed, such as isomorphism equivalence, trace equivalence, simulation equivalence and bisimulation equivalence. Internal, unobservable actions are also considered and many behavioral equivalences are suitably adapted for this case.

2.1 Modeling a Reactive System

A reactive system, such as the vending machine in Figure 2.1, is a rather complex system with many features, some externally visible, such as the buttons, which the user can interact with, and some internal, such as the supply of water, that may influence the actual functioning of the machine but that are not under the control of the user. In particular, the picture shows a slot where coins can be inserted, two buttons for selecting different kinds of coffee and an opening where the beverage is removed.

In order to reason about the behavior of this vending machine, we need to create a model of it, that is, an abstraction where only the pieces of information that are considered relevant are explicitly represented; such an abstraction should be analyzable by means of mathematically based techniques.

So, in order to create a model of such a vending machine, what abstractions do we need? First, we have to single out what *actions* can take place (e.g., inserting a coin) and how the internal *state* of the machine evolves in response to those actions (e.g., a beverage can now be chosen). Actions are usually assumed to be *atomic*: they are either performed completely or not at all, i.e., they enjoy the *all-or-nothing* property; actions are also assumed to be *instantaneous*, so that two actions cannot occur at the same time.

Second, we have to decide how many internal states are necessary to model correctly all the possible variations due to the effect of the actions performed. We assume that the reactive system is *discrete*, i.e., the number of its states is count-

Fig. 2.1 A simple vending machine

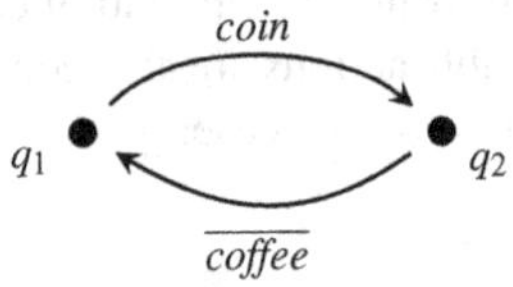

Fig. 2.2 The first simple model for the vending machine

able. This assumption may be unrealistic, e.g., when one needs to model real time-dependent behavior, where some action can occur exactly at some prescribed time instant or within a predetermined time interval. For instance, in some vending machines, after selecting the beverage, the user has one second to choose the level of sugar; each time instant in the interval $[0, 1]$ is appropriate for such a selection, hence there may be the need for uncountably many states to accurately model such a situation. However, for the purpose of the theories described in this book, countable sets of states suffice.

Hence, the models we use are very similar to (possibly infinite) edge-labeled directed graphs, where the nodes are the *states* and the edges are the *transitions*, each one labeled with one action, describing how the system evolves upon the occurrence of that action. The passing of time is not explicitly represented, as we model only the temporal ordering of actions, not their actual timing. We can say that, starting from some state q, some action a can be performed and afterwards another action, say b, can be performed, but we do not model the precise time instant when a or b have been performed.

As a first, tiny example, the vending machine of Figure 2.1 can be modeled very abstractly by the labeled directed graph depicted in Figure 2.2; we have two states q_1 and q_2, in which the machine can let time pass, and two transitions, describing how the system evolves when some input or output occurs. For instance, the transition

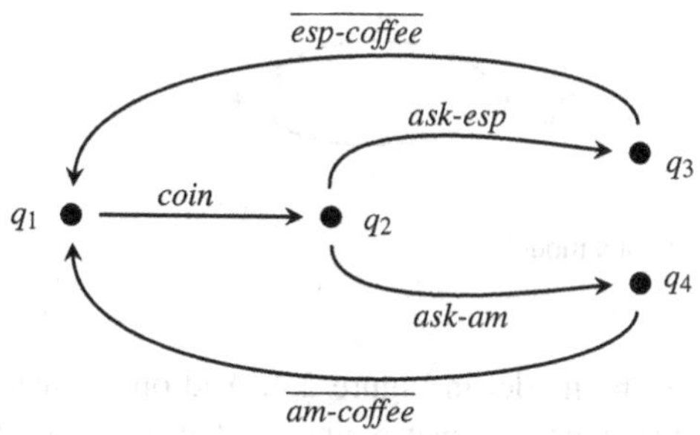

Fig. 2.3 The second simple model for the vending machine

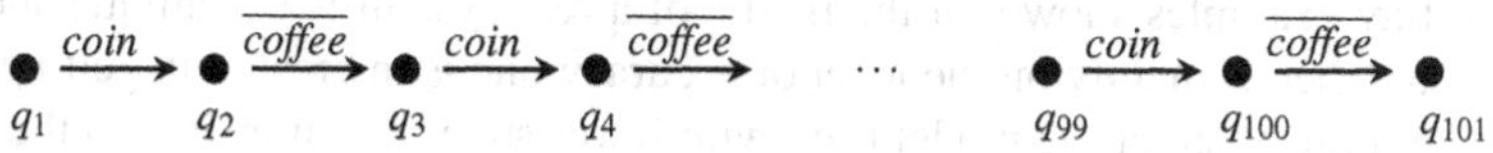

Fig. 2.4 The third simple model for the vending machine

$q_1 \xrightarrow{coin} q_2$ indicates that when the machine is in state q_1, it can react to the input of a *coin* and move to the new state q_2. Action $\overline{coffee}$ indicates that the machine outputs a cup of coffee: for simplicity's sake, we have abstracted the two kinds of beverages in one single (abstract) beverage. (We use the convention that output actions are overlined.) The behavior described by this model is that whenever a coin is inserted, then a cup of coffee is delivered. Note that many possible behaviors are forbidden by this model; for instance, it is not possible to insert two coins in a row: the insertion of the coin and the delivery of the coffee are to be performed alternately.

This model is making one important simplification: as more kinds of coffee are offered, the user should be able to choose the desired one. So, it is necessary to include more actions in the model. We can have input action *ask-esp* to select typical Italian espresso coffee, and input action *ask-am* to select American coffee. The new, more detailed model is shown in Figure 2.3. This model is built according to the so-called *black-box view*: only the interaction interface of the vending machine is relevant (i.e., the slot for coins, the selection buttons and the drinks in the opening, each one modeled by a specific action), while what the machine performs internally (e.g., preparation of the ground coffee) is unobservable.

The very abstract model in Figure 2.2 is making another, rather unrealistic simplification: when the coin is inserted, the beverage is always available. Indeed, whenever a coffee is delivered, the level of water and ground coffee inside the machine diminishes, so that after some time the machine stops working. A more realistic model, taking into account that after, say, 50 coffees the machine cannot supply further coffee, is sketched in Figure 2.4, where state q_{101} is a *deadlock*: no further action is possible.

Exercise 2.1. Model a limited coffee machine, according to the description of Figure 2.3, that cannot offer anything after two beverages have been delivered. □

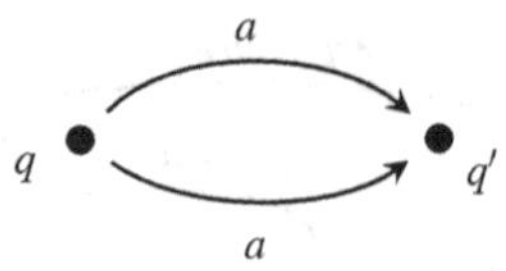

Fig. 2.5 A graph which is not a model

Exercise 2.2. Consider the model in Figure 2.4. Add one transition, labeled *refill*, that models the fact that when service is unavailable, the only action that can be taken (by an operator) is refilling the machine of water and ground coffee. □

These tiny examples show that the model of a real system is not unique and its description depends mainly on the level of accuracy one is interested in getting. Of course, the more abstract the model (i.e., with fewer states and transitions) the less accurate it is, but the easier it is to analyze; on the contrary, the more concrete the model (i.e., with more states and transitions) the more accurate the model, but the more complex its analysis (it may even be impossible when the number of states is infinite). For instance, the first model of the vending machine is detailed enough for studying the basic input/output interaction between a user and the machine, while the third model more accurately takes into account also the special case of unavailable service, which can occur in real life.

At first sight, edge-labeled directed graphs may be considered a natural model for reactive systems. However, they are a bit too concrete, meaning that some graphs have no interpretation as systems. For instance, consider the graph in Figure 2.5 where we have two edges from state q to state q', both labeled with action a.[1] As a transition is completely determined by the change of state and the interaction due to the action performed, there is no observable reason for not identifying the two edges. In other words, whenever two edges have the same source state, the same target state and the same label, then the two are to be identified. Hence, the class of models for reactive systems we are looking for is a slight abstraction of edge-labeled directed graphs.

Classical automata (see Section 1.3.4 for a short overview, or, e.g., [HMU01] for more details), which are largely studied in parsing theory, are also very similar to edge-labeled directed graphs, but with the important abstraction that transitions are defined by means of a transition relation: a transition is just a triple (q,a,q'), usually denoted with $q \xrightarrow{a} q'$. Indeed, the model we want to define — called *labeled transition systems* and originally introduced by Keller in [Kel76] — are essentially automata (with possibly infinitely many states); the only difference is that we do not consider the additional distinction between accepting (or final) and non-accepting states (as discussed in Section 2.3.2). The graphs in Figures 2.2–2.4 are actually also labeled transition systems, while the graph in Figure 2.5 is not.

[1] A graph which is permitted to have *multiple* edges, that is, many edges that have the same end nodes, is sometimes called a *multigraph*.

2.2 Labeled Transition Systems

Definition 2.1. (Actions) Let $\mathscr{L}$ be a countable set of *input* actions, ranged over by $a, b, \ldots$. Let $\overline{\mathscr{L}}$ be the set of co-actions, ranged over by $\overline{a}, \overline{b}, \ldots$, usually called the *outputs*. The set $\mathscr{L} \cup \overline{\mathscr{L}}$, ranged over by $\alpha, \beta, \ldots$, is the set of *visible actions*.

Let $Act = \mathscr{L} \cup \overline{\mathscr{L}} \cup \{\tau\}$, such that $\tau \notin \mathscr{L} \cup \overline{\mathscr{L}}$, be the set of *actions* (or *labels*), ranged over by μ. Action τ denotes an invisible, internal activity. □

Definition 2.2. (Labeled transition systems) A labeled transition system (LTS for short) is a triple $TS = (Q, A, \rightarrow)$ where

- Q is the nonempty, countable set of *states*, ranged over by q (possibly indexed);
- $A \subseteq Act$ is the countable set of *labels* (or *actions*), ranged over by μ (possibly indexed);
- $\rightarrow \subseteq Q \times A \times Q$ is the *transition relation*.

Given a transition $(q, \mu, q') \in \rightarrow$, q is called the *source*, q' the *target* and μ the *label* of the transition. A *rooted* labeled transition system is a pair (TS, q_0) where $TS = (Q, A, \rightarrow)$ is an LTS and $q_0 \in Q$ is the *initial state* (or *root*). Sometimes we write $TS = (Q, A, \rightarrow, q_0)$ for a rooted LTS. □

Given the formal definition of an LTS as above, there is an obvious, associated graphical representation: each state is represented as a node, labeled with the (name of the) state, and each transition (q, a, q') as an edge from the source node/state q to the target node/state q', labeled a. Of course, the graphical representation for a given LTS definition is not unique, because of the obvious freedom in choosing the actual disposition of states and shape of edges.

Remark 2.1. **(Minimal definition of a graphical representation)** It is also possible to do the reverse: given a graphical representation, one can derive the set Q and the transition relation $\rightarrow$. For instance, the labeled transition system depicted in Figure 2.2 gives rise to the triple $TS = (Q, A, \rightarrow)$, where $Q = \{q_1, q_2\}$, $A = \{coin, \overline{coffee}\}$ and $\rightarrow = \{(q_1, coin, q_2), (q_2, \overline{coffee}, q_1)\}$.

However, also this reverse operation is not unique. Strictly speaking, if we extend the LTS TS above with an extra action, say a, which is not used in any transition, then the LTS $TS' = (Q, A \cup \{a\}, \rightarrow)$ is still represented as in Figure 2.2. When deriving a formal definition of an LTS from a graphical representation, we usually assume that the set of labels contains only labels that are used in some transitions, hence TS is a good definition for the vending machine in Figure 2.2, while TS' is not. Under the assumption that in a formal definition an action a is in A only if there is a transition labeled with such an action a, there is only one *minimal formal definition* for a graphical representation of an LTS. □

Exercise 2.3. Provide the minimal definition for the LTS depicted in Figure 2.3. □

Notation: In the following, given an LTS $(Q, A, \rightarrow)$, we denote $(q, \mu, q') \in \rightarrow$ by $q \xrightarrow{\mu} q'$. Moreover,

$$q \xrightarrow{\mu} \text{ if and only if } \exists q'.q \xrightarrow{\mu} q'$$
$$q \stackrel{\mu}{\nrightarrow} \text{ if and only if } \nexists q'.q \xrightarrow{\mu} q'$$
$$q \rightarrow \text{ if and only if } \exists \mu \in A, q \xrightarrow{\mu}$$
$$q \nrightarrow \text{ if and only if } \forall \mu \in A, q \stackrel{\mu}{\nrightarrow}$$

Definition 2.3. Given an LTS $TS = (Q, A, \rightarrow)$, and two states $q, q' \in Q$, a *path* (or *computation*) of length n from q to q' is a sequence of transitions $q_1 \xrightarrow{\mu_1} q'_1\ q_2 \xrightarrow{\mu_2} q'_2 \ldots q_n \xrightarrow{\mu_n} q'_n$ such that $q = q_1$, $q' = q'_n$ and $q'_i = q_{i+1}$ for $i = 1, \ldots, n$, usually denoted as

$$q_1 \xrightarrow{\mu_1} q_2 \xrightarrow{\mu_2} \ldots q_n \xrightarrow{\mu_n} q_{n+1}.$$

When $n = 0$, the path is *empty* and $q = q' = q_1$. If $q_i \neq q_j$ for all $i \neq j$ ($i, j \in \{1, \ldots, n+1\}$), then the path is *acyclic*, otherwise it is *cyclic*. The rooted LTS (TS, q_0) is *acyclic* if it contains no cyclic path starting from q_0. The LTS TS is *acyclic* if it contains no cyclic path. We say that q' is *reachable* from q if a path exists from q to q'; we denote by Q_q the set of all the states in Q reachable from q. A computation may also be infinite: the infinite sequence $q_1, q_2, q_3, \ldots$, such that $q_i \xrightarrow{\mu_i} q_{i+1}$ for each $i \in \mathbb{N}$, yields the infinite path $q_1 \xrightarrow{\mu_1} q_2 \xrightarrow{\mu_2} q_3 \ldots$. □

The LTS in Figure 2.2 is not acyclic because a cyclic path is

$$q_1 \xrightarrow{coin} q_2 \xrightarrow{\overline{coffee}} q_1$$

The LTS in Figure 2.6(a) is not acyclic as there is an obvious cyclic path starting from q_3. However, if we consider it as rooted in q_0, then it is acyclic, because no cyclic path is possible from the initial state q_0, as q_3 is not reachable from q_0.

Definition 2.4. (Reachability relation) Let A^*, ranged over by σ, be the set of all the strings on A, including the empty string ε. The concatenation of strings σ_1 and σ_2 yields $\sigma_1\sigma_2$, with the proviso that $\varepsilon\sigma = \sigma = \sigma\varepsilon$. We define the *reachability relation* $\rightarrow^* \subseteq Q \times A^* \times Q$ as the reflexive and transitive closure of $\rightarrow$, i.e., as the least relation induced by the following axiom and rules:

$$\frac{}{q \xrightarrow{\varepsilon}{}^* q} \qquad \frac{q_1 \xrightarrow{\mu} q_2}{q_1 \xrightarrow{\mu}{}^* q_2} \qquad \frac{q_1 \xrightarrow{\sigma_1}{}^* q_2 \quad q_2 \xrightarrow{\sigma_2}{}^* q_3}{q_1 \xrightarrow{\sigma_1\sigma_2}{}^* q_3}$$

We simply write $q_1 \rightarrow^* q_2$ to state that q_2 is *reachable* from q_1 when there exists a string σ such that $q_1 \xrightarrow{\sigma}{}^* q_2$. □

Exercise 2.4. Let $\sigma = \mu_1 \ldots \mu_n$ with $n \geq 0$. Prove, by induction on n, that $q \xrightarrow{\sigma}{}^* q'$ if and only if there exist $q_1, \ldots, q_{n+1}$ such that $q = q_1$, $q' = q_{n+1}$ and

$$q_1 \xrightarrow{\mu_1} q_2 \xrightarrow{\mu_2} \ldots q_n \xrightarrow{\mu_n} q_{n+1}.$$

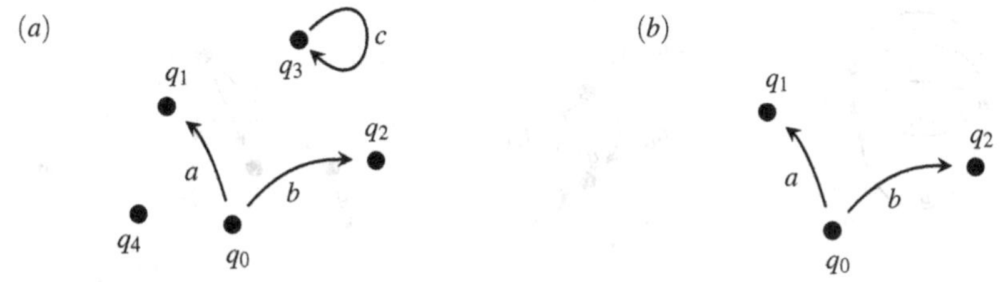

Fig. 2.6 Reachable LTS

This implies that, when $n \geq 1$, $q \xrightarrow{\sigma}{}^* q'$ if and only if a state q'' exists such that $q \xrightarrow{\sigma'}{}^* q'' \xrightarrow{\mu_n} q'$, with $\sigma' = \mu_1 \dots \mu_{n-1}$.

This exercise shows that the definition of reachable state in Definition 2.3 is equivalent to the one based on reachability relation $\longrightarrow^*$ of Definition 2.4. □

Definition 2.5. Given an LTS $TS = (Q, A, \rightarrow)$ and a state $q \in Q$, we define the sort of q as the set $sort(q) = \{\mu \in A \mid \exists q'. q \rightarrow^* q' \xrightarrow{\mu}\}$. We define the rooted LTS $TS_q = (Q_q, sort(q), \rightarrow_q, q)$, called the *reachable LTS from q*, where

- Q_q is the set of the states reachable from q, i.e., $Q_q = \{q' \in Q \mid q \rightarrow^* q'\}$, and
- $\rightarrow_q$ is the restriction of $\rightarrow$ on $Q_q \times sort(q) \times Q_q$. □

Let us consider Figure 2.6. It is not difficult to see that the reachable LTS from state q_0 of the LTS (a) on the left is indeed the LTS (b) on the right.

Definition 2.6. A rooted LTS $TS = (Q, A, \rightarrow, q_0)$ is *reduced* if TS is exactly the reachable LTS from q_0, i.e., $TS = TS_{q_0}$. □

The LTS in Figure 2.6(a) is not reduced, while the one in (b) is reduced. Note that a rooted LTS $TS = (Q, A, \rightarrow, q_0)$ is reduced when all the states are reachable from the initial state (i.e., when $Q = Q_{q_0}$), under the proviso of Remark 2.1, i.e., that the formal definition is minimal.

We briefly introduce some classes of LTSs we will use in the following.

Definition 2.7. (Classes of LTSs) An LTS $TS = (Q, A, \rightarrow)$ is:

- *finite* if it is acyclic and Q and A are finite sets;
- *finite-state* if Q and A are finite sets;
- *boundedly-branching* if $\exists k \in \mathbb{N}$ such that $\forall q \in Q$ the set $T_q = \{(q, \mu, q') \mid \exists \mu \in A$ $\exists q' \in Q.\ q \xrightarrow{\mu} q'\}$ has cardinality at most k; the least k satisfying the above condition is called the *branching-degree* of the LTS;
- *finitely-branching* if the set $T_q = \{(q, \mu, q') \mid \exists \mu \in A\ \exists q' \in Q.\ q \xrightarrow{\mu} q'\}$ is finite for all $q \in Q$; if this is not the case, the LTS is *infinitely-branching*;
- *image-finite* if the set $T_{q,\mu} = \{(q, \mu, q') \mid \exists q' \in Q.\ q \xrightarrow{\mu} q'\}$ is finite for all $q \in Q$ and for all $\mu \in A$;

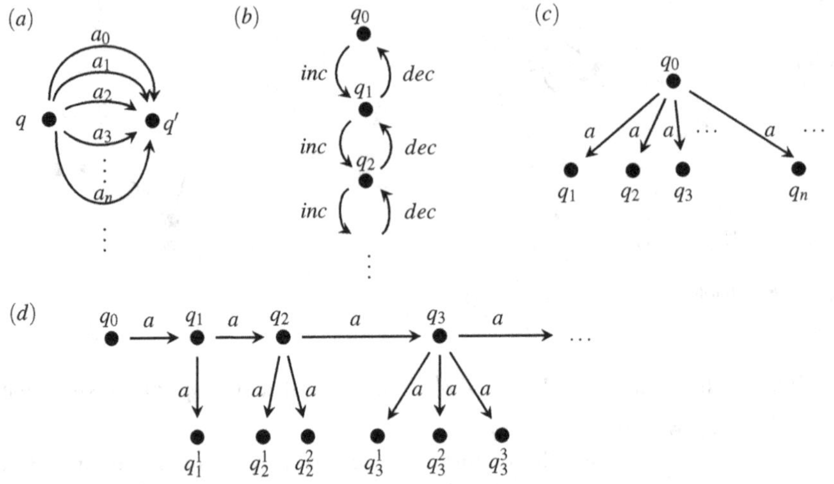

Fig. 2.7 Some labeled transition systems

- *deterministic* if $q \xrightarrow{\mu} q'$ and $q \xrightarrow{\mu} q''$ imply that $q' = q''$, for all $q \in Q$ and for all $\mu \in A$. □

The LTS in Figure 2.4 is finite, as is the LTS in Figure 2.6(b). Note that a finite LTS may offer only finitely many different paths.

Exercise 2.5. Demonstrate that for any finite LTS $TS = (Q, A, \rightarrow)$ there exists $k \in \mathbb{N}$ such that the longest path of TS has length less than k. (*Hint:* Take $k = |Q|$) □

Of course, any finite LTS is also finite-state. However, the LTS in Figure 2.2 (as well as the LTS in Figure 2.6(a)) is finite-state but not finite, because it is not acyclic. Note that a finite-state LTS may offer infinitely many different paths. Note also that the transitions are necessarily finitely many when both Q and A are finite.

In Figure 2.7(a), an LTS is depicted that has finitely many states but that is not finite-state because A is infinite; note that this LTS is not even finitely branching. Formally, this LTS is the triple $(Q, A, \rightarrow)$, where $Q = \{q, q'\}$, $A = \{a_i \mid i \in \mathbb{N}\}$ and $\rightarrow = \{(q, a_i, q') \mid a_i \in A\}$.

Exercise 2.6. (i) Prove that a finite-state LTS is boundedly-branching. Does the converse hold? (ii) Prove that a boundedly-branching LTS is finitely-branching. Does the converse hold? □

As a hint for Exercise 2.6(i), observe that Figure 2.7(b) depicts a boundedly-branching LTS (with branching-degree $k = 2$), which is not finite-state: formally, it is a triple $(Q, A, \rightarrow)$, where $Q = \{q_i \mid i \in \mathbb{N}\}$, $A = \{inc, dec\}$ and $\rightarrow = \{(q_i, inc, q_{i+1}) \mid i \in \mathbb{N}\} \cup \{(q_{i+1}, dec, q_i) \mid i \in \mathbb{N}\}$. This LTS represents a counter (but without the ability to test for zero – see Section 3.4.4 and 3.4.6 for a discussion about various kinds of counters) that can increment (action *inc*) or decrement (action *dec*) the value n stored in the counter state q_n.

As a hint for Exercise 2.6(ii), look at the LTS in Figure 2.7(d). It depicts an LTS $(Q,A,\rightarrow)$, where $Q = \{q_i \mid i \in \mathbb{N}\} \cup \{q^j_{i+1} \mid i \in \mathbb{N}, 1 \leq j \leq i+1\}$, $A = \{a\}$ and $\rightarrow = \{(q_i, a, q_{i+1}) \mid i \in \mathbb{N}\} \cup \{(q_{i+1}, a, q^j_{i+1}) \mid i \in \mathbb{N}, 1 \leq j \leq i+1\}$. Clearly, for all $i \in \mathbb{N}$, exactly $i+1$ a-labeled transitions start from state q_i, hence, this LTS is not boundedly-branching, even if it is finitely-branching.

Exercise 2.7. (i) Prove that a finitely-branching LTS with finitely many states is boundedly-branching. Might it be not finite-state? (*Hint:* Consider the set of labels, that may be infinite.) (ii) Prove that an LTS that is not boundedly-branching but that it is finitely-branching cannot be finite-state. □

Exercise 2.8. Prove that a finitely-branching LTS is image-finite. Does the converse hold? (*Hint:* Just read the following line.) □

The LTS in Figure 2.7(a) is image-finite but not finitely-branching. The LTS in Figure 2.7(c) is not image-finite.

Exercise 2.9. Define the LTS in Figure 2.7(c). □

Examples of nondeterministic LTSs are shown in Figure 2.7(c) and (d). A finite-state nondeterministic LTS is depicted in Figure 2.10 (Section 2.3.2).

Exercise 2.10. Show that a deterministic LTS is image-finite. Does the converse hold? Can you find, among the examples above, a deterministic LTS which is not finitely-branching? □

2.3 Behavioral Equivalences

It is of paramount importance to develop suitable behavioral equivalence relations over LTSs. One good reason for developing such equivalences is *interchangeability*: if two systems offer the same behavior, we can safely replace one of the two with the other one. Another very good reason is that a behavioral equivalence relation supports the verification technique called *equivalence-checking*: a complex, detailed model is correct if it is behaviorally equivalent to some simpler model that is self-evidently correct. In Chapter 3, especially in Sections 3.4.2–3.4.6, we will see many examples of use of the equivalence-checking technique.

When are two systems to be considered equivalent? There is not an obvious answer to this question: many different notions of behavioral equivalence have been proposed in the literature over LTSs, some of which are motivated by peculiar technical reasons. Here we present only the main ones. The reader interested in a comprehensive overview may consult [vGl01, vGl93, San12].

In this section, we consider the internal action τ as observable as any other action. Equivalences of this kind are called *strong* to reflect this strict requirement. In the next section we will discuss the problem of abstracting from the internal action τ. Equivalences discussed there are called *weak* as they turn out to be less discriminating than the corresponding strong ones, discussed in this section.

We start our investigation from the most discriminating equivalence, i.e. *isomorphism*, inspired by graph theory; then, we consider *trace equivalence*, inspired by automata theory; we will argue that both are not convincing as behavioral equivalences in the setting of reactive systems. Then, a radically different definition, called *simulation*, is presented, which is a sufficient condition for trace equivalence; finally, *bisimulation* equivalence is presented as a further refinement of simulation equivalence.

2.3.1 Isomorphism

As LTSs are basically edge-labeled directed graphs (even if with possibly infinitely many states), we can consider graph isomorphism as our first choice.

Definition 2.8. Let $TS_1 = (Q_1, A_1, \rightarrow_1)$ and $TS_2 = (Q_2, A_2, \rightarrow_2)$ be two labeled transition systems. An *isomorphism* is a bijection $f : Q_1 \rightarrow Q_2$ that preserves transitions:

$$q \xrightarrow{\mu}_1 q' \quad \text{if and only if} \quad f(q) \xrightarrow{\mu}_2 f(q')$$

for all $q, q' \in Q_1$ and $\mu \in A_1 \cup A_2$. If there exists an isomorphism between TS_1 and TS_2 then we say that TS_1 and TS_2 are *isomorphic*, denoted $TS_1 \cong TS_2$.

This definition can be applied to rooted labeled transition systems by requiring that the isomorphism f preserve also the initial states, i.e., $f(q_1) = q_2$, if q_1 and q_2 are the initial states of TS_1 and TS_2, respectively. □

Remark 2.2. Observe that if $TS_1 = (Q_1, A_1, \rightarrow_1)$ and $TS_2 = (Q_2, A_2, \rightarrow_2)$ are isomorphic, then they are also isomorphic to $TS_1' = (Q_1, A, \rightarrow_1)$ and $TS_2' = (Q_2, A, \rightarrow_2)$, where $A = A_1 \cap A_2$. □

Exercise 2.11. Prove that LTSs isomorphism $\cong$ is an equivalence relation, i.e., reflexive, symmetric and transitive. □

Exercise 2.12. Prove that a finitely-branching LTS with finitely many states is (isomorphic to) a finite-state LTS. □

Two isomorphic LTSs are of course indistinguishable by any observer: whatever is done on the first can be replicated on the second, and vice versa.[2] However, isomorphism equivalence is too discriminating as it distinguishes between systems that are to be equated intuitively. For instance, consider the two LTSs in Figure 2.8, which are clearly not isomorphic because there is no bijection between the two sets

[2] According to the *black-box view*, an observer can only observe the actions that a system may perform, but not its states or internal structure.

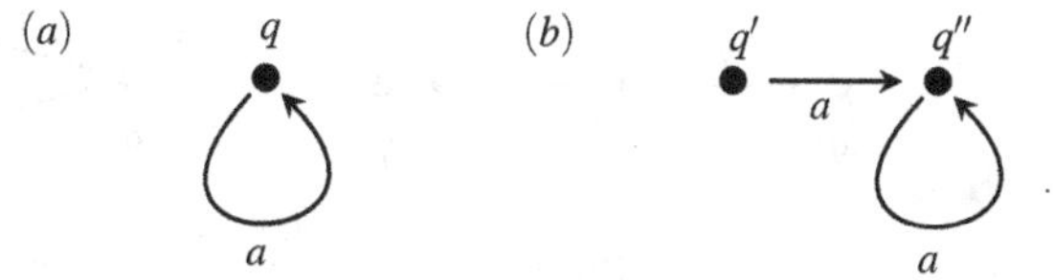

Fig. 2.8 Two equivalent, yet not isomorphic LTSs

of states. Nonetheless, no observer can tell them apart as, after all, both can only do any sequence of a's. Hence, isomorphism equivalence is not an appropriate behavioral equivalence relation. A more suitable behavioral equivalence, weaker than isomorphism, that equates these two LTSs may just compare them on the basis of the sequences of actions they can perform. This is the subject of the next section.

Moreover, checking graph isomorphism over finite-state LTSs is an NP problem that is not known to be in P, and the best known algorithm is exponential in the number of states [BKL83]; therefore, verification based on isomorphism equivalence is, in general, not viable in practice.

2.3.2 Traces

Labeled transition systems are also very similar to automata (see Section 1.3.4), and so we can take inspiration from the notion of equivalence defined over automata. An automaton, besides an initial state, has a designated set of *accepting* (or *final*) states. Any directed arc connecting two states is labeled with one symbol of an alphabet (while for LTSs the transition labels are the actions). A sequence of symbols (a *string* or *word*, in automata terminology) is recognized if there is a path in the automaton starting from the initial state and ending in one of its final states by reading that string. Two automata are equivalent if they recognize the same strings (*language equivalence*).

Intuitively, LTSs differ in one important aspect from automata: while we check if string σ can drive the automaton to a final state, for LTSs we check if it is able to perform that string *interactively*. Hence, if the LTS performs a string σ, necessarily it is also able to perform any prefix of σ; this implies that we should consider an analogous definition of equivalence over LTSs, which implicitly assumes that all the states are final.

Definition 2.9. (Trace equivalence) Let $(Q, A, \to)$ be an LTS and let $q \in Q$. A *trace* of q is a sequence of actions $\mu_1 \mu_2 \dots \mu_n$ (possibly empty, when $n = 0$) such that there exists a path

$$q \xrightarrow{\mu_1} q_1 \xrightarrow{\mu_2} \dots q_{n-1} \xrightarrow{\mu_n} q_n.$$

In other words, according to Exercise 2.4, the set of *traces* of q is

$$Tr(q) = \{\sigma \in A^* \mid \exists q' \in Q.\, q \xrightarrow{\sigma}{}^{*} q'\}.$$

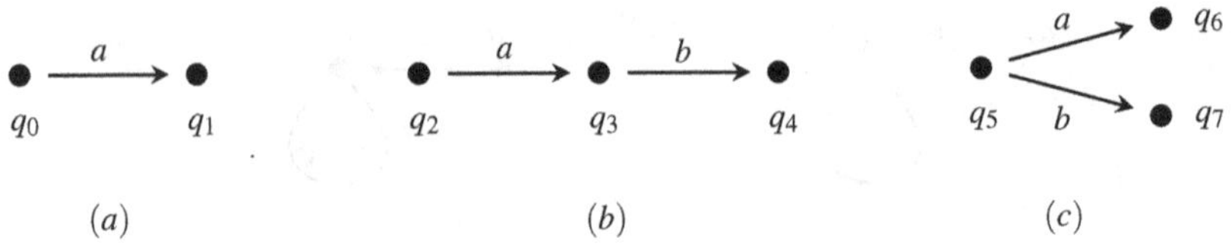

Fig. 2.9 Three not trace equivalent LTSs

Two states $q_1, q_2 \in Q$ are *trace equivalent* if $Tr(q_1) = Tr(q_2)$, and this is denoted as $q_1 =_{tr} q_2$. This definition can be extended to rooted LTSs as follows. The set $Tr(TS)$ of traces of the rooted LTS $TS = (Q, A, \rightarrow, q_0)$ is $Tr(q_0)$. Two rooted LTSs, TS_1 and TS_2, are *trace equivalent* if $Tr(TS_1) = Tr(TS_2)$. □

Example 2.1. Consider Figure 2.8(a). We want to show that $Tr(q) = \{a^n \mid n \in \mathbb{N}\}$, where $a^0 = \varepsilon$ and $a^{n+1} = aa^n$. This can be proven by induction on n. The base case is $n = 0$ and, since $q \xrightarrow{\varepsilon}{}^* q$, $\varepsilon \in Tr(q)$. Now, the inductive case: assume $a^n \in Tr(q)$; this means that a state q' exists such that $q \xrightarrow{a^n}{}^* q'$; however, q' must be q, as this is the only state of the LTS. It is easy to observe that also $q \xrightarrow{a^n}{}^* q \xrightarrow{a} q$, hence $q \xrightarrow{a^{n+1}}{}^* q$, i.e., $a^{n+1} \in Tr(q)$. It is an easy exercise then to check that the two transition systems of Figure 2.8 are trace equivalent, i.e., $Tr(q) = Tr(q')$. □

Exercise 2.13. Compute the set of traces for states q_0, q_2 and q_5 in Figure 2.9 and observe that they are not trace equivalent. □

Exercise 2.14. A nonempty set L of traces over the alphabet A is *prefix closed* if whenever $\sigma_1\sigma_2 \in L$, then $\sigma_1 \in L$ for all $\sigma_1, \sigma_2 \in A^*$. Hence, a prefix closed, nonempty set L is such that $\varepsilon \in L$. Prove that, for any rooted LTS $TS = (Q, A, \rightarrow, q_0)$, the set $Tr(TS)$ of its traces is nonempty and prefix closed. □

Exercise 2.15. (*i*) Prove that if $TS = (Q, A, \rightarrow)$ is a finite LTS, then $Tr(q)$ is a finite set for all $q \in Q$. (Hint: Look at Exercise 2.5.) (*ii*) Prove also that for any finite, nonempty, prefix closed $L \subseteq A^*$, it is possible to define a finite, deterministic LTS TS' such that $Tr(TS') = L$. □

Exercise 2.16. (Trace preorder) The relation of *trace preorder* $\leq_{tr} \subseteq Q \times Q$ is defined as follows: $q \leq_{tr} q'$ if and only if $Tr(q) \subseteq Tr(q')$. Prove that $\leq_{tr}$ is a preorder, i.e., it is reflexive and transitive. Observe that $q =_{tr} q'$ if and only if $q \leq_{tr} q'$ and $q' \leq_{tr} q$. Prove that trace equivalence $=_{tr} \subseteq Q \times Q$ is an equivalence relation. As an example, consider Figure 2.9. Observe that $q_0 \leq_{tr} q_2$ but that $q_2 \not\leq_{tr} q_0$. Similarly, $q_0 \leq_{tr} q_5$ but $q_5 \not\leq_{tr} q_0$. □

Exercise 2.17. (Nondeterminism vs Determinism for trace equivalence) Given a nondeterministic, finite-state, rooted LTS $TS_1 = (Q, A, \rightarrow_1, q_0)$, one can build a finite-state rooted LTS $TS_2 = (R, A, \rightarrow_2, \{q_0\})$, where $R = \mathcal{P}(Q) \setminus \{\emptyset\}$ is the set of all nonempty subsets of Q and the transition relation $\rightarrow_2 \subseteq R \times A \times R$ is defined

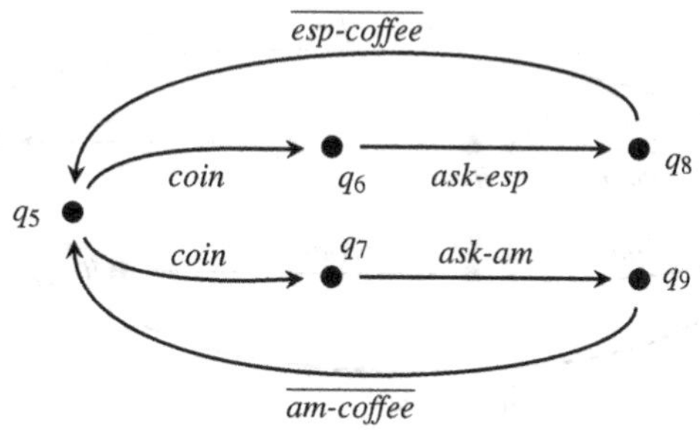

Fig. 2.10 An impolite vending machine

as follows: for any $P \in R$ and for any $a \in A$, we have that $P \xrightarrow{a}_2 P'$ if $P' = \{q' \in Q \mid \exists q \in P. q \xrightarrow{a}_1 q'\}$ is nonempty.[3] Prove that TS_2 is deterministic and that the two LTSs are trace equivalent, i.e., $Tr(TS_1) = Tr(TS_2)$. Apply the construction to the nondeterministic LTS of Figure 2.10, rooted in q_5, and compare the result with the deterministic LTS in Figure 2.3.

This construction is a slight variant of the famous *Rabin-Scott subset construction* for nondeterministic finite-state automata [RS59] (see, e.g., [Sip06, HMU01] for a gentle introduction). In principle, it may be extended also to LTSs with infinitely many states, but the resulting deterministic LTSs may have uncountably many states (which is excluded in Definition 2.2), as the powerset of a countable set may be uncountable.

This exercise proves that, from an expressiveness point of view, nondeterminism is inessential for trace equivalence, as, given a nondeterministic LTS, it is always possible to find a deterministic one exhibiting the same set of traces. (Observe that, on the contrary, it is not possible to find a deterministic LTS isomorphic to a nondeterministic LTS.) □

But is trace equivalence useful for reactive systems? Consider the model of an impolite vending machine in Figure 2.10 and compare it with the model in Figure 2.3. It is not difficult to see that the two are trace equivalent, as both offer the same traces for interaction with a user. However, we cannot declare them equivalent, as an observer can really detect some difference in their behavior. Observe that in the model of Figure 2.3, after inserting a *coin*, the user can choose between asking for an espresso or for an American coffee. On the contrary, in Figure 2.10, upon insertion of the coin, the machine nondeterministically chooses to reach either q_6, where only *ask-esp* is possible, or q_7, where only *ask-am* is possible; hence, in this case it is the machine that makes the choice and not the user! This example explains that in concurrency theory the timing of a choice can be a crucial aspect of the behavior and cannot be neglected, as in trace equivalence.

[3] This simple construction is not optimal, as it may generate states that are unreachable from the initial state $\{q_0\}$. It is an easy exercise (try it!) to optimize it in order to get only reachable states.

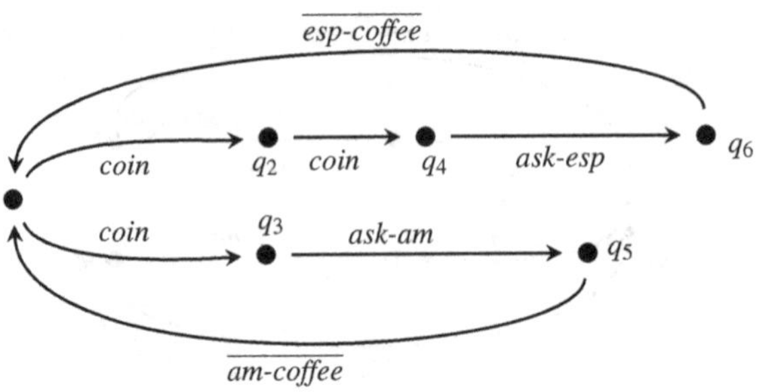

Fig. 2.11 Another impolite vending machine

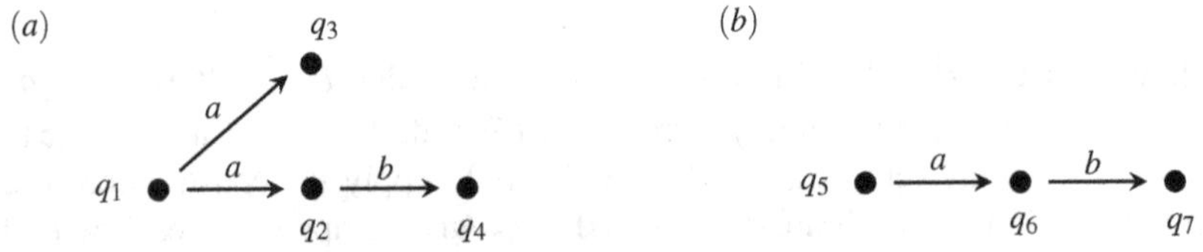

Fig. 2.12 Two trace equivalent LTSs with different deadlock behavior

Exercise 2.18. (*i*) Define a model for a vending machine that differs from the one in Figure 2.3 because two coins are needed to get an espresso. (*ii*) Elaborate on the model to allow the selection of American coffee also after two coins have been inserted, keeping credit of one coin. (*iii*) Are the two LTSs trace equivalent? (*Hint*: A solution, in CCS, can be found in Example 3.4, Section 3.4.2.) □

Exercise 2.19. Consider the vending machine in Figure 2.11. Argue that it is trace equivalent to the first model of the previous exercise. Is it reasonable to consider equivalent these two models? In particular, what may happen to a user with only one coin (and willing to get American coffee with it)? □

Another good reason for rejecting trace equivalence is that it equates LTSs with different deadlock behavior.

Definition 2.10. (Deadlock) A state q is a *deadlock* if there is no transition starting from it, i.e., $\nexists q' \in Q, \nexists \mu \in A$ such that $q \xrightarrow{\mu} q'$, usually abbreviated as $q \not\rightarrow$. An LTS $TS = (Q, A, \rightarrow)$ is *deadlock-free* if for all $q \in Q$, q is not a deadlock. □

Indeed, trace equivalence is not sensitive to deadlock. Consider the two LTSs in Figure 2.12. Both can perform traces $\{\varepsilon, a, ab\}$, so they are trace equivalent. However, the LTS on the left is nondeterministic and, after a, it can reach the deadlock q_3, while q_5, after a, reaches q_6, which is not a deadlock. We can slightly refine trace equivalence in order to get an equivalence that is sensitive to deadlock.

Definition 2.11. (Completed trace equivalence) Let $TS = (Q, A, \rightarrow)$ be a transition system. A *completed trace* for state $q \in Q$ is a (possibly empty) sequence of actions

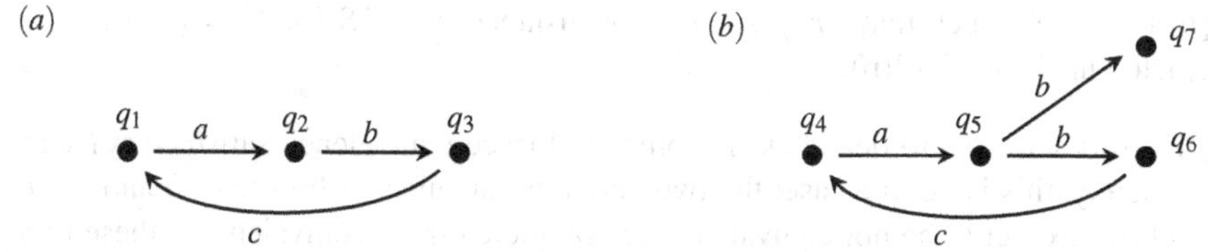

Fig. 2.13 Two not completed trace equivalent systems

$\mu_1 \ldots \mu_n$ such that there exists a path $q_1 \xrightarrow{\mu_1} \ldots q_n \xrightarrow{\mu_n} q_{n+1}$ such that $q_1 = q$ and q_{n+1} is a deadlock. Hence, the set of *completed traces* of a state $q \in Q$ is

$$CTr(q) = \{\sigma \in A^* \mid \exists q' \in Q.\, q \xrightarrow{\sigma}{}^{*} q' \wedge q' \nrightarrow\}.$$

Two states, $q_1, q_2 \in Q$, are completed trace equivalent if $Tr(q_1) = Tr(q_2)$ and $CTr(q_1) = CTr(q_2)$, and this is denoted as $q_1 =_{ctr} q_2$. □

The two states q_1 and q_5 in Figure 2.12 are not completed trace equivalent, as the set of the completed traces of q_1 is $\{a, ab\}$, while for q_5 it is $\{ab\}$. This example shows that nondeterminism cannot be always removed when considering completed trace semantics, as the nondeterministic LTS in Figure 2.12(a) is not completed trace equivalent to its deterministic counterpart in (b), which is isomorphic to the LTS (reachable from $\{q_1\}$) obtained by the construction in Exercise 2.17.

Example 2.2. Let us consider the LTSs in Figure 2.13. It is not difficult to observe that q_1 and q_4 are trace equivalent. However, they are not completed trace equivalent. As a matter of fact, the LTS in (a) is clearly deadlock-free, hence $CTr(q_1) = \emptyset$. On the contrary, the LTS in (b) is such that $CTr(q_4) = \{ab, abcab, abcabcab, \ldots\}$, which can be represented by the regular expression $(abc)^*ab$. □

Remark 2.3. Observe that two states may have the same set of completed traces but not be trace equivalent (and so not even completed trace equivalent). For instance, consider the LTS $TS = (\{q_1, q_2\}, \{a, b\}, \{(q_1, a, q_1), (q_2, b, q_2)\})$. Clearly $CTr(q_1) = CTr(q_2) = \emptyset$, but $Tr(q_1) \neq Tr(q_2)$; hence q_1 and q_2 are not (completed) trace equivalent. □

Exercise 2.20. (Any finite language is representable by a finite LTS) Prove that for any nonempty, finite $L \in \{\varepsilon\} \cup \wp(A^+)$, where $A^+ = A^* \setminus \{\varepsilon\}$, it is possible to define a finite rooted LTS $(Q, A, \rightarrow, q_0)$ such that $CTr(q_0) = L$.[4] For instance, a finite LTS for $L = \{a, ab\}$ is depicted in Figure 2.12(a).

In Remark 2.7 (Section 2.4.1), we will show that all *regular* languages (and only regular languages) can be represented by *finite-state* LTSs.

This exercise is a companion of Exercise 2.15, where we asked to prove that for any finite, *prefix-closed* $L' \subseteq A^*$, we can define a finite, *deterministic* rooted LTS

[4] Observe that if $\varepsilon \in CTr(q)$, then q is a deadlock, i.e., $\sigma \notin CTr(q)$ for all $\sigma \in A^+$. On the contrary, if $\sigma \in CTr(q)$ and $\sigma \neq \varepsilon$, then $\varepsilon \notin CTr(q)$. Hence, a finite LTS can represent strongly only a finite, nonempty language in $\{\varepsilon\} \cup \wp(A^+)$.

$(Q', A, \to', q'_0)$ such that $Tr(q'_0) = L'$. For instance, an LTS for $L' = \{\varepsilon, a, ab\}$ is depicted in Figure 2.12(b). □

Even if sensitive to deadlock, is completed trace equivalence satisfactory? Unfortunately, this is not the case: the two vending machines of Figures 2.3 and 2.10, which we expect to be not equivalent, are completed trace equivalent: as these two LTSs are deadlock-free, it turns out that $CTr(q_1) = CTr(q_5) = \emptyset$. So it is necessary to find some finer notion of equivalence that is able to capture the timing of choices.

Moreover, language equivalence over automata is PSPACE-complete [SM73, HRS76], and so is also trace equivalence (as well as completed trace equivalence) for finite-state LTSs. It is therefore advisable to find a finer notion of equivalence that is more easily checkable. One possibility is the simulation preorder and equivalence (originally proposed in [Park81]) we introduce in the next section.

Nonetheless, trace equivalence is useful for the verification of so-called *safety* properties. A safety property is a property that states that "something bad never happens". If an LTS TS_1 (the *specification* of the system) satisfies a safety property and the set of traces of TS_2 (the *implementation*) is included in the set of traces of TS_1, then also TS_2 satisfies that safety property. An example of a safety property for the vending machine may be "coffee must not be deliverable if at least one coin has not been inserted". On the contrary, trace equivalence cannot be used for checking so-called *liveness* properties, which require some progress: "something good will happen eventually". In the specific case of Figure 2.12, the liveness property is "action b will happen eventually", which holds for q_5 but not for q_1.

2.3.3 Simulation

Definition 2.12. Let $TS = (Q, A, \to)$ be a transition system. A *simulation* is a relation $R \subseteq Q \times Q$ such that if $(q_1, q_2) \in R$ then for all $\mu \in A$

- $\forall q'_1$ such that $q_1 \xrightarrow{\mu} q'_1$, $\exists q'_2$ such that $q_2 \xrightarrow{\mu} q'_2$ and $(q'_1, q'_2) \in R$.

State q is simulated by q', denoted $q \lesssim q'$, if there exists a simulation R such that $(q, q') \in R$. Two states q and q' are *simulation equivalent*, denoted $q \simeq q'$, if $q \lesssim q'$ and $q' \lesssim q$. □

Remark 2.4. The definition above comprises also the case of a simulation between two LTSs, say, $TS_1 = (Q_1, A_1, \to_1)$ and $TS_2 = (Q_2, A_2, \to_2)$ with $Q_1 \cap Q_2 = \emptyset$.[5] In such a case, we may consider just one single LTS $TS = (Q_1 \cup Q_2, A_1 \cup A_2, \to_1 \cup \to_2)$. A simulation $R \subseteq Q_1 \times Q_2$ is also a simulation on $(Q_1 \cup Q_2) \times (Q_1 \cup Q_2)$. We say that the rooted LTS $TS_1 = (Q_1, A_1, \to_1, q_1)$ is simulated by the rooted LTS $TS_2 = (Q_2, A_2, \to_2, q_2)$ if there exists a simulation $R \subseteq Q_1 \times Q_2$ containing the pair (q_1, q_2). □

[5] If this is not the case, we can take an isomorphic LTS such that the intersection of the sets of states is empty.

The definition of simulation is such that whatever action q_1 performs, reaching q_1', q_2 can respond with the same action reaching a state q_2' that is still able to simulate q_1'. So q_2 can do whatever sequence q_1 can execute.

Proposition 2.1. *Let $TS = (Q, A, \to)$ be a transition system. For any $q, q' \in Q$, if $q \lesssim q'$ then $q \leq_{tr} q'$, i.e., $Tr(q) \subseteq Tr(q')$.*

Proof. *If $q \lesssim q'$, then there exists a simulation $R \subseteq Q \times Q$ such that $(q, q') \in R$. We then prove, by induction on the length of traces, a sligthly stronger result: if $q \xrightarrow{\sigma}{}^* q_1$, then $q' \xrightarrow{\sigma}{}^* q_1'$ with $(q_1, q_1') \in R$; this implies the thesis $Tr(q) \subseteq Tr(q')$. The base case is when $\sigma = \varepsilon$ and is trivial as $q \xrightarrow{\varepsilon}{}^* q$ and $q' \xrightarrow{\varepsilon}{}^* q'$ with $(q, q') \in R$. Now if $q \xrightarrow{\sigma}{}^* q_1$ with $|\sigma| = n+1$, then by Exercise 2.4, there exist a state $\overline{q}$, a trace σ' and an action μ such that $q \xrightarrow{\sigma'}{}^* \overline{q} \xrightarrow{\mu} q_1$ with $\sigma = \sigma'\mu$. Hence, induction can be applied to conclude that a state $\overline{q}'$ exists such that $q' \xrightarrow{\sigma'}{}^* \overline{q}'$ with $(\overline{q}, \overline{q}') \in R$. As R is a simulation, a state q_1' exists such that transition $\overline{q} \xrightarrow{\mu} q_1$ is to be matched by $\overline{q}' \xrightarrow{\mu} q_1'$ with $(q_1, q_1') \in R$, and so $q' \xrightarrow{\sigma}{}^* q_1'$ by Exercise 2.4. Summing up, transition $q \xrightarrow{\sigma}{}^* q_1$ is matched by $q' \xrightarrow{\sigma}{}^* q_1'$ with $(q_1, q_1') \in R$, as required.* □

As a trivial corollary, we have that if $q \simeq q'$ then $q =_{tr} q'$, i.e., $Tr(q) = Tr(q')$. However, the inverse implication does not hold. Consider again the two trace equivalent vending machines of Figure 2.3 and Figure 2.10. We have that $q_5 \lesssim q_1$, but $q_1 \not\lesssim q_5$. To demonstrate that q_5 is simulated by q_1 it is enough to exhibit a simulation relation R containing the pair (q_5, q_1). Relation R is $\{(q_5, q_1), (q_6, q_2), (q_7, q_2), (q_8, q_3), (q_9, q_4)\}$. In order to check that R is a simulation, we have to show that, for any pair $(q_i, q_j) \in R$, for any action a, for all q_i' such that $q_i \xrightarrow{a} q_i'$, there exists q_j' such that $q_j \xrightarrow{a} q_j'$ and $(q_i', q_j') \in R$. For instance, for the first pair (q_5, q_1), we have that transition $q_5 \xrightarrow{coin} q_6$ can be simulated by $q_1 \xrightarrow{coin} q_2$ with (q_6, q_2) in R, as well as that transition $q_5 \xrightarrow{coin} q_7$ can be simulated by $q_1 \xrightarrow{coin} q_2$ with $(q_7, q_2) \in R$. Hence, the first pair is OK. Now consider pair (q_6, q_2). State q_6 can perform only transition $q_6 \xrightarrow{ask\text{-}esp} q_8$, which is matched by $q_2 \xrightarrow{ask\text{-}esp} q_3$, with $(q_8, q_3) \in R$. So, also the second pair is OK.

Exercise 2.21. Complete the check that R is a simulation, i.e., perform the check above for all the other pairs in R. □

However, we can prove that it is not possible to find a simulation R' that contains the pair (q_1, q_5). Suppose, towards a contradiction, we have a simulation R' such that $(q_1, q_5) \in R'$. By Definition 2.12, to transition $q_1 \xrightarrow{coin} q_2$, q_5 has to respond with either $q_5 \xrightarrow{coin} q_6$ or with $q_5 \xrightarrow{coin} q_7$; this means that either pair (q_2, q_6) or pair (q_2, q_7) must belong to simulation R'. But this is impossible: if $(q_2, q_6) \in R'$ then q_6 cannot respond to transition $q_2 \xrightarrow{ask\text{-}am} q_4$, invalidating the assumption that R' is a simulation; similarly, if $(q_2, q_7) \in R'$, then q_7 cannot respond to transition $q_2 \xrightarrow{ask\text{-}esp} q_3$. Therefore, the two vending machines are not simulation equivalent.

Example 2.3. **(How to build a simulation)** We have seen above how to check if a relation is a simulation; however, given two finite-state LTSs, how do we build a simulation relation? As a first, extremely simple case, consider the two LTSs in Figure 2.8 and assume we want to prove that $q \lesssim q'$. The candidate simulation relation S must contain the pair (q,q') and such a pair has to satisfy the simulation condition, i.e., for all $a \in A$ and for all q_1 such that $q \xrightarrow{a} q_1$, there exists a state q_2 such that $q' \xrightarrow{a} q_2$ and $(q_1,q_2) \in S$. From q we have only one transition, namely $q \xrightarrow{a} q$, hence $q_1 = q$; this transition can be matched by the only transition from q', namely $q' \xrightarrow{a} q''$, hence $q_2 = q''$. This means that for the pair (q,q') it is necessary that also the pair (q,q'') belong to the candidate simulation relation S. But then we have to check that the pair (q,q'') satisfies the simulation condition; from q only transition $q \xrightarrow{a} q$ is executable, and similarly only $q'' \xrightarrow{a} q''$ is executable from q''. Therefore, for the pair (q,q'') to be justified in the simulation S, it is necessary that ... (q,q'') itself be a part of the simulation relation! Hence, the simulation relation S we were looking for is just $\{(q,q'),(q,q'')\}$. □

Exercise 2.22. Considering Figure 2.8, check that only one of the following relations over $Q = \{q,q',q''\}$ is not a simulation:

- $S_0 = \{(q,q)\}$
- $S_1 = \{(q,q'),(q,q''),(q'',q)\}$
- $S_2 = \{(q,q'),(q,q''),(q'',q')\}$
- $S_3 = \emptyset$. □

Exercise 2.23. (*i*) Considering again Figure 2.8, build a simulation relation that proves that $q' \lesssim q$. (*ii*) How many different simulation relations over $Q = \{q,q',q''\}$ can you build to prove $q' \lesssim q$? (*iii*) Is $S' = \{(q,q),(q',q'),(q'',q''),(q',q),(q'',q),(q,q'),(q,q''),(q',q''),(q'',q')\}$ one of such simulation relations? If so, is it the largest one? □

Exercise 2.24. Given a simulation $S \subseteq Q \times Q$, argue that, for any pair $(q_1,q_2) \in S$, if q_2 is a deadlock, then also q_1 is a deadlock, i.e., if $q_2 \nrightarrow$ then $q_1 \nrightarrow$.

Moreover, argue that if we add to simulation S a pair (q,q') such that q is a deadlock, then $S \cup \{(q,q')\}$ is still a simulation. In particular, as $S = \emptyset$ is a simulation, also $\{(q,q')\}$ is a simulation. □

Exercise 2.25. Consider Figure 2.9. Show that $q_0 \lesssim q_2$ but $q_2 \not\lesssim q_0$. Similarly, show that $q_0 \lesssim q_5$ but $q_5 \not\lesssim q_0$. Finally, note that $q_2 \not\lesssim q_5$ and $q_5 \not\lesssim q_2$. □

Example 2.4. **(How to build a simulation 2)** Continuing Example 2.3, let us consider the less obvious case depicted in Figure 2.14. We would like to build a simulation relation S containing the pair (q_0,q_4).

For any move of q_0, we have to find a match among the transitions of q_4 so that the state reached from q_0 is simulated by the state reached from q_4. To transition $q_0 \xrightarrow{a} q_1$, state q_4 can respond with three different transitions, reaching q_5, q_6 and q_7, respectively; it is necessary that at least one of these match the simulation request. Among the three, transition $q_4 \xrightarrow{a} q_7$ is not OK because the pair we

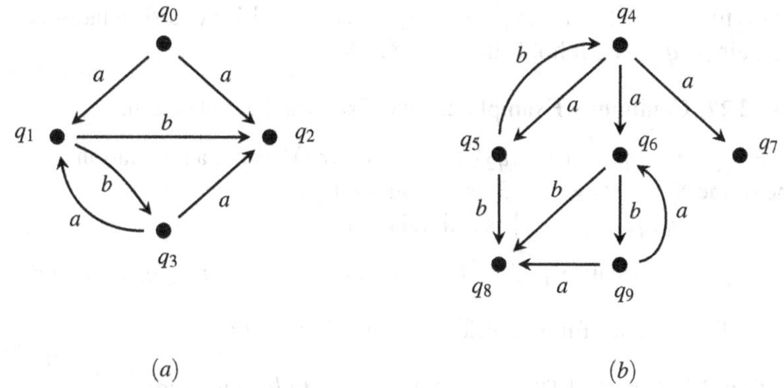

Fig. 2.14 Two simulation equivalent systems

should add to S is (q_1, q_7), but q_7 cannot simulate q_1 because q_7 is a deadlock. So, we may try to add to S one of the other two pairs, say, (q_1, q_5), which must then be checked. But before examining this new pair, let us continue exploring the conditions for the first pair (q_0, q_4). To transition $q_0 \xrightarrow{a} q_2$, state q_4 can respond with three different transitions, reaching q_5, q_6 and q_7, respectively; as q_2 is a deadlock, any state can simulate it, so we are free to choose, e.g., q_7, and so S is now $\{(q_0, q_4), (q_1, q_5), (q_2, q_7), \ldots\}$. We have completed the analysis of the first pair: if the other two pairs satisfy the simulation condition, then (q_0, q_4) is really a simulation pair.

Let us now examine pair (q_1, q_5). Transition $q_1 \xrightarrow{b} q_2$ reaches a deadlock, so any of the b-labeled transitions from q_5 is OK; take, e.g., $q_5 \xrightarrow{b} q_8$ and add (q_2, q_8) to S. The other transition from q_1 is $q_1 \xrightarrow{b} q_3$. This transition cannot be matched by $q_5 \xrightarrow{b} q_8$ because q_3 is not a deadlock, while q_8 is. So the only possibility left is that $q_5 \xrightarrow{b} q_4$ and so S is now $\{(q_0, q_4), (q_1, q_5), (q_2, q_7), (q_2, q_8), (q_3, q_4) \ldots\}$. So the second pair has also been analyzed.

Pairs (q_2, q_7) and (q_2, q_8) vacuously satisfy the simulation condition because q_2 is a deadlock and so there is no transition to match.

Finally, pair (q_3, q_4) is OK because transition $q_3 \xrightarrow{a} q_1$ can be matched by $q_4 \xrightarrow{a} q_5$ and (q_1, q_5) is already in S; and also the other transition $q_3 \xrightarrow{a} q_2$ can be matched by $q_4 \xrightarrow{a} q_7$ and (q_2, q_7) is already in S. Hence, the simulation S we were looking for, proving that $q_0 \lesssim q_4$, is $\{(q_0, q_4), (q_1, q_5), (q_2, q_7), (q_2, q_8), (q_3, q_4)\}$. □

Exercise 2.26. Continuing Example 2.4, check that the following are simulation relations:

- $S_1 = \{(q_0, q_4), (q_1, q_5), (q_2, q_6), (q_2, q_4), (q_3, q_4)\}$
- $S_2 = \{(q_0, q_4), (q_1, q_6), (q_2, q_6), (q_2, q_9), (q_3, q_9)\}$
- $S_3 = S \cup \{(q_2, q_6)\}$

Check also that $S_4 = \{(q_0,q_4),(q_1,q_6),(q_2,q_8),(q_0,q_9)\}$ is not a simulation, even if for each pair $(q,q') \in S_4$ it holds that $q \lesssim q'$. □

Exercise 2.27. Continuing Example 2.4 and Exercise 2.26, check that

- $S^{-1} = \{(q_4,q_0),(q_5,q_1),(q_7,q_2),(q_8,q_2),(q_4,q_3)\}$ is not a simulation,
- None of the S_i^{-1}, for $i = 1,2,3$, is a simulation.
- $R = S^{-1} \cup \{(q_6,q_1),(q_9,q_3)\}$ is a simulation.

Therefore, $q_0 \simeq q_4$ because $q_0 \lesssim q_4$ (due to relation S), and $q_4 \lesssim q_0$ (due to R). □

We now list some useful properties of simulation relations.

Proposition 2.2. *For any LTS $TS = (Q,A,\rightarrow)$, the following hold:*

1. *the identity relation $\mathscr{I} = \{(q,q) \mid q \in Q\}$ is a simulation;*
2. *the relational composition $R_1 \circ R_2 = \{(q,q'') \mid \exists q'.(q,q') \in R_1 \wedge (q',q'') \in R_2\}$ of two simulations R_1 and R_2 is a simulation.*
3. *the union $\bigcup_{i \in I} R_i$ of simulations R_i is a simulation.*

Proof. *The proof of (1) is immediate: $(q,q) \in \mathscr{I}$ is a simulation pair because whatever transition q performs (say, $q \xrightarrow{\mu} q'$), the other q in the pair does exactly the same transition $q \xrightarrow{\mu} q'$ with $(q',q') \in \mathscr{I}$.*

The proof of (2) is also easy: given a pair $(q,q'') \in R_1 \circ R_2$, there exists a state q' such that $(q,q') \in R_1$ and $(q',q'') \in R_2$; as $(q,q') \in R_1$, if $q \xrightarrow{\mu} q_1$, there exists q_2 such that $q' \xrightarrow{\mu} q_2$ with $(q_1,q_2) \in R_1$. But as $(q',q'') \in R_2$, we have also that there exists q_3 such that $q'' \xrightarrow{\mu} q_3$ with $(q_2,q_3) \in R_2$. Summing up, for any pair $(q,q'') \in R_1 \circ R_2$, if $q \xrightarrow{\mu} q_1$, then there exists a state q_3 such that $q'' \xrightarrow{\mu} q_3$ with $(q_1,q_3) \in R_1 \circ R_2$, as required.

The proof of (3) is trivial, too: assume $(q,q') \in \bigcup_{i \in I} R_i$; then, there exists $j \in I$ such that (q,q') belongs to simulation R_j. If $q \xrightarrow{\mu} q_1$, then there must exist q_2 such that $q' \xrightarrow{\mu} q_2$ with $(q_1,q_2) \in R_j$. Hence, $(q_1,q_2) \in \bigcup_{i \in I} R_i$ as $R_j \subseteq \bigcup_{i \in I} R_i$. So $\bigcup_{i \in I} R_i$ is a simulation, too. □

Remember that $q \lesssim q'$ if there exists a simulation containing the pair (q,q'). This means that $\lesssim$ is the union of all simulations, i.e.,

$$\lesssim = \bigcup \{R \subseteq Q \times Q \mid R \text{ is a simulation}\}.$$

By Proposition 2.2(3), $\lesssim$ is also a simulation, hence the largest such relation.

Proposition 2.3. *For any LTS $TS = (Q,A,\rightarrow)$, relation $\lesssim \subseteq Q \times Q$ is the largest simulation relation.* □

Observe that all the simulation relations we have presented so far are neither reflexive, nor transitive (with the exception of relation S' in Exercise 2.23(*iii*)). Nonetheless, the largest simulation relation $\lesssim$ is a preorder.

Proposition 2.4. *For any LTS $TS = (Q, A, \to)$, relation $\lesssim$ is a preorder, i.e., reflexive and transitive, while relation $\simeq$ is an equivalence relation.*

Proof. *By Proposition 2.2.1, the identity relation $\mathscr{I} = \{(q,q) \mid q \in Q\}$ is a simulation; since $\lesssim$ is the union of all simulations, $\mathscr{I} \subseteq \lesssim$ and so $\lesssim$ is reflexive.*

Transitivity states that if $q_1 \lesssim q_2$ and $q_2 \lesssim q_3$ then $q_1 \lesssim q_3$. If $q_1 \lesssim q_2$ then there exists a simulation R_1 such that $(q_1,q_2) \in R_1$; similarly, if $q_2 \lesssim q_3$ then there exists a simulation R_2 such that $(q_2,q_3) \in R_2$; by Proposiiton 2.2.2, $R_1 \circ R_2$ is a simulation containing the pair (q_1,q_3). Since $\lesssim$ is the union of all simulations, $R_1 \circ R_2 \subseteq \lesssim$ and so $q_1 \lesssim q_3$, i.e. $\lesssim$ is transitive.

Remember that $q \simeq q'$ if $q \lesssim q'$ and $q' \lesssim q$. So, $\simeq$ is trivially reflexive and transitive because $\lesssim$ is reflexive and transitive. Relation $\simeq$ is also trivially symmetric: $q \lesssim q'$ and $q' \lesssim q$ implies $q \simeq q'$, *but then also $q' \simeq q$ because, of course, we can reorder the two premises of the implication in the logical* and. □

Exercise 2.28. (Maximum and minimum elements for the simulation preorder) (*i*) Prove that a one-state LTS exists that can be simulated by any other rooted LTS (i.e., a *minimum* element in the simulation preorder $\lesssim$). (*ii*) Prove also that, for a given set A of labels, a one-state LTS exists that can simulate any other rooted LTS over A (i.e., a *maximum* element in the simulation preorder $\lesssim$). (*iii*) Are these one-state LTSs minima or maxima also for the trace preorder $\leq_{tr}$ defined in Exercise 2.16? (*iv*) Show that there exist many different maxima (or minima) rooted LTSs for the simulation preorder, but any two of them are simulation equivalent. (*v*) Show a maximum element for the trace preorder which is not a maximum element for the simulation preorder. □

As pointed out in [vGP08], the simulation preorder is the coarsest preorder included in the trace preorder that is known to be decidable in polynomial time and polynomial space (see, e.g., [GPP03, RT07, vGP08, CRT11] for some performant algorithms), hence establishing a simulation between two processes is an efficient way of showing that they are related by trace inclusion.

However, simulation equivalence is not sensitive to deadlock. Consider again the two LTSs in Figure 2.12. It is easy to see that the two are simulation equivalent. Relation $R_1 = \{(q_5,q_1),(q_6,q_2),(q_7,q_4)\}$ is a simulation proving that $q_5 \lesssim q_1$. Similarly, $R_2 = \{(q_1,q_5),(q_2,q_6),(q_3,q_6),(q_4,q_7)\}$ is a simulation proving that $q_1 \lesssim q_5$.

Exercise 2.29. Consider Figure 2.13. Show that $q_1 \simeq q_4$ by providing two suitable simulation relations. □

One can slightly refine the definition of simulation to obtain such a sensitivity.

Definition 2.13. (Completed simulation) Let $TS = (Q, A, \to)$ be a transition system. A *completed simulation* is a simulation relation $R \subseteq Q \times Q$ such that, whenever $(q_1,q_2) \in R$, if $q_1 \not\to$ then $q_2 \not\to$.[6]

State q is completely simulated by q', denoted $q \lesssim_c q'$, if there exists a completed simulation R such that $(q,q') \in R$. Two states q and q' are *completed simulation equivalent*, denoted $q \simeq_c q'$, if $q \lesssim_c q'$ and $q' \lesssim_c q$ □

[6] Observe that, by Exercise 2.24, this condition can be expressed as $q_1 \not\to$ if and only if $q_2 \not\to$.

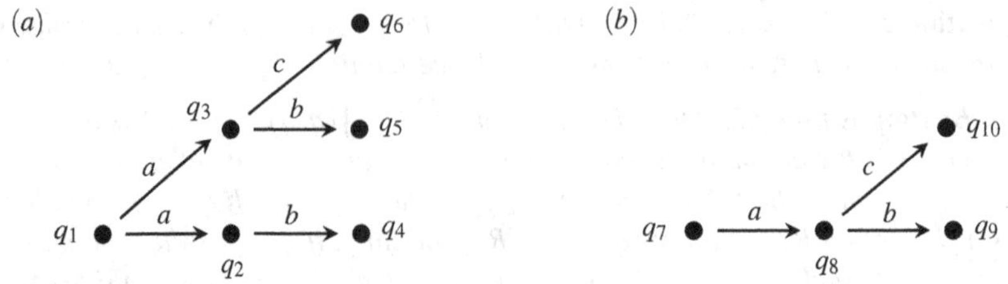

Fig. 2.15 Two completed simulation equivalent systems

Exercise 2.30. Continuing Exercise 2.25, observe that $q_0 \not\lesssim_c q_2$ in Figure 2.9, even if $q_0 \lesssim q_2$. On the contrary, $q_0 \lesssim_c q_5$ because relation $S = \{(q_0, q_5), (q_1, q_6)\}$ is a completed simulation. Prove also that the two LTSs in Figure 2.12, as well as those in Figure 2.13, are not completed simulation equivalent. □

Exercise 2.31. Following the same idea of the proof of Proposition 2.1, demonstrate that if q and q' are completed simulation equivalent, then q and q' are completed trace equivalent. The reverse implication does not hold: consider the usual two vending machines of Figures 2.3 and 2.10 that are completed trace equivalent but not (completed) simulation equivalent. □

Exercise 2.32. Following Exercise 2.28, show that there is no minimum LTS for the completed simulation preorder, nor a maximum element (*Hint:* Consider an LTS composed of a single deadlock state and the LTS of Figure 2.9(a)). □

However, we are not yet completely satisfied because the ability to sense the timing of choices is not captured in a complete way. Consider the two transition systems in Figure 2.15. These two systems are completed simulation equivalent (check this!), but the left one, after reacting to a, can reach a state (namely q_2) where only b is available, while this situation is not possible for the system at the right: after a, state q_8 is reached, which offers both b and c. Is this a good reason for rejecting (completed) simulation equivalence? It is sometimes argued that this is a matter of taste. However, we think that these two systems can be observationally distinguished by an observer who can repeat experiments on the two systems and can realize that in any case, after a, the system on the right can always react to b and c, while the system on the left can sometimes refuse to react to c.

Summing up, the main advantages of the simulation preorder are

- a simple proof principle (simply exhibit a relation and check easily that it is a simulation),
- sufficient condition for trace inclusion, more easily checkable, hence useful when verifying safety properties;

and the main disadvantage is that it is still too abstract, i.e., it identifies machines that should be distinguished. An obvious enhancement would be to require that the simulation game is played step by step not only in one direction but in both directions, leading to the notion of *bisimulation* (see below). As a further step towards bisimulation, consider the following exercise.

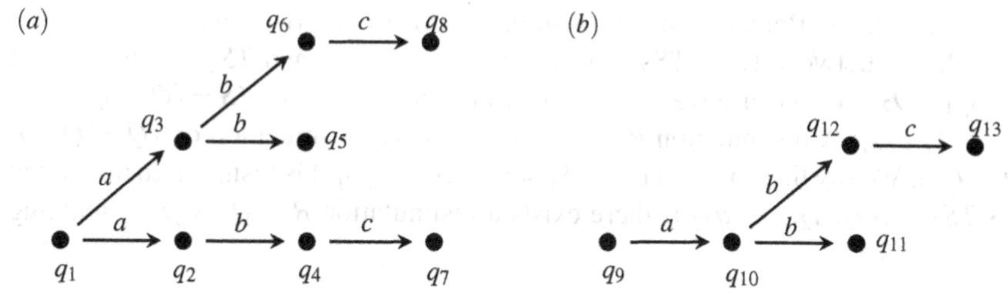

Fig. 2.16 Two ready simulation equivalent systems

Exercise 2.33. (Ready simulation) Let $TS = (Q, A, \rightarrow)$ be a labeled transition system. A *ready simulation*[7] is a simulation relation $R \subseteq Q \times Q$ such that whenever $(q_1, q_2) \in R$, if $q_2 \xrightarrow{\mu}$ then $q_1 \xrightarrow{\mu}$. This means that, not only q_1 is simulated by q_2, but that q_1 and q_2 have the same menu of possible initial moves. State q is ready simulated by q', denoted $q \lesssim_{rs} q'$, if there exists a ready simulation R such that $(q, q') \in R$. Two states q and q' are ready simulation equivalent, denoted $q \simeq_{rs} q'$, if $q \lesssim_{rs} q'$ and $q' \lesssim_{rs} q$.

- Prove that $\lesssim_{rs}$ is a preorder, and that $\simeq_{rs}$ is an equivalence relation.
- Prove also that if $q \lesssim_{rs} q'$, then $q \lesssim_{c} q'$; does the converse implication hold? (*Hint*: Consider states q_0 and q_5 of Figure 2.9)
- Show that the two LTSs in Figure 2.15 are not ready simulation equivalent.
- Show that the two LTSs in Figure 2.16 are ready simulation equivalent; argue why, according to the discussion above, they should not be considered equivalent.
- Show that there is no rooted LTS that can ready simulate any other rooted LTS, nor a rooted LTS that can be ready simulated by any other rooted LTS (i.e., the ready simulation preorder has no maximum and no minimum elements, *cf.* Exercises 2.28 and 2.32). □

2.3.4 Bisimulation

Definition 2.14. Let $TS = (Q, A, \rightarrow)$ be a transition system. A *bisimulation* (originated in [Park81, Mil89]) is a relation $R \subseteq Q \times Q$ such that R and its inverse R^{-1} are both simulation relations. More explicitly, a bisimulation is a relation R such that if $(q_1, q_2) \in R$ then for all $\mu \in A$

- $\forall q_1'$ such that $q_1 \xrightarrow{\mu} q_1'$, $\exists q_2'$ such that $q_2 \xrightarrow{\mu} q_2'$ and $(q_1', q_2') \in R$,
- $\forall q_2'$ such that $q_2 \xrightarrow{\mu} q_2'$, $\exists q_1'$ such that $q_1 \xrightarrow{\mu} q_1'$ and $(q_1', q_2') \in R$.

Two states q and q' are *bisimilar* (or *bisimulation equivalent*), denoted $q \sim q'$, if there exists a bisimulation R such that $(q, q') \in R$. □

[7] Ready simulation was originally proposed in [LS91] under the name of 2/3 bisimulation, and further studied in [BIM95], which gave it its current name.

In the light of Remark 2.4, the definition above comprises also the case of a bisimulation between two LTSs, say, $TS_1 = (Q_1, A_1, \rightarrow_1)$ and $TS_2 = (Q_2, A_2, \rightarrow_2)$ with $Q_1 \cap Q_2 = \emptyset$, because we may consider just one single LTS $TS = (Q_1 \cup Q_2, A_1 \cup A_2, \rightarrow_1 \cup \rightarrow_2)$: A bisimulation $R \subseteq Q_1 \times Q_2$ is also a bisimulation on $(Q_1 \cup Q_2) \times (Q_1 \cup Q_2)$. We say that a rooted LTS $TS_1 = (Q_1, A_1, \rightarrow_1, q_1)$ is bisimilar to the rooted LTS $TS_2 = (Q_2, A_2, \rightarrow_2, q_2)$ if there exists a bisimulation $R \subseteq Q_1 \times Q_2$ containing the pair (q_1, q_2).

Remark 2.5. **(Symmetric simulation and bisimulation)** A relation $R \subseteq Q \times Q$ is symmetric if for all $q, q' \in Q$, $(q, q') \in R$ iff $(q', q) \in R$. Therefore, if R is symmetric, then $R = R^{-1} = \{(q', q) \mid (q, q') \in R\}$. As a consequence, any symmetric simulation relation S is also a bisimulation, because if S is a simulation, then also $S^{-1} = S$ is a simulation. However, a bisimulation relation need not to be symmetric, as we will see in the following examples, even if it has to be a simulation. □

Two states q and q' are bisimilar, $q \sim q'$, if there exists a relation R containing the pair (q, q') such that both R and its inverse R^{-1} are simulations. On the contrary, we have that q and q' are simulation equivalent, $q \simeq q'$, if we have two simulation relations R_1 and R_2 such that $(q, q') \in R_1$ and $(q', q) \in R_2$, but R_2 may be different from R_1^{-1}. Moreover, if $(q, q') \in R$ for some bisimulation R, then q and q' offer the same menu of initial moves. Therefore, the following implications trivially hold.

Proposition 2.5. *$q \sim q'$ implies $q \simeq_{rs} q'$ implies $q \simeq_c q'$ implies $q \simeq q'$.* □

All these three implications above are strict. We will see in Figure 2.20 a summary of the counterexamples for the reverse implications. The following exercise completes the comparison with other equivalences.

Exercise 2.34. (Isomorphism equivalence implies bisimilarity) Given two labeled transition systems $TS_1 = (Q_1, A_1, \rightarrow_1)$ and $TS_2 = (Q_2, A_2, \rightarrow_2)$, prove that if TS_1 and TS_2 are isomorphic via bijection f (see Definition 2.8), then $R = \{(q_1, q_2) \in Q_1 \times Q_2 \mid f(q_1) = q_2\}$ is a bisimulation. (Moreover, if $Q_1 \cap Q_2 = \emptyset$, then R is a bisimulation over the union of the two transition systems $TS = (Q_1 \cup Q_2, A_1 \cup A_2, \rightarrow_1 \cup \rightarrow_2)$.) □

As for the simulation preorder, proving two states q and q' bisimilar is often an easy task: just exhibit a bisimulation R such that $(q, q') \in R$. For instance, consider the two LTSs in Figure 2.8; they are bisimulation equivalent because $R = \{(q, q'), (q, q'')\}$ is a bisimulation relation. Checking that R is indeed a bisimulation means checking that each pair $(q_1, q_2) \in R$ satisfies the bisimulation condition

$$\forall q_1' \text{ such that } q_1 \xrightarrow{a} q_1',\ \exists q_2' \text{ such that } q_2 \xrightarrow{a} q_2' \text{ and } (q_1', q_2') \in R$$

and conversely if q_2 moves first. Note that this is not the only bisimulation relation between these two LTSs: for instance, $R' = \{(q, q'')\}$ is another bisimulation. However, $R'' = \{(q, q')\}$ is not a bisimulation, because q' can respond to transition $q \xrightarrow{a} q$ only with $q' \xrightarrow{a} q''$, but $(q, q'') \notin R''$.

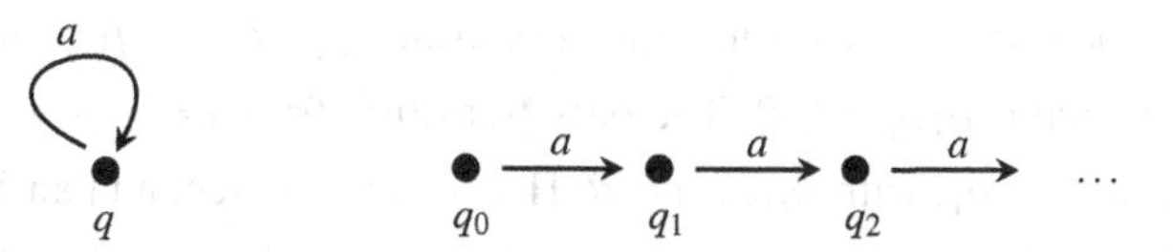

Fig. 2.17 Two bisimulation equivalent processes

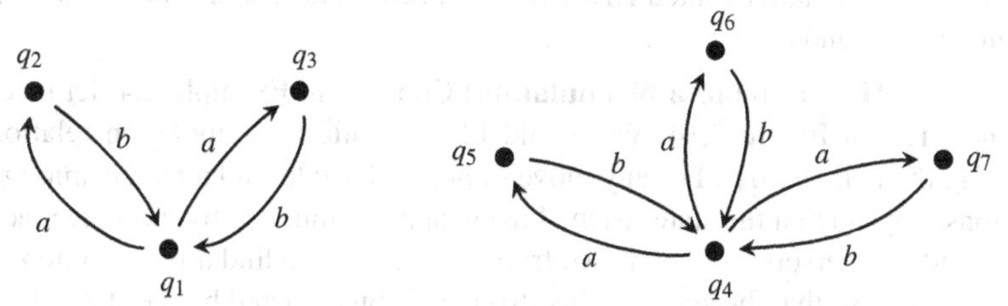

Fig. 2.18 Other two bisimulation equivalent processes

Exercise 2.35. Check that $R = \{(q,q),(q,q'),(q,q''),(q',q''),(q'',q'')\}$ is a bisimulation over the (union of the) LTSs in Figure 2.8. Compute all the bisimulations over $\{q,q',q''\} \times \{q,q',q''\}$. Is there any difference w.r.t. the set of all simulations computed in Exercise 2.23? How is the largest (w.r.t. set inclusion) such relation? Is it reflexive, symmetric and transitive? □

Exercise 2.36. (i) Observe that, for any LTS $(Q,A,\to)$, the empty relation $R = \emptyset$ is a bisimulation. (ii) Single out an LTS $TS_1 = (Q_1,A,\to_1)$ such that the universal relation $R' = Q_1 \times Q_1$ is a bisimulation, and another LTS $TS_2 = (Q_2,A,\to_2)$ such that the universal relation $R'' = Q_2 \times Q_2$ is not a bisimulation. □

Observe also that checking bisimilarity between infinite-state systems may be not more difficult. Consider, for instance, the two LTSs in Figure 2.17. They are bisimulation equivalent, as $R = \{(q,q_i) \mid i \in \mathbb{N}\}$ is a bisimulation. To check that R is a bisimulation, it is enough to observe that, for any i, the pair $(q,q_i) \in R$ evolves into the pair (q,q_{i+1}) still in R.

Example 2.5. Consider the two LTSs in Figure 2.18 and take the relation

$$R = \{(q_1,q_4),(q_2,q_5),(q_3,q_5),(q_2,q_6),(q_2,q_7)\}.$$

We want to show that R is a bisimulation. Take the first pair (q_1,q_4). We have to take all the transitions from q_1 and find a suitable match among the transitions from q_4; and symmetrically, for all the transitions from q_4, we have to find a suitable match among the transitions from q_1. Let us start with the transitions from q_1. We have $q_1 \xrightarrow{a} q_2$ and $q_1 \xrightarrow{a} q_3$. Both transitions can be matched by $q_4 \xrightarrow{a} q_5$ with $(q_2,q_5) \in R$ as well as $(q_3,q_5) \in R$. Now consider the transitions from q_4; they are: $q_4 \xrightarrow{a} q_5$, $q_4 \xrightarrow{a} q_6$ and $q_4 \xrightarrow{a} q_7$. All the three transitions can be matched by $q_1 \xrightarrow{a} q_2$ with $\{(q_2,q_5),(q_2,q_6),(q_2,q_7)\} \subseteq R$. Hence, the first pair is OK. Now consider the pair (q_2, q_5). The only transition from q_2 is $q_2 \xrightarrow{b} q_1$, which is matched by $q_5 \xrightarrow{b} q_4$, with $(q_1, q_4) \in R$. The only transition from q_5 is $q_5 \xrightarrow{b} q_4$, which is matched by $q_2 \xrightarrow{b} q_1$, with $(q_1, q_4) \in R$. Hence, also the second pair is OK. □

Exercise 2.37. (*i*) Complete the check that R above is a bisimulation. (*ii*) How many different bisimulation relations containing the pair (q_1, q_4) can you find? (*iii*) Is it possible to find a smaller rooted LTS (i.e., with fewer states and transitions) which is bisimilar to q_1 and q_4? □

Example 2.6. **(How to build a bisimulation)** Continuing Example 2.4, let us consider the LTSs in Figure 2.14. We would like to build a bisimulation relation R containing the pair (q_0, q_4). For any move of q_0, we have to find a match among the transitions of q_4 so that the state reached from q_0 is bisimulated by the state reached from q_4 and, conversely, for any move from q_4, we have to find a match among the transitions of q_0 so that the state reached from q_0 is bisimulated by the state reached from q_4.

To transition $q_0 \xrightarrow{a} q_1$, state q_4 can respond with three different transitions, reaching q_5, q_6 and q_7, respectively; it is necessary that at least one of these match the bisimulation request. Among the three, transition $q_4 \xrightarrow{a} q_7$ is not OK because the pair we should add to R is (q_1, q_7), but q_7 cannot bisimulate q_1 because q_7 is a deadlock. So, we may try to add to R one of the other two pairs, say, (q_1, q_5), which must then be checked. But before examining this new pair, let us continue exploring the conditions for the first pair (q_0, q_4). To transition $q_0 \xrightarrow{a} q_2$, state q_4 can respond with three different transitions, reaching q_5, q_6 and q_7, respectively; as q_2 is a deadlock, the only suitable match is $q_4 \xrightarrow{a} q_7$ and (q_2, q_7) is to be added to R. In order to complete the analysis of the first pair, we now have to consider all the moves from q_4. The suitable match for transition $q_4 \xrightarrow{a} q_7$ is transition $q_0 \xrightarrow{a} q_2$, and the pair (q_2, q_7) is already in R. To transition $q_4 \xrightarrow{a} q_5$, q_0 can respond only with $q_0 \xrightarrow{a} q_1$, and the pair (q_1, q_5) is already in R. To transition $q_4 \xrightarrow{a} q_6$, q_0 can respond only with $q_0 \xrightarrow{a} q_1$, and the pair (q_1, q_6) is to be added to R, which is now $\{(q_0, q_4), (q_1, q_5), (q_2, q_7), (q_1, q_6), \ldots\}$ We have completed the analysis of the first pair: if the other three pairs satisfy the bisimulation condition, then (q_0, q_4) is really a bisimulation pair.

Let us examine pair (q_1, q_5) now. Transition $q_1 \xrightarrow{b} q_2$ reaches a deadlock, so only transition $q_5 \xrightarrow{b} q_8$ can match and we add (q_2, q_8) to R. The other transition from q_1 is $q_1 \xrightarrow{b} q_3$. This transition cannot be matched by $q_5 \xrightarrow{b} q_8$ because q_3 is not a deadlock, while q_8 is. So the only possibility left is that $q_5 \xrightarrow{b} q_4$ and so R is now $\{(q_0, q_4), (q_1, q_5), (q_2, q_7), (q_1, q_6), (q_2, q_8), (q_3, q_4) \ldots\}$. In order to complete the analysis of the second pair, we have now to consider all the moves from q_5. But this can be done as above: transition $q_5 \xrightarrow{b} q_8$ can be matched by $q_1 \xrightarrow{b} q_2$, and transition $q_5 \xrightarrow{b} q_4$ can be matched by $q_1 \xrightarrow{b} q_3$, and so no new pairs are added to R. So also the second pair has been analyzed.

Pairs (q_2, q_7) and (q_2, q_8) satisfy vacuously the bisimulation condition because q_2, q_7 and q_8 are deadlocks.

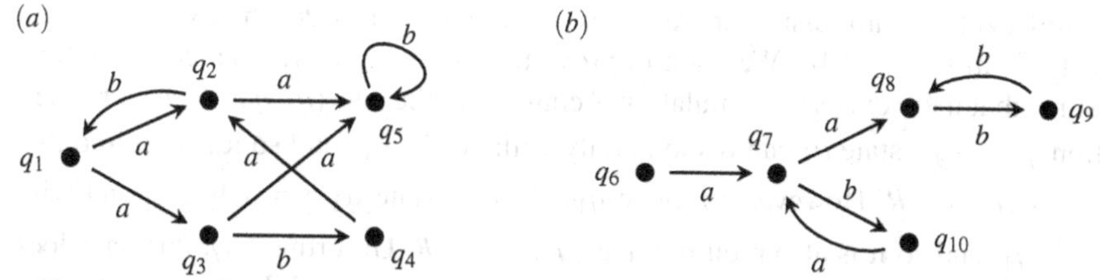

Fig. 2.19 Are q_1 and q_6 bisimulation equivalent?

Let us consider pair (q_1,q_6). Transition $q_1 \xrightarrow{b} q_2$ reaches a deadlock, so only transition $q_6 \xrightarrow{b} q_8$ can match, and the pair (q_2,q_8) is already in R. The other transition from q_1 is $q_1 \xrightarrow{b} q_3$. This transition can be matched only by $q_6 \xrightarrow{b} q_9$, and (q_3,q_9) is to be added to R. In order to complete the analysis of the second pair, we have now to consider all the moves from q_6. But this is as above, hence omitted. So, R is now $\{(q_0,q_4),(q_1,q_5),(q_2,q_7),(q_1,q_6),(q_2,q_8),(q_3,q_4),(q_3,q_9),\ldots\}$.

Pair (q_3,q_4) is OK because transition $q_3 \xrightarrow{a} q_1$ can be matched by $q_4 \xrightarrow{a} q_5$ and (q_1,q_5) is already in R; and also the other transition $q_3 \xrightarrow{a} q_2$ can be matched by $q_4 \xrightarrow{a} q_7$, and (q_2,q_7) is already in R.

Finally, we have to examine pair (q_3,q_9). Transition $q_3 \xrightarrow{a} q_1$ can be only matched by $q_9 \xrightarrow{s} q_6$, and (q_1,q_6) is already in R. Transition $q_3 \xrightarrow{a} q_2$ can be only matched by $q_9 \xrightarrow{s} q_8$, and (q_2,q_8) is already in R. Similarly, if q_9 moves first.

Hence, the bisimulation R we were looking for, proving that $q_0 \sim q_4$, is $\{(q_0,q_4),(q_1,q_5),(q_2,q_7),(q_1,q_6),(q_2,q_8),(q_3,q_4),(q_3,q_9)\}$. □

Exercise 2.38. Continuing Example 2.6, check which of the following are bisimulation relations:

- $R_1 = R \cup \{(q_7,q_8),(q_1,q_1),(q_0,q_9)\}$
- $R_2 = R \cup \{(q_0,q_3),(q_3,q_6))\}$
- $R_3 = R \cup \{(q_5,q_6),(q_4,q_9),(q_8,q_8),(q_7,q_8)\}$
- $R_4 = \{(q_5,q_6),(q_4,q_9),(q_8,q_8),(q_8,q_7)\}$

□

Exercise 2.39. Check that the two states q_1 and q_6 in Figure 2.19 are bisimulation equivalent. □

Example 2.7. (**How to prove that two states are not bisimilar**) Given a finite-state LTS, a simple-minded strategy to prove that two states, say q and q', are not bisimilar would consist of enumerating all the possible binary relations (over the *finite* set of states) containing the pair (q,q'), and checking that none of them is a bisimulation: clearly a very inefficient, exponential algorithm.

A much better way to prove nonbisimilarity of two (initial) states would consist in finding a sequence of moves that inevitably leads to two (final) states that are self-evidently not bisimilar. As an example, let us consider the two ready similar LTSs in Figure 2.16. We want to prove that $q_1 \not\sim q_9$. Towards a contradiction, assume that there exists a bisimulation R containing the pair (q_1, q_9). Hence, to transition $q_1 \xrightarrow{a} q_2$, state q_9 can respond only with $q_9 \xrightarrow{a} q_{10}$ and so it is required that also $(q_2, q_{10}) \in R$. However, to move $q_{10} \xrightarrow{b} q_{11}$, state q_2 can only respond with $q_2 \xrightarrow{b} q_4$, and so it is also required that $(q_4, q_{11}) \in R$. Unfortunately, this pair does not satisfy the bisimulation condition, as q_{11} is a deadlock, while q_4 is not. Hence, we have singled out a sequence of moves from the (initial) states q_1 and q_9 that has inevitably led to inequivalent (final) states q_4 and q_{11}. Therefore, we can conclude that $q_1 \not\sim q_9$. □

Consider again the two vending machines in Figures 2.3 and 2.10. The argument we have used after Exercise 2.21 to prove that q_1 cannot be simulated by q_5 also proves that $q_1 \not\sim q_5$.

Exercise 2.40. Prove that q_1 and q_7 in Figure 2.15 are not bisimilar. □

Exercise 2.41. (String bisimulation) Let $TS = (Q, A, \rightarrow)$ be an LTS. A *string bisimulation* is a relation $R \subseteq Q \times Q$ such that if $(q_1, q_2) \in R$ then for all $\sigma \in A^*$

- $\forall q_1'$ such that $q_1 \xrightarrow{\sigma}{}^* q_1'$, $\exists q_2'$ such that $q_2 \xrightarrow{\sigma}{}^* q_2'$ and $(q_1', q_2') \in R$,
- $\forall q_2'$ such that $q_2 \xrightarrow{\sigma}{}^* q_2'$, $\exists q_1'$ such that $q_1 \xrightarrow{\sigma}{}^* q_1'$ and $(q_1', q_2') \in R$.

Two states q and q' are string bisimilar, denoted $q \sim^* q'$, if there exists a string bisimulation R such that $(q, q') \in R$.

(*i*) Prove that bisimilarity $\sim$ and string bisimilarity $\sim^*$ coincide, i.e., $q_1 \sim q_2$ if and only if $q_1 \sim^* q_2$. (*ii*) Prove that an easy consequence of the above is that if $q_1 \sim q_2$ then $Tr(q_1) = Tr(q_2)$, as well as $CTr(q_1) = CTr(q_2)$. □

We now list some useful properties of bisimulation relations.

Proposition 2.6. *For any LTS* $TS = (Q, A, \rightarrow)$*, the following hold:*

1. *the identity relation* $\mathscr{I} = \{(q,q) \mid q \in Q\}$ *is a bisimulation;*
2. *the inverse relation* $R^{-1} = \{(q',q) \mid (q,q') \in R\}$ *of a bisimulation R is a bisimulation;*
3. *the relational composition* $R_1 \circ R_2 = \{(q,q'') \mid \exists q'.(q,q') \in R_1 \wedge (q',q'') \in R_2\}$ *of two bisimulations* R_1 *and* R_2 *is a bisimulation.*
4. *the union* $\bigcup_{i \in I} R_i$ *of bisimulations* R_i *is a bisimulation.*

Proof. Left as an exercise to the reader (it is very similar to the proof of Proposition 2.2). □

Remember that $q \sim q'$ if there exists a bisimulation containing the pair (q, q'). This means that $\sim$ is the union of all bisimulations, i.e.,

$$\sim = \bigcup \{R \subseteq Q \times Q \mid R \text{ is a bisimulation}\}.$$

By Proposition 2.6(4), $\sim$ is also a bisimulation, hence the largest such relation.

Proposition 2.7. *For any LTS $TS = (Q, A, \to)$, relation $\sim \subseteq Q \times Q$ is the largest bisimulation relation.* □

Observe that most bisimulation relations we have presented so far are not reflexive, not symmetric, and not transitive. Nonetheless, the largest bisimulation relation $\sim$ is an equivalence relation. As a matter of fact, as the identity relation $\mathscr{I}$ is a bisimulation by Proposition 2.6(1), we have that $\mathscr{I} \subseteq \sim$, and so $\sim$ is reflexive. Symmetry derives from the following argument. For any $(q,q') \in \sim$, there exists a bisimulation R such that $(q,q') \in R$; by Proposition 2.6(2), relation R^{-1} is a bisimulation containing the pair (q',q); hence, $(q',q) \in \sim$ because $R^{-1} \subseteq \sim$. Transitivity also holds for $\sim$. Assume $(q,q') \in \sim$ and $(q',q'') \in \sim$; hence, there exist two bisimulations R_1 and R_2 such that $(q,q') \in R_1$ and $(q',q'') \in R_2$; by by Proposition 2.6(3), relation $R_1 \circ R_2$ is a bisimulation containing the pair (q,q''); hence, $(q,q'') \in \sim$, because $R_1 \circ R_2 \subseteq \sim$. Summing up, we have the following:

Proposition 2.8. *For any LTS $TS = (Q, A, \to)$, relation $\sim \subseteq Q \times Q$ is an equivalence relation.* □

Let us now define *recursively* a new behavioral relation $\sim' \subseteq Q \times Q$ as follows: $q_1 \sim' q_2$ *if and only if* for all $\mu \in A$

- $\forall q_1'$ such that $q_1 \xrightarrow{\mu} q_1'$, $\exists q_2'$ such that $q_2 \xrightarrow{\mu} q_2'$ and $q_1' \sim' q_2'$,
- $\forall q_2'$ such that $q_2 \xrightarrow{\mu} q_2'$, $\exists q_1'$ such that $q_1 \xrightarrow{\mu} q_1'$ and $q_1' \sim' q_2'$.

We would like to adopt this as *the* definition of our behavioral equivalence; but, this does not identify a unique relation, as many different relations satisfy this recursive definition. As an example, consider the LTS in Figure 2.8(b): then it is not difficult to check that, e.g., $R_1 = \emptyset$, $R_2 = \{(q',q'),(q'',q'')\}$ and $R_3 = \{(q',q'),(q'',q''),(q',q''),(q'',q')\}$ are solutions of this recursive definition. It is important to observe that a relation R which is a solution of the recursive definition above must be a bisimulation because the condition "$q_1 \sim' q_2$ *if and only if* for all $\mu \in A$ ..." above implies the condition "$q_1 \sim' q_2$ *implies* for all $\mu \in A$ ..." of the definition of bisimulation (implication from left to right). However, not all bisimulations are solutions; consider again Figure 2.8(b): the bisimulation relation $R = \{(q'',q'')\}$ is not a solution because the implication from right to left requires that the solution contain also the pair (q',q').

Exercise 2.42. Find an LTS such that the empty bisimulation $R = \emptyset$ is not a solution of the recursive definition above. (*Hint:* Consider an LTS composed of one deadlocked state only.) □

Now we want to prove that $\sim$, the largest bisimulation, is a solution of this recursive definition, hence proving that bisimulation equivalence may be seen as a fixed point of this recursive definition. A more formal study of bisimulation as a fixed point is postponed to Section 2.5, where we show that $\sim$ is the largest fixed point of a suitable relation transformer. Here we give a more elementary account of this fact.

Proposition 2.9. *For any LTS, bisimulation equivalence* $\sim$ *is such that* $q_1 \sim q_2$ if and only if *for all* $\mu \in A$

- $\forall q_1'$ *such that* $q_1 \xrightarrow{\mu} q_1'$, $\exists q_2'$ *such that* $q_2 \xrightarrow{\mu} q_2'$ *and* $q_1' \sim q_2'$,
- $\forall q_2'$ *such that* $q_2 \xrightarrow{\mu} q_2'$, $\exists q_1'$ *such that* $q_1 \xrightarrow{\mu} q_1'$ *and* $q_1' \sim q_2'$.

Proof. Note that in Definition 2.14, we have "implies" *instead of* "if and only if". *Hence, the implication from left to right is due to the fact that* $\sim$ *is itself a bisimulation. For the implication from right to left, we follow the proof in [Mil89]. First, define a new relation* $\sim''$ *in terms of* $\sim$ *as follows:*

$q_1 \sim'' q_2$ if and only if *for all* $\mu \in A$

- $\forall q_1'$ *such that* $q_1 \xrightarrow{\mu} q_1'$, $\exists q_2'$ *such that* $q_2 \xrightarrow{\mu} q_2'$ *and* $q_1' \sim q_2'$,
- $\forall q_2'$ *such that* $q_2 \xrightarrow{\mu} q_2'$, $\exists q_1'$ *such that* $q_1 \xrightarrow{\mu} q_1'$ *and* $q_1' \sim q_2'$.

Now we want to prove that $\sim = \sim''$, *hence proving the property above.*

First, if $q_1 \sim q_2$, *then (as* $\sim$ *is a bisimulation)*

- $\forall q_1'$ *such that* $q_1 \xrightarrow{\mu} q_1'$, $\exists q_2'$ *such that* $q_2 \xrightarrow{\mu} q_2'$ *and* $q_1' \sim q_2'$,
- $\forall q_2'$ *such that* $q_2 \xrightarrow{\mu} q_2'$, $\exists q_1'$ *such that* $q_1 \xrightarrow{\mu} q_1'$ *and* $q_1' \sim q_2'$

and so (by using the implication from right to left in the definition of $\sim''$*) we have that* $q_1 \sim'' q_2$. *It remains to prove that* $q_1 \sim'' q_2$ *implies* $q_1 \sim q_2$. *To obtain this, we prove that* $\sim''$ *is a bisimulation. Assume that* $q_1 \xrightarrow{\mu} q_1'$ *(the symmetric case when* q_2 *moves first is analogous, hence omitted). By definition of* $\sim''$, *we have that there exists a state* q_2' *such that* $q_2 \xrightarrow{\mu} q_2'$ *and* $q_1' \sim q_2'$; *but, by what we just proved, we have also that* $q_1' \sim'' q_2'$, *and we are done.* □

It is sometimes convenient to write compactly a bisimulation, by removing those pairs that differ from others only for the use of bisimulation equivalent alternatives. The resulting relation is *not* a bisimulation, rather a bisimulation up to $\sim$. We denote by $\sim R \sim$ the relational composition $\sim \circ R \circ \sim$; in other words, by $q \sim R \sim q'$ we mean that there exist two states q_1 and q_2 such that $q \sim q_1$, $(q_1, q_2) \in R$ and $q_2 \sim q'$.

Definition 2.15. (Bisimulation up to $\sim$) A bisimulation up to $\sim$ is a relation $R \subseteq Q \times Q$ such that if $(q_1, q_2) \in R$ then for all $\mu \in A$

- $\forall q_1'$ such that $q_1 \xrightarrow{\mu} q_1'$, $\exists q_2'$ such that $q_2 \xrightarrow{\mu} q_2'$ and $q_1' \sim R \sim q_2'$,
- $\forall q_2'$ such that $q_2 \xrightarrow{\mu} q_2'$, $\exists q_1'$ such that $q_1 \xrightarrow{\mu} q_1'$ and $q_1' \sim R \sim q_2'$. □

Lemma 2.1. *If* R *is a bisimulation up to* $\sim$, *then* $\sim R \sim$ *is a bisimulation.*

Proof. Assume $q \sim R \sim q'$, *i.e., there exist* q_1 *and* q_2 *such that* $q \sim q_1$, $(q_1, q_2) \in R$ *and* $q_2 \sim q'$. *We have to prove that for any* $q \xrightarrow{\mu} q^1$ *there exists* $q' \xrightarrow{\mu} q^2$ *such that* $q^1 \sim R \sim q^2$ *(the symmetric case when* q' *moves first is omitted). Since* $q \sim q_1$, *there exists* q_1' *such that* $q_1 \xrightarrow{\mu} q_1'$ *with* $q^1 \sim q_1'$. *Since* $(q_1, q_2) \in R$, *there exists* q_2' *such that* $q_2 \xrightarrow{\mu} q_2'$ *with* $q_1' \sim R \sim q_2'$. *Since* $q_2 \sim q'$, *there exists* q^2 *such that* $q' \xrightarrow{\mu} q^2$

with $q_2' \sim q^2$. Summing up, $q^1 \sim q_1'$ and $q_1' \sim R \sim q_2'$ and $q_2' \sim q^2$ can be shortened to $q^1 \sim R \sim q^2$, because $\sim \circ \sim = \sim$ by Proposition 2.6. Hence, we have proved that if $q \sim R \sim q'$ then for any q^1 such that $q \xrightarrow{\mu} q^1$ there exists q^2 such that $q' \xrightarrow{\mu} q^2$ with $q^1 \sim R \sim q^2$, as required by the definition of bisimulation. □

Proposition 2.10. *If R is a bisimulation up to $\sim$, then $R \subseteq \sim$.*

Proof. *By Lemma 2.1, $\sim R \sim$ is a bisimulation, hence $\sim R \sim \subseteq \sim$ by definition of $\sim$. As the identity relation $\mathscr{I} \subseteq \sim$ by Proposition 2.6(1), we have that relation $R = \mathscr{I} \circ R \circ \mathscr{I} \subseteq \sim R \sim$, hence $R \subseteq \sim$ by transitivity.* □

The above proposition states the correctness of the proof principle based on the above up-to technique: the fact that $R \subseteq \sim$ ensures that no erroneous equalities are introduced. Hence, in order to prove that $p \sim q$, it is enough to exhibit a bisimulation up to $\sim$ that contains the pair (p,q).

Exercise 2.43. Consider Figure 2.19. Assume to have already proved that $q_2 \sim q_3$ and $q_8 \sim q_9$. Then, prove that $q_1 \sim q_6$ by showing that relation $R = \{(q_1,q_6), (q_2,q_7), (q_5,q_8), (q_1,q_{10})\}$ is a bisimulation up to $\sim$. Note that R is not a bisimulation: some missing pairs are (q_3,q_7) and (q_4,q_{10}). □

The up-to technique is very useful when we consider LTSs generated by CCS processes, as we will see in Chapter 3. In that setting, states are CCS processes, and on CCS processes we will prove in Chapter 4 that many algebraic properties hold (e.g., associativity and commutativity of parallel composition) for bisimulation equivalence; hence, we can economize on the number of pairs in R by replacing one CCS process by some other equivalent one, according to some algebraic laws. Instances of application of this principle are outlined, e.g., in Example 3.9 (Section 3.4.3) and Example 3.12 (Section 3.4.4).

Bisimulation equivalence over a finite-state LTS with n states and m transitions can be computed in $O(m \log n)$ time [PT87] (see [DPP01] for an upgraded version of this algorithm and [FV99] for an experimental overview of the merits of various algorithms for bisimulation equivalence; see also Section 2.5 for an intuitive, non optimal algorithm). Differently from all the other equivalences, it is even decidable over some classes of infinite-state systems we will introduce in the next chapters (notably over BPP in Section 3.4.4 and BPA in Section 5.4.4). The reader interested in a gentle introduction to decidability of bisimilarity over these classes of infinite-state systems may consult [AIS12].

We summarize the various equivalence relations we have introduced so far in the diagram in Figure 2.20. An arrow from an equivalence relation to another means that the former is finer than the latter; e.g., completed simulation equivalence is finer than completed trace equivalence. A dashed arrow labeled with an index of a figure means that such an implication does not hold because of the counterexample discussed in that figure; for instance, the vending machines in Figures 2.3 and 2.10 are completed trace equivalent, but not completed simulation equivalent.

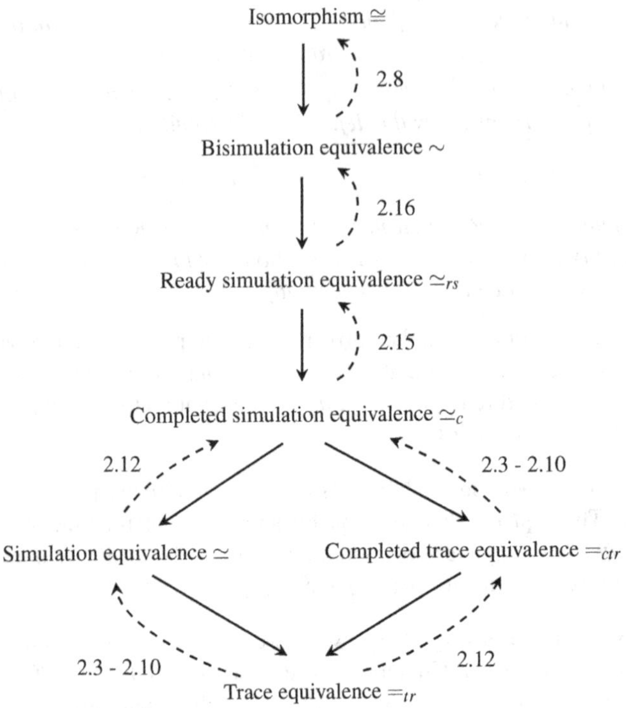

Fig. 2.20 Summary of (strong) behavioral equivalences

Exercise 2.44. (Determinism) Prove that trace equivalence and bisimulation equivalence coincide over deterministic LTSs. (*Hint:* Given a deterministic LTS $(Q, A, \to)$, show that $R = \{(q_1, q_2) \mid q_1, q_2 \in Q \text{ and } q_1 =_{tr} q_2\}$ is a strong bisimulation.)

This means that the diagram above collapses to a two-node diagram, one node for isomorphism and one node for all the other equivalences. Moreover, this observation offers a simple algorithm to check trace equivalence between two finite-state rooted LTSs $TS_i = (Q_i, A_i, \to_i, q_i)$ for $i = 1, 2$: first, transform TS_1 and TS_2 into trace equivalent, yet deterministic, rooted LTSs DTS_1 and DTS_2, respectively, by means of the procedure described in Exercise 2.17; then, check if the finite-state, rooted, deterministic LTSs DTS_1 and DTS_2 are strongly bisimilar. □

2.4 Abstracting from Invisible Actions

All the equivalences and preorders we have discussed so far are sometimes called *strong* to denote that all the actions of the LTS are equally observable. In fact, in real life this is not the case. A lot of activities of a system are completely internal and cannot be influenced by any interacting user and so, to some extent, are not

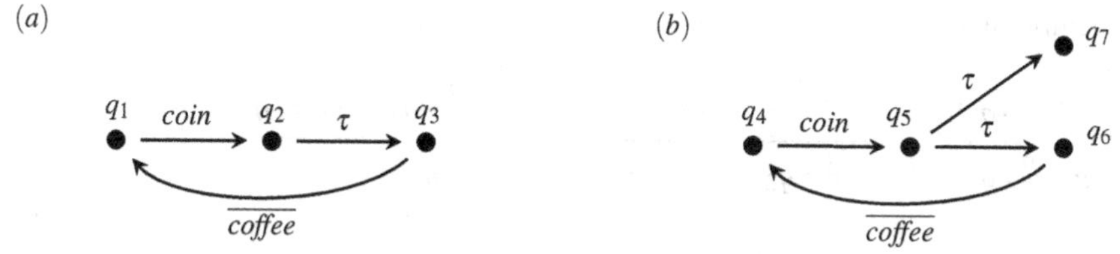

Fig. 2.21 Two models for the vending machine

observable. For instance, the first model in Figure 2.2 of the vending machine is not representing the internal activity of the preparation of the coffee, an activity that the machine performs but that the user cannot interact with. It is usually assumed that all the internal activities are equally represented by the same action τ, to express that the actual content of what an internal action does is unobservable. A model that more realistically represents the vending machine is outlined in Figure 2.21(a). An action τ can also be the result of a synchronization between two compound subsystems, as we will see in the next chapter. So, when specifying a system, it is often the case that the resulting LTS is rich of τ-labeled transitions.

Weak behavioral equivalences are those equivalences that compare systems by considering unobservable the occurrences of action τ. But we have to be precise in this respect. For instance, consider the LTSs in Figures 2.2 and 2.21(a). One can easily convince oneself that there is no observable difference between the two, as both can only perform the same traces of observable actions:

$$\{\varepsilon, coin, coin\,\overline{coffee}, coin\,\overline{coffee}\,coin, \ldots\}$$

Now, compare the two LTSs in Figure 2.21. If we are only interested in looking at the observable traces, we should say that the two are equivalent. However, the two differ for the important aspect of deadlock: the left one cannot deadlock, while the right one has always the possibility to deadlock after inserting a coin. Hence, τ transitions are not completely unobservable, because some choices can be taken internally by the machine and may have an observable effect by preempting other possible behaviors of the machine.

In the following, we briefly recall the main definitions that are needed to define weak equivalences suitable for our aims. Also in the case of weak semantics, many variant equivalences can be defined. The reader may consult the survey [vGl93] or the book [San12] for a comprehensive discussion on the merits of many of them.

2.4.1 Weak Traces

Definition 2.16. For any LTS $TS = (Q, A \cup \{\tau\}, \rightarrow)$, where $\tau \notin A$, define relation $\Longrightarrow \subseteq Q \times A^* \times Q$ as the *weak* reflexive and transitive closure of $\rightarrow$ (*cf.* Definition 2.4), i.e., as the least relation induced by the following axiom and rules, where ε is the empty trace:

$$\frac{q_1 \xrightarrow{\alpha} q_2}{q_1 \stackrel{\alpha}{\Longrightarrow} q_2} \qquad \frac{q_1 \xrightarrow{\tau} q_2}{q_1 \stackrel{\varepsilon}{\Longrightarrow} q_2} \qquad \frac{}{q \stackrel{\varepsilon}{\Longrightarrow} q} \qquad \frac{q_1 \stackrel{\sigma_1}{\Longrightarrow} q_2 \quad q_2 \stackrel{\sigma_2}{\Longrightarrow} q_3}{q_1 \stackrel{\sigma_1\sigma_2}{\Longrightarrow} q_3}$$

□

Note that a path $q_1 \xrightarrow{\tau} q_2 \xrightarrow{\tau} \dots q_n \xrightarrow{\tau} q_{n+1}$ (with $n \geq 0$) yields that $q_1 \stackrel{\varepsilon}{\Longrightarrow} q_{n+1}$. Moreover, it can be proved that $q \stackrel{\alpha}{\Longrightarrow} q'$ if and only if there exist two states q_1 and q_2 such that $q \stackrel{\varepsilon}{\Longrightarrow} q_1 \xrightarrow{\alpha} q_2 \stackrel{\varepsilon}{\Longrightarrow} q'$. Finally, if $\sigma = \alpha_1\alpha_2\dots\alpha_n$, then $q_1 \stackrel{\sigma}{\Longrightarrow} q_{n+1}$ if and only if there exist $q_2,\dots,q_n$ such that $q_1 \stackrel{\alpha_1}{\Longrightarrow} q_2 \stackrel{\alpha_2}{\Longrightarrow} \dots q_n \stackrel{\alpha_n}{\Longrightarrow} q_{n+1}$.

Definition 2.17. (Weak trace equivalence) Let $(Q, A \cup \{\tau\}, \rightarrow)$ be an LTS, where $\tau \notin A$. A *weak trace* of $q \in Q$ is a sequence $\sigma \in A^*$ such that $q \stackrel{\sigma}{\Longrightarrow} q'$ for some q'. Hence, the set $WTr(q)$ of weak traces of q is

$$WTr(q) = \{\sigma \in A^* \mid \exists q' \in Q.\, q \stackrel{\sigma}{\Longrightarrow} q'\}.$$

Two states $q_1, q_2 \in Q$ are *weak trace equivalent* if $WTr(q_1) = WTr(q_2)$, and this is denoted $q_1 =_{wtr} q_2$. This definition can be adapted to rooted LTSs: the set $WTr(TS)$ of weak traces of the rooted LTS $TS = (Q, A \cup \{\tau\}, \rightarrow, q_0)$ is $WTr(q_0)$. Two rooted LTSs, TS_1 and TS_2, are *weak trace equivalent* if $WTr(TS_1) = WTr(TS_2)$. □

Exercise 2.45. (Weak Trace preorder) The *weak trace preorder* $\leq_{wtr} \subseteq Q \times Q$ is defined as follows: $q \leq_{wtr} q'$ if and only if $WTr(q) \subseteq WTr(q')$. Prove that $\leq_{wtr}$ is a preorder, i.e., reflexive and transitive. Note that $q =_{wtr} q'$ iff $q \leq_{wtr} q'$ and $q' \leq_{wtr} q$. Prove that weak trace equivalence $=_{wtr}$ is an equivalence relation. □

Exercise 2.46. (Strong traces vs weak traces) Following Definition 2.9, given an LTS $TS = (Q, A \cup \{\tau\}, \rightarrow)$, a (strong) trace is any $\sigma \in (A \cup \{\tau\})^*$ such that $q \xrightarrow{\sigma}{}^* q'$ for some $q' \in Q$. Prove that two (strong) trace equivalent states are also weak trace equivalent, i.e., $Tr(q_1) = Tr(q_2)$ implies $WTr(q_1) = WTr(q_2)$ for any pair of states q_1 and q_2. (*Hint*: Prove that if $q \xrightarrow{\sigma}{}^* q'$, according to Definition 2.4, then $q \stackrel{\sigma'}{\Longrightarrow} q'$, according to Definition 2.16, where σ' is σ with all occurrences of action τ been removed.) □

It is easy to see that the LTS in Figure 2.2 and the LTS in Figure 2.21(a) are weak trace equivalent, but not (strongly) trace equivalent: e.g., *coin* τ is a strong trace only for the latter. Moreover, the two vending machines in Figure 2.21 are (strong and) weak trace equivalent; as discussed above, these two machines should not be considered equivalent, as they behave differently w.r.t. deadlock; hence, as expected, also weak trace equivalence is not sensitive to deadlock.

Definition 2.18. (Weak completed traces) Given an LTS $TS = (Q, A \cup \{\tau\}, \rightarrow)$, where $\tau \notin A$, and a state $q \in Q$, the set of the *weak completed traces* of q is

$$WCTr(q) = \{\sigma \in A^* \mid \exists q' \in Q.\, q \stackrel{\sigma}{\Longrightarrow} q' \wedge q' \stackrel{\alpha}{\not\Longrightarrow} \text{ for all observable } \alpha \in A\}.$$

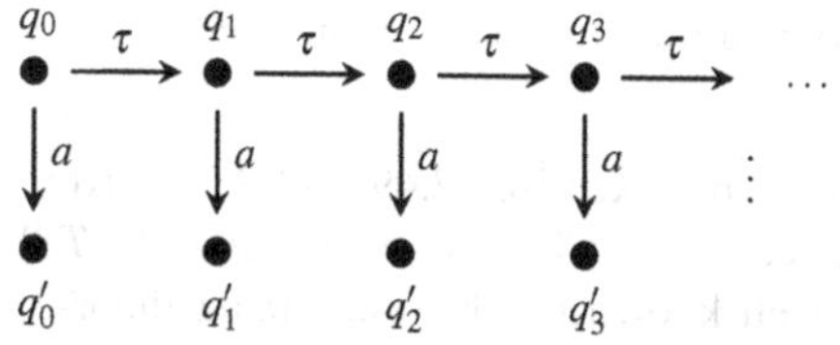

Fig. 2.22 An image finite LTS, whose τ-abstracted LTS is not image-finite

Note that state q' above need not be a deadlock state, as it may still perform silent, τ-labeled transitions. Two states $q_1, q_2 \in Q$ are *weak completed trace equivalent* if $WTr(q_1) = WTr(q_2)$ and $WCTr(q_1) = WCTr(q_2)$, denoted as $q_1 =_{wctr} q_2$.

This definition can be extended to rooted LTSs: the set $WCTr(TS)$ of weak completed traces of the rooted LTS $TS = (Q, A \cup \{\tau\}, \rightarrow, q_0)$ is $WCTr(q_0)$. Two rooted LTSs ,TS_1 and TS_2, are weak completed trace equivalent if $WTr(TS_1) = WTr(TS_2)$ and $WCTr(TS_1) = WCTr(TS_2)$. □

Exercise 2.47. Show that the two vending machines in Figure 2.21 are not weak completed trace equivalent. Furthermore, prove that for any pair of states q_1 and q_2, $CTr(q_1) = CTr(q_2)$ implies $WCTr(q_1) = WCTr(q_2)$. □

Exercise 2.48. (Tau-free LTS) An LTS is τ-free if it is labeled only with observable actions in A. Show that there is a τ-free LTS weak (completed) trace equivalent to the LTS in Figure 2.21(b). □

The exercise above can be generalized to show that, given an LTS with τ transitions, we can always build an associated τ-free LTS which is weak (completed) trace equivalent.

Exercise 2.49. (Tau-abstracted LTS) For a rooted LTS $TS = (Q, A \cup \{\tau\}, \rightarrow, q_0)$, where $\tau \notin A$, define its associated τ-abstracted rooted LTS $\tau ATS = (Q, A, \Rightarrow', q_0)$, where $\Rightarrow' = \{(q, \alpha, q') \mid \alpha \in A \wedge (q, \alpha, q') \in \Rightarrow\}$. Prove that:

- TS is finite-state if and only if τATS is finite-state.
- If τATS is finitely branching, then TS is finitely branching; does the converse implication hold? (*Hint:* Look at Figure 2.22).
- $WTr(TS) = Tr(\tau ATS)$.
- *Either* $WCTr(TS) = \{\varepsilon\} = CTr(\tau ATS)$ *or* $WCTr(TS) \setminus \{\varepsilon\} = CTr(\tau ATS)$.[8] □

A consequence of (the last two items of) this exercise is that, for finite-state LTSs, one can compute (completed) weak trace equivalence by means of strong (completed) trace equivalence: one has first to derive the τ-abstracted LTS by computing the (partial) transitive closure $\Rightarrow'$ of the transition relation $\rightarrow$ (a procedure, based on the classic Floyd-Warshall algorithm [Flo62], which takes time at most $O(n^3)$ with

[8] Observe that, in the *or* case, we removed ε from $WCTr(TS)$ because τATS is τ-free. For instance, the LTS $TS' = (\{q_0, q_1, q_2\}, \{a, \tau\}, \{(q_0, a, q_1), (q_0, \tau, q_2)\})$ is such that $WCTr(TS') = \{a, \varepsilon\}$, while $CTr(\tau ATS') = \{a\}$.

n the number of states), and then to check (completed) trace equivalence, which is PSPACE-complete.

Remark 2.6. Observe that, by Exercises 2.49 and 2.17, given an LTS TS, we can construct a deterministic, τ-free LTS TS' such that $WTr(TS) = Tr(TS')$. This is closely related to the well-known result in automata theory (see, e.g., [HMU01, Sip06]) that a nondeterministic finite automaton (NFA) can be transformed into a language equivalent, deterministic finite automaton (DFA). □

Exercise 2.50. Compute the τ-abstracted LTSs associated to the LTSs in Figure 2.21(a) and (b), and show that they are (strong) trace equivalent. □

Remark 2.7. **(All regular languages – and only these – are representable by finite state LTSs)** In automata theory (see Section 1.3.4), the languages recognized by finite automata are called *regular* languages. We want to show that *all the regular languages* can be represented by finite-state LTSs.

Let L be a regular language. Then, there exists a DFA $M = (Q, A, \delta, F, q_0)$ — where $Q = \{q_0, q_1, \ldots, q_n\}$, and δ has type $\delta : Q \times A \to Q$ — such that $L = L[M]$. Starting from M, we can build a rooted LTS $TS_M = (Q \cup F', A \cup \{\tau\}, \to, q_0)$, where $F' = \{q_i' \mid q_i \in F\}$ is a set of copies of the final states of M, and the transition relation is defined as the minimal relation generated by the following rules:

$$\frac{\delta(q_i,a) = q_j}{q_i \xrightarrow{a} q_j} \qquad \frac{\delta(q_i,a) = q_j \wedge q_j \in F}{q_i \xrightarrow{a} q_j'} \qquad \frac{q_0 \in F}{q_0 \xrightarrow{\tau} q_0'}$$

Note that for a transition $\delta(q_i,a) = q_j$ with $q_j \in F$ in M, we have two transitions in TS_M: $q_i \xrightarrow{a} q_j$ and $q_i \xrightarrow{a} q_j'$. Hence, TS_M is a nondeterministic LTS, even if M is a deterministic automaton. Note that the states in F' are deadlocks.

It is not difficult to prove that $(q_0, w) \longrightarrow^* (q_k, \varepsilon)$ in M iff $q_0 \xrightarrow{w}{}^* q_k$ in TS_M, for all $q_k \in Q$ and $w \in A^*$. The proof is by induction on the length of w; the base case is $w = \varepsilon$; in such a case, $(q_0, \varepsilon) \longrightarrow^* (q_0, \varepsilon)$ as well as $q_0 \xrightarrow{\varepsilon}{}^* q_0$, as required. Now, assume $w = va$. In M, $(q_0, va) \longrightarrow^* (q_k, \varepsilon)$ is derivable iff there exists a state $q_i \in Q$ such that $(q_0, va) \longrightarrow^* (q_i, a)$ and $\delta(q_i, a) = q_k$. It can be easily proved that $(q_0, va) \longrightarrow^* (q_i, a)$ in M iff $(q_0, v) \longrightarrow^* (q_i, \varepsilon)$; hence, by induction, we can conclude that $q_0 \xrightarrow{v}{}^* q_i$ in TS_M and, by definition of the transition relation, also $q_i \xrightarrow{a} q_k$; therefore, $q_0 \xrightarrow{va}{}^* q_k$ by Definition 2.4, as required.

As a consequence, we can easily prove that $w \in WCTr(TS_M)$ if and only if $w \in L[M]$ for all $w \in A^*$. The empty trace ε belongs to $WCTr(TS_M)$ iff transition $q_0 \xrightarrow{\tau} q_0'$ is present in TS_M; in turn, this is possible iff $q_0 \in F$, and so iff $\varepsilon \in L[M]$. Now, let us assume that $va \in WCTr(TS_M)$. This is possible iff there exist $q_i \in Q$ and $q_j' \in F'$ such that $q_0 \xrightarrow{v}{}^* q_i \xrightarrow{a} q_j'$. By the argument above, we have that $(q_0, v) \longrightarrow^* (q_i, \varepsilon)$; moreover, by definition of $\to$, we have $\delta(q_i, a) = q_j$ with $q_j \in F$. It can be easily proved that $(q_0, v) \longrightarrow^* (q_i, \varepsilon)$ in M iff $(q_0, va) \longrightarrow^* (q_i, a)$; therefore, $(q_0, va) \longrightarrow^* (q_j, \varepsilon)$ in M, i.e., $va \in L[M]$.

Conversely, we can also show that *only regular languages* can be represented by finite-state LTSs. Let us consider a rooted LTS $TS = (Q, A \cup \{\tau\}, \to, q_0)$. Starting

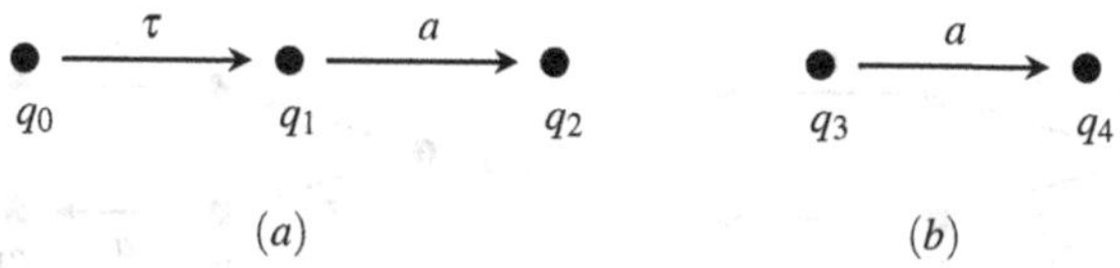

Fig. 2.23 Two weakly similar systems

from *TS*, we can build an NFA $M = (Q, A, \delta, F, q_0)$, where the set of final states is $F = \{q \in Q \mid q \text{ is a deadlock}\}$, and $\delta \subseteq Q \times (A \cup \{\varepsilon\}) \times Q$ is defined as the minimal relation generated by the following rules:

$$\frac{q_i \xrightarrow{a} q_j}{(q_i, a, q_j) \in \delta} \qquad \frac{q_i \xrightarrow{\tau} q_j}{(q_i, \varepsilon, q_j) \in \delta}$$

It is easy to prove, with a reasoning similar to the above, that $w \in WCTr(TS)$ if and only if $w \in L[M]$ for all $w \in A^*$; this because the two automata-like structures are essentially isomorphic and a state is final for M iff it is a deadlock for *TS*.

Summing up, we have shown that a language L is regular if and only if there exists a finite-state, rooted LTS *TS* such that $L = WCTr(TS)$. □

2.4.2 Weak Simulation and Weak Bisimulation

Definition 2.19. For any LTS $TS = (Q, A \cup \{\tau\}, \rightarrow)$, where $\tau \notin A$, a *weak simulation* is a relation $R \subseteq Q \times Q$ such that if $(q_1, q_2) \in R$ then for all $\alpha \in A$

- $\forall q_1'$ such that $q_1 \xrightarrow{\alpha} q_1'$, $\exists q_2'$ such that $q_2 \overset{\alpha}{\Longrightarrow} q_2'$ and $(q_1', q_2') \in R$,
- $\forall q_1'$ such that $q_1 \xrightarrow{\tau} q_1'$, $\exists q_2'$ such that $q_2 \overset{\varepsilon}{\Longrightarrow} q_2'$ and $(q_1', q_2') \in R$.

State q is *weakly simulated* by q', denoted $q \lessapprox q'$, if there exists a weak simulation R such that $(q, q') \in R$. Two states q and q' are *weakly simulation equivalent*, denoted $q \approxeq q'$, if $q \lessapprox q'$ and $q' \lessapprox q$. □

In other words, the weak simulation preorder $\lessapprox$ is the union of all the weak simulations:

$$\lessapprox = \bigcup \{R \subseteq Q \times Q \mid R \text{ is a weak simulation}\}.$$

Example 2.8. Note that in the weak simulation game, to a transition $q_1 \xrightarrow{\tau} q_1'$, q_2 can reply also by idling, as, among the possible q_2' such that $q_2 \overset{\varepsilon}{\Longrightarrow} q_2'$, we have also q_2 itself. This is necessary in order to get that the LTS in Figure 2.23(a) is simulated by the LTS in Figure 2.23(b), as we naturally would expect: both LTSs can only perform a single action a. The weak simulation relation proving this is $S_1 = \{(q_0, q_3), (q_1, q_3), (q_2, q_4)\}$. Note that to transition $q_0 \xrightarrow{\tau} q_1$, q_3 responds by idling:

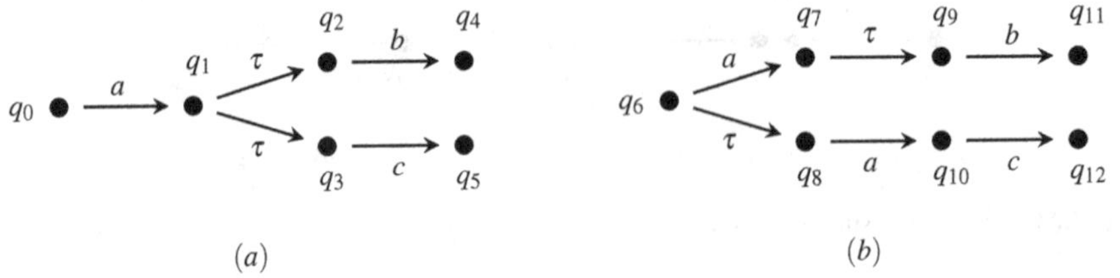

Fig. 2.24 Two non-weakly similar systems

$q_3 \stackrel{\varepsilon}{\Longrightarrow} q_3$. It is also possible to prove that the LTS in Figure 2.23(b) is simulated by the one in Figure 2.23(a), and a possible weak simulation proving this is $S_2 = \{(q_3, q_0), (q_4, q_2)\}$. Note that to transition $q_3 \stackrel{a}{\longrightarrow} q_4$, q_0 responds with $q_0 \stackrel{a}{\Longrightarrow} q_2$. Summing up, these two LTSs are weakly simulation equivalent, i.e., $q_0 \approxeq q_3$. □

Exercise 2.51. (Strong simulation vs weak simulation) Following Definition 2.12 for any LTS $TS = (Q, A \cup \{\tau\}, \rightarrow)$, a (strong) simulation is a relation $R \subseteq (Q \times Q)$ such that if $(q_1, q_2) \in R$ then for all $\mu \in A \cup \{\tau\}$

- $\forall q_1'$ such that $q_1 \stackrel{\mu}{\longrightarrow} q_1'$, $\exists q_2'$ such that $q_2 \stackrel{\mu}{\longrightarrow} q_2'$ and $(q_1', q_2') \in R$,

and we denote by $q \lesssim q'$ that there exists a strong simulation R containing the pair (q, q'). Prove that $q \lesssim q'$ implies $q \lessapprox q'$ by showing that a strong simulation is also a weak simulation. Show that the inverse implication does not hold by providing a suitable counterexample. (*Hint:* Consider the LTSs in Figure 2.23.) □

Exercise 2.52. Following the proof of Proposition 2.1, prove that weak simulation equivalence is finer than weak trace equivalence, i.e., $q \approxeq q'$ implies $q =_{wtr} q'$. □

In order to show that the inverse implication of Exercise 2.52 does not hold, we can consider the two LTSs in Figure 2.24 as a suitable counterexample. As a matter of fact, $WTr(q_0) = WTr(q_6) = \{\varepsilon, a, ab, ac\}$, while $q_0 \not\approxeq q_6$ because, even if $q_6 \lessapprox q_0$ (check this!), we have that $q_0 \not\lessapprox q_6$. The latter holds because to transition $q_0 \stackrel{a}{\longrightarrow} q_1$, q_6 can respond with either $q_6 \stackrel{a}{\Longrightarrow} q_7$ or $q_6 \stackrel{a}{\Longrightarrow} q_9$ or $q_6 \stackrel{a}{\Longrightarrow} q_{10}$; however, (q_1, q_7) is not a simulation pair, as to move $q_1 \stackrel{\tau}{\longrightarrow} q_3$, q_7 can respond either idling or moving to q_9, but $q_3 \stackrel{c}{\longrightarrow}$, while q_7 and q_9 cannot perform c weakly. This explains that also (q_1, q_9) is not a simulation pair. Similarly, (q_1, q_{10}) is not a simulation pair, as to move $q_1 \stackrel{\tau}{\longrightarrow} q_2$, q_{10} can only respond by idling, but $q_2 \stackrel{b}{\longrightarrow}$ while q_{10} cannot do b. Hence, q_0 cannot be weakly simulated by q_6.

Exercise 2.53. For any LTS $TS = (Q, A \cup \{\tau\}, \rightarrow)$, where $\tau \notin A$, given a weak simulation $R \subseteq Q \times Q$, prove that if $(q_1, q_2) \in R$ and $q_1 \stackrel{\delta}{\Longrightarrow} q_1'$, then there exists q_2' such that $q_2 \stackrel{\delta}{\Longrightarrow} q_2'$ with $(q_1', q_2') \in R$, for $\delta \in A \cup \{\varepsilon\}$. (*Hint:* By induction on the proof of $q_1 \stackrel{\delta}{\Longrightarrow} q_1'$, according to Definition 2.16.) □

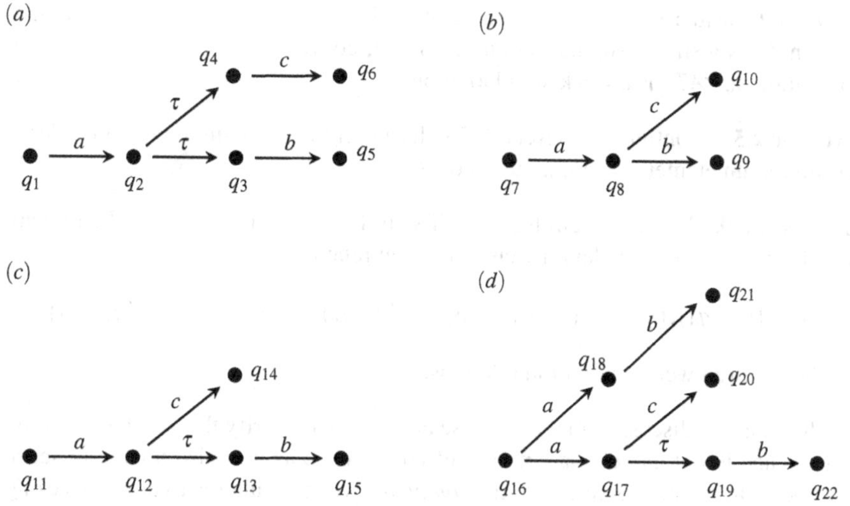

Fig. 2.25 Some labeled transition systems

Exercise 2.54. Prove that, for any LTS $(Q, A \cup \{\tau\}, \rightarrow)$, relation $\lesssim \subseteq Q \times Q$ is a preorder, while relation $\approx \subseteq Q \times Q$ is an equivalence relation. (*Hint:* Follow the same steps of Proposition 2.2 and Proposition 2.4. In proving that the relational composition of weak simulations is a weak simulation, you need also the result of Exercise 2.53.) □

Exercise 2.55. Consider the LTSs in Figure 2.25. Prove that relation

$$R = \{(q_1, q_7), (q_2, q_8), (q_3, q_8), (q_4, q_8), (q_5, q_9), (q_6, q_{10})\}$$

is a weak simulation proving that the LTS in Figure 2.25(a) is weakly simulated by the LTS in Figure 2.25(b). Prove also that the LTS in (b) is weakly simulated by the LTS in (a) by providing a suitable weak simulation relation.

Prove also that the LTS in Figure 2.25(c) is weakly simulation equivalent to the LTS in Figure 2.25(b), as well as to the LTS in Figure 2.25(d). □

Exercise 2.56. In the light of Exercise 2.55, by continuing Exercise 2.53, show that, for any LTS $TS = (Q, A \cup \{\tau\}, \rightarrow)$, where $\tau \notin A$, and for any $q_1, q_2 \in Q$, if $q_1 \lesssim q_2$ and $q_1 \stackrel{\varepsilon}{\Longrightarrow} q_1'$, then $q_1' \lesssim q_2$. This means that if we consider the largest simulation $\lesssim$, to a silent τ-move of q_1, q_2 can respond simply by idling. (*Hint:* Prove first that if $q_1 \stackrel{\varepsilon}{\Longrightarrow} q_1'$, then $q_1' \lesssim q_1$; then, the thesis follows by transitivity of $\lesssim$.) □

Exercise 2.57. (Can τ's be removed safely?) In the light of Exercise 2.55, by continuing Exercise 2.49 about the construction of the τ-abstracted LTS, show that two states q and q' are weak simulation equivalent in TS if and only if they are strong simulation equivalent in τATS.

(*Hint*: Consider Exercises 2.53 and 2.56 and show that the largest weak simulation on *TS* is a strong simulation on τATS and, conversely, that the largest strong simulation on τATS is a weak simulation on *TS*.) □

Exercise 2.58. Continuing Exercise 2.28, discuss if the weak simulation preorder $\lesssim$ has minimum elements and maximum ones. □

Exercise 2.59. Consider again the two LTSs in Figure 2.21. Show that the two are weakly simulation equivalent, by checking that relations

$$S_1 = \{(q_1,q_4),(q_2,q_5),(q_3,q_6)\} \quad S_2 = \{(q_4,q_1),(q_5,q_2),(q_6,q_3),(q_7,q_3)\}$$

are the required weak simulation relations. □

The example discussed in the exercise above should clarify that weak simulation equivalence is not an adequate notion of equivalence as it is also unable to sense deadlock. Of course, one might define *completed* weak simulation as done for strong simulation.

Exercise 2.60. (Weak completed simulation) A *weak completed simulation R* is a weak simulation such that for all $(q_1,q_2) \in R$ if $q_1 \not\rightarrow$ then $q_2 \stackrel{\alpha}{\not\Rightarrow}$ for all observable α (but q_2 may still perform silent transitions). State q_1 is weakly completed simulated by q_2, denoted $q_1 \lesssim_c q_2$, if there exists a weak completed simulation R such that $(q_1,q_2) \in R$. States q_1 and q_2 are *weakly completed simulation equivalent*, $q_1 \approxeq_c q_2$, if $q_1 \lesssim_c q_2$ and $q_2 \lesssim_c q_1$.

Show that the two weakly simulation equivalent vending machines of Figure 2.21 are actually not weakly completed simulation equivalent. □

Even if sensitive to deadlock, such a variant of weak simulation would be unable to sense the timing of choices in a complete way, as illustrated in Figure 2.15 for the strong case. As we have seen for the strong behavioral equivalences, a natural strengthening is obtained by bisimulation.

Definition 2.20. (Weak bisimulation) For any LTS $TS = (Q, A \cup \{\tau\}, \rightarrow)$, where $\tau \notin A$, a *weak bisimulation* is a relation $R \subseteq (Q \times Q)$ such that both R and its inverse R^{-1} are weak simulations. More expicitly, a weak bisimulation is a relation R such that if $(q_1,q_2) \in R$ then for all $\alpha \in A$

- $\forall q_1'$ such that $q_1 \xrightarrow{\alpha} q_1'$, $\exists q_2'$ such that $q_2 \stackrel{\alpha}{\Longrightarrow} q_2'$ and $(q_1',q_2') \in R$,
- $\forall q_1'$ such that $q_1 \xrightarrow{\tau} q_1'$, $\exists q_2'$ such that $q_2 \stackrel{\varepsilon}{\Longrightarrow} q_2'$ and $(q_1',q_2') \in R$,

and, symmetrically,

- $\forall q_2'$ such that $q_2 \xrightarrow{\alpha} q_2'$, $\exists q_1'$ such that $q_1 \stackrel{\alpha}{\Longrightarrow} q_1'$ and $(q_1',q_2') \in R$,
- $\forall q_2'$ such that $q_2 \xrightarrow{\tau} q_2'$, $\exists q_1'$ such that $q_1 \stackrel{\varepsilon}{\Longrightarrow} q_1'$ and $(q_1',q_2') \in R$.

States q and q' are *weakly bisimilar* (or *weak bisimulation equivalent*), denoted with $q \approx q'$, if there exists a weak bisimulation R such that $(q,q') \in R$. □

In other words, weak bisimulation equivalence is the union of all weak bisimulations:

$$\approx = \bigcup\{R \subseteq Q \times Q \mid R \text{ is a weak bisimulation}\}.$$

Example 2.9. Continuing Example 2.8, the LTS in Figure 2.23(a) is actually weakly bisimilar to the LTS in Figure 2.23(b). The weak bisimulation relation proving this is: $S_1 = \{(q_0,q_3),(q_1,q_3),(q_2,q_4)\}$. Indeed, S_1 is a weak simulation, as well as $S_1^{-1} = \{(q_3,q_0),(q_3,q_1),(q_4,q_2)\}$. Summing up, these two LTSs are weak bisimulation equivalent, i.e., $q_0 \approx q_3$. □

Exercise 2.61. (Strong vs weak bisimulation) Following Definition 2.14, for any LTS $TS = (Q, A \cup \{\tau\}, \rightarrow)$, a (strong) bisimulation is a relation $R \subseteq Q \times Q$ such that if $(q_1,q_2) \in R$ then for all $\mu \in A \cup \{\tau\}$

- $\forall q_1'$ such that $q_1 \xrightarrow{\mu} q_1'$, $\exists q_2'$ such that $q_2 \xrightarrow{\mu} q_2'$ and $(q_1',q_2') \in R$,
- $\forall q_2'$ such that $q_2 \xrightarrow{\mu} q_2'$, $\exists q_1'$ such that $q_1 \xrightarrow{\mu} q_1'$ and $(q_1',q_2') \in R$.

Two states q and q' are bisimilar, denoted $q \sim q'$, if there exists a strong bisimulation R such that $(q,q') \in R$.

Prove that $q \sim q'$ implies $q \approx q'$ by showing that a strong bisimulation is also a weak bisimulation. Show that the reverse implication does not hold by providing a suitable counterexample. □

Exercise 2.62. Prove that, for any LTS $(Q, A \cup \{\tau\}, \rightarrow)$, relation $\approx \subseteq Q \times Q$ is an equivalence relation. (*Hint:* Follow the same steps of Propositions 2.6 and 2.8, in turn based on the proofs of Propositions 2.2 and 2.4. In proving that the relational composition of weak bisimulations is a weak bisimulation, you need also the result of Exercise 2.53.) Prove also that $\approx$ is the largest weak bisimulation. □

Exercise 2.63. Prove that $q \approx q'$ implies $q \approxeq q'$. Show that the inverse implication does not hold by providing a suitable counterexample. (*Hint*: If you do not find it, read below.) □

Consider again Figure 2.25. In Exercise 2.55 we stated that all the four LTSs are weak simulation equivalent. Now, to show that the LTSs (c) and (d) are weakly bisimilar, it is enough to exhibit a suitable weak bisimulation, e.g.,

$$R = \{(q_{11},q_{16}),(q_{12},q_{17}),(q_{13},q_{19}),(q_{13},q_{18}),(q_{14},q_{20}),(q_{15},q_{21}),(q_{15},q_{22})\}.$$

Observe that when $q_{16} \xrightarrow{a} q_{18}$ in (d), then $q_{11} \overset{a}{\Longrightarrow} q_{13}$ in (c) and the reached states can only react to b.

It is not difficult to see that no weak bisimulation can relate the LTS in Figure 2.25(b) with the LTS in (c): move $q_{11} \xrightarrow{a} q_{12}$ in (c) can be matched by move $q_7 \xrightarrow{a} q_8$ in (b), but now if in (c) we have move $q_{12} \xrightarrow{\tau} q_{13}$, in (b) we can respond only by idling $q_8 \overset{\varepsilon}{\Longrightarrow} q_8$, and then q_{13} and q_8 are not bisimilar because the former cannot execute c while the latter can.

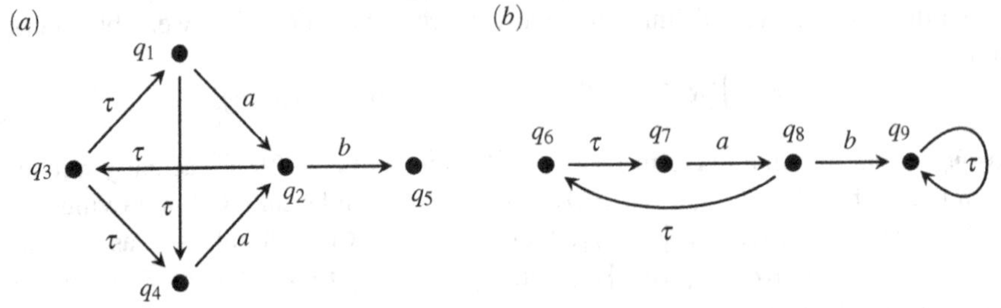

Fig. 2.26 Are states q_1 and q_6 weakly bisimilar?

Exercise 2.64. Prove that any other pair of LTSs in Figure 2.25 does not belong to bisimulation equivalence $\approx$. In particular, observe that (a) and (b), even if weakly simulation equivalent (see Exercise 2.55), are not weakly bisimilar. □

Exercise 2.65. Continuing Exercise 2.59, show that the two LTSs in Figure 2.21 are not weakly bisimilar. This supports our claim that weak bisimilarity is sensitive to deadlock. □

Exercise 2.66. Show that states q_1 and q_6 in Figure 2.26 are weakly bisimilar. □

Exercise 2.67. (Tau-free LTS, again) Let us continue Exercise 2.48 about τ-free LTSs. (i) Prove that no τ-free LTS can be weakly bisimilar to the LTS in Figure 2.25(a). This is in contrast with weak simulation equivalence, as the τ-free LTS in Figure 2.25(b) is weakly similar to that in Figure 2.25(a). (ii) Prove also that there exists no τ-free LTS weakly bisimilar to those in Figure 2.26. (*Hint:* Look at state q_8.) (iii) Show an example of an LTS with τ transitions which is weakly bisimilar to a τ-free LTS. □

The exercise above shows that it is not possible to abstract completely from τ-labeled transitions when considering weak bisimulation equivalence. Hence, given an LTS $TS = (Q, A \cup \{\tau\}, \rightarrow)$, the associated τ-abstracted LTS, as outlined in Exercise 2.49, is not preserving all the needed information in this setting.

Exercise 2.68. (Abstract LTS) Continuing Exercise 2.49, given an LTS $TS = (Q, A \cup \{\tau\}, \rightarrow)$, where $\tau \notin A$, define its associated abstract LTS $ATS = (Q, A \cup \{\varepsilon\}, \Rightarrow'')$, where $\Rightarrow'' = \{(q, \delta, q') \mid \delta \in A \cup \{\varepsilon\} \wedge (q, \delta, q') \in \Rightarrow\}$, Prove that $q \approx q'$ in TS if and only if $q \sim q'$ in ATS. □

A consequence of Exercise 2.68 is that it is possible to offer an alternative, yet equivalent, definition of weak bisimulation as follows: A weak bisimulation is a relation R such that if $(q_1, q_2) \in R$ then for all $\delta \in A \cup \{\varepsilon\}$

- $\forall q_1'$ such that $q_1 \stackrel{\delta}{\Longrightarrow} q_1'$, $\exists q_2'$ such that $q_2 \stackrel{\delta}{\Longrightarrow} q_2'$ and $(q_1', q_2') \in R$,
- $\forall q_2'$ such that $q_2 \stackrel{\delta}{\Longrightarrow} q_2'$, $\exists q_1'$ such that $q_1 \stackrel{\delta}{\Longrightarrow} q_1'$ and $(q_1', q_2') \in R$,

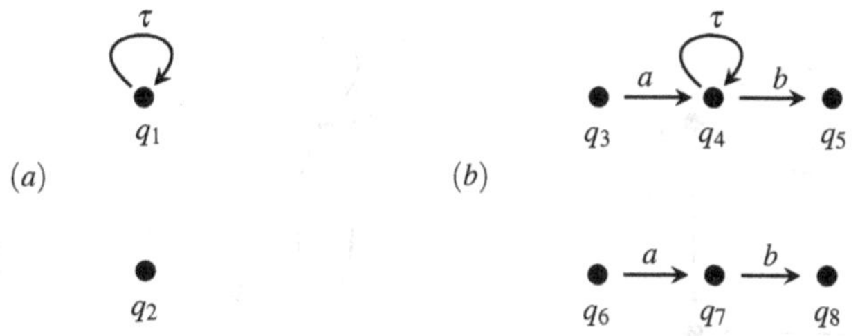

Fig. 2.27 Comparing some divergent systems

which is exactly the definition of strong bisimulation on the abstract LTS (cf. also Exercise 2.53).

Exercise 2.69. As a consequence of the observation above, prove the analogous of Proposition 2.9 where $\approx$ replaces $\sim$. Hence, also weak bisimulation equivalence can be seen as the largest fixed point of a suitable recursive definition. □

Another consequence of Exercise 2.68 is that, from a complexity point of view, computing weak bisimulation equivalence over finite-state LTSs is just a bit harder than computing (strong) bisimulation equivalence: as mentioned above, one has first to derive the abstract LTS by computing the (partial) transitive closure $\Rightarrow''$ of the transition relation $\rightarrow$ (by means of the classic Floyd-Warshall algorithm [Flo62], which runs in $O(n^3)$, where n is the number of states) and then to check (strong) bisimulation equivalence, which is in $O(m \log n)$ time [PT87], where m is the number of transitions.

Exercise 2.70. Compute the abstract LTSs associated to the LTSs in Figure 2.21(a) and (b), and show that they are not (strong) bisimulation equivalent. Compare these abstract LTSs with those τ-abstracted LTSs computed in Exercise 2.50. □

A peculiar aspect of weak bisimulation equivalence, as well as of all the weak behavioral equivalences we have introduced so far, is that it equates systems with different divergent behavior.

Definition 2.21. (Divergent state and livelock) A state q is *divergent*, denoted $q \Uparrow$, if there exist an infinite path $q_1 \xrightarrow{\tau} q_2 \xrightarrow{\tau} \dots$ of τ-labeled transitions with $q_1 = q$. An LTS $TS = (Q, A \cup \{\tau\}, \rightarrow)$ is *divergence-free* if no state $q \in Q$ is divergent.

A state q is a *livelock* if for all reachable q', i.e., for all q' such that $q \longrightarrow^* q'$, q' can do at least one τ-labeled transition and cannot do any observable transitions, i.e., $q' \xrightarrow{\tau}$ and $q' \stackrel{\alpha}{\nrightarrow}$ for all observable α. □

Clearly, a livelock state is also divergent. Consider the LTSs in Figure 2.27. States q_1 and q_4 are divergent, even if only q_1 is a livelock. Similarly, states q_1, q_2, q_5 and q_6 of Figure 2.28 are divergent, even if only q_6 is a livelock. In general, if the LTS contains infinitely many states, a divergent state (as well as a livelock) need not

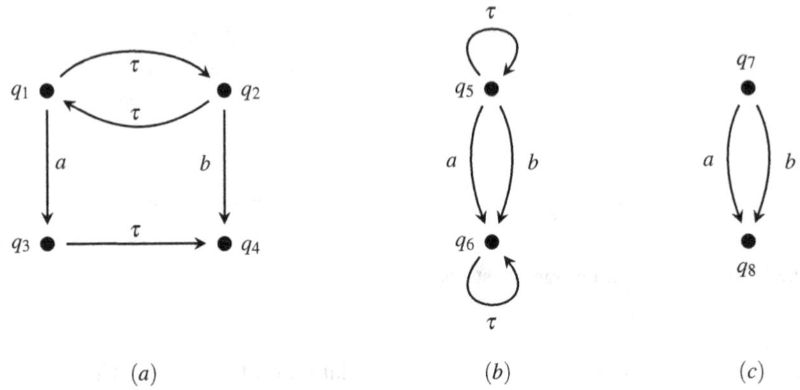

Fig. 2.28 Comparing other divergent systems

belong to a cyclic τ-labeled path; for instance, the states q_i, for $i \in \mathbb{N}$, of Figure 2.22 are all divergent.

Technically, the subset of divergent states can be characterized as the largest fixed point of the functional $F : \wp(Q) \to \wp(Q)$ defined as:

$$F(S) = \{q_1 \mid \exists q_2 \text{ such that } q_1 \xrightarrow{\tau} q_2 \text{ and } q_2 \in S\}$$

This set can be computed iteratively over finite-state LTSs as follows: start initially by considering all the states in Q as potentially divergent, and then compute $F(Q)$, which is obtained by removing all those states that cannot perform a τ-labeled transition initially. If $Q = F(Q)$, then halt; otherwise, continue by computing $F(F(Q))$, which is obtained by removing all those states that cannot perform a τ-labeled transition initially, reaching a state in $F(Q)$, i.e., by removing those states that cannot perform two consecutive τ-labeled transitions initially. If $F(F(Q)) = F(Q)$, then halt. Otherwise, continue the algorithm by computing $F(F(F(Q)))$ and proceed until the set stabilizes. Following this algorithm, one can conclude that the divergent states of the LTS in Figure 2.28(a) are exactly q_1 and q_2. (See Section 2.5 for another instance on how to compute the largest fixed point of a functional).

Exercise 2.71. Characterize the set of livelocks of an LTS $TS = (Q, A, \to)$ as the largest fixed point of a suitable functional $G : \wp(Q) \to \wp(Q)$. □

Now we discuss in what sense weak bisimilarity is not sensitive to divergence. Consider the two LTSs in Figure 2.27(a). They are weak trace equivalent and also weak bisimulation equivalent; hence, these weak behavioral equivalences do not distinguish a divergent state (as well as a livelock) from a deadlock! So, weak behavioral equivalences intend to abstract not only from finite amounts of internal work, but also from infinite amounts (i.e., divergences).

Similarly, the two LTSs in Figure 2.27(b) are equated, despite the fact that the upper one, in principle, may diverge and never execute b. The intuition behind this identification is that τ-cycles cannot be taken forever when an alternative is present, i.e., weak bisimilarity assumes that any computation is *fair*: if b is possible infinitely often, then b will be eventually chosen and executed.

Exercise 2.72. Show that the two LTSs in Figure 2.27(a) are weak completed trace equivalent, by showing that $WCTr(q_1) = \{\varepsilon\} = WCTr(q_2)$. □

Exercise 2.73. Consider Figure 2.28. Show that $q_1 \approx q_2$, as well as $q_1 \approx q_5$ and $q_1 \approx q_7$. Show that $q_3 \approx q_6$ as well as $q_3 \approx q_8$. □

As we have done for strong bisimulation (cf. Definition 2.15), it is sometimes convenient to write compactly a weak bisimulation, by removing those pairs that differ from others only for the use of equivalent alternatives. The resulting relation is *not* a weak bisimulation, but rather a weak bisimulation up to $\approx$.

Definition 2.22. (Weak bisimulation up to $\approx$) Given an LTS $TS = (Q, A \cup \{\tau\}, \rightarrow)$, where $\tau \notin A$, a weak bisimulation up to $\approx$ is a relation $R \subseteq Q \times Q$ such that if $(q_1, q_2) \in R$ then for all $\alpha \in A$

- $\forall q_1'$ such that $q_1 \xrightarrow{\alpha} q_1'$, $\exists q_2'$ such that $q_2 \stackrel{\alpha}{\Longrightarrow} q_2'$ and $q_1' \sim R \approx q_2'$,
- $\forall q_1'$ such that $q_1 \xrightarrow{\tau} q_1'$, $\exists q_2'$ such that $q_2 \stackrel{\varepsilon}{\Longrightarrow} q_2'$ and $q_1' \sim R \approx q_2'$,

and, symmetrically,

- $\forall q_2'$ such that $q_2 \xrightarrow{\alpha} q_2'$, $\exists q_1'$ such that $q_1 \stackrel{\alpha}{\Longrightarrow} q_1'$ and $q_1' \approx R \sim q_2'$,
- $\forall q_2'$ such that $q_2 \xrightarrow{\tau} q_2'$, $\exists q_1'$ such that $q_1 \stackrel{\varepsilon}{\Longrightarrow} q_1'$ and $q_1' \approx R \sim q_2'$. □

Exercise 2.74. Prove that if R is a weak bisimulation up to $\approx$, then $R \subseteq \approx$. (*Hint:* Prove first that $\sim R \approx$ and $\approx R \sim$ are both weak bisimulations, following the similar proof of Lemma 2.1.) □

The exercise above proves the correctness of this up-to technique: as $R \subseteq \approx$, we are not inducing erroneous equalities. So, in order to prove that $p \approx q$, it is enough to exhibit a weak bisimulation up to $\approx$ which contains (p, q). This principle will be useful when considering LTSs generated by CCS processes, because weak bisimilarity enjoys some useful algebraic laws. Instances of application of this proof principle are outlined, e.g., in Examples 3.17 and 3.25.

Exercise 2.75. Let us relax Definition 2.22 by replacing the occurrences of $\sim$ with $\approx$, i.e., by replacing $\sim R \approx$ with $\approx R \approx$ and, symmetrically, $\approx R \sim$ with $\approx R \approx$, so that the new definition is perfectly symmetric. Then check that R may *not* be a subset of $\approx$! That is, this new definition of weak bisimulation up to $\approx$ is incorrect. (*Hint:* Consider the LTS in Figure 2.23(a), performing the strong trace τa and composed of the states q_0, q_1 and q_2; consider also a deadlock state, called q_3. Then, relation $R = \{(q_0, q_3)\}$ satisfies the new definition of weak bisimulation up to $\approx$, because $q_0 \approx q_1$; however, q_0 is not weakly bisimilar to q_3.) □

In the following chapters, we will see that weak bisimilarity is not a compositional semantics w.r.t. the CCS operator of alternative composition $+$. In this respect, a useful weak equivalence is *rooted weak bisimilarity*, which we will prove to be the coarsest congruence for CCS contained in weak bisimilarity (Theorem 4.5). In its definition, we use an auxiliary relation $q \stackrel{\tau}{\Longrightarrow} q'$ defined as $q \stackrel{\varepsilon}{\Longrightarrow} q_1 \stackrel{\tau}{\longrightarrow} q_2 \stackrel{\varepsilon}{\Longrightarrow} q'$.

Definition 2.23. (Rooted weak bisimilarity) Given an LTS $(Q, A \cup \{\tau\}, \rightarrow)$, two states q_1 and q_2 are rooted weak bisimilar, denoted $q_1 \approx^c q_2$, if for all $\mu \in A \cup \{\tau\}$

- $\forall q_1'$ such that $q_1 \stackrel{\mu}{\longrightarrow} q_1'$, $\exists q_2'$ such that $q_2 \stackrel{\mu}{\Longrightarrow} q_2'$ and $q_1' \approx q_2'$,
- $\forall q_2'$ such that $q_2 \stackrel{\mu}{\longrightarrow} q_2'$, $\exists q_1'$ such that $q_1 \stackrel{\mu}{\Longrightarrow} q_1'$ and $q_1' \approx q_2'$. □

Observe that, in the definition of $\approx^c$, the possible initial silent transition $q_1 \stackrel{\tau}{\longrightarrow} q_1'$ is to be matched by a weak transition $q_2 \stackrel{\tau}{\Longrightarrow} q_2'$ (so, at least one τ transition must be performed by q_2). Afterwards, as only $\approx$ is required, any subsequent τ-labeled transition $\stackrel{\tau}{\longrightarrow}$ performed by q_1' is to be matched by a weak transition $\stackrel{\varepsilon}{\Longrightarrow}$ from q_2', hence possibly also by idling. (Symmetrically, if $q_2 \stackrel{\tau}{\longrightarrow} q_2'$.) The term "*rooted* weak bisimilarity" emphasizes this distinction about τ-labeled transitions of the roots.

Exercise 2.76. Prove that $q_1 \approx^c q_2$ implies $q_1 \approx q_2$. Show that the inverse implication does not hold. □

As a hint for the solution to the previous exercise, observe that the LTSs in Figure 2.27(b) are rooted weak bisimilar, while the LTSs in Figure 2.27(a) are only weakly bisimilar. Consider now the LTSs in Figure 2.28: $q_3 \approx^c q_6$ because $q_4 \approx q_6$; on the contrary, $q_3 \not\approx^c q_8$, even if $q_3 \approx q_8$.

Exercise 2.77. Prove that $\approx^c$ is an equivalence relation. □

Exercise 2.78. Show that if $p \stackrel{\tau}{\Longrightarrow} q$ and $q \stackrel{\tau}{\Longrightarrow} p$, then $p \approx^c q$. □

2.4.3 Branching Bisimulation

To conclude this overview of weak bisimulation-based equivalences, we present another one, originally proposed in [vGW96] under the label of *branching bisimulation* equivalence. The name reflects the argument that weak bisimulation equivalence $\approx$ is not completely respecting the timing of choices (the so-called *branching structure* of systems). For instance, consider the two rooted weakly bisimilar systems in Figure 2.29. In the LTS on the left, in each computation the choice between b and c is made *after* the a-labeled transition, while in the LTS on the right there is a computation where c is already discarded after a. Hence, it may be argued that the two LTSs should not be equivalent. A finer notion of equivalence that distinguishes between these two systems is as follows.

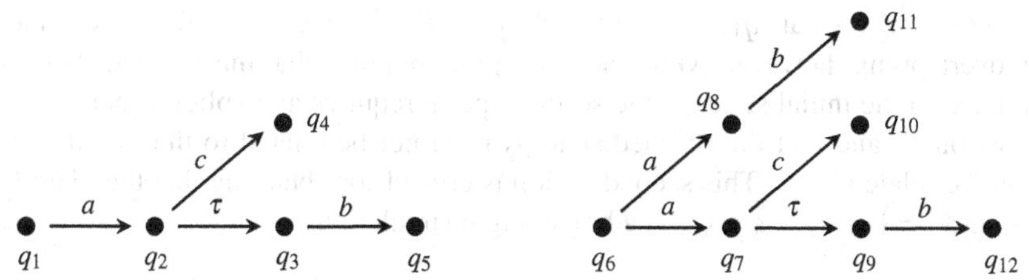

Fig. 2.29 Is the timing of choices respected?

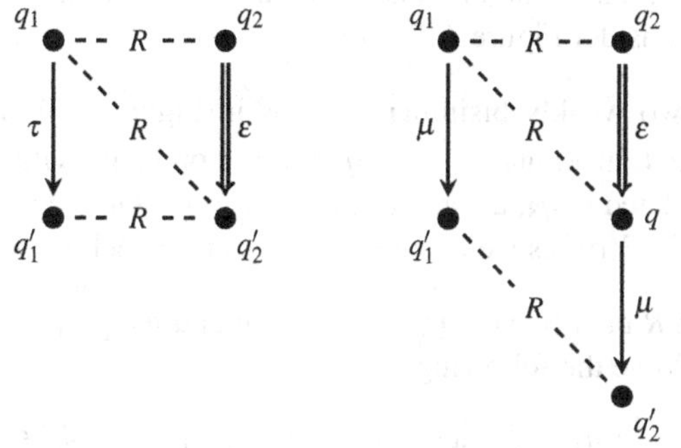

Fig. 2.30 Graphical representation of branching bisimulation *either-or* conditions

Definition 2.24. Given an LTS $TS = (Q, A \cup \{\tau\}, \rightarrow)$, a *branching bisimulation* is a relation $R \subseteq Q \times Q$ such that if $(q_1, q_2) \in R$, for all $\mu \in A \cup \{\tau\}$

- $\forall q_1'.\ q_1 \xrightarrow{\mu} q_1'$,
  - either $\mu = \tau$ and $\exists q_2'.\ q_2 \stackrel{\varepsilon}{\Longrightarrow} q_2'$ with $(q_1, q_2') \in R$ and $(q_1', q_2') \in R$,
  - or $\exists q, q_2'.\ q_2 \stackrel{\varepsilon}{\Longrightarrow} q \xrightarrow{\mu} q_2'$ with $(q_1, q) \in R$ and $(q_1', q_2') \in R$

  (such conditions are pictorially represented in Figure 2.30), and symmetrically
- $\forall q_2'.\ q_2 \xrightarrow{\mu} q_2'$,
  - either $\mu = \tau$ and $\exists q_1'.\ q_1 \stackrel{\varepsilon}{\Longrightarrow} q_1'$ with $(q_1', q_2) \in R$ and $(q_1', q_2') \in R$,
  - or $\exists q, q_1'.\ q_1 \stackrel{\varepsilon}{\Longrightarrow} q \xrightarrow{\mu} q_1'$ with $(q, q_2) \in R$ and $(q_1', q_2') \in R$.

States q_1 and q_2 are *branching bisimilar* (or *branching bisimulation equivalent*), denoted $q_1 \approx_{br} q_2$, if there exists a branching bisimulation R that relates them. □

Hence, branching bisimilarity $\approx_{br}$ is the union of all branching bisimulations:

$$\approx_{br} = \bigcup \{R \subseteq Q \times Q \mid R \text{ is a branching bisimulation}\}.$$

Remark 2.8. **(Either-or conditions)** In Definition 2.24, to a silent move $q_1 \xrightarrow{\tau} q_1'$, q_2 can reply *either* with $q_2 \stackrel{\varepsilon}{\Longrightarrow} q_2'$ so that $(q_1, q_2') \in R$ and $(q_1', q_2') \in R$, *or* with

$q_2 \stackrel{\varepsilon}{\Longrightarrow} q \stackrel{\tau}{\longrightarrow} q_2'$ so that $(q_1, q) \in R$ and $(q_1', q_2') \in R$. The two possibilities are somehow overlapping; however, while the first option requires that the reached state q_2' be related to the initial state q_1, the second option requires an explicit execution of at least one τ and that the reached state q_2' need not be related to the initial state q_1, but be related to q_1'. This second option is crucial for obtaining that the identity relation $\mathscr{I} = \{(q,q) \mid q \in Q\}$ is a branching bisimulation. □

Exercise 2.79. Prove that $q_1 \approx_{br} q_2$ implies $q_1 \approx q_2$ by observing that a branching bisimulation is also a weak bisimulation, but that a weak bisimulation may fail to satisfy the branching bisimulation conditions. Is it true that if $q_1 \approx_{br} q_2$ then also $q_1 \approx^c q_2$? (*Hint:* Consider Figure 2.27(a).) □

Note that the two weakly bisimilar systems in Figure 2.29 are not branching bisimilar because to transition $q_6 \stackrel{a}{\longrightarrow} q_8$, q_1 can respond only with $q_1 \stackrel{\varepsilon}{\Longrightarrow} q_1 \stackrel{a}{\longrightarrow} q_2$, but q_2 cannot be related to q_8, as only q_2 can perform c. On the other hand, the two systems in Figure 2.27(a), as well as those in (b), are branching bisimilar.

Exercise 2.80. Let R be a branching bisimulation and let q_1, q_2 be two states such that $(q_1, q_2) \in R$. Prove the following:

- $\forall q_1'$ such that $q_1 \stackrel{\varepsilon}{\Longrightarrow} q_1'$, $\exists q_2'$ such that $q_2 \stackrel{\varepsilon}{\Longrightarrow} q_2'$ with $(q_1', q_2') \in R$;

and symmetrically for q_2. (*Hint:* By induction on the proof of $q_1 \stackrel{\varepsilon}{\Longrightarrow} q_1'$, according to Definition 2.16.) □

Exercise 2.81. Following Proposition 2.6 and Proposition 2.8 as well as Exercise 2.62, prove that branching bisimilarity $\approx_{br}$ is an equivalence relation. (*Hint:* In proving that the relational composition of two branching bisimulations is a branching bisimulation you need the result of Exercise 2.80.) Prove also that the union of branching bisimulations is a branching bisimulation, hence deriving that $\approx_{br}$ is the largest branching bisimulation. □

Exercise 2.82. (Stuttering Lemma)
Prove that, given a τ-labeled path $q_1 \stackrel{\tau}{\longrightarrow} q_2 \stackrel{\tau}{\longrightarrow} \ldots q_n \stackrel{\tau}{\longrightarrow} q_{n+1}$, if $q_1 \approx_{br} q_{n+1}$, then $q_i \approx_{br} q_j$ for all $i, j = 1, \ldots n+1$. This is sometimes called the *stuttering property*. Prove that this property holds also for weak bisimilarity $\approx$. □

Remark 2.9. **(Stuttering implies no change of equivalence class)** The exercise above justifies the following observation. As $\approx_{br}$ is a branching bisimulation, it satisfies the conditions in Definition 2.24. We start by taking $q_1 \approx_{br} q_2$. Then, suppose $q_1 \stackrel{\tau}{\longrightarrow} q_1'$ and that q_2 responds by performing $q_2 \stackrel{\varepsilon}{\Longrightarrow} q_2'$ with $q_1 \approx_{br} q_2'$ and $q_1' \approx_{br} q_2'$. By transitivity, we have that also $q_2 \approx_{br} q_2'$. Hence, by the stuttering lemma, q_1 is branching bisimilar to each state in the path from q_2 to q_2'. Similarly, assume $q_1 \stackrel{\mu}{\longrightarrow} q_1'$ and that q_2 responds by performing $q_2 \stackrel{\varepsilon}{\Longrightarrow} q \stackrel{\mu}{\longrightarrow} q_2'$ with $q_1 \approx_{br} q$ and $q_1' \approx_{br} q_2'$. By transitivity, $q_2 \approx_{br} q$, hence, by the stuttering lemma, q_1 is branching bisimilar to each state in the path from q_2 to q. (Symmetrically, the requirement $q_1' \approx_{br} q_2$ in the third item of Definition 2.24 implies that q_2 is related to each state

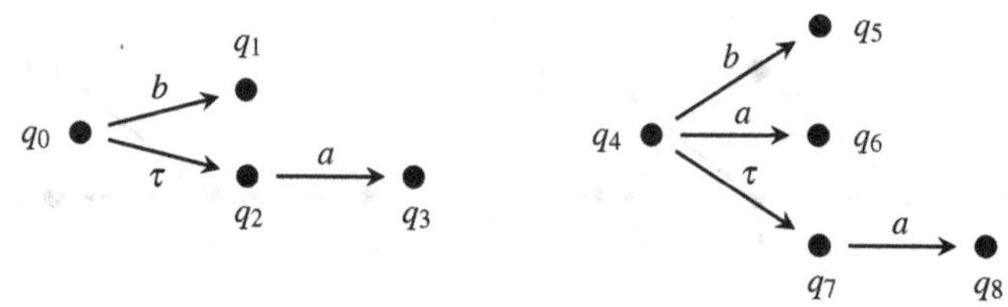

Fig. 2.31 Two weakly bisimilar LTSs that are not branching bisimilar

in the path from q_1 to q_1', while $q \approx_{br} q_2$ in the fourth item ensures that q_2 is related to each state in the path from q_1 to q.)

These constraints are not required by weak bisimilarity: given $q_1 \approx q_2$, when matching a transition $q_1 \xrightarrow{\mu} q_1'$ with $q_2 \stackrel{\varepsilon}{\Longrightarrow} q \xrightarrow{\mu} q' \stackrel{\varepsilon}{\Longrightarrow} q_2'$, weak bisimilarity only requires that $q_1' \approx q_2'$, but does not impose any condition on the intermediate states; in particular, it is not required that $q_1 \approx q$, or that $q_1' \approx q'$. On the contrary, given $q_1 \approx_{br} q_2$, when matching a transition $q_1 \xrightarrow{\mu} q_1'$ with $q_2 \stackrel{\varepsilon}{\Longrightarrow} q \xrightarrow{\mu} q_2'$, branching bisimilarity requires that $q_1 \approx_{br} q$ (i.e., while performing initial τ moves from q_2 to q, equivalent states only are traversed), and that $q_1' \approx_{br} q_2'$.

For instance, consider the two LTSs in Figure 2.31. Not surprisingly, we have that $q_0 \approx q_4$. On the contrary, $q_0 \not\approx_{br} q_4$. Indeed, to move $q_4 \xrightarrow{a} q_6$, state q_0 can respond only with $q_0 \stackrel{\varepsilon}{\Longrightarrow} q_2 \xrightarrow{a} q_3$, but $q_2 \not\approx_{br} q_4$ because only q_4 can perform action b. □

As done for strong and weak bisimulations, it is sometimes convenient to write compactly a branching bisimulation, by removing those pairs that differ from others only for their use of branching equivalent alternatives. The resulting relation is *not* a branching bisimulation, but rather a branching bisimulation up to $\approx_{br}$.

Definition 2.25. (Branching bisimulation up to $\approx_{br}$) A branching bisimulation up to $\approx_{br}$ is a relation $R \subseteq Q \times Q$ such that if $(q_1, q_2) \in R$ then for all $\mu \in A \cup \{\tau\}$

- $\forall q_1'.\ q_1 \xrightarrow{\mu} q_1'$,
  - either $\mu = \tau$ and $\exists q_2'.\ q_2 \stackrel{\varepsilon}{\Longrightarrow} q_2'$ with $q_1 \sim R \approx_{br} q_2'$ and $q_1' \sim R \approx_{br} q_2'$,
  - or $\exists q_2', q_2''$ such that $q_2 \stackrel{\varepsilon}{\Longrightarrow} q_2' \xrightarrow{\mu} q_2''$ with $q_1 \sim R \approx_{br} q_2'$ and $q_1' \sim R \approx_{br} q_2''$;

  and, symmetrically
- $\forall q_2'.\ q_2 \xrightarrow{\mu} q_2'$,
  - either $\mu = \tau$ and $\exists q_1'.\ q_1 \stackrel{\varepsilon}{\Longrightarrow} q_1'$ with $q_1' \approx_{br} R \sim q_2$ and $q_1' \approx_{br} R \sim q_2'$,
  - or $\exists q_1', q_1''.\ q_1 \stackrel{\varepsilon}{\Longrightarrow} q_1' \xrightarrow{\mu} q_1''$ with $q_1' \approx_{br} R \sim q_2$ and $q_1' \approx_{br} R \sim q_2''$ □

Proposition 2.11. *If R is a branching bisimulation up to $\approx_{br}$, then $R \subseteq \approx_{br}$.*

Proof. The proof is based on the fact that both $\sim R \approx_{br}$ and $\approx_{br} R \sim$ are branching bisimulations (check this!),[9] *from which it follows that $\approx_{br}$ includes both $\sim R \approx_{br}$*

[9] To prove this fact, one has to use the result of Exercise 2.80.

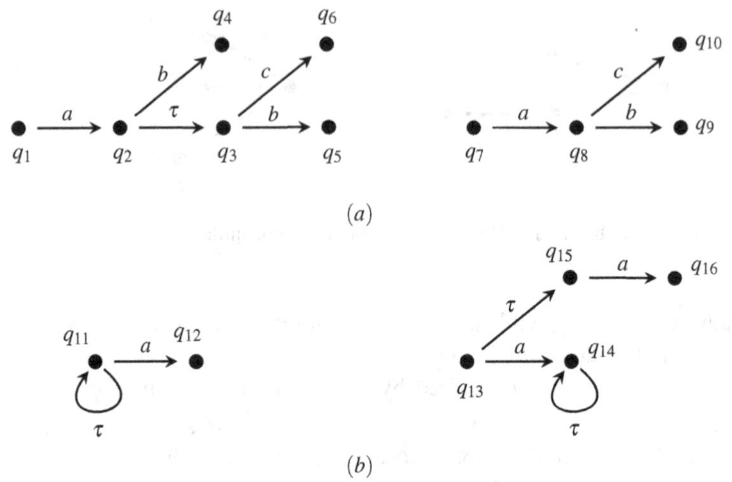

Fig. 2.32 Are these rooted branching bisimilar?

and $\approx_{br} R \sim$. *The thesis* $R \subseteq \approx_{br}$ *follows because the identity relation* $\mathscr{I}$ *is a subset of both* $\sim$ *and* $\approx_{br}$, *hence* $R = \mathscr{I} R \mathscr{I} \subseteq \sim R \approx_{br} \subseteq \approx_{br}$. □

As a consequence of this proposition, in order to prove that $p \approx_{br} q$, it is enough to exhibit a branching bisimulation up to $\approx_{br}$ containing the pair (p,q).

Exercise 2.83. Continuing Exercise 2.75, show that if in Definition 2.25 we replace $\sim$ with $\approx_{br}$ in order to get a completely symmetric definition, then it might happen that a branching bisimulation up to $\approx_{br}$ is not included in $\approx_{br}$. □

In the next chapters, we will see that branching bisimilarity is not a compositional semantics w.r.t. the CCS operator of alternative composition $+$. In this respect, a useful variant of branching bisimulation equivalence is *rooted branching bisimilarity*, denoted $\approx_{br}^c$, which we will prove to be the coarsest congruence for CCS contained in branching bisimilarity $\approx_{br}$. The peculiar feature of rooted branching bisimilarity is that initial moves are matched as in strong bisimulation, while subsequent moves are matched as for branching bisimilarity.

Definition 2.26. (Rooted branching bisimilarity) Given an LTS $(Q, A \cup \{\tau\}, \rightarrow)$, q_1 and q_2 are rooted branching bisimilar, denoted $q_1 \approx_{br}^c q_2$, if for all $\mu \in A \cup \{\tau\}$

- $\forall q_1'$ such that $q_1 \xrightarrow{\mu} q_1'$, $\exists q_2'$ such that $q_2 \xrightarrow{\mu} q_2'$ and $q_1' \approx_{br} q_2'$,
- $\forall q_2'$ such that $q_2 \xrightarrow{\mu} q_2'$, $\exists q_1'$ such that $q_1 \xrightarrow{\mu} q_1'$ and $q_1' \approx_{br} q_2'$. □

Exercise 2.84. Prove the following statements:

- $\approx_{br}^c$ is an equivalence relation,
- $\approx_{br}^c$ is finer than $\approx_{br}$, as well as finer than $\approx^c$,
- $\sim$ is finer than $\approx_{br}^c$. □

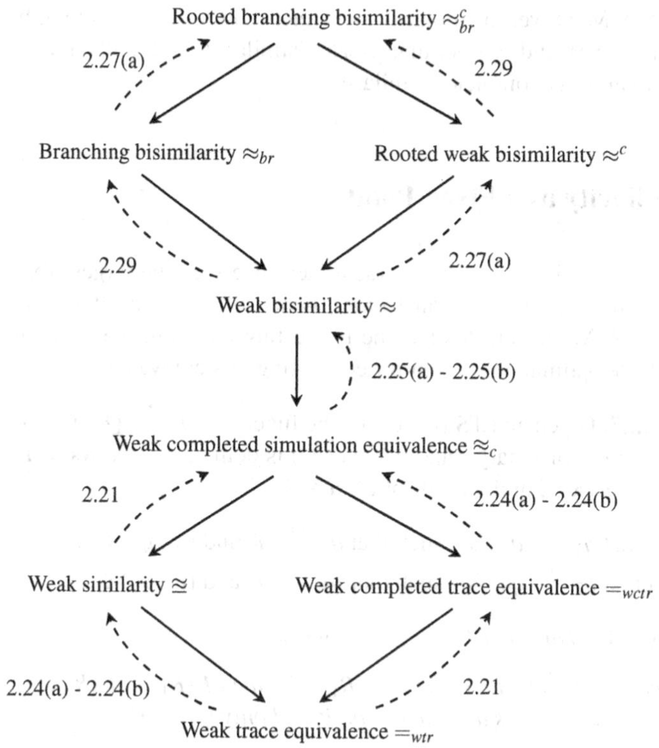

Fig. 2.33 Summary of weak behavioral equivalences

Exercise 2.85. Check if the pairs of LTSs in Figure 2.32(a) and (b) are branching bisimilar by providing suitable branching bisimulation relations. Are they rooted branching bisimilar? □

From a complexity point of view, branching bisimilarity is the easiest weak equivalence to decide. According to [vGW96], it can be checked on finite-state LTSs with time complexity $O(l+nm)$ and space complexity $O(n+m)$, where $l = |A|$, n the number of states and m the number of transitions.

We summarize the various weak equivalence relations we have introduced so far in the diagram in Figure 2.33. An arrow from an equivalence relation to another one means that the former is finer than the latter; e.g., weak simulation equivalence is finer than weak trace equivalence. A dashed arrow labeled with an index of a figure means that such an implication does not hold because of the counterexample discussed in that figure; for instance, the LTS in Figure 2.24(a) is weak trace equivalent to the LTS in Figure 2.24(b), but the two are not weakly simulation equivalent.

In comparing this summary with the one for strong equivalences in Figure 2.20, we should recall that all the strong equivalences are finer than their corresponding weak ones (e.g., strong similarity $\simeq$ is finer than weak similarity $\lessapprox$, as discussed in

Exercise 2.51). Moreover, bisimilarity $\sim$ is finer than rooted branching bisimilarity $\approx_{br}^{c}$ (Exercise 2.84), and so finer than weak bisimilarity $\approx$. Finally, ready similarity $\simeq_{rs}$ is finer than weak completed similarity $\approxeq_c$.

2.5 Bisimilarity as a Fixed Point

Bisimulation equivalence can be characterized nicely as the largest fixed point of a suitable monotone relation transformer, hence offering an alternative proof of Proposition 2.9. Moreover, this has the interesting consequence of defining a natural, even if not optimal, algorithm for computing this equivalence.

Definition 2.27. Given an LTS $(Q, A, \rightarrow)$, the functional $F : \wp(Q \times Q) \rightarrow \wp(Q \times Q)$ (i.e., a transformer of binary relations over Q) is defined as follows. If $R \subseteq Q \times Q$, then $(q_1, q_2) \in F(R)$ if and only if for all $\mu \in A$

- $\forall q_1'$ such that $q_1 \xrightarrow{\mu} q_1'$, $\exists q_2'$ such that $q_2 \xrightarrow{\mu} q_2'$ and $(q_1', q_2') \in R$,
- $\forall q_2'$ such that $q_2 \xrightarrow{\mu} q_2'$, $\exists q_1'$ such that $q_1 \xrightarrow{\mu} q_1'$ and $(q_1', q_2') \in R$. □

Proposition 2.12. *For any LTS $(Q, A, \rightarrow)$, we have that:*

1. *The functional F is monotone, i.e., if $R_1 \subseteq R_2$ then $F(R_1) \subseteq F(R_2)$.*
2. *A relation $R \subseteq Q \times Q$ is a bisimulation if and only if $R \subseteq F(R)$.*

Proof. *The proof of (1) derives immediately form the definition of F: if $(q_1, q_2) \in F(R_1)$ then for all $\mu \in A$*

- *$\forall q_1'$ such that $q_1 \xrightarrow{\mu} q_1'$, $\exists q_2'$ such that $q_2 \xrightarrow{\mu} q_2'$ and $(q_1', q_2') \in R_1$,*
- *$\forall q_2'$ such that $q_2 \xrightarrow{\mu} q_2'$, $\exists q_1'$ such that $q_1 \xrightarrow{\mu} q_1'$ and $(q_1', q_2') \in R_1$.*

Since $R_1 \subseteq R_2$, the above implies that for all $\mu \in A$

- *$\forall q_1'$ such that $q_1 \xrightarrow{\mu} q_1'$, $\exists q_2'$ such that $q_2 \xrightarrow{\mu} q_2'$ and $(q_1', q_2') \in R_2$,*
- *$\forall q_2'$ such that $q_2 \xrightarrow{\mu} q_2'$, $\exists q_1'$ such that $q_1 \xrightarrow{\mu} q_1'$ and $(q_1', q_2') \in R_2$,*

which means that $(q_1, q_2) \in F(R_2)$.

The proof of (2) is also easy: if R is a bisimulation, then if $(q_1, q_2) \in R$ then for all $\mu \in A$

- *$\forall q_1'$ such that $q_1 \xrightarrow{\mu} q_1'$, $\exists q_2'$ such that $q_2 \xrightarrow{\mu} q_2'$ and $(q_1', q_2') \in R$,*
- *$\forall q_2'$ such that $q_2 \xrightarrow{\mu} q_2'$, $\exists q_1'$ such that $q_1 \xrightarrow{\mu} q_1'$ and $(q_1', q_2') \in R$,*

and, by using the reverse implication, this means that $(q_1, q_2) \in F(R)$, i.e., $R \subseteq F(R)$. Similarly, if $R \subseteq F(R)$, then the condition holding for $F(R)$ holds also for all the elements of R, hence R is a bisimulation. □

A *fixed point* for F is a relation R such that $R = F(R)$. Knaster-Tarski's fixed point theorem (see, e.g., [DP02]) ensures that the largest fixed point of the monotone functional F is

$$\bigcup\{R \subseteq Q \times Q \mid R \subseteq F(R)\}.$$

We want to show that this largest fixed point is $\sim$. A *post-fixed point* of F is a relation R such that $R \subseteq F(R)$. By Proposition 2.12(2), we know that bisimulations are the post-fixed points of F. Bisimilarity $\sim$ is the union of all the bisimulations:

$$\sim = \bigcup\{R \subseteq Q \times Q \mid R \text{ is a bisimulation}\}.$$

Hence, we also conclude that $\sim$ is the largest fixed point of F, i.e.:

$$\sim = \bigcup\{R \subseteq Q \times Q \mid R \subseteq F(R)\}.$$

Here we provide a direct proof of this fact.

Theorem 2.1. *Strong bisimilarity $\sim$ is the largest fixed point of F.*

Proof. *We first prove that $\sim$ is a fixed point, i.e., $\sim = F(\sim)$, by proving that $\sim \subseteq F(\sim)$ and that $F(\sim) \subseteq \sim$. Since $\sim$ is a bisimulation, $\sim \subseteq F(\sim)$ by Proposition 2.12(2). As F is monotonic, by Proposition 2.12(1) we have that $F(\sim) \subseteq F(F(\sim))$, i.e., also $F(\sim)$ is a post-fixed point of F i.e., a bisimulation. Since we know that $\sim$ is the union of all bisimulation relations (as well as the largest post-fixed point of F), $F(\sim) \subseteq \sim$ follows.*

Now we want to show that $\sim$ is the largest fixed point. Assume T is another fixed point of F, i.e. $T = F(T)$. Then, in particular, we have that $T \subseteq F(T)$, i.e., T is a bisimulation by Propostion 2.12(2), hence $T \subseteq \sim$. □

Exercise 2.86. (Simulation preorder as a fixed point) Given an LTS $(Q, A, \rightarrow)$, functional $G : \wp(Q \times Q) \rightarrow \wp(Q \times Q)$ is defined as follows. If $R \subseteq Q \times Q$, then $(q_1, q_2) \in G(R)$ if and only if $\forall \mu \in A$

- $\forall q_1'$ such that $q_1 \xrightarrow{\mu} q_1'$, $\exists q_2'$ such that $q_2 \xrightarrow{\mu} q_2'$ and $(q_1', q_2') \in R$.

Show that:

- functional G is monotone, i.e., if $R_1 \subseteq R_2$ then $G(R_1) \subseteq G(R_2)$;
- relation $R \subseteq Q \times Q$ is a simulation if and only if $R \subseteq G(R)$;
- the simulation preorder $\lesssim$ is the largest fixed point of G. □

There is a natural iterative way of approximating $\sim$ (over image-finite LTSs) by means of a descending (actually, nonincreasing) chain of relations indexed on the natural numbers. We will see that there is a strict relation between this chain of relations and the functional F above.

Definition 2.28. Given an LTS $TS = (Q, A, \rightarrow)$, for each natural $i \in \mathbb{N}$, we define relation $\sim_i$ over Q as follows:

- $\sim_0 = Q \times Q$.

- $q_1 \sim_{i+1} q_2$ if and only if for all $\mu \in A$
  - $\forall q_1'$ such that $q_1 \xrightarrow{\mu} q_1'$, $\exists q_2'$ such that $q_2 \xrightarrow{\mu} q_2'$ and $q_1' \sim_i q_2'$
  - $\forall q_2'$ such that $q_2 \xrightarrow{\mu} q_2'$, $\exists q_1'$ such that $q_1 \xrightarrow{\mu} q_1'$ and $q_1' \sim_i q_2'$.

We denote by $\sim_\omega$ the relation $\bigcap_{i \in \mathbb{N}} \sim_i$. □

Intuitively, $q_1 \sim_i q_2$ if and only if the two states are bisimilar up to paths of length at most i. Hence, all the states are in the relation $\sim_0$. Considering Figure 2.15, it is easy to see that the $q_1 \sim_1 q_7$, while $q_1 \not\sim_2 q_7$, because there is no state, reachable from q_7, $\sim_1$-equivalent to q_2.

Proposition 2.13. *For each $i \in \mathbb{N}$:*

1. *relation $\sim_i$ is an equivalence relation,*
2. *$\sim_{i+1} \subseteq \sim_i$,*
3. *$\sim_i = F^i(Q \times Q)$*

Moreover, $\sim_\omega = \bigcap_{i \in \mathbb{N}} \sim_i$ is an equivalence relation.

Proof. The three proofs are by induction on i.

(1) The base case is obvious: $\sim_0$ is an equivalence relation because so is the universal relation. Assuming that $\sim_i$ is an equivalence relation, we show that also $\sim_{i+1}$ is an equivalence relation. Reflexivity is trivial: as for all $\mu \in A$

- *$\forall q'$ such that $q \xrightarrow{\mu} q'$, $\exists q'$ such that $q \xrightarrow{\mu} q'$ and $q' \sim_i q'$*

it follows that also $q \sim_{i+1} q$. Symmetry is also trivial: we have to prove that if $q_1 \sim_{i+1} q_2$ then $q_2 \sim_{i+1} q_1$. We know that $q_1 \sim_{i+1} q_2$ if and only if for all $\mu \in A$

- *$\forall q_1'$ such that $q_1 \xrightarrow{\mu} q_1'$, $\exists q_2'$ such that $q_2 \xrightarrow{\mu} q_2'$ and $q_1' \sim_i q_2'$*
- *$\forall q_2'$ such that $q_2 \xrightarrow{\mu} q_2'$, $\exists q_1'$ such that $q_1 \xrightarrow{\mu} q_1'$ and $q_1' \sim_i q_2'$.*

We also know, by inductive hypothesis, that $q_2' \sim_i q_1'$, hence, also that

- *$\forall q_2'$ such that $q_2 \xrightarrow{\mu} q_2'$, $\exists q_1'$ such that $q_1 \xrightarrow{\mu} q_1'$ and $q_2' \sim_i q_1'$.*
- *$\forall q_1'$ such that $q_1 \xrightarrow{\mu} q_1'$, $\exists q_2'$ such that $q_2 \xrightarrow{\mu} q_2'$ and $q_2' \sim_i q_1'$*

which means that $q_2 \sim_{i+1} q_1$. Transitivity is similar, hence left to the reader.

(2) and (3) are left as exercises for the reader. We simply explain what we mean by the ith power of F: $F^0(R) = R$ and $F^{n+1}(R) = F(F^n(R))$.

Observe that $\sim_\omega = \bigcap_{i \in \mathbb{N}} \sim_i$ is an equivalence relation, as $\sim_i$ is an equivalence relation for all $i \in \mathbb{N}$. As a matter of fact, the identity relation $\mathscr{I}$ is a subset of all the $\sim_i$'s, hence $\mathscr{I} \subset \sim_\omega$, i.e., $\sim_\omega$ is reflexive. Relation $\sim_\omega$ is also symmetric because, if $(q_1, q_2) \in \sim_\omega$, then $(q_1, q_2) \in \sim_i$ for all $i \in \mathbb{N}$. Since each $\sim_i$ is symmetric, $(q_2, q_1) \in \sim_i$ for all $i \in \mathbb{N}$, so $(q_2, q_1) \in \sim_\omega$, hence $\sim_\omega$ is symmetric. Transitivity of $\sim_\omega$ can be proved similarly. □

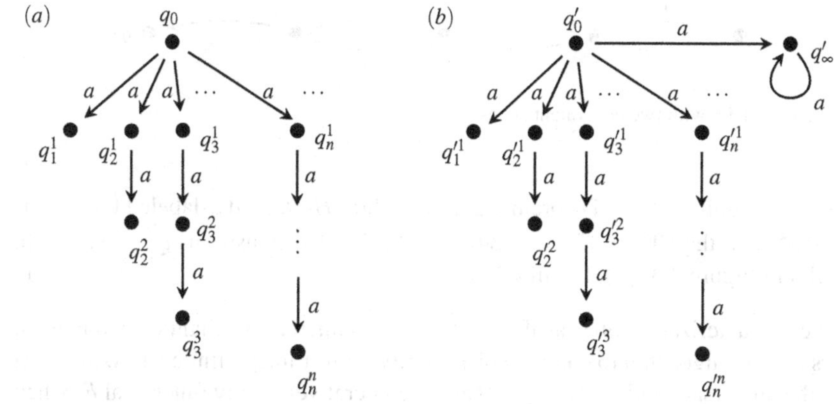

Fig. 2.34 Two not image-finite LTSs

Hence, we have a nonincreasing chain of equivalence relations,

$$\sim_0 = F^0(Q \times Q) \supseteq \sim_1 = F^1(Q \times Q) \supseteq \ldots \supseteq \sim_i = F^i(Q \times Q) \supseteq \ldots \supseteq \sim_\omega,$$

with relation $\sim_\omega$ as its limit. Interestingly, this limit, for image-finite LTSs, coincides with strong bisimilarity $\sim$.

Theorem 2.2. *If the LTS $TS = (Q, A, \rightarrow)$ is image-finite, then $\sim = \sim_\omega$.*

Proof. We prove first that $\sim \subseteq \sim_i$ for all i by induction on i. Indeed, $\sim \subseteq \sim_0$ (the universal relation); moreover, assuming $\sim \subseteq \sim_i$, by monotonicity of F and the fact that $\sim$ is a fixed point for F, we get $\sim = F(\sim) \subseteq F(\sim_i) = \sim_{i+1}$. Hence, $\sim \subseteq \sim_\omega$.

Now we prove that $\sim_\omega \subseteq \sim$, by proving that relation $R = \{(q_1, q_2) \mid q_1 \sim_\omega q_2\}$ is a bisimulation. Assume $(q_1, q_2) \in R$, hence $q_1 \sim_i q_2$ for all $i \in \mathbb{N}$. If $q_1 \xrightarrow{\mu} q'_1$, then for all i, there exists q_{2_i} such that $q_2 \xrightarrow{\mu} q_{2_i}$ with $q'_1 \sim_i q_{2_i}$. Since the LTS is image-finite, the set $K = \{q_{2_k} \mid q_2 \xrightarrow{\mu} q_{2_k} \wedge q'_1 \sim_k q_{2_k} \wedge k \in \mathbb{N}\}$ is finite; hence, there is at least one $q_{2_n} \in K$ such that $q'_1 \sim_i q_{2_n}$ for infinitely many i. But since if $q \sim_i q'$ then $q \sim_j q'$ for any $j < i$, we can conclude that $q'_1 \sim_i q_{2_n}$ for all i, hence $q'_1 \sim_\omega q_{2_n}$, and so $(q'_1, q_{2_n}) \in R$. The symmetric case when q_2 moves first is analogous, and hence omitted. So $R = \sim_\omega$ is a bisimulation, therefore $\sim_\omega \subseteq \sim$. □

The hypothesis of image-finiteness is crucial for the correctness of the Theorem above, as the following exercise shows.

Exercise 2.87. Consider the two LTSs in Figure 2.34; these are not image-finite. Prove, by induction on i, that the two roots q_0 and q'_0 are related by $\sim_i$ for any i, but they are not related by $\sim$. □

Exercise 2.88. Show a pair of distinct LTSs such that they are both not image-finite, but such that the two roots are related by $\sim_i$ for any i, and also by $\sim$. This shows that

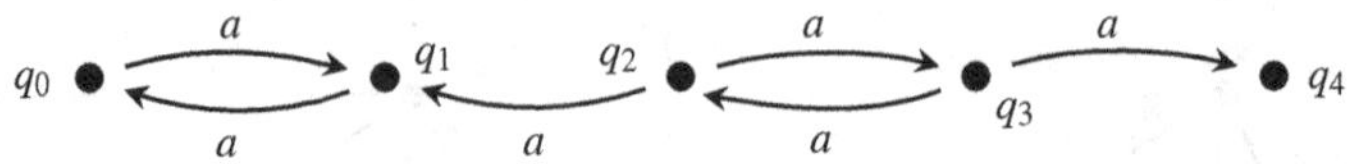

Fig. 2.35 An LTS with two equivalent states

the reverse implication of Theorem 2.2 is not valid. (*Hint:* Add a-labeled transitions $q_k^k \xrightarrow{a} q_{k+1}^{k+1}$ to the LTS in Figure 2.34(a) and a-labeled transitions $q_k'^k \xrightarrow{a} q_{k+1}'^{k+1}$ to the LTS in Figure 2.34(b), for all $k \geq 1$.) □

The characterization of $\sim$ as the limit of the nonincreasing chain of relations $\sim_i$ offers an easy algorithm to compute bisimilarity $\sim$ over image-finite LTSs: just start from the universal relation $R_0 = Q \times Q$ and then iteratively apply functional F; when $R_{i+1} = F(R_i) = R_i$ then stop and take R_i as the bisimilarity relation. In other words, the algorithm can be expressed by the following sequence of instructions:

$R := Q \times Q$;
$S := F(R)$;
while $R \neq S$ **do** $\{R := S; S := F(S)\}$;
return R

Of course, this algorithm may not terminate for LTSs with infinitely many states, but it always terminates for finite-state LTSs.

As an example, take the LTS (actually, the union of the two) in Figure 2.18, where $Q = \{q_1, q_2, \ldots, q_7\}$. Initially, $R_0 = Q \times Q$, which is composed by one equivalence class only: all the states are equivalent. Then, we compute $F(R_0) = R_1$, which is given by $R \cup R^{-1} \cup \mathscr{I}$, with $R = \{(q_1,q_4),(q_2,q_3),(q_2,q_5),(q_2,q_6),(q_2,q_7),(q_3,q_5),(q_3,q_6),(q_3,q_7),(q_5,q_6),(q_5,q_7),(q_6,q_7)\}$ and $\mathscr{I} = \{(q_i,q_i) \mid 1 \leq i \leq 7\}$. Observe that the only pairs of states that have been removed from R_0 are those whose states do not offer the same menu of possible initial moves, e.g., (q_1,q_2). Relation R_1 defines two equivalence classes only: $\{q_1,q_4\}$ and $\{q_2,q_3,q_5,q_6,q_7\}$. As $F(R_1) = R_1$, we can conclude that strong bisimilarity $\sim$ on such a transition system is given by relation R_1. Hence, we have a *minimum* (with respect to strong bisimulation) realization of this LTS: just take one state for each equivalence class! In our example, we can call q the state for the equivalence class $\{q_1,q_4\}$ and q' the state for $\{q_2,q_3,q_5,q_6,q_7\}$. The resulting minimum LTS is the triple $(Q', A, \rightarrow)$, where $Q' = \{q, q'\}$, $A = \{a,b\}$ and $\rightarrow = \{(q,a,q'),(q',b,q)\}$.

Exercise 2.89. Compute relation $\sim$ for the LTS in Figure 2.35 by applying the iterative algorithm described above. Check that the resulting relation $\sim$ is $\mathscr{I} \cup \{(q_0,q_1),(q_1,q_0)\}$. □

Definition 2.29. (LTS minimization w.r.t. $\sim$) Given an LTS $TS = (Q, A, \rightarrow)$, let $\sim$ be the largest bisimulation relation over $Q \times Q$. Then, we can define the *minimum* LTS $TS_\sim = (Q_\sim, A, \rightarrow_\sim)$, where:

- $Q_\sim = \{[q]_\sim \mid q \in Q\}$, where $[q]_\sim = \{q' \in Q \mid q \sim q'\}$,
- $\rightarrow_\sim = \{([q]_\sim, \mu, [q']_\sim) \mid (q, \mu, q') \in \rightarrow\}$. □

Observe that in the definition of the minimum LTS $TS_{\sim}$, any state $[q]_{\sim}$ is an equivalence class of states of TS: for all $q, q' \in Q$, $q \sim q'$ if and only if $[q]_{\sim} = [q']_{\sim}$. Moreover, if $([q]_{\sim}, \mu, [q']_{\sim})$ is a transition in $TS_{\sim}$, then for all $q_1 \in Q$ such that $q \sim q_1$, there exists a $q_2 \in Q$ such that $q_1 \xrightarrow{\mu} q_2$ and $q' \sim q_2$, and so $([q_1]_{\sim}, \mu, [q_2]_{\sim}) = ([q]_{\sim}, \mu, [q']_{\sim})$. In other words, the definition of $TS_{\sim}$ is independent of the choice of the representative state q for its equivalence class $[q]_{\sim}$.

Proposition 2.14. *Given an LTS $TS = (Q, A, \rightarrow)$ and its associated minimum LTS $TS_{\sim} = (Q_{\sim}, A, \rightarrow_{\sim})$, the following hold:*

- *$q \sim [q]_{\sim}$ for all $q \in Q$ and $[q]_{\sim} \in Q_{\sim}$, i.e., $TS_{\sim}$ is a correct realization of TS;*
- *for all $[q]_{\sim}, [q']_{\sim} \in Q_{\sim}$ we have that if $[q]_{\sim} \sim [q']_{\sim}$ then $[q]_{\sim} = [q']_{\sim}$, i.e., $TS_{\sim}$ is the minimum (up to isomorphism).*

Proof. *For the proof of the first item, consider relation $R \subseteq Q \times Q_{\sim}$ defined as follows: $R = \{(q, [q]_{\sim}) \mid q \in Q\}$. It is easy to see that R is a bisimulation.*

For the proof of the second item, we have that $q \sim [q]_{\sim}$ as well as $q' \sim [q']_{\sim}$ by the previous item. Therefore, if $[q]_{\sim} \sim [q']_{\sim}$, then by transitivity we also have that $q \sim q'$ and so, by construction of $TS_{\sim}$, we have that $[q]_{\sim} = [q']_{\sim}$. □

Remark 2.10. (**Minimality of the minimum LTS w.r.t.** $\sim$) The fact that $TS_{\sim} = (Q_{\sim}, A, \rightarrow_{\sim})$ is *the* minimum LTS (up to isomorphism) directly follows from the fact that in $Q_{\sim}$ there are no two different states that are bisimilar, as stated by the second item of Proposition 2.14. As a matter of fact, assume, towards a contradiction, that there exists an LTS $TS' = (Q', A, \rightarrow')$, such that $|Q'| < |Q_{\sim}|$ and for all $[q]_{\sim} \in Q_{\sim}$ there exists (at least) one corresponding bisimilar state $q' \in Q'$. Then, by cardinality, there must exist two distinct states $[q_1]_{\sim}, [q_2]_{\sim} \in Q_{\sim}$ and one state $q_3 \in Q'$ such that $[q_1]_{\sim} \sim q_3$ and $[q_2]_{\sim} \sim q_3$; hence, by transitivity, $[q_1]_{\sim} \sim [q_2]_{\sim}$, which is impossible.

To be precise, one further optimization is possible for *rooted* LTSs. Given the rooted LTS $TS = (Q, A, \rightarrow, q_0)$, the associated minimum LTS $TS_{\sim} = (Q_{\sim}, A, \rightarrow_{\sim}, [q_0]_{\sim})$ might be further reduced by removing all the states (and relevant transitions) that are not reachable from the initial state $[q_0]_{\sim}$. □

Remark 2.11. (**Minimum LTS w.r.t. trace equivalence?**) Given a rooted LTS $TS = (Q, A, \rightarrow, q_0)$, we can obtain the minimum *deterministic* LTS w.r.t. trace equivalence $=_{tr}$ (which is unique up to isomorphism) by means of the following three-step algorithm. First, transform TS into a deterministic LTS, which we call dTS, according to the construction sketched in Exercise 2.17. Then, compute relation $\sim$ over dTS by means of the algorithm above (remember that $\sim$ and $=_{tr}$ coincide on deterministic LTS, as mentioned in Exercise 2.44). Finally, compute the minimum LTS $dTS_{\sim}$, according to Definition 2.29. We are sure that $dTS_{\sim}$ is the minimum *deterministic* LTS w.r.t. trace equivalence for TS. However, it is possible that there exists a smaller (i.e., with fewer states) nondeterministic LTS $ndTS$ trace equivalent to TS, as illustrated in Exercise 2.91. We conjecture that finding a general polynomial-time algorithm to minimize a nondeterministic LTS into a nondeterministic LTS is impossible, as in automata theory a similar conjecture is expressed in [HMU01] for the analogous problem of nondeterministic minimization of NFA. □

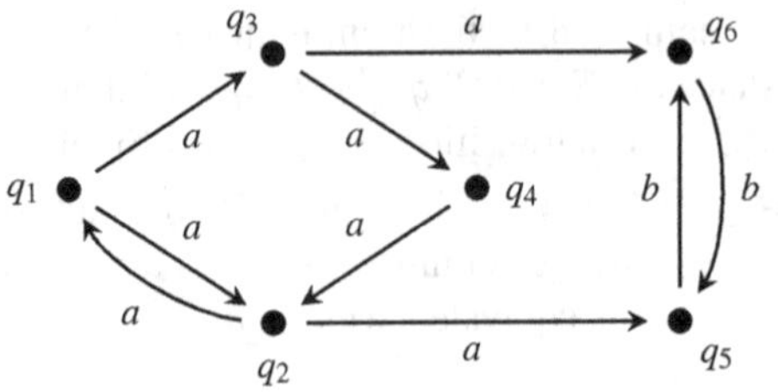

Fig. 2.36 An LTS with three pairs of equivalent states

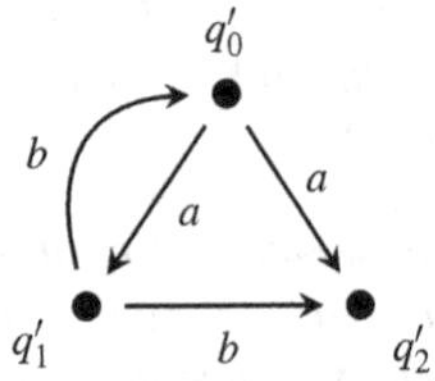

Fig. 2.37 The minimum LTS w.r.t. $\sim$ for those in Figure 2.14

Exercise 2.90. Continuing Exercise 2.89, compute the minimum LTS w.r.t. bisimilarity of the LTS in Figure 2.35. Then, compute the minimum deterministic LTS w.r.t. trace equivalence for the same LTS, assuming q_2 as its initial state. □

Exercise 2.91. Compute the minimum LTS w.r.t. bisimulation equivalence of the LTS in Figure 2.36. Then, compute the minimum deterministic LTS w.r.t. trace equivalence for the same LTS, assuming q_1 as its initial state. Observe that the minimum LTS w.r.t. $\sim$ is nondeterministic and smaller than the minimum deterministic LTS w.r.t. trace equivalence. □

Exercise 2.92. Check that the minimum LTS w.r.t. $\sim$ for those outlined in Figure 2.14 is isomorphic to the LTS in Figure 2.37. □

Exercise 2.93. (Approximating the simulation preorder $\lesssim$) Given an LTS $TS = (Q, A, \to)$, for each natural $i \in \mathbb{N}$, define the relations $\lesssim_i$ over Q as follows:

- $\lesssim_0 = Q \times Q$.
- $q_1 \lesssim_{i+1} q_2$ if and only if for all $\mu \in A$
    - $\forall q'_1$ such that $q_1 \xrightarrow{\mu} q'_1$, $\exists q'_2$ such that $q_2 \xrightarrow{\mu} q'_2$ and $q'_1 \lesssim_i q'_2$.

We denote by $\lesssim_\omega$ the relation $\bigcap_{i \in \mathbb{N}} \lesssim_i$. Prove that, for each $i \in \mathbb{N}$:

1. the relation $\lesssim_i$ is a preorder,
2. $\lesssim_{i+1} \subseteq \lesssim_i$,
3. $\lesssim_i = G^i(Q \times Q)$, where G is defined in Exercise 2.86.

Moreover, prove that if TS is image-finite, then $\lesssim = \lesssim_\omega$. □

Hence, also the simulation preorder $\lesssim$ can be computed by the easy iterative algorithm we have proposed for computing bisimulation equivalence, where the occurrences of functional F are to be replaced by functional G.

Also, the weak simulation preorder $\lessapprox$ can be computed for finite-state LTSs with this iterative algorithm. As a matter of fact, Exercise 2.57 ensures that the weak simulation preorder on an LTS *TS* can be computed as the (strong) simulation preorder $\lesssim$ over its associated tau-abstracted LTS τATS (Definition 2.49).

Finally, weak bisimilarity $\approx$ can also be defined as the largest fixed point of a suitable relation transformer H, following the same steps as above because, as a consequence of Exercise 2.68, it is possible to offer an alternative, yet equivalent, definition of weak bisimulation on an LTS *TS* (with transition relation $\longrightarrow$) as a strong bisimulation on its associated abstract LTS *ATS* (with transition relation $\Longrightarrow''$). Even if weak bisimilarity $\approx$ can be characterized by means of the approximations $\approx_i$ on the natural numbers for finite-state LTSs, such a characterization is sometimes prevented for LTSs with infinitely many states, because it may happen that an image-finite LTS may originate a not image-finite $\Longrightarrow''$ relation (see, e.g., Figure 2.22, which is the relation used in the definition of H.

Exercise 2.94. LTS minimization can be performed also w.r.t. weak bisimilarity $\approx$. Consider the LTSs in Figure 2.26. First compute their associated abstract LTSs, according to Exercise 2.68, and then minimize these abstract LTSs w.r.t. strong bisimulation. The resulting minimum LTS has three states only. □

Chapter 3
CCS: A Calculus of Communicating Systems

Abstract The process calculus CCS for describing reactive systems is introduced. Its syntax is defined, as well as its operational semantics in terms of labeled transition systems. Some subcalculi are singled out that possess some specific interesting expressiveness properties. It is shown that CCS is Turing-complete by offering an encoding of Counter Machines into CCS. As a byproduct, all the behavioral equivalences of interest are undecidable over the class of CCS processes, even if they are decidable over some subcalculi.

3.1 A Language for Describing Reactive Systems

In this chapter we present the process calculus CCS (acronym of *Calculus of Communicating Systems*), proposed by Milner in 1980 in his seminal book [Mil80] and revisited in [Mil89, Mil99]. Many other process calculi have been proposed in the literature to describe reactive systems, e.g., ACP [BK84a, BW90, BBR10], CSP [Hoa85, Ros98], and Lotos [BoBr87, BLV95]. The main reason for choosing CCS is its deep simplicity, due to a good choice of intuitive operators, and also because it has been extended smoothly to include other features, such as mobility with the π-calculus [MPW92, Mil99, SW01] and atomicity with Multi-CCS (see Chapter 6).

Why do we want a language for expressing reactive systems? There are at least four good reasons for defining a process calculus: linguistic support for describing succinctly objects of the semantic model, support for compositional modeling and reasoning, support for equational reasoning, as well as support for early prototyping of the model.

- **Linguistic support:** In the previous chapter we have shown models of reactive systems based on a labeled transition system representation. However, these models are not given very easily, as we have to draw explicitly the whole state space. Except for relatively small systems, a graphical representation by LTS is laborious; indeed, real-world systems easily have thousands (sometimes millions or even billions) of states, which makes drawing them practically impossible. In order to be able to describe a complex system, we would prefer a textual (i.e., linear), implicit representation as a term in some process language. We would like to develop a language, and an associated semantics in terms of labeled transition systems, so that, given a term of the language, the semantics produces automatically an LTS representation of the behavior of the system described by that term. For instance, the LTS in Figure 2.2 can be represented textually as the following process definition:

$$A \stackrel{def}{=} coin.\overline{coffee}.A$$

where A is a process *constant* (i.e., a name for a process), $\stackrel{def}{=}$ is the special symbol that means "is defined as" and $coin.\overline{coffee}.A$ is the code meaning that first A can accept a coin in input, then it can deliver a coffee in output, and in doing so, it returns to its initial state A. (The operator $a.p$ is called *action prefixing*, and determines a sequentialization of action a before process p starts.)
In some cases the state space of a reactive system may be even infinite and so it cannot be built in a finite amount of time: see, e.g., the semi-counter in Figure 2.7(b). We will see how to model in CCS such a system with a parametric definition that, however, involves an unbounded number of process constants (see Example 3.1); interestingly, such a process can be proved bisimulation equivalent to a simple CCS process defined by means of just one constant (see Example 3.12 in Section 3.4.4), so that the infinite LTS of a semi-counter is succinctly represented, up to bisimilarity, by a finite term.

- **Compositionality:** Another good reason to develop a language for describing reactive systems is that it opens the way to compositional modeling and reasoning. When a complex system is to be modeled, it is often convenient to identify the sequential subcomponents of the system, to model these subcomponents separately in order to reason on each of them independently of the other components, and then, possibly, to generate the complete model by composition of the individual models in order to analyze global properties of the system.
As an instance of the kind of analyzes that are possible by compositionality, suppose a system is described by the CCS process term $p_1 \mid p_2$, where p_1 and p_2 are two processes that are composed in parallel by means of the parallel operator $\mid$. It often happens that the compound system $p_1 \mid p_2$ satisfies a certain property if this property holds for the constituents p_1 and p_2. As an instance of a property of this form, consider "the process always terminates". Hence, instead of checking this property against the large state space of the compound system, we can just check if this holds for the two smaller state spaces of the constituents. In general, assuming that the state space of each process p_i (for $1 \leq i \leq n$) is composed of ten states, we will see that the state space of $p_1 \mid p_2 \mid \dots \mid p_n$ is composed of 10^n states: hence, checking the property on the compound system is exponential in the number of components, while with a compositional reasoning, it reduces to checking n times, the property on a system with ten states (linear complexity).
For instance, when checking the equivalence between two composite systems, say $p_1 \mid p_2$ and $q_1 \mid q_2$, it might be more convenient to check separately if p_1 is

Fig. 3.1 The three LTSs for **0**, $a.\mathbf{0}$, and $a.b.\mathbf{0}$

equivalent to q_1, as well as if p_2 is equivalent to q_2, instead of considering the two large global state spaces, because when the equivalence is preserved by parallel composition (which is indeed the case for most equivalences, as we will see in Chapter 4), then we are sure that $p_1 \mid p_2$ and $q_1 \mid q_2$ are equivalent, too.

- **Equational reasoning:** One further good reason for developing a process calculus to describe reactive systems is that it allows for the definition of equational theories (usually called *axiomatizations*) for behavioral equivalences.
  So far, two reactive systems can be proved bisimulation equivalent by checking that there exists a bisimulation over the state spaces of the two LTSs; but now, as the two LTSs are actually generated by suitable terms, say p_1 and p_2, of the same language, this can be also proved with a purely syntactical argument by showing that the process term p_1 can be equated with the process term p_2 by means of some equational deductive proof. This is the subject of Section 4.3.
- **Early prototyping:** Since CCS is executable, we can perform simple forms of analysis of the model by "simulating" its behavior. This is the so-called *early prototyping*: the *specification* (i.e., the model) is executable and so it can be analyzed before being implemented. As a matter of fact, by running the CCS process representing the model, we can extract useful pieces of information about the model, e.g., by testing its behavior under specific circumstances. Model testing is a widespread technique in software engineering and this technique is available also for CCS-based model descriptions because CCS is executable.

3.1.1 An Informal Overview of CCS Operators

Now we give an informal, intuitive overview of CCS operators, as well as of the semantics of CCS process terms, by showing examples of LTSs that are associated by the semantics to such terms. The semantics generates LTSs such that the states are the terms of the language.

Inaction: The simplest process one can define in CCS is **0** (called *nil*), the empty process. Semantically, it is represented by an LTS with one state only: the deadlocked process term **0** itself (see Figure 3.1(a)).

Action prefixing: If we prefix **0** with an action a, we get $a.\mathbf{0}$, which is a process that can perform a and then stops. Its associated LTS is depicted in Figure 3.1(b).

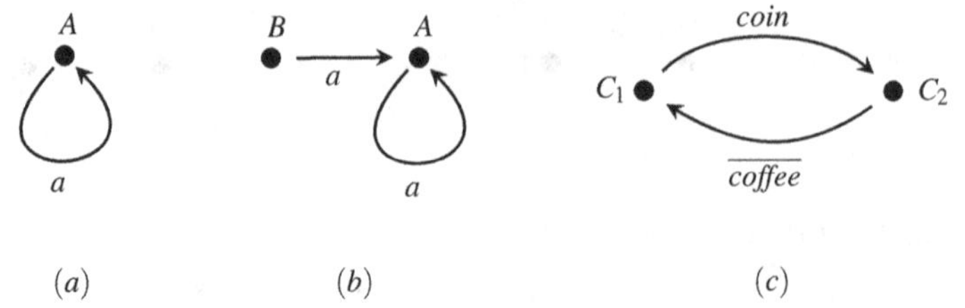

Fig. 3.2 Three simple recursive LTSs

In general, the action prefixing operator takes one action μ (which can be an input a, an output $\bar{a}$, or the invisible τ) and a process p to build the process $\mu.p$ that can execute μ first and then continue with the residual p. This is the basic form of sequentialization of actions available in CCS.

Exercise 3.1. Define a process term generating an LTS isomorphic to the one in Figure 2.12(b). (*Hint:* See Figure 3.1(c).) □

With the empty process and the action prefixing operator, one can only build processes that always terminate. For instance, it is not difficult to see that an LTS isomorphic to the one in Figure 2.4 can be described by a term built with only these two operators:

$$coin.\overline{coffee}.coin.\overline{coffee}\ldots coin.\overline{coffee}.\mathbf{0}$$

To describe nonterminating processes we need the following construct.

Process constant: If we want to give a name to some process p, we may use a process constant (usually denoted by a capital letter such as $A, B, \ldots$) equipped with its defining equation $A \stackrel{def}{=} p$. For instance, $B \stackrel{def}{=} a.b.\mathbf{0}$ denotes the finite process B that can perform the sequence ab, and $C \stackrel{def}{=} \mathbf{0}$ denotes another empty process. Given $A \stackrel{def}{=} p$, we sometimes say that p is the *body* of the defined constant A.

Process constants are very important because they provide the means for expressing nonterminating behavior when the constant to be defined occurs inside its body. For instance,

$$A \stackrel{def}{=} a.A$$

denotes the one-state cyclic LTS in Figure 3.2(a). Of course, one can define mutually recursive constants such as $C_1 \stackrel{def}{=} coin.C_2$ and $C_2 \stackrel{def}{=} \overline{coffee}.C_1$ (see Figure 3.2(c) and the original Figure 2.2). More generally, one can define a system of mutually recursive constants (which may even be infinitely many).

Exercise 3.2. Provide the definition of the process constant B corresponding to the LTS in Figure 3.2(b). □

Choice: With the empty process, the action prefixing operator and the process constants, we can only build terms whose associated LTSs are such that each state

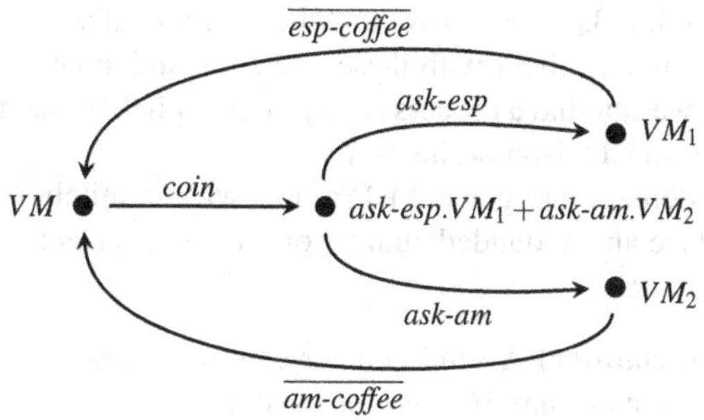

Fig. 3.3 A polite vending machine

has at most one outgoing transition. When we need to express the choice among some alternative actions, we use the choice operator $+$. For instance, the CCS term $a.A + b.B$, where $A \stackrel{def}{=} \mathbf{0}$ and $B \stackrel{def}{=} \mathbf{0}$, would generate an LTS isomorphic to the one in Figure 2.9(c).

Exercise 3.3. Try to figure out what is the LTS associated to $a.\mathbf{0} + b.\mathbf{0}$ (which is not isomorphic to the one in Figure 2.9(c), as it has only two states), as well as to $a.b.\mathbf{0} + b.a.\mathbf{0}$. □

In general, $p_1 + p_2$ is a process that can perform either an action from p_1 (and then continues with the residual of p_1), or an action from p_2 (and then continues with the residual of p_2): in any case, the other alternative is discarded. As a further example, let us consider the LTS in Figure 3.3, which is a CCS analogous to the LTS in Figure 2.3. The CCS process term for it is *VM*, defined as follows:

$$VM \stackrel{def}{=} coin.(ask\text{-}esp.VM_1 \; + \; ask\text{-}am.VM_2)$$
$$VM_1 \stackrel{def}{=} \overline{esp\text{-}coffee}.VM \qquad\qquad VM_2 \stackrel{def}{=} \overline{am\text{-}coffee}.VM$$

Exercise 3.4. Try to figure out what is the LTS corresponding to the vending machine specified by the term $VM' \stackrel{def}{=} coin.(ask\text{-}esp.\overline{coffee}.VM' + ask\text{-}am.\overline{coffee}.VM')$. (*Hint:* Note that after the selection of the beverage, the residual $\overline{coffee}.VM'$ is the same.) (*Remark:* Note that when a constant B is defined as $B \stackrel{def}{=} a.(b.B + c.B)$, we cannot factorize as $B \stackrel{def}{=} a.(b + c).B$, because $b + c$ is not an action and so not a prefix.) □

Exercise 3.5. Define a CCS term corresponding to the LTS in Figure 2.10 (impolite vending machine). (*Hint:* Solution is in Exercise 3.9.) □

Exercise 3.6. Discuss whether a CCS term for the LTS in Figure 2.13(b) may be $A \stackrel{def}{=} a.(b.\mathbf{0} + b.c.A)$, with A corresponding to state q_4. □

With $\mathbf{0}$, action prefixing, choice and a finite number of process constants, we can define any *finite-state* LTS (i.e., with finitely many states and transitions). To be more precise, in Section 3.4.2, we prove that the operational semantics associates a finite-state LTS to any term built with these operators and, conversely, that for any finite-state LTS *TS* we can find a process term p built with these operators only, such that its semantics is an LTS isomorphic to *TS*.

If one wants to define more general LTSs (i.e., with infinitely many states), one has either to introduce an unbounded number of process constants, or to add further operators to the calculus.

Example 3.1. (**Semi-counter**) A semi-counter can be represented by means of an unbounded number of constants $SCount_i$ for $i = 0, 1, \ldots$:

$$SCount_0 \stackrel{def}{=} inc.SCount_1$$
$$SCount_n \stackrel{def}{=} inc.SCount_{n+1} + dec.SCount_{n-1} \quad n > 0$$

The LTS for $SCount_0$ is isomorphic to the one (with infinitely many states) reported in Figure 2.7(b), where each process constant $SCount_i$ is mapped to its corresponding state q_i, for any $i \in \mathbb{N}$. □

CCS offers two important additional operators: parallel composition and restriction. Parallel composition is a binary, infix operator $|$ that applied to p_1 and p_2 gives the term $p_1 \,|\, p_2$. Restriction is an operator that takes an input action, say a, and a process, say p, and generates the term $(\nu a)p$. These operators are called *static* because, as opposed to the *dynamic* ones (namely, action prefixing and choice), they do not disappear, but rather persist, while computation proceeds. These operators are particularly useful when modeling a complex system as a suitable composition of many simpler components.

Parallel composition: Two processes p_1 and p_2 can run in parallel when we compose them with the operator of parallel composition: $p_1 \,|\, p_2$. This means that the two can run asynchronously (at unpredictable relative speed) or interact by performing complementary input/output actions synchronously (so-called *handshake synchronization*). For instance, the LTS for $a.\mathbf{0} \,|\, \bar{a}.\mathbf{0}$ is depicted in Figure 3.4(a), where a stands for an input and $\bar{a}$ for its complementary output. When an interaction takes place, the visible effect is null, that is no further interaction is possible: we represent this fact by means of the invisible action τ. Hence, synchronization is strictly binary in CCS. However, note that synchronization between the two subprocesses is not mandatory: action a can also be performed asynchronously to represent its willingness to be an input for some output coming from the external environment.

Exercise 3.7. Guess the expected LTS for $a.b.\mathbf{0} \,|\, \bar{a}.\mathbf{0}$. Try to draw also the LTS for $(a.b.\mathbf{0} + a.c.\mathbf{0}) \,|\, \bar{a}.\mathbf{0}$ as well as the one for $A \,|\, B$, where $A \stackrel{def}{=} a.b.A$ and $B \stackrel{def}{=} \bar{a}.B$. □

Restriction: Suppose one wants to make action a private for a process p; this can be obtained by means of the restriction operator: $(\nu a)p$ declares that action a cannot be offered for interaction to any process in parallel with p, and that — within p — a can only be used for internal synchronization. For instance, the LTS for $(\nu a)(a.\mathbf{0} \,|\, \bar{a}.\mathbf{0})$

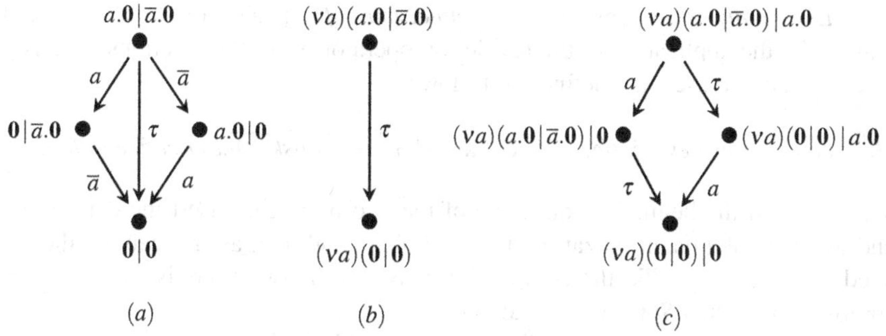

Fig. 3.4 The three LTSs for $a.\mathbf{0}\,|\,\bar{a}.\mathbf{0}$, $(\nu a)(a.\mathbf{0}\,|\,\bar{a}.\mathbf{0})$ and $(\nu a)(a.\mathbf{0}\,|\,\bar{a}.\mathbf{0})\,|\,a.\mathbf{0}$

is depicted in Figure 3.4(b), where the only available transition is due to an internal synchronization on a. Note that actions not under the scope of restriction can be freely executed: the LTS for process $(\nu a)(a.\mathbf{0}\,|\,\bar{a}.\mathbf{0})\,|\,a.\mathbf{0}$ is depicted in Figure 3.4(c). Now action a executed by the rightmost subprocess $a.\mathbf{0}$ is semantically different from the restricted action a executed by the leftmost $a.\mathbf{0}$! The latter can synchronize with the restricted action $\bar{a}$, the former cannot.

Exercise 3.8. Continuing Exercise 3.7, draw the expected LTSs for the three CCS terms $(\nu a)(a.b.\mathbf{0}\,|\,\bar{a}.\mathbf{0})$, $(\nu a)((a.b.\mathbf{0}+a.c.\mathbf{0})\,|\,\bar{a}.\mathbf{0})$ and $(\nu a)(A\,|\,B)$, where $A \stackrel{def}{=} a.b.A$ and $B \stackrel{def}{=} \bar{a}.B$. □

Remark 3.1. **(Static operators and infinite state space)** When the operator of parallel composition or restriction occurs inside the body q of a recursively defined constant A, that is $A \stackrel{def}{=} q$, then the state space for A may contain infinitely many states. For instance, the LTS for constant A, defined as $A \stackrel{def}{=} (\nu b)(a.A)$, is isomorphic to the one on the right of Figure 2.17, where A corresponds to state q_0, $(\nu b)A$ to state q_1, $(\nu b)((\nu b)A)$ to state q_2, and so on. On the contrary, note that the LTS for $(\nu b)B$ with $B \stackrel{def}{=} a.B$ (i.e., restriction does not occur in the body of B) is isomorphic to the one on the left of Figure 2.17, where $(\nu b)B$ corresponds to q. □

Remark 3.2. **(Competition)** With parallel composition and restriction we can express *competition* for the use of some shared resource, hence expressing a form of choice. For instance, in process $(\nu d)(a.(d.b.\mathbf{0}\,|\,d.c.\mathbf{0})\,|\,\bar{d}.\mathbf{0})$ the shared output $\bar{d}$ can be synchronized with only one of the two inputs d. The resulting LTS is isomorphic to the one in Figure 2.25(a), where the system, after a, makes an *internal choice* (i.e., not driven by the environment) between b and c. □

Exercise 3.9. As a further example, if we want to model the fact that two users, say U_1 and U_2, are competing for the use of the vending machine VM specified as in Figure 3.3, we can write

$$Sys \stackrel{def}{=} (\nu L)((U_1\,|\,U_2)\,|\,VM)$$

where $L = \{coin, ask\text{-}esp, ask\text{-}am, esp\text{-}coffee,\ am\text{-}coffee\}$ and (νL) is a shorthand notation for the application of the restriction operator over all the actions in L. Assume that the two users are defined as follows:

$$U_1 \stackrel{def}{=} \overline{coin}.\overline{ask\text{-}esp}.esp\text{-}coffee.milk.U_1 \text{ and } U_2 \stackrel{def}{=} \overline{coin}.\overline{ask\text{-}am}.am\text{-}coffee.milk.U_2.$$

Argue that, at the beginning, only one of the two users can insert the coin in the vending machine (synchronization-driven choice), and that, after receiving the selected beverage and milk, the competition between the two users is still possible. Moreover, argue that *Sys* cannot deadlock.

Now consider the impolite vending machine of Figure 2.10 which is represented by the CCS process term:

$$IVM \stackrel{def}{=} coin.ask\text{-}esp.IVM_1 + coin.ask\text{-}am.IVM_2$$
$$IVM_1 \stackrel{def}{=} \overline{esp\text{-}coffee}.IVM \qquad IVM_2 \stackrel{def}{=} \overline{am\text{-}coffee}.IVM.$$

Argue that the compound system $Sys' \stackrel{def}{=} (\nu L)((U_1 \,|\, U_2) \,|\, IVM)$ can deadlock. □

Continuing Exercise 3.9, suppose that now U_1 wants to have private access to *VM* so that U_2 is not allowed to interact with *VM*. This can be obtained by restricting the scope of the actions that are used for interaction:

$Sys'' \stackrel{def}{=} (\nu L)(U_1 \,|\, VM) \,|\, U_2$.

Of course, U_2 is allowed to interact with any other vending machine in order to get American coffee. For instance,

$Sys''' \stackrel{def}{=} (\nu L)(U_1 \,|\, VM) \,|\, (\nu L)(U_2 \,|\, VM)$

represents a system where each user has a private coffee machine.

Remark 3.3. (**Flow-graph or interface**) The LTS associated to a CCS term denotes its dynamic bahavior. However, we may be interested also in a representation of the static interconnection architecture of a CCS term, called *flow-graph* [Mil89] or *interface*. The flow graph for a process term p is completely determined by the static operators occurring in p.

Each sequential subprocess p_i is represented by a circle with the name p_i inside, and its interaction ports are represented by small blobs on its circumference, each labeled with the name of an action p_i can perform. For instance, the interface for the vending machine *VM* is represented in Figure 3.5(a).

When composing in parallel two sequential subprocesses, say p_1 and p_2, we put together the two interfaces and, additionally, we connect their complementary ports to represent that a corresponding communication channel has been established. For instance, when composing *VM* in parallel with U_1, we get the interface in Figure 3.5(b). Similarly, Figure 3.5(c) describes the interface of $(U_1 \,|\, VM) \,|\, U_2$. Note that a port, e.g., *coin*, may be connected to several different complementary ports, e.g., all those ports labeled with $\overline{coin}$.

When restricting action a over the parallel composition of p_1 and p_2 (i.e., $(\nu a)(p_1 \,|\, p_2)$), we take the interface of $p_1 \,|\, p_2$ and then hide the ports for a and

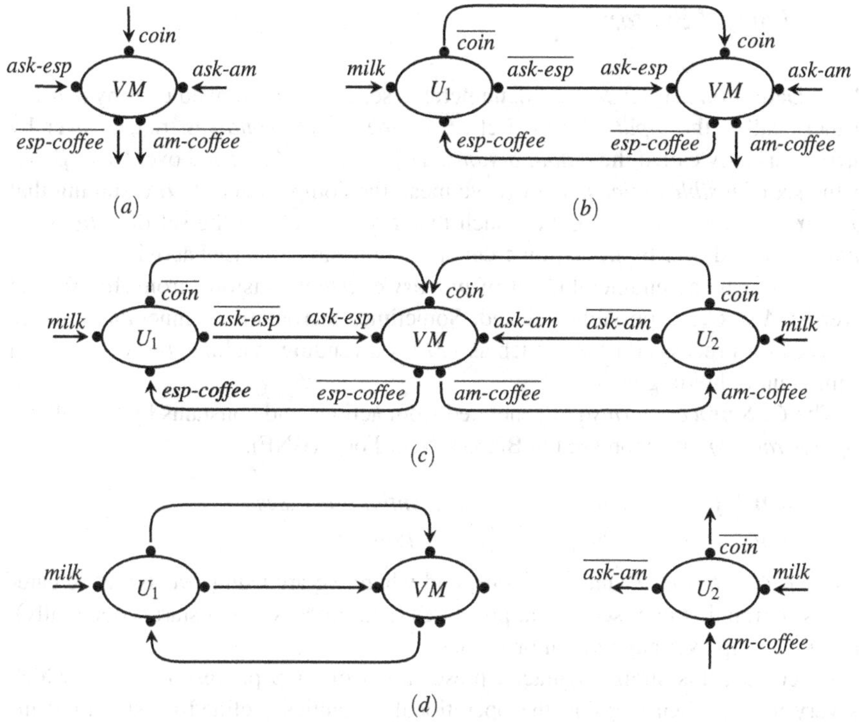

Fig. 3.5 The flow-graphs for the vending machine *VM* in (a), for $U_1 \mid VM$ in (b), for $(U_1 \mid VM) \mid U_2$ in (c), and for the private use of the *VM* by user U_1 in (d)

$\bar{a}$, to represent these ports as no longer being available for interaction with the external environment; nonetheless, the already established communication channels are preserved. For instance, the interface for $Sys'' \stackrel{def}{=} (\nu L)(U_1 \mid VM) \mid U_2$ is depicted in Figure 3.5(d), where all the channels between U_1 and *VM* are made private by deleting the names of the ports involved. From the point of view of U_2, the only visible activity of U_1 is on the port *milk*. □

Remark 3.4. (**Sequentialization**) Restriction and parallel composition can be used together to express also some form of sequentialization, more general than action prefixing. For instance, process $(\nu d)(a.(b.d.\mathbf{0} + c.d.\mathbf{0}) \mid \bar{d}.q)$ prescribes that q be activated by means of a synchronization on d only after the left subprocess $a.(b.d.\mathbf{0} + c.d.\mathbf{0})$ has performed either ab or ac. The example can be adapted to the case where the process that has to terminate in order to activate q is composed of some parallel components; e.g., take $(\nu d)(a.(b.d.\mathbf{0} \mid c.d.\mathbf{0}) \mid \bar{d}.\bar{d}.q)$, where q is activated only after both components have terminated properly by synchronizing on d. This basic idea can be generalized to define a sequentialization operator $p \cdot q$ which states that q can start only after p has properly terminated its execution. Full details about this additional operator will be given in Section 5.4. □

3.1.2 Formal Syntax

As in Definition 2.1, let $\mathscr{L}$ be a denumerable set of *names*, ranged over by $a, b, \ldots$, usually called the *input actions*. Let $\overline{\mathscr{L}}$ be the set of *co-names*, ranged over by $\overline{a}, \overline{b}, \ldots$, usually called the *output actions*. The set $\mathscr{L} \cup \overline{\mathscr{L}}$, ranged over by $\alpha, \beta, \ldots$, is the set of *visible actions*. With $\overline{\alpha}$ we mean the complement of α, assuming that $\overline{\overline{\alpha}} = \alpha$. Let $Act = \mathscr{L} \cup \overline{\mathscr{L}} \cup \{\tau\}$, such that $\tau \notin \mathscr{L} \cup \overline{\mathscr{L}}$, be the set of *actions* (or *labels*), ranged over by μ. Action τ denotes an invisible, internal activity.

Let $\mathscr{C}ons$ be a denumerable set of *process constants*, disjoint from Act, ranged over by $A, B, C, \ldots$, possibly indexed. Sometimes, acronyms or mnemonic names are used for process constants, such as *VM* for a vending machine, or $SCount_i$ for a semi-counter holding value of i.

The *CCS process terms* are generated from actions and constants by the following *abstract syntax*, expressed in Backus-Naur Form (BNF):

$$\begin{array}{ll} p ::= \mathbf{0} \;\mid\; \mu.q \;\mid\; p + p & \textit{sequential processes} \\ q ::= p \;\mid\; q | q \;\mid\; (\nu a)q \;\mid\; C & \textit{processes} \end{array}$$

where $\mu \in Act$, $a \in \mathscr{L}$ and $C \in \mathscr{C}ons$, and where we are using two syntactic categories: p, ranging over sequential processes (i.e., processes that start sequentially), and q, ranging over any kind of processes.

Of course, this simple syntactic presentation of CCS process terms — which is very useful when defining the operational semantics (defined by structural induction on the abstract syntax) — is ambiguous, and so we make use of brackets whenever necessary. For instance, the derivable term $a.b.\mathbf{0} + c.\mathbf{0}$ is an ambiguous linear description as it may represent either $(a.(b.\mathbf{0})) + (c.\mathbf{0})$ or $a.((b.\mathbf{0}) + (c.\mathbf{0}))$, which originate two very different abstract syntax trees.

Remark 3.5. **(Notational convention)** In order to avoid the use of too many parentheses when writing a process in an unambiguous way, we assume that the operators have a different binding strength, by giving them a different priority in parsing; in particular, choice has lowest priority, then parallel composition, followed by action prefixing and finally by restriction:

$$(\nu a)\_ \;>\; \mu.\_ \;>\; \_|\_ \;>\; \_+\_$$

Therefore, $a.b.\mathbf{0} + c.\mathbf{0}$ denotes the term $(a.(b.\mathbf{0})) + (c.\mathbf{0})$; analogously, $(\nu a)b.c.\mathbf{0} \,|\, a.\mathbf{0}$ denotes the term $((\nu a)(b.(c.\mathbf{0}))) \,|\, (a.\mathbf{0})$. Similarly, $b.(\nu a)c.\mathbf{0} + a.b.\mathbf{0}$ represents the term $(b.((\nu a)(c.\mathbf{0}))) + (a.(b.\mathbf{0}))$.

As we will prove in Section 4.1.1 that both parallel composition and choice are associative w.r.t. all the behavioral equivalences we have studied in Chapter 2, we sometimes use the n-ary version of these operators. So we write $p_1 + p_2 + \ldots + p_n$, shortened as $\Sigma_{i=1}^n p_i$. Similarly, we write $p_1 \,|\, p_2 \,|\, \ldots \,|\, p_n$, shortened as $\Pi_{i=1}^n p_i$. We also write $\Sigma_{i \in I} p_i$ as well as $\Pi_{i \in I} p_i$, meaning that they denote the empty process $\mathbf{0}$ when I is empty. We will also prove that $\mathbf{0}$ is the neutral element for $+$ w.r.t. all the behavioral equivalences; hence, we often omit occurrences of $\mathbf{0}$ summands; e.g.,

$(\mathbf{0}+p)+\mathbf{0}$ is simply denoted as p. Using this alternative n-ary choice operator and the assumption of absorption of useless $\mathbf{0}$ summands, the CCS syntax can be more succinctly denoted as

$$p ::= \Sigma_{j\in J}\mu_j.p_j \mid p|p \mid (\nu a)p \mid C$$

with the assumption that J is finite and $\Sigma_{j\in J}\mu_j.p_j = \mathbf{0}$ when $J=\emptyset$. The sequential process terms are those of the form $\Sigma_{j\in J}\mu_j.p_j$.

Another syntactic simplification is for restriction: $(\nu a_1)(\nu a_2)\dots(\nu a_n)p$ is more succinctly denoted as $(\nu a_1 a_2 \dots a_n)p$ or even as $(\nu L)p$ where $L=\{a_1,a_2,\dots,a_n\}$. Finally, we sometimes omit trailing $\mathbf{0}$'s, so that, for instance, $a.\mathbf{0}\,|\,b.\mathbf{0}$ is sometimes denoted as $a\,|\,b$. □

Let us examine in detail the syntax of sequential processes (i.e., processes that start sequentially).

Process $\mu.q$ is sequential because it starts with a unique possible initial action, even if, after μ, process q can be a parallel one.

Note that $p+p'$ is the sequential process obtained by the alternative composition of *sequential* processes p and p'; hence we are restricting the use of $+$ to the so-called *guarded sum*. For instance, a term of the form $(a.\mathbf{0}\,|\,b.\mathbf{0})+c.\mathbf{0}$ is not legal because the first summand is not sequential. Similarly, the illegal term $(a.\mathbf{0}\,|\,b.\mathbf{0})+(c.\mathbf{0}\,|\,d.\mathbf{0})$ represents a *distributed choice*, where the execution of a should be able, in a single atomic transition, also to kill the two parallel processes $c.\mathbf{0}$ and $d.\mathbf{0}$, while keeping the sibling $b.\mathbf{0}$. The limitation to guarded sum only is inessential for expressiveness, as we are able to describe all LTSs anyway.[1] Moreover, from a practical point of view, it seems that the distributed choice mechanism is not physically implementable. Indeed, we consider it much more realistic that a choice can be performed only by sequential processes, locally.

We have listed a constant C in the category q not only because C can be used as a component of a parallel system, but also to indicate that the body q' in the defining equation for C, $C \stackrel{def}{=} q'$, is a generic process of type q.

With abuse of notation, in the following, when clear from the context, we will often use p to denote any kind of process term, not only sequential ones.

Exercise 3.10. Check that none of the following terms is a CCS process term:

$$\begin{array}{llll} a.(A+B) & (b.c.\mathbf{0}+d.\mathbf{0}).a.\mathbf{0} & (\nu a)(b.\mathbf{0})+c.\mathbf{0} & (\nu\tau)(a.\tau.\mathbf{0}) \\ (A\,|\,c.\mathbf{0})+a.B & A+(a.B) & \mathbf{0}.a.\mathbf{0} & (a.A+\bar{a}.\mathbf{0}).B \end{array}$$ □

We want to compute the set $Const(p)$ of process constants used in a CCS process term p. This is important as we usually restrict our attention to those process terms that use only finitely many constants.

[1] To be precise, we will see that all finitely-branching LTSs can be defined by CCS terms with guarded sum, possibly with the help of infinitely many constants (see Theorem 3.2 in Section 3.4.2 and Exercise 3.66 in Section 3.4.6).

Definition 3.1. (*Const*(*p*): **Set of constants used by** *p*) With $Const(p)$ we denote the *least* set of process constants such that the following equations are satisfied:

$$
\begin{array}{rlrl}
Const(\mathbf{0}) = \emptyset & & Const(p_1 + p_2) = Const(p_1) \cup Const(p_2) & \\
Const(\mu.p) = Const(p) & & Const(p_1 \,|\, p_2) = Const(p_1) \cup Const(p_2) & \\
Const((\nu a)p) = Const(p) & & Const(A) = \begin{cases} \{A\} & \text{if } A \text{ undef.} \\ \{A\} \cup Const(q) & \text{if } A \overset{def}{=} q \end{cases} &
\end{array}
$$

A term p such that $Const(p)$ is finite is called *finitary*, as it uses finitely many process constants only. □

Example 3.2. Let us consider the simple vending machine specified as follows:

$$C_1 \overset{def}{=} coin.C_2 \quad C_2 \overset{def}{=} \overline{coffee}.C_1$$

It is not difficult to see that

$$
\begin{aligned}
Const(C_1) &= \{C_1\} \cup Const(coin.C_2) \\
&= \{C_1\} \cup Const(C_2) \\
&= \{C_1, C_2\} \cup Const(\overline{coffee}.C_1) \\
&= \{C_1, C_2\} \cup Const(C_1)
\end{aligned}
$$

so that the *least* set $Const(C_1)$ satisfying this recursive equation is $\{C_1, C_2\}$.

As a further example, consider process A_0, with the family of process constants $A_i \overset{def}{=} a_i.A_{i+1}$ for $i \in \mathbb{N}$; it is easy to observe that

$$
\begin{aligned}
Const(A_0) &= \{A_0\} \cup Const(a_0.A_1) &&= \{A_0\} \cup Const(A_1) \\
&= \{A_0, A_1\} \cup Const(a_1.A_2) &&= \{A_0, A_1\} \cup Const(A_2) \\
&= \{A_0, A_1, A_2\} \cup Const(a_2.A_3) &&= \{A_0, A_1, A_2\} \cup Const(A_3) \\
&= \ldots
\end{aligned}
$$

so that the limit of this increasing sequence of finite sets is $Const(A_0) = \{A_i \mid i \in \mathbb{N}\}$, which is an infinite set. □

The computation of $Const(p)$ may be much more intricate than in the examples above; a formal treatment is outside the scope of this introductory text. However, observe that, when p is finitary, there is an obvious algorithm to compute $Const(p)$: it is enough to remember all the constants that have been already found while scanning p, in order to avoid applying again function *Const* over their bodies. This can be achieved by the following auxiliary function δ, which has, as an additional parameter, a set I of already known constants:

$$
\begin{array}{ll}
\delta(\mathbf{0}, I) = \emptyset & \delta(p_1 + p_2, I) = \delta(p_1, I) \cup \delta(p_2, I) \\
\delta(\mu.p, I) = \delta(p, I) & \delta(p_1 \,|\, p_2, I) = \delta(p_1, I) \cup \delta(p_2, I) \\
\delta((\nu a)p, I) = \delta(p, I) & \delta(A, I) = \begin{cases} \emptyset & A \in I, \\ \{A\} & A \notin I \wedge A \text{ undef.} \\ \{A\} \cup \delta(p, I \cup \{A\}) & A \notin I \wedge A \overset{def}{=} p \end{cases}
\end{array}
$$

Then, for any finitary term p, $Const(p) = \delta(p, \emptyset)$, where the additional parameter is the empty set because so is the set of process constants we assume to know at the beginning of the scanning of p.

Definition 3.2. (Defined constant and fully defined term) A process constant A is *defined* if it possesses a defining equation: $A \stackrel{def}{=} p$. A process term q is *fully defined* if all the constants in $Const(q)$ are defined. □

The request that a term is fully defined is due to the fact that its semantics cannot be fully given otherwise. For instance, term $a.A$ can execute a, but after that we do not know what to do if A is not equipped with a defining equation.

Definition 3.3. (Guarded constant) An occurrence of a constant A in a process q is *strongly guarded* in q if such occurrence of A occurs within a prefixed subprocess $\mu.q'$ of q. Constant A, defined as $A \stackrel{def}{=} p$, is *guarded* if each occurrence of A in p is strongly guarded in p and each occurrence of any other constant B occurring in p is either strongly guarded in p or simply guarded. □

Constant B, defined as $B \stackrel{def}{=} b.B$, is guarded because B is strongly guarded in its body $b.B$; similarly, constant A, defined as $A \stackrel{def}{=} a.A \,|\, B$, is guarded because A occurs strongly guarded in its body $a.A \,|\, B$ and B, defined above, is guarded. On the contrary, constant $C \stackrel{def}{=} a.\mathbf{0} \,|\, C$ is not guarded.

A bit of care is needed because the definition of guarded constant is recursive. For $A \stackrel{def}{=} p$, it may happen that some of the not strongly guarded occurrences of B in p may have not strongly guarded occurrences of A in their definition body, so that we cannot derive that A (as well as B) is guarded. For instance, constants D and F defined as $D \stackrel{def}{=} F, F \stackrel{def}{=} D$ are not guarded, and a naïve attempt to prove that they are guarded would loop forever.

Exercise 3.11. Given $A \stackrel{def}{=} a.A \,|\, B$ and $B \stackrel{def}{=} b.B \,|\, A$, argue that neither A nor B is guarded. □

Exercise 3.12. Check which of the following constants is guarded:

$$A \stackrel{def}{=} (\nu a)(b.\mathbf{0} \,|\, A) \qquad B \stackrel{def}{=} b.B + a.A$$
$$C \stackrel{def}{=} c.C + d.D \qquad D \stackrel{def}{=} \mathbf{0} + c.C$$

□

The request that a term is guarded is not always necessary, and is done mainly for simplicity. Some good reasons for using this restriction are listed in Section 3.3. In particular, Proposition 3.3 states that a guarded process term generates a finitely-branching LTS; moreover, Section 3.3.2 shows that guardedness is necessary to get uniqueness of solutions of process equations modulo bisimulation equivalence. As an instance of such arguments, we will see in Section 3.3.1 that $C \stackrel{def}{=} a.\mathbf{0} \,|\, C$ generates an infinitely-branching (actually, not even image-finite) LTS, i.e., a form of LTS that may be considered pathological.

Definition 3.4. (CCS processes) The set $\mathscr{P}$ of CCS processes contains those process terms p where each constant $A \in Const(p)$ is defined and guarded. [2] With abuse of notation, $\mathscr{P}$ will be ranged over by p, q, possibly indexed. A CCS process p is *finitary* if $Const(p)$ is finite. □

Consider Exercise 3.12. Observe that B is a guarded constant, but not a CCS process, because $A \in Const(B)$ is not guarded. Consider Example 3.2: VM'' is a finitary process, while $b.A_0$ is not. We will usually restrict our attention to finitary processes, as only these are equipped with a finite syntactic definition.

Exercise 3.13. Check which of the following CCS terms is a CCS process in $\mathscr{P}$, assuming that $A \stackrel{def}{=} a.A$, $B \stackrel{def}{=} b.\mathbf{0} \,|\, B$ and $C \stackrel{def}{=} a.B$:

$$\begin{array}{ccc} a.\mathbf{0}+\mathbf{0} & a.C\,|\,\mathbf{0} & a.A\,|\,b.B \\ (\nu a)(\bar{a}.\mathbf{0}\,|\,A) & a.(\nu a)(b.a.\mathbf{0}\,|\,\bar{a}.\bar{b}.B) & (\nu a)C \end{array}$$

□

3.2 Structural Operational Semantics

The next step is to define the labeled transition system for the whole language CCS. This is, of course, an infinite LTS, for the number of states, the number of labels and the number of transitions. The states are actually the CCS processes: a state p performs the transitions corresponding to the actions process p prescribes. A finite implicit representation of this infinite, countable set is given by means of the grammar (in Backus-Naur Form) outlined in the previous section. More difficult is finding a finite implicit representation of the infinite, countable set of transitions. For this, we resort to Plotkin's technique called *Structural Operational Semantics* (SOS for short) [Plo04a, Plo04b], according to which the transitions are defined by means of an inference system composed of axioms and rules whose definition is syntax-driven.

A typical SOS operational rule has the form $\dfrac{\textit{premises}}{\textit{conclusion}}$ *side condition*

where *premises* is the conjunction of zero (and in such a case the rule is called an *axiom*) or more transitions, *conclusion* is one transition and *side condition* is a predicate that must be true for rule applicability.

Definition 3.5. The CCS labeled transition system $\mathscr{C}$ is the triple $(\mathscr{P}, Act, \rightarrow)$ where $\rightarrow \subseteq \mathscr{P} \times Act \times \mathscr{P}$ is the least transition relation generated by the axiom and rules in Table 3.1. □

Let us comment on the rules in Table 3.1.

[2] Note that a guarded constant is also defined, so that we can more simply require that any constant be guarded.

(Pref)	$\dfrac{}{\mu.p \xrightarrow{\mu} p}$	(Cons)	$\dfrac{p \xrightarrow{\mu} p'}{C \xrightarrow{\mu} p'}$ $C \stackrel{def}{=} p$
(Sum$_1$)	$\dfrac{p \xrightarrow{\mu} p'}{p+q \xrightarrow{\mu} p'}$	(Sum$_2$)	$\dfrac{q \xrightarrow{\mu} q'}{p+q \xrightarrow{\mu} q'}$
(Par$_1$)	$\dfrac{p \xrightarrow{\mu} p'}{p \mid q \xrightarrow{\mu} p' \mid q}$	(Par$_2$)	$\dfrac{q \xrightarrow{\mu} q'}{p \mid q \xrightarrow{\mu} p \mid q'}$
(Com)	$\dfrac{p \xrightarrow{\alpha} p' \quad q \xrightarrow{\overline{\alpha}} q'}{p \mid q \xrightarrow{\tau} p' \mid q'}$	(Res)	$\dfrac{p \xrightarrow{\mu} p'}{(\nu a)p \xrightarrow{\mu} (\nu a)p'}$ $\mu \neq a, \bar{a}$

Table 3.1 Structural Operational Semantics: syntax-driven axiom and inference rules

(Pref) is an axiom, as it has no premises; it declares that for any action μ and for any process p, state $\mu.p$ can perform a transition labeled μ reaching state p. Note that the occurrence of the prefix disappears in the reached state (it is "consumed"); for this reason this operator is called *dynamic*.

Rule (Sum$_1$) clarifies why we say that the rules are defined by induction on the structure of terms; this rule states that, in order to derive a transition from state $p+q$, we have to solve first the simpler problem of finding a transition from p: if $p \xrightarrow{\mu} p'$, then $p+q \xrightarrow{\mu} p'$. Rule (Sum$_2$) is symmetric. The two rules for choice state that $p+q$ can do whatever p or q can do. Note that a summand, say p, can be chosen only if p can move. For instance, in $a.b+\mathbf{0}$, summand $\mathbf{0}$ cannot be chosen as it does not contribute with any behavior. Note that, when a choice is made, the alternative is disregarded: indeed, the occurrence of the choice operator disappears in the reached state (it is "consumed"); for this reason this operator is called *dynamic*.

(Cons) is a rule that states that a constant C can do what is prescribed by the body of its defining equation: if $C \stackrel{def}{=} p$ and $p \xrightarrow{\mu} p'$, then $C \xrightarrow{\mu} p'$. This is the only rule which is not defined by structural induction.

(Par$_1$) and (Par$_2$) are the rules describing the asynchronous execution of an action by one of the two subcomponents of a parallel process. Specifically, (Par$_1$) states that if $p \xrightarrow{\mu} p'$, then $p \mid q \xrightarrow{\mu} p' \mid q$. Note that q is not discarded by the transition, and for this reason the operator is called *static*. (Par$_2$) is symmetric. These two rules together state that $p \mid q$ can do whatever p and q can do, possibly interleaving their executions. Rule (Com) describes how interaction can take place: if the two subcomponents can execute complementary input/output actions, then a synchronization is possible and the resulting transition, labeled τ, cannot be used for further interaction with another parallel component, as the premise of rule (Com) requires that the transition be labeled with a visible action. Hence, synchronization in CCS is strictly binary (point-to-point communication).

Rule (Res) explains that the role of a restriction is to bind a name, so that it is not freely available for the external environment. Clearly, $(\nu a)p$ impedes any transition labeled a or $\bar{a}$ that p might produce, while having no effect on the other transitions of p, in particular on possible synchronizations (labeled τ) along channel a between parallel subcomponents of p. Restriction is a *static* operator, too.

Observe that there is no operational rule for the nil process $\mathbf{0}$, hence no transition can have $\mathbf{0}$ as its source. In other words, $\mathbf{0}$ is a deadlock.

Example 3.3. **(How to compute transitions?)** How can we compute, by means of the inference rules, the transitions that can be performed by a process? As a simple example, consider the CCS process $a.b.\mathbf{0}+b.a.\mathbf{0}$. The two possible initial transitions are $a.b.\mathbf{0}+b.a.\mathbf{0} \xrightarrow{a} b.\mathbf{0}$ and $a.b.\mathbf{0}+b.a.\mathbf{0} \xrightarrow{b} a.\mathbf{0}$, which can be proved as follows:

$$
(\mathrm{Sum}_1)\ \dfrac{(\mathrm{Pref})\ \dfrac{}{a.b.\mathbf{0} \xrightarrow{a} b.\mathbf{0}}}{a.b.\mathbf{0}+b.a.\mathbf{0} \xrightarrow{a} b.\mathbf{0}}
\qquad
(\mathrm{Sum}_2)\ \dfrac{(\mathrm{Pref})\ \dfrac{}{b.a.\mathbf{0} \xrightarrow{b} a.\mathbf{0}}}{a.b.\mathbf{0}+b.a.\mathbf{0} \xrightarrow{b} a.\mathbf{0}}
$$

These are proof-trees, with (instantiations of) axioms as leaves and the theorems (i.e., the transitions to be proved) as their roots. Observe that the proof is by induction on the shape of the term: as the top-level operator in $a.b.\mathbf{0}+b.a.\mathbf{0}$ is $+$, we have to resort to either (Sum_1) or (Sum_2). Consider the first proof-tree, which makes use of rule (Sum_1). We have first to match the source state in the conclusion of that rule, namely $p+q$, with $a.b.\mathbf{0}+b.a.\mathbf{0}$, hence binding p to $a.b.\mathbf{0}$ and q to $b.a.\mathbf{0}$. Then, in order to satisfy the premise of that rule, namely $p \xrightarrow{\mu} p'$, we are asked to find a transition with source $p = a.b.\mathbf{0}$. Since in $a.b.\mathbf{0}$ the top-level operator is action prefixing, we must use axiom (Pref), stating that $\mu.p \xrightarrow{\mu} p$. If we bind μ to a and p to $b.\mathbf{0}$ (where the new instances of μ and p for this application of the rule are not to be confused with the previous ones), then, by this instance of axiom (Pref), we can derive transition $a.b.\mathbf{0} \xrightarrow{a} b.\mathbf{0}$. Therefore, the premise of rule (Sum_1) we were looking for is now ready — $a.b.\mathbf{0} \xrightarrow{a} b.\mathbf{0}$ — binding not only p to $a.b.\mathbf{0}$, but also μ to a and p' to $b.\mathbf{0}$. Finally, the conclusion — $a.b.\mathbf{0}+b.a.\mathbf{0} \xrightarrow{a} b.\mathbf{0}$ — can be derived by the application of the rule, because the various p, q, μ and p' occurring in the rule have been all suitably instantiated.

As a more complex example, see the following proof tree:

$$
(\mathrm{Res})\ \dfrac{(\mathrm{Com})\ \dfrac{(\mathrm{Pref})\ \dfrac{}{a.c.\mathbf{0} \xrightarrow{a} c.\mathbf{0}} \qquad (\mathrm{Sum}_1)\ \dfrac{(\mathrm{Pref})\ \dfrac{}{\bar{a}.\mathbf{0} \xrightarrow{\bar{a}} \mathbf{0}}}{\bar{a}.\mathbf{0}+c.\mathbf{0} \xrightarrow{\bar{a}} \mathbf{0}}}{a.c.\mathbf{0}\,|\,(\bar{a}.\mathbf{0}+c.\mathbf{0}) \xrightarrow{\tau} c.\mathbf{0}\,|\,\mathbf{0}}}{(\nu a)(a.c.\mathbf{0}\,|\,(\bar{a}.\mathbf{0}+c.\mathbf{0})) \xrightarrow{\tau} (\nu a)(c.\mathbf{0}\,|\,\mathbf{0})}
$$

where induction is performed on the structure of the more elaborate process term $(\nu a)(a.c.\mathbf{0}\,|\,(\bar{a}.\mathbf{0}+c.\mathbf{0}))$. First, rule (Res) is used, as restriction is the top-level operator, with the effect of binding the (formal) restricted action a in the rule to the (actual) restricted action a, and the formal p in the rule to the actual $a.c.\mathbf{0}\,|\,(\bar{a}.\mathbf{0}+c.\mathbf{0})$.

Since the premise of rule (Res) requires a transition from $p = a.c.\mathbf{0} \mid (\bar{a}.\mathbf{0} + c.\mathbf{0})$, we have to single out the top operator of p, which is parallel composition, and use one of the three SOS rules for it. In this proof, we use rule (Com), hence binding p (the new one for this application of the rule) to $a.c.\mathbf{0}$ and q to $\bar{a}.\mathbf{0} + c.\mathbf{0}$. Rule (Com) requires two premises: one transition from $p = a.c.\mathbf{0}$ and one from $q = \bar{a}.\mathbf{0} + c.\mathbf{0}$. For p, the top-level operator is action prefixing and so we can apply axiom (Pref), as already discussed in the previous example, deriving $a.c.\mathbf{0} \xrightarrow{a} c.\mathbf{0}$. For q, the top-level operator is $+$ and so we can use two different rules; in this proof we use (Sum$_1$), and the derivation of $\bar{a}.\mathbf{0} + c.\mathbf{0} \xrightarrow{\bar{a}} \mathbf{0}$ is similar to the case discussed in the previous example. Summing up, the two premises of rule (Com) have been derived; the conclusion of that rule, $a.c.\mathbf{0} \mid (\bar{a}.\mathbf{0} + c.\mathbf{0}) \xrightarrow{\tau} c.\mathbf{0} \mid \mathbf{0}$, can be derived as well and used as the premise for rule (Res), thus deriving the "theorem" $(\nu a)(a.c.\mathbf{0} \mid (\bar{a}.\mathbf{0} + c.\mathbf{0})) \xrightarrow{\tau} (\nu a)(c.\mathbf{0} \mid \mathbf{0})$, because the side condition of rule (Res) is satisfied (τ is different from the restricted action a).

Observe that if we had tried with rule (Par$_1$) instead of (Com), we would end in failure because $p = a.c.\mathbf{0}$ can only execute a, which is then restricted, hence the mandatory rule (Res) cannot be applied. On the contrary, if we had tried with rule (Par$_2$), then we may end in failure as above, if (Sum$_1$) is used to select $\bar{a}.\mathbf{0}$; or we can derive another transition, namely $(\nu a)(a.c.\mathbf{0} \mid (\bar{a}.\mathbf{0} + c.\mathbf{0})) \xrightarrow{c} (\nu a)(a.c.\mathbf{0} \mid \mathbf{0})$, if (Sum$_2$) is used instead. □

Exercise 3.14. Use the operational rules of Table 3.1 to prove that the following transitions are derivable:

$$(\nu a)(a.c.\mathbf{0} \mid (\bar{a}.\mathbf{0} + c.\mathbf{0})) \xrightarrow{c} (\nu a)(a.c.\mathbf{0} \mid \mathbf{0}) \qquad (a.\mathbf{0} \mid \bar{a}.\mathbf{0}) \mid a.\mathbf{0} \xrightarrow{\tau} (\mathbf{0} \mid \mathbf{0}) \mid a.\mathbf{0}$$
$$(\nu a)(c.\mathbf{0} \mid \mathbf{0}) \xrightarrow{c} (\nu a)(\mathbf{0} \mid \mathbf{0}) \qquad (a.\mathbf{0} \mid \bar{a}.\mathbf{0}) \mid a.\mathbf{0} \xrightarrow{\tau} (a.\mathbf{0} \mid \mathbf{0}) \mid \mathbf{0}$$

□

Exercise 3.15. Use the operational rules of Table 3.1 to prove that the following transitions are *not* derivable:

$$(\nu c)(a.c.\mathbf{0} \mid b.\mathbf{0}) \xrightarrow{\tau} (\nu c)(c.\mathbf{0} \mid \mathbf{0}) \qquad (\nu a)(a.\mathbf{0} \mid \bar{a}.\mathbf{0}) \xrightarrow{a} (\nu a)(\mathbf{0} \mid \bar{a}.\mathbf{0})$$
$$(\nu c)(a.c.\mathbf{0} \mid b.c.\mathbf{0}) \xrightarrow{b} (\nu a)(c.\mathbf{0} \mid c.\mathbf{0}) \qquad (\nu a)(a.b.\mathbf{0} \mid \bar{a}.\mathbf{0}) \xrightarrow{\tau} (\nu a)(\mathbf{0} \mid \mathbf{0})$$

□

Exercise 3.16. Use the operational rules to derive all the possible initial transitions for the following CCS processes:

$$(\nu a)((a.c.\mathbf{0} + a.b.\mathbf{0}) \mid \bar{a}.\mathbf{0}) \qquad (a.\mathbf{0} \mid \bar{a}.\mathbf{0}) \mid a.\mathbf{0} \qquad (\nu a)(a.\mathbf{0} \mid \bar{a}.\mathbf{0}) \mid a.\mathbf{0}$$

□

It is clear that the SOS rules of Table 3.1 can be easily implemented in a declarative programming language such as Prolog (see, e.g., [ClMe03]), in order to obtain a simple sequential interpreter for CCS. As CCS is an *executable* specification language, one can do early prototyping of CCS specifications in order to perform simple analyses of the expected behavior. This can be done with the help of some semi-automated tools, such as the *Concurrency Workbench* [CPS93, CWB].

We then observe that the operational semantics is well behaved, meaning that if we move from a CCS process p in one step, reaching p', then we are sure that also p' is a CCS process.

Proposition 3.1. *For any $p \in \mathscr{P}$, if $p \xrightarrow{\mu} p'$, then $p' \in \mathscr{P}$.*

Proof. *It is enough to observe, by induction on the proof of $p \xrightarrow{\mu} p'$, that if p is a CCS process term, then also p' is a CCS process term. Additionally, we can prove that $Const(p') \subseteq Const(p)$, so that if p satisfies the guardedness condition, the same holds for p'.* □

Remark 3.6. **(Notation)** For any $p \in \mathscr{P}$, the reachable LTS from p (Definition 2.5) is $\mathscr{C}_p = (\mathscr{P}_p, sort(p), \rightarrow_p, p)$, where $\mathscr{P}_p$ is the set of states reachable from p, $sort(p)$ is the set of actions that can be performed by p (formally, $sort(p) = \{\mu \in Act \mid \exists p'.p \longrightarrow^* p' \xrightarrow{\mu}\}$) and $\rightarrow_p$ is the restriction of the transition relation to $\mathscr{P}_p \times sort(p) \times \mathscr{P}_p$. (Note that, by Proposition 3.1, any state q in $\mathscr{P}_p$ is a CCS process.) E.g., the reachable LTS from $a.\mathbf{0} \,|\, \overline{a}.\mathbf{0}$ is depicted in Figure 3.4(a). □

The following obvious fact follows.

Proposition 3.2. *For any $p \in \mathscr{P}$, the LTS $\mathscr{C}_p = (\mathscr{P}_p, sort(p), \rightarrow_p, p)$ reachable from p is a reduced rooted LTS.* □

Exercise 3.17. Use the operational rules to derive the portion of the CCS transition system reachable from $(a.\mathbf{0} \,|\, \bar{a}.\mathbf{0})$, $(\nu a)(a.\mathbf{0} \,|\, \bar{a}.\mathbf{0})$ and $(\nu a)(a.\mathbf{0} \,|\, \bar{a}.\mathbf{0}) \,|\, a.\mathbf{0}$. Compare the resulting LTSs with those in Figure 3.4. □

Exercise 3.18. Continuing Exercise 3.4, derive the LTS for the vending machine VM'. Is it strongly bisimilar to the one in Figure 3.3? □

Exercise 3.19. Use the operational rules to derive the part of the CCS transition system reachable from the following six states:

$$a.b.\mathbf{0}+\mathbf{0} \qquad a.(b.\mathbf{0}+c.d.\mathbf{0}) \qquad A \stackrel{def}{=} a.(b.A+c.\mathbf{0})$$
$$a.(b.\mathbf{0}+c.d.\mathbf{0}) \,|\, \bar{b}.\mathbf{0} \qquad (\nu b)(a.(b.\mathbf{0}+c.d.\mathbf{0}) \,|\, \bar{b}.\mathbf{0}) \qquad (\nu a)(B \,|\, C)$$

where $B \stackrel{def}{=} a.B + b.\mathbf{0}$ and $C \stackrel{def}{=} \bar{a}.C + c.\mathbf{0}$. Observe that $a.b.\mathbf{0}+\mathbf{0}$ originates an LTS isomorphic to the one in Figure 3.1(c). □

Exercise 3.20. Consider again Exercise 3.7. Now that the operational rules have been provided, draw the LTSs for the CCS terms $a.b.\mathbf{0} \,|\, \bar{a}.\mathbf{0}$, $(a.b.\mathbf{0}+a.c.\mathbf{0}) \,|\, \bar{a}.\mathbf{0}$, $A \,|\, B$, where $A \stackrel{def}{=} a.b.A$ and $B \stackrel{def}{=} \bar{a}.B$. Similarly, consider again Exercise 3.8. Draw the LTSs for $(\nu a)(a.b.\mathbf{0} \,|\, \bar{a}.\mathbf{0})$, $(\nu a)((a.b.\mathbf{0}+a.c.\mathbf{0}) \,|\, \bar{a}.\mathbf{0})$ and $(\nu a)(A \,|\, B)$. □

Exercise 3.21. Consider Remark 3.1. Draw the LTS for constant A, defined as $A \stackrel{def}{=} (\nu b)(a.A)$, and compare it with the one for $(\nu b)B$ with $B \stackrel{def}{=} a.B$. □

Exercise 3.22. (Counting states) Recall that $\mathscr{P}_p$ denotes the set of processes reachable from p and that $|J|$ is the cardinality of set J. Looking at the SOS rules, argue that, if $|\mathscr{P}_{p_1}| = k_1$ and $|\mathscr{P}_{p_2}| = k_2$, then $|\mathscr{P}_{p_1+p_2}| \leq k_1+k_2+1$[3] and $|\mathscr{P}_{p_1 | p_2}| = k_1 \times k_2$. Moreover, if $q = (\nu a)p_1$, conclude that $|\mathscr{P}_q| \leq k_1$. □

[3] As an example showing that the bound k_1+k_2+1 can be really reached, consider the processes $p_1 = a.A$ and $p_2 = b.B$ where $A \stackrel{def}{=} a.a.A$ and $B \stackrel{def}{=} b.b.B$.

Remark 3.7. **(State space explosion problem)** The exercise above explains that if we have a compound process $p_1 \mid p_2 \mid \ldots \mid p_n$, where each p_i generates an LTS with ten states, then the LTS for $p_1 \mid p_2 \mid \ldots \mid p_n$ has 10^n states, i.e., the state space of a compound system grows exponentially w.r.t. the number of components. This phenomenon is sometimes called the *state space explosion problem.* □

Exercise 3.23. (Deadlock-freeness) Consider the definition of deadlock-free LTS in Definition 2.10. (*i*) Argue that, if $\mathscr{C}_{p_1}$ is deadlock-free, then also $\mathscr{C}_{p_1 \mid p_2}$ is deadlock-free, for any $p_2 \in \mathscr{P}$. (*ii*) Argue also that, for any deadlock-free LTS $\mathscr{C}_p$, the LTS $\mathscr{C}_{(\nu a)p}$ is deadlock-free if and only if $\forall p', p''. p \longrightarrow^* p' \xrightarrow{\alpha} p''$ with $\alpha \in \{a, \bar{a}\}, \exists q\ \exists \beta \notin \{a, \bar{a}\}. p' \xrightarrow{\beta} q$. □

Exercise 3.24. (Reachable deadlock) (*i*) Argue that, for any $p_1, p_2 \in \mathscr{P}$, $\mathscr{C}_{p_1 \mid p_2}$ may reach a deadlock if and only if both $\mathscr{C}_{p_1}$ and $\mathscr{C}_{p_2}$ may reach a deadlock. (*ii*) Show an example of a process p such that p may reach a deadlock, while $\mathscr{C}_{(\nu a)p}$ cannot reach a deadlock; show an example of a process q such that q cannot reach a deadlock, while $\mathscr{C}_{(\nu a)q}$ may reach a deadlock. □

Exercise 3.25. (Divergence) Consider the definition of divergent state in Definition 2.21. (*i*) Argue that, if $\mathscr{C}_{p_1}$ has a divergent state, then $\mathscr{C}_{p_1 \mid p_2}$ also has a divergent state, for any $p_2 \in \mathscr{P}$. (*ii*) Argue that if $\mathscr{C}_{p_1 \mid p_2}$ has a divergent state, then it may happen that both $\mathscr{C}_{p_1}$ and $\mathscr{C}_{p_2}$ do not have any divergent state. (*iii*) Argue also that, for any p, if $\mathscr{C}_{(\nu a)p}$ has a divergent state, then the LTS $\mathscr{C}_p$ has a divergent state. Show an example of a process p such that the LTS $\mathscr{C}_p$ has a divergent state while the LTS $\mathscr{C}_{(\nu a)p}$ has no divergent state. □

Exercise 3.26. (Livelock) Consider the definition of livelock in Definition 2.21. Remember that a livelock state is also a divergent state, but not the converse. (*i*) Argue that, if $\mathscr{C}_{p_1}$ has a livelock, then $\mathscr{C}_{p_1 \mid p_2}$ may not have a livelock, but has a divergent state for sure. (*ii*) Argue also that, for any p, if $\mathscr{C}_{(\nu a)p}$ has a livelock, we can only conclude that $\mathscr{C}_p$ has a divergent state. Show an example of a process p such that the LTS $\mathscr{C}_p$ has a livelock but the LTS $\mathscr{C}_{(\nu a)p}$ has no livelock. □

Remark 3.8. Example 3.1 shows a process p, with an infinite set $Const(p)$, that generates an LTS with infinitely many states. However, it is not true that if $Const(p)$ is infinite, then the LTS reachable from p has necessarily infinitely many states. Consider A defined as $A \stackrel{def}{=} a.A + b.(\nu c)(c.A_0)$, where $A_i \stackrel{def}{=} c_i.A_{i+1}$ for $i \in \mathbb{N}$: $Const(A)$ is infinite, but A generates a finite-state LTS. This is because set $Const(p)$ gives a syntactic upper bound to the number of constants that can be unwound during the execution of p; in the case above, the constant that A can unwind is only A itself, because restriction on c prevents unveiling A_0 (see also Exercise 3.74 in Section 3.5.3). □

3.3 About Guardedness

By Definition 3.4, a CCS process p is a CCS term such that each $A \in Const(p)$ is guarded. We discuss two good reasons for restricting our attention to guarded terms only. The first reason is that the SOS rules generate a finitely-branching LTS for a CCS term with guarded constants. The second one is that guardedness ensures uniqueness of the solution (up to $\sim$) of equations defined over processes. These are the subject of the following subsections.

3.3.1 Guardedness Implies Finite Branching

The operational rules in Table 3.1 can be applied also to unguarded process terms. Let us consider a typical example of an unguarded constant: $C \stackrel{def}{=} a.\mathbf{0} \mid C$. According to rule (Cons), C can perform any transition derivable by its body $a.\mathbf{0} \mid C$. By rule (Par$_1$), $a.\mathbf{0} \mid C$ can do what $a.\mathbf{0}$ can do. As $a.\mathbf{0} \xrightarrow{a} \mathbf{0}$ by axiom (Pref), we can conclude the proof that $C \xrightarrow{a} \mathbf{0} \mid C$.

$$
\text{(Cons)}\ \dfrac{\text{(Par}_1\text{)}\ \dfrac{\text{(Pref)}\ \dfrac{}{a.\mathbf{0} \xrightarrow{a} \mathbf{0}}}{a.\mathbf{0} \mid C \xrightarrow{a} \mathbf{0} \mid C}}{C \xrightarrow{a} \mathbf{0} \mid C}
$$

However, this is not the only transition derivable by $a.\mathbf{0} \mid C$. By rule (Par$_2$), $a.\mathbf{0} \mid C$ can do what C can do, hence deriving a circular proof that C can do what C can do, plus some context. The two proof trees

$$
\text{(Cons)}\ \dfrac{\text{(Par}_2\text{)}\ \dfrac{\text{(Cons)}\ \dfrac{\text{(Par}_1\text{)}\ \dfrac{\text{(Pref)}\ \dfrac{}{a.\mathbf{0} \xrightarrow{a} \mathbf{0}}}{a.\mathbf{0} \mid C \xrightarrow{a} \mathbf{0} \mid C}}{C \xrightarrow{a} \mathbf{0} \mid C}}{a.\mathbf{0} \mid C \xrightarrow{a} a.\mathbf{0} \mid (\mathbf{0} \mid C)}}{C \xrightarrow{a} a.\mathbf{0} \mid (\mathbf{0} \mid C)}
$$

$$
\text{(Cons)}\ \dfrac{\text{(Par}_2\text{)}\ \dfrac{\text{(Cons)}\ \dfrac{\text{(Par}_2\text{)}\ \dfrac{\text{(Cons)}\ \dfrac{\text{(Par}_1\text{)}\ \dfrac{\text{(Pref)}\ \dfrac{}{a.\mathbf{0} \xrightarrow{a} \mathbf{0}}}{a.\mathbf{0} \mid C \xrightarrow{a} \mathbf{0} \mid C}}{C \xrightarrow{a} \mathbf{0} \mid C}}{a.\mathbf{0} \mid C \xrightarrow{a} a.\mathbf{0} \mid (\mathbf{0} \mid C)}}{C \xrightarrow{a} a.\mathbf{0} \mid (\mathbf{0} \mid C)}}{a.\mathbf{0} \mid C \xrightarrow{a} a.\mathbf{0} \mid (a.\mathbf{0} \mid (\mathbf{0} \mid C))}}{C \xrightarrow{a} a.\mathbf{0} \mid (a.\mathbf{0} \mid (\mathbf{0} \mid C))}
$$

give evidence that C has infinitely many transitions, all labeled a, each one reaching a different state. Hence C originates an infinitely-branching LTS, which is actually not even image-finite.

Exercise 3.27. Consider the unguarded constant $A \stackrel{def}{=} a.\mathbf{0} + A$. Show that there are infinitely many different proofs for its unique transition $A \xrightarrow{a} \mathbf{0}$. □

As a matter of fact, it is not difficult to see that guarded process terms (i.e., CCS processes) have a finite number of outgoing transitions.

Proposition 3.3. *For any $q \in \mathscr{P}$, set $T_q = \{(q, \mu, q') \mid \exists \mu \in Act, \exists q' \in \mathscr{P}. q \xrightarrow{\mu} q'\}$ is finite.*

Proof. *We can define an upper bound $\gamma(q)$ on the number of transitions leaving a given state/process q. Function $\gamma : \mathscr{P} \to \mathbb{N}$ is as follows:*

$$\begin{array}{ll} \gamma(\mathbf{0}) = 0 & \gamma(\mu.p) = 1 \\ \gamma(p_1 + p_2) = \gamma(p_1) + \gamma(p_2) & \gamma(A) = \gamma(p) \;\; \text{if } A \stackrel{def}{=} p \\ \gamma((\nu a)p) = \gamma(p) & \gamma(p_1 \mid p_2) = \gamma(p_1) + \gamma(p_2) + \gamma(p_1) \times \gamma(p_2) \end{array}$$

By guardedness, we are sure that $\gamma(A)$ will not call itself recursively, and so it is guaranteed that $\gamma(A)$ is always a finite number. It is not difficult then to check — by reasoning on the shape of the SOS inference rules — that indeed $\gamma(p)$ is an upper bound on the number of transitions leaving p. □

Corollary 3.1. *For any $p \in \mathscr{P}$, the LTS $\mathscr{C}_p = (\mathscr{P}_p, sort(p), \to_p)$ reachable from p is finitely-branching.*

Proof. *All the states reachable from $p \in \mathscr{P}$ are CCS processes by Proposition 3.1. Hence $\mathscr{C}_p$ is a finitely-branching LTS by Proposition 3.3.* □

Indeed, even the whole LTS $\mathscr{C}$ for CCS (see Definition 3.5) is finitely-branching.

Remark 3.9. **(Alternative rule for constants)** It is possible to define an alternative operational rule for process constants as follows:

$$(\text{ACon}) \; \frac{}{C \xrightarrow{\tau} p} \; C \stackrel{def}{=} p$$

Observe that with this alternative semantics, there is no need to restrict process terms to have guarded constants, as a constant would produce exactly one single (τ-labeled) transition. However, there are good reasons to reject this approach. First, C and its body p are no longer strongly bisimilar, but only weakly bisimilar. Moreover, the resulting LTS is much richer of τ-labeled transitions. A further negative aspect of this alternative operational rule is that it would prevent the possibility to represent, up to isomorphism, all the (finite-state) LTSs by means of CCS process terms (see Theorem 3.2 in Section 3.4.2 and Exercise 3.66 in Section 3.4.6). □

3.3.2 Unique Solution of Equations

There is another very good reason for restricting attention to guarded process terms only: it ensures uniqueness of the solution (up to $\sim$) of equations defined over processes. Consider the following process equation:

$$X \sim E(X)$$

where X is a *process variable* and $E(X)$ is some *open* expression built with the CCS operators which may contain some occurrences of X itself (for instance, $E(X) = a.\mathbf{0} + X$ or $E(X) = a.\mathbf{0} + \tau.X$).[4]

This equation defines a set of possible solutions: all the CCS processes p such that $p \sim E\{p/X\}$, where $E\{p/X\}$ is term $E(X)$ with each occurrence of X replaced by process p.

If we consider equation $X \sim a.\mathbf{0} + X$, it is clear that processes $p = a.\mathbf{0}$ and $q = a.\mathbf{0} + b.\mathbf{0}$ are both solutions of the equation, even if $p \not\sim q$. Actually it is not difficult to realize that infinitely many, not bisimilar processes are solutions of the equation: it is enough that a process possesses a summand $a.\mathbf{0}$ to be a solution. The distinguishing feature of this equation is that the process variable X does not occur strongly guarded in the right-hand side component $E(X)$.[5]

Exercise 3.28. How many not bisimilar solutions does the equation $X \sim a.\mathbf{0} | X$ possess? (*Hint:* Consider first the simple solution $A \stackrel{def}{=} a.A$.) □

On the contrary, if we consider $X \sim a.\mathbf{0} + \tau.X$, which is a typical instance of a process equation where the process variable X occurs strongly guarded in the right-hand side, we can easily get convinced that A, where $A \stackrel{def}{=} a.\mathbf{0} + \tau.A$, is a solution, and that any other solution should be very similar to this. In fact, one can prove that if p and q are two solutions for this process equation, then $p \sim q$, i.e., the solution is unique up to $\sim$. In the proof of this fact, following [Mil89], we make use of the following lemma.

Lemma 3.1. *Let X be a process variable strongly guarded in $E(X)$. If $E\{p/X\} \xrightarrow{\mu} p'$, then $p' = E'\{p/X\}$ for some $E'(X)$, and moreover, for any q, $E\{q/X\} \xrightarrow{\mu} E'\{q/X\}$.*

Proof. *By induction on the proof of $E\{p/X\} \xrightarrow{\mu} p'$. We examine the possible shapes of $E(X)$.*

- *$E(X) = X$. But X is assumed guarded in $E(X)$, hence this case is impossible.*
- *$E(X) = \mu'.F(X)$: in such a case $E\{p/X\} = \mu'.(F\{p/X\}) \xrightarrow{\mu'} F\{p/X\}$. Hence, $\mu = \mu'$ and $p' = F\{p/X\}$. Similarly $E\{q/X\} = \mu'.(F\{q/X\}) \xrightarrow{\mu'} F\{q/X\}$; hence the thesis follows by choosing $E'(X) = F(X)$.*
- *$E(X) = E_1(X) + E_2(X)$: in this case $E\{p/X\} = E_1\{p/X\} + E_2\{p/X\} \xrightarrow{\mu} p'$ must be due to either $E_1\{p/X\} \xrightarrow{\mu} p'$ or $E_2\{p/X\} \xrightarrow{\mu} p'$. Without loss of generality, assume the transition is due to $E_1\{p/X\} \xrightarrow{\mu} p'$. By inductive hypothesis, $p' = E_1'\{p/X\}$ for some $E_1'(X)$, and moreover, for any q, $E_1\{q/X\} \xrightarrow{\mu} E_1'\{q/X\}$.*

---

[4] A precise definition of *open* CCS terms is given in Section 4.2.2.

[5] A process *variable* X is strongly guarded in $E(X)$ if each occurrence of X in $E(X)$ is within a subterm of the form $\mu.F$. This definition is essentially the same we have given for process *constants* in Definition 3.3.

By rule (Sum_1), we can derive $E\{p/X\} = E_1\{p/X\} + E_2\{p/X\} \xrightarrow{\mu} E'_1\{p/X\}$ as well as $E\{q/X\} = E_1\{q/X\} + E_2\{q/X\} \xrightarrow{\mu} E'_1\{q/X\}$ for any q. Hence, the thesis follows by choosing $E'(X) = E'_1(X)$.

- $E(X) = E_1(X) \mid E_2(X)$: in this case $E\{p/X\} = E_1\{p/X\} \mid E_2\{p/X\} \xrightarrow{\mu} p'$ must be due to three different possibilities.
  First: $E_1\{p/X\} \xrightarrow{\mu} p_1$ and $p' = p_1 \mid E_2\{p/X\}$. Hence, by inductive hypothesis, p_1 is of the form $E'_1\{p/X\}$ for some $E'_1(X)$, and for any q, $E_1\{q/X\} \xrightarrow{\mu} E'_1\{q/X\}$. By rule ($Par_1$), we have that $E_1\{p/X\} \mid E_2\{p/X\} \xrightarrow{\mu} E'_1\{p/X\} \mid E_2\{p/X\}$ as well as $E_1\{q/X\} \mid E_2\{q/X\} \xrightarrow{\mu} E'_1\{q/X\} \mid E_2\{q/X\}$ for any q. Hence, $E\{p/X\} = E_1\{p/X\} \mid E_2\{p/X\} \xrightarrow{\mu} E'_1\{p/X\} \mid E_2\{p/X\} = (E'_1 \mid E_2)\{p/X\}$, as well as $E\{q/X\} = E_1\{q/X\} \mid E_2\{q/X\} \xrightarrow{\mu} E'_1\{q/X\} \mid E_2\{q/X\} = (E'_1 \mid E_2)\{q/X\}$ for any q. Hence, the thesis follows by choosing $E'(X) = E'_1(X) \mid E_2(X)$.
  Second: $E_2\{p/X\} \xrightarrow{\mu} p_2$ and $p' = E_1\{p/X\} \mid p_2$. This case is symmetric to the one above, and hence omitted.
  Third: $\mu = \tau$, $E_1\{p/X\} \xrightarrow{\alpha} p_1$ and $E_2\{p/X\} \xrightarrow{\overline{\alpha}} p_2$. By inductive hypothesis, $p_1 = E'_1\{p/X\}$ for some $E'_1(X)$, $p_2 = E'_2\{p/X\}$ for some $E'_2(X)$, as well as, for any q, $E_1\{q/X\} \xrightarrow{\alpha} E'_1\{q/X\}$ and $E_2\{q/X\} \xrightarrow{\bar{\alpha}} E'_2\{q/X\}$. By rule ($Com$), $E\{p/X\} = E_1\{p/X\} \mid E_2\{p/X\} \xrightarrow{\tau} E'_1\{p/X\} \mid E'_2\{p/X\} = (E'_1 \mid E'_2)\{p/X\}$, as well as $E\{q/X\} = E_1\{q/X\} \mid E_2\{q/X\} \xrightarrow{\tau} E'_1\{q/X\} \mid E'_2\{q/X\} = (E'_1 \mid E'_2)\{q/X\}$ for any q. The thesis follows by choosing $E'(X) = E'_1(X) \mid E'_2(X)$.
- $E(X) = (\nu a)F(X)$: very similar to the first subcase of the case above, hence left as an exercise.
- $E(X) = C$: in this case, since no X occurs in E, $E\{p/X\} = C \xrightarrow{\mu} p'$, which is possible only if $C \stackrel{def}{=} p_1$ and $p_1 \xrightarrow{\mu} p'$. Clearly, also $E\{q/X\} = C \xrightarrow{\mu} p'$. Hence, we can simply take $E'(X) = p'$. □

Proposition 3.4. *Let X be a process variable strongly guarded in $E(X)$. Then, if $p \sim E\{p/X\}$ and $q \sim E\{q/X\}$, then $p \sim q$.*

Proof. Consider relation

$$R = \{(E'\{p/X\}, E'\{q/X\}) \mid E'(X) \text{ is a CCS term using } X\} \cup \mathscr{I}$$

where $E'(X)$ is any CCS process term which may use only X as a variable, and $\mathscr{I} = \{(p,p) \mid p \in \mathscr{P}\}$ is the identity relation. Note that when $E'(X)$ is X, we get $(p,q) \in R$. If we prove that R is a strong bisimulation up to $\sim$ (see Definition 2.15), then we get the required thesis $p \sim q$ by Proposition 2.10 (stating the correctness of this proof principle).

By symmetry, it is enough to prove that if $E'\{p/X\} \xrightarrow{\mu} p'$, then $E'\{q/X\} \xrightarrow{\mu} q'$ with $p' \sim R \sim q'$. The proof proceeds by induction of the proof of $E'\{p/X\} \xrightarrow{\mu} p'$. We examine the possible shapes of $E'(X)$.

- $E'(X) = X$. *Then, we have* $E'\{p/X\} = p$ *and* $E'\{q/X\} = q$. *We want to prove that for any transition from* $E'\{p/X\} = p$, *e.g.,* $E'\{p/X\} \xrightarrow{\mu} p'$, *there exists a process* q' *such that* $E'\{q/X\} = q \xrightarrow{\mu} q'$ *with* $p' \sim R \sim q'$. *By hypothesis,* $p \sim E\{p/X\}$, *hence there exists* p'' *such that* $E\{p/X\} \xrightarrow{\mu} p''$ *with* $p' \sim p''$. *But as X is guarded in E, Lemma 3.1 ensures that* $p'' = E''\{p/X\}$ *for some* E'', *and that* $E\{q/X\} \xrightarrow{\mu} E''\{q/X\}$. *But* $E'\{q/X\} = q \sim E\{q/X\}$, *so there exists a* q' *such that* $E'\{q/X\} = q \xrightarrow{\mu} q' \sim E''\{q/X\}$. *Hence* $p' \sim R \sim q'$ *because* $(E''\{p/X\}, E''\{q/X\}) \in R$.
- $E'(X) = \mu.F(X)$. *In this case,* $E'\{p/X\} = \mu.F\{p/X\} \xrightarrow{\mu} F\{p/X\}$. *Similarly,* $E'\{q/X\} = \mu.F\{q/X\} \xrightarrow{\mu} F\{q/X\}$. *Note that* $(F\{p/X\}, F\{q/X\}) \in R$, *and so trivially* $F\{p/X\} \sim R \sim F\{q/X\}$, *as required.*
- $E'(X) = E_1(X) + E_2(X)$. *In this case,* $E'\{p/X\} = E_1\{p/X\} + E_2\{p/X\}$. *A transition from* $E'\{p/X\}$, *namely* $E_1\{p/X\} + E_2\{p/X\} \xrightarrow{\mu} p'$ *is derivable only if* $E_i\{p/X\} \xrightarrow{\mu} p'$ *for some* $i = 1, 2$. *Without loss of generality, assume the transition is due to* $E_1\{p/X\} \xrightarrow{\mu} p'$. *We can apply induction (as we are shortening the proof tree) and conclude that* $E_1\{q/X\} \xrightarrow{\mu} q'$ *with* $p' \sim R \sim q'$. *By rule (Sum*$_1$*), transition* $E_1\{q/X\} \xrightarrow{\mu} q'$ *ensures that also* $E'\{q/X\} = E_1\{q/X\} + E_2\{q/X\} \xrightarrow{\mu} q'$, *with* $p' \sim R \sim q'$, *as required.*
- $E'(X) = E_1(X) \,|\, E_2(X)$ *and* $E'(X) = (\nu a)E_1(X)$, *left as exercises for the reader.*
- $E'(X) = C$. *Since no instance of X occurs in* E', $E'\{p/X\} = C \xrightarrow{\mu} p'$, *which is possible only if* $C \stackrel{def}{=} p_1$ *and* $p_1 \xrightarrow{\mu} p'$. *Clearly, also* $E'\{q/X\} = C \xrightarrow{\mu} p'$ *and* $(p', p') \in R$. □

The proposition above works for a single equation, $X \sim E(X)$, exploiting CCS terms of the form $E(X)$, which use at most one single process variable X. However, the generalization to a system of equations of this form

$$\begin{aligned} X_1 &\sim E_1(X_1, \ldots, X_n) \\ X_2 &\sim E_2(X_1, \ldots, X_n) \\ &\ldots \\ X_n &\sim E_n(X_1, \ldots, X_n) \end{aligned}$$

where each $E_i(X_1, \ldots, X_n)$, for $i = 1, \ldots n$, is a CCS term that may use the variables $X_1, \ldots X_n$ is quite obvious, even if technically more involved.

More interesting is the case of process equations with respect to weak bisimulation. Consider $X \approx \tau.X + a.\mathbf{0}$: even if X is strongly guarded, we have infinitely many different non-weakly bisimilar solutions! E.g., $p = a.\mathbf{0}$ and $q = a.\mathbf{0} + b.\mathbf{0}$ (as well as any process with a summand $a.\mathbf{0}$). It can be proved (see [Mil89] pages 157-160) that the solution of an equation $X \approx E(X)$ is unique (up to $\approx$) only if X is *observationally guarded* and *sequential* in $E(X)$.

A variable X is observationally guarded in $E(X)$ if each occurrence of X is within a subterm of the form $\alpha.F$, where α is a visible action. Note that X in $E(X) = \tau.X + a.\mathbf{0}$ is not observationally guarded.

A variable X is sequential in $E(X)$ if every subexpression of $E(X)$ which contains X (apart from X itself) is of the form $\mu.F(X)$ or $E_1(X)+E_2(X)$. For instance, X in $E(X)=a.(b.\mathbf{0}\,|\,a.X)$ is not sequential because the subexpression $b.\mathbf{0}\,|\,a.X$ containing X is not of the required form.

As an example, consider the equation $X \approx a.\tau.X + b.\mathbf{0}$. Note that X in $E(X) = a.\tau.X + b.\mathbf{0}$ occurs observationally guarded and sequential. Hence, it admits a unique solution (up to $\approx$), e.g., constant B with $B \stackrel{def}{=} a.\tau.B + b.\mathbf{0}$.

3.4 Some Subclasses of CCS Processes

In this section we introduce a hierarchy of six subclasses of CCS — of increasing expressive power — that are important for various theoretical and practical reasons.

The first subclass, called *finite* CCS, is obtained by forbidding the use of process constants; hence, finite CCS processes are, from a practical point of view, of limited interest as they cannot afford any cyclic behavior: indeed, a finite CCS process originates a finite LTS. However, finite CCS is expressive enough to describe some simple, useful examples, e.g., for discriminating between competing behavioral equivalences (as illustrated in Chapter 2, where some examples of finite LTSs are presented). Moreover, for didactical reasons, we will restrict ourselves to finite CCS processes when describing how to prove equationally that two processes are equivalent (as discussed in Section 4.3).

The second subclass of CCS, called *finite-state* CCS, is obtained by restricting the syntax to allow only for the following operators: $\mathbf{0}$, action prefixing, choice and process constants. The name well represents the distinguishing feature of this class of processes. Any syntactical term p belonging to this subcalculus generates a finite-state LTS $\mathscr{C}_p$, but also we have a *representability theorem*: for any finite-state, reduced LTS *TS* there exists a finite-state CCS process p such that the LTS $\mathscr{C}_p$ generated by p is isomorphic to *TS*. Since finite-state LTSs are very important for verification purposes (because all the interesting properties and behavioral equivalences are decidable for them), this subclass of CCS — generating all and only such finite models — is very important in practice.

The third subclass is called *regular* CCS, whose processes still generate finite-state LTSs; nonetheless, regular CCS allows for limited use of parallel composition and restriction. This class is the most useful from an applicative point of view, as it allows for the modeling of finite-state systems in a modular way.[6]

Then we will present *Basic Parallel Processes* (BPP for short), obtained from CCS by removing the restriction operator and by ignoring rule (Com) in Table 3.1 (i.e., no communication is possible). The distinguishing feature of BPP is that a

---

[6] In the literature, the sublanguage we have called finite-state CCS is sometimes called *regular* CCS, while the processes we have called regular are sometimes called *networks of regular processes*.

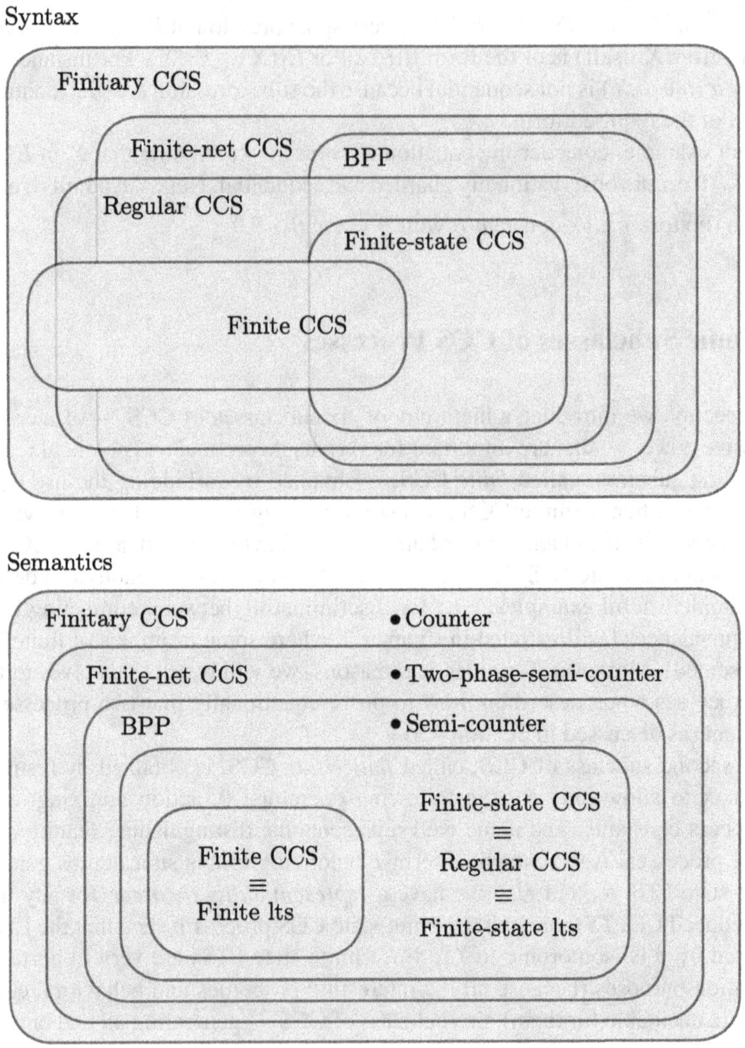

Fig. 3.6 Syntactic and semantic classifications of the six CCS subcalculi

BPP process may generate an infinite-state LTS, but still bisimulation equivalence is decidable on processes in this class.

The fifth subclass is called *finite-net* CCS, obtained from CCS by preventing restriction from occurring inside the body of recursively defined constants. This class is interesting because it has been proved in [Gor15, GV10] that a finite-net CCS process generates a finite P/T Petri net (see, e.g., the textbooks [Pet81, Rei85] or the survey [DesRei98] for an introduction to this model of distributed computation).[7]

[7] Petri nets were invented by Carl Adam Petri in 1939, at the age of 13, for the purpose of describing chemical processes, and were made popular by his Ph.D. dissertation [Petri62] in 1962.

Even if bisimulation equivalence is undecidable over finite-net processes, we might take advantage of the many decidable properties over finite P/T Petri nets, which are hence also decidable for these CCS processes. Moreover, in [Gor15, GV10] a representability theorem is presented: for any finite P/T net N, we can find a finite-net process p in a proper extension of CCS with multiway synchronization, called Multi-CCS (described in Chapter 6), such that the Petri net of p is isomorphic to N. Hence, finite-net Multi-CCS is for Petri nets as fundamental as finite-state CCS is for LTSs.

Finally, we present *finitary* CCS, where the only syntactical constraint is that, for any $p \in \mathscr{P}$, the set $Const(p)$ is finite. Actually, finitary CCS is the real language we will use throughout the book, as we consider feasible processes only those that can be described finitely, i.e., with the help of finitely many constants. Finitary CCS is proved to be Turing-complete in Section 3.5, where it is outlined how to model faithfully *Counter Machines* [Min67], a well-known Turing-complete formalism.

Summing up, the six subclasses of CCS we will discuss in this section can be classified as in Figure 3.6, where in the upper part the classification is based on *syntactic containment* (e.g., regular CCS processes are a proper superclass of finite-state CCS processes), while in the lower part the classification is based on *semantic expressiveness* with respect to the class of LTSs that can be generated by such subcalculi (e.g., regular CCS processes and finite-state CCS processes are equally expressive, as they both can represent all and only finite-state LTSs). The semantic classification contains also three distinct processes, namely:

- the *semi-counter* (discussed in Example 3.12 of Section 3.4.4), which is a typical example of a BPP process that is not in finite-state CCS;
- the *two-phase-semi-counter* (discussed in Example 3.15 of Section 3.4.5), which is a typical example of a finite-net CCS process that is not BPP;
- and the *counter* (discussed in Example 3.17 of Section 3.4.6), which is a typical process of a finitary CCS process that is not in finite-net CCS.

3.4.1 Finite CCS

Let $\mathscr{P}_{fin}$ be the set of *finite CCS processes*, i.e. of processes without any occurrence of any constant. Formally, a finite CCS process is a process generated by the following restricted abstract syntax

$$\begin{array}{l} p ::= \mathbf{0} \mid \mu.q \mid p+p \\ q ::= p \mid q|q \mid (\nu a)q \end{array}$$

which, under the assumptions of Remark 3.5, can be more succinctly denoted as

$$p ::= \Sigma_{j \in J} \mu_j.p_j \mid p|p \mid (\nu a)p$$

with the assumption that J is finite and $\Sigma_{j\in J}\mu_j.p_j = \mathbf{0}$ when $J = \emptyset$. The peculiar feature of a finite CCS process is that its reachable LTS $\mathscr{C}_p$ is finite, i.e., with a finite number of states and with an acyclic transition relation.

Proposition 3.5. (Finiteness) *For any $p \in \mathscr{P}_{fin}$, the reachable labeled transition system $\mathscr{C}_p = (\mathscr{P}_p, sort(p), \rightarrow_p, p)$ is a finite LTS.*

Proof. *The fact that $\mathscr{P}_p$ is finite, $sort(p)$ is finite and the LTS is acyclic derives trivially from the fact that the calculus is lacking any form of recursive construct (i.e., process constants).* □

Exercise 3.29. (Size of a finite CCS process) The *size* of a finite CCS process p, denoted *size*(p), is the number of prefixes occurring in p. Formally:

$$\begin{aligned} size(\mathbf{0}) &= 0 \\ size(\mu.p) &= 1 + size(p) \\ size(p_1 + p_2) &= size(p_1) + size(p_2) \\ size(p_1 \,|\, p_2) &= size(p_1) + size(p_2) \\ size((\nu a)p) &= size(p) \end{aligned}$$

Observe that, for any $p \in \mathscr{P}_{fin}$, $size(p)$ is a natural number and that if $size(p) = 0$, then $p \nrightarrow$. Prove that, for any $p \in \mathscr{P}_{fin}$, if $p \xrightarrow{\mu} p'$, then $size(p') < size(p)$. This proof can be done by induction on the proof of transition $p \xrightarrow{\mu} p'$. This observation offers an alternative proof of Proposition 3.5: the transition relation must be acyclic because any path

$$q_1 \xrightarrow{\mu_1} q_2 \xrightarrow{\mu_2} \ldots q_n \xrightarrow{\mu_n} q_{n+1}$$

determines a descending chain $size(q_1) > size(q_2) > \ldots > size(q_n) > size(q_{n+1})$ that has 0 as lower bound. Hence, if $size(q_1) = k$, the path cannot be longer than k (actually, it may be shorter). (See also Exercise 4.9 for the extension of function *size* to finitary CCS processes.) □

Exercise 3.30. (i) Observe that $\mathbf{0}$, $(\nu a)\mathbf{0}$ and $(\nu a)a.\mathbf{0}$ are all strongly bisimilar. (ii) Show that $\mathbf{0}$, $\tau.\mathbf{0}$, $\tau.\tau.\mathbf{0}$ and $\tau.\mathbf{0} + \tau.(\nu a)\mathbf{0}$ are all weakly bisimilar, but not all of them are rooted weakly bisimilar. (iii) Show that $a.\mathbf{0}$, $\tau.a.\mathbf{0}$, $a.\tau.\mathbf{0}$ and $a.\mathbf{0} + \tau.a.\mathbf{0}$ are all branching bisimilar, but not all of them are rooted branching bisimilar. (iv) Show that $a.\mathbf{0} + b.\mathbf{0} + \tau.b.\mathbf{0}$ and $a.\mathbf{0} + \tau.b.\mathbf{0}$ are weakly bisimilar but not branching bisimilar. (v) Show that $\tau.(a.\mathbf{0} + \tau.\mathbf{0}) + \tau.(b.\mathbf{0} + \tau.\mathbf{0})$ and $\tau.(a.\mathbf{0} + \tau.\mathbf{0}) + \tau.(b.\mathbf{0} + \tau.\mathbf{0}) + \tau.\mathbf{0}$ are weakly bisimilar but not branching bisimilar. □

Exercise 3.31. Show that $a.(b.\mathbf{0} \,|\, c.d.\mathbf{0})$ is weakly bisimilar, but not strongly bisimilar, to $(\nu e)(a.e.\mathbf{0} \,|\, \overline{e}.(b.c.d.\mathbf{0} + c.(b.d.\mathbf{0} + d.b.\mathbf{0})))$. □

Exercise 3.32. Prove that the finite CCS process $a.((\nu b)\mathbf{0}) + a.b.((\nu a)\mathbf{0})$ generates an LTS isomorphic to the one in Figure 2.12(a). Build the LTS for $a.\mathbf{0} + a.b.\mathbf{0}$ and show that the two are strongly bisimilar. □

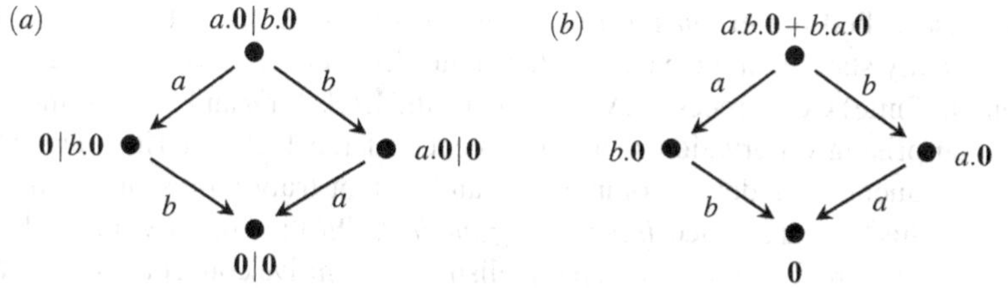

Fig. 3.7 Interleaving law: two isomorphic LTSs

Exercise 3.33. Define two finite CCS processes generating the two LTSs in Figure 2.15. Show that they are not strongly bisimilar. □

Exercise 3.34. Define two finite CCS processes generating the LTSs in Figure 2.29. Show that they are weakly bisimilar, but not branching bisimilar. □

Exercise 3.35. (Representability) Generalize the exercises above, by showing that, for any reduced, finite, rooted LTS *TS*, there exists a finite CCS process p such that the reachable LTS $\mathscr{C}_p = (\mathscr{P}_p, sort(p), \to_p, p)$ is isomorphic to *TS*.

(*Hint:* Start from the deadlock states, each one represented by a process of the form $(\nu d)\mathbf{0}$ for some new name d, as exemplified in Exercise 3.32, whose semantics is indeed a (distinct) deadlock state. A state q such that $T(q) = \{(q,\mu,q_k) \mid \exists \mu \in A, \exists q_k \in Q.\ q \xrightarrow{\mu} q_k\}$, which reaches only states q_k that are already represented by suitable CCS terms p_k, originates the finite CCS process $\Sigma_{(q,\mu,q_k)\in T(q)}\mu.p_k$.) □

Remark 3.10. **(Graphs vs LTSs)** Figure 2.5 depicts a directed graph that is not a labeled transition system. We want to observe that no finite CCS process can generate such a graph, indeed. The only natural candidate process is $a.\mathbf{0}+a.\mathbf{0}$, because we have two different branches $a.\mathbf{0}$, and the reached state, after a, is the deadlocked state $\mathbf{0}$ for both. However, if we use the SOS operational rules, we discover that the only derivable transition is $a.\mathbf{0}+a.\mathbf{0} \xrightarrow{a} \mathbf{0}$, for which we have two different proofs, so that its associated LTS is isomorphic to the one in Figure 3.1(b) for $a.\mathbf{0}$. □

Exercise 3.36. Prove that the finite processes $a.\mathbf{0} \mid b.\mathbf{0}$ and $a.b.\mathbf{0}+b.a.\mathbf{0}$ are strongly bisimilar (see Figure 3.7). Check that the required bisimulation relation is actually an LTS isomorphism. □

Remark 3.11. **(Sequentiality vs Concurrency)** The exercise above shows that, even if isomorphism is the most discriminating (i.e., most concrete) equivalence one can define over LTSs, it is unfortunately already abstract enough to be unable to distinguish *parallelism* (or *concurrency*) from *sequentiality*! Indeed, the parallel execution of two actions a and b, denoted in CCS by the parallel process term $a.\mathbf{0} \mid b.\mathbf{0}$, generates an LTS isomorphic to the one for the choice between their two possible sequential orderings, in CCS denoted by the sequential process term $a.b.\mathbf{0}+b.a.\mathbf{0}$. This example is an instance of a more general law for CCS, called the *interleaving* *law* — also called the *expansion law* – (see also Section 4.1.1, Proposition 4.3), which roughly states that, given a parallel (finite) CCS process, say q, we can find a sequential (finite) CCS process, say p, such that the LTSs for q and p are isomorphic.

This surprising observation is a consequence of the fact that LTSs are intrinsically a sequential model of computation and cannot truly represent parallelism; models of this sort are called *interleaving models*. On the contrary, in models of concurrency where distribution and parallelism are primitive concepts, such as Petri nets [Petri62, Pet81, Rei85, DesRei98], the interleaving law does not hold, e.g., the Petri net for the parallel CCS process $a.\mathbf{0} \mid b.\mathbf{0}$ and the Petri net for sequential CCS process $a.b.\mathbf{0}+b.a.\mathbf{0}$ are not isomorphic, as shown in [DDM88, Gol90, Old91]. (See also Section 6.3.2, Example 6.10 in particular, for a discussion about interleaving models and non-interleaving (or *truly concurrent*) ones.) □

3.4.2 Finite-State CCS

The *finite-state CCS processes* are generated from actions and constants as described by the following abstract syntax:

$$\begin{aligned} p &::= \mathbf{0} \mid \mu.q \mid p+p \\ q &::= p \mid C \end{aligned}$$

where we assume that all constants are defined and guarded[8] and, more importantly, that the set of constants $Const(q)$ used for any finite-state process q is finite. Under the assumptions listed in Remark 3.5, sometimes the syntax of finite-state CCS processes is more succinctly given as:

$$p ::= \Sigma_{j\in J}\mu_j.p_j \mid C$$

where J is finite and $\Sigma_{j\in J}\mu_j.p_j = \mathbf{0}$ when $J = \emptyset$.

As a simple example of a finite-state CCS process, consider the polite vending machine VM defined as follows:

$$\begin{aligned} VM &\stackrel{def}{=} coin.(ask\text{-}esp.VM_1 + ask\text{-}am.VM_2) \\ VM_1 &\stackrel{def}{=} \overline{esp\text{-}coffee}.VM \qquad\qquad VM_2 \stackrel{def}{=} \overline{am\text{-}coffee}.VM \end{aligned}$$

whose associated LTS is depicted in Figure 3.3.

We now study the properties of this subcalculus: first we prove that any finite-state CCS process originates, via the operational semantics, a finite-state LTS, i.e., an LTS with finitely many states and labels; then, we provide a representability theorem, stating that any finite-state LTS can be represented by a finite-state CCS process.

[8] Guardedness is not strictly necessary for this subcalculus: even allowing for unguarded constants, the generated LTSs would still be finite-state; however, guardedness prevents unwanted phenomena, such as those described in Exercise 3.27.

Theorem 3.1. (Finite number of reachable states) *For any finite-state CCS process p, the set $\mathscr{P}_p$ of its reachable states is finite.*

Proof. *We define the number $\chi(p,\emptyset)$ as an upper bound on the number of possible states reachable from p:*

$$\chi(\mathbf{0},I) = 1 \quad \chi(p_1+p_2,I) = \chi(p_1,I)+\chi(p_2,I)+1 \quad \chi(\mu.p,I) = 1+\chi(p,I)$$

$$\chi(A,I) = \begin{cases} 0 & \textit{if } A \in I, \\ 1+\chi(p,I \cup \{A\}) & \textit{if } A \stackrel{def}{=} p \textit{ and } A \notin I \end{cases}$$

As the set Const(p) of process constants used in p is finite, it follows that the calculation of $\chi(p,\emptyset)$ always terminates (it cannot loop, because of its second argument recording the constants already found), returning a finite number. □

Corollary 3.2. (Finite-state LTS) *For any finite-state CCS process p, the LTS reachable from p, $\mathscr{C}_p = (\mathscr{P}_p, sort(p), \rightarrow_p, p)$, is finite-state.*

Proof. *By Theorem 3.1, the set of reachable states $\mathscr{P}_p$ is finite. By Corollary 3.1, $\mathscr{C}_p$ is finitely-branching. Hence, also sort(p) must be finite, as required by Definition 2.7.* □

Theorem 3.2. (Representability) *For any reduced, finite-state, rooted LTS TS, there exists a finite-state CCS process p such that the reachable LTS $\mathscr{C}_p = (\mathscr{P}_p, sort(p), \rightarrow_p, p)$ is isomorphic to TS.*

Proof. *Let $TS = (Q, A, \rightarrow_1, q_0)$, with $Q = \{q_0, q_1, \ldots, q_n\}$. We define a process constant C_i in correspondence with state q_i, for $i = 0, 1, , \ldots, n$, defined as follows: if q_i is a deadlock, then $C_i \stackrel{def}{=} \mathbf{0}$; if $T(q_i) = \{(q_i, \mu, q_k) \mid \exists \mu \in A, \exists q_k \in Q.\ q_i \stackrel{\mu}{\longrightarrow}_1 q_k\}$, then $C_i \stackrel{def}{=} \Sigma_{(q_i,\mu,q_k) \in T(q_i)} \mu.C_k$. Let us consider $\mathscr{C}_{C_0} = (\mathscr{P}_{C_0}, sort(C_0), \rightarrow_2, C_0)$. It is not difficult to see that $\mathscr{P}_{C_0} = \{C_0, C_1 \ldots, C_n\}$ because TS is reduced. Hence, the bijection we are looking for is $f : Q \rightarrow \mathscr{P}_{C_0}$, defined as $f(q_i) = C_i$. It is also easy to observe that the two conditions of isomorphism are satisfied, namely:*

- $C_0 = f(q_0)$, *and*
- $q \stackrel{\mu}{\longrightarrow}_1 q'$ *iff* $f(q) \stackrel{\mu}{\longrightarrow}_2 f(q')$

Hence, f is indeed an LTS isomorphism. □

As a simple application of this theorem, the finite-state LTS in Figure 2.2 can be represented by the two mutually recursive constants $C_1 \stackrel{def}{=} coin.C_2$ and $C_2 \stackrel{def}{=} \overline{coffee}.C_1$; their associated LTS is depicted in Figure 3.2(c).

Exercise 3.37. Take the finite-state LTS in Figure 2.14(a). Build a finite-state CCS process p such that the LTS reachable from p is isomorphic to it. □

Definition 3.6. (Finite-state CCS language) A language $L \subseteq (\mathscr{L} \cup \overline{\mathscr{L}})^*$ is a *finite-state CCS language* if there exists a finite state CCS process p such that the set of its weak completed traces is L, i.e., $WCTr(p) = L$. □

Proposition 3.6. (Finite-state CCS languages are regular languages) *The class of finite-state CCS languages coincides with the class of regular languages.*

Proof. *By Remark 2.7, all, and only, the regular languages can be represented by finite-state LTSs, up to weak completed trace equivalence. By Corollary 3.2 and Theorem 3.2, finite-state CCS processes generate all, and only, finite-state LTSs. Hence, the class of finite-state CCS languages coincides with the class of regular languages.* □

Exercise 3.38. (From right-linear grammars to finite-state CCS processes) In Remark 2.7 we showed that all the regular languages can be represented by finite-state LTSs by means of finite automata. An alternative proof of this fact can be obtained by means of *right-linear grammars* (see Section 1.3.3). Show a simple translation from a reduced[9] right-linear grammar $G = (N, T, S, P)$ to a finite-state CCS p such that $L(G) = WCTr(p)$, i.e., the language generated by the grammar G is the same as the set of weak completed traces of p. (*Hint*: Consider the grammar nonterminal symbols as process constants, take $p = S$, and generate for each nonterminal, say, A with productions, say, $A \to bA \mid b \mid \varepsilon$, the corresponding constant definition $A \overset{def}{=} b.A + b.\mathbf{0} + \tau.\mathbf{0}$.) □

Exercise 3.39. (Vending machines) Prove that the vending machines *VM* — defined at the beginning of this section and whose LTS is depicted in Figure 3.3 — and *IVM* — defined in Exercise 3.9 — are not bisimilar. □

Example 3.4. **(Vending machines, again)** Let us consider the finite-state LTSs for the two vending-machines discussed in Exercise 2.18. A possible finite-state process for case (i) is the following:

$$AVM \overset{def}{=} coin.(coin.ask\text{-}esp.\overline{esp\text{-}coffee}.AVM \; + \; ask\text{-}am.\overline{am\text{-}coffee}.AVM)$$

A possible finite-state process for the case (ii) is the following:

$$0VM \overset{def}{=} coin.1AVM$$
$$1VM \overset{def}{=} ask\text{-}am.\overline{am\text{-}coffee}.0VM + coin.2VM$$
$$2VM \overset{def}{=} ask\text{-}esp.\overline{esp\text{-}coffee}.0VM + ask\text{-}am.\overline{am\text{-}coffee}.1VM$$

where the constants $0VM, 1VM, 2VM$ stand for the number of coins collected by the machine, i.e., if the machine is in state iVM, then its credit is i, for $i = 0, 1, 2$. □

Exercise 3.40. Modify the vending machine $0VM$ of Example 3.4, so that it returns the coins inserted in case something goes wrong (e.g., no more water available), by modeling this situation by an internal transition to an erroneous state that handles the exception by returning the coins and entering a deadlock state. □

---

[9] A grammar $G = (N, T, S, P)$ is reduced if for any nonterminal $A \in N$ there is a production in P with A in its left-hand side, e.g., $A \to aB$.

Exercise 3.41. Consider the following vending machine:

$$0VM' \stackrel{def}{=} coin.1VM' + coin.0VM'$$
$$1VM' \stackrel{def}{=} ask\text{-}am.(\overline{am\text{-}coffee}.0VM' + \tau.1VM') + coin.2VM'$$
$$2VM' \stackrel{def}{=} ask\text{-}esp.\overline{esp\text{-}coffee}.0VM' + ask\text{-}am.(\overline{am\text{-}coffee}.1VM' + \tau.2VM')$$

Comment on the expected unfair bahavior of such a machine. □

Exercise 3.42. Define a finite-state CCS process for a vending machine that accepts in input coins of one or two euros, sells espresso for two euros and american coffee for one euro, keeps credit up to four euros (it refuses to input coins otherwise), does not steal money and does not allow a beverage to be dispensed if the previously delivered one has not been collected. □

We recall that bisimulation equivalence over a finite-state LTS with n states and m transitions can be computed in $O(m \log n)$ time [PT87] — see Section 2.5 for an intuitive, non-optimal algorithm. On the contrary, as language equivalence over automata is PSPACE-complete [SM73, HRS76], so is also trace equivalence (as well as completed trace equivalence) for finite-state LTSs.

3.4.3 Regular CCS

Finite-state CCS is rather powerful, as it may represent any finite-state LTS; also it is very tractable algorithmically, as all the behavioral equivalences we have defined in Chapter 2 are decidable for finite-state LTSs. However, from a modeling point of view, it lacks the basic operators of parallel composition and restriction, that are very useful when modeling a complex system in a compositional way. Regular CCS aims at overcoming such a problem, by allowing restricted use of these additional static operators; however, this is achieved without increasing the expressive power of the language, so that this subcalculus is as tractable as finite-state CCS. For these reasons, regular CCS is considered the most appropriate calculus for applications.

The regular CCS processes are generated from actions and constants by the following abstract syntax:

$$\begin{aligned} s &::= \mathbf{0} \mid \mu.p \mid s+s \\ p &::= s \mid C \\ q &::= p \mid (\nu a)q \mid q|q \end{aligned}$$

where, as usual, we assume that process constants are always defined and guarded and, more importantly, that the set of constants *Const(q)* used for any process q is finite (see Definition 3.1). Under the assumptions listed in Remark 3.5, sometimes the syntax of regular CCS processes is more succinctly given as

$$\begin{aligned} p &::= \Sigma_{j \in J} \mu_j.p_j \mid C \\ q &::= p \mid (\nu a)q \mid q|q \end{aligned}$$

which more clearly expresses the fact that a regular process is obtained as the parallel composition (or restriction) of a finite number of finite-state CCS processes. Hence, from a syntactical point of view, regular CCS is a proper superclass of finite-state CCS.

Remark 3.12. **(Notational convention and extension)** For simplicity's sake, in the following we let p and q denote any regular CCS process term, while in the syntax definition above, p ranges only over finite-state CCS process terms. Additionally, for convenience, in order to give a name to every process, we will sometimes make use of constants that are defined nonrecursively over the general class of regular processes. For instance, we consider acceptable the constant definition $A \stackrel{def}{=} (\nu a)(p \,|\, q)$ when A does not belong to $Const(p) \cup Const(q)$. This convention is extensively adopted in the following, e.g., for constant $2Sem$ in Example 3.5 or constant $Protocol$ in Example 3.6. □

From a semantical point of view, regular CCS is as expressive as finite-state CCS.

Proposition 3.7. (Regular CCS processes originate finite-state LTSs) *For any regular CCS process p, the LTS* $\mathscr{C}_p = (\mathscr{P}_p,\ sort(p),\ \rightarrow_p, p)$ *reachable from p is finite-state.*

Proof. *By the syntactic definition of regular CCS, a regular CCS process is the parallel composition (or restriction) of a finite number of finite-state CCS processes. By Corollary 3.2 each finite-state CCS process generates a finite-state LTS. The thesis then follows by Exercise 3.22, which proves that the parallel composition of two CCS processes, generating finite-state LTSs, generates a finite-state LTS,*[10] *as well as that the restriction of a finite-state process remains finite-state.* □

Hence, from an expressiveness point of view, it seems that we gain nothing by the richer regular CCS syntax. However, from a modeling point of view, we gain a lot, as it is quite useful to define a complex system by describing its parallel subcomponents separately and how they are connected to one another. Indeed, this allows for compositional reasoning as discussed in the introduction of this chapter and in the following example.

Example 3.5. **(Semaphore)** A binary semaphore is an abstract data type — realized by means of a variable that can hold value 0 or 1 — that is used for controlling access, by multiple processes, to a common resource in a parallel environment [Dij68]: if the stored value is 1, then request operation p is executable by a process on the semaphore, which has the effect of decrementing the stored value to 0 and of granting private access to the shared resource; when the process wants to release the shared resource, it will perform operation v on the semaphore with the effect of incrementing the stored value to 1. Abstractly, in our process algebra setting, we can model a binary (or one-position) semaphore by the finite-state CCS process

[10] The generalization to an arbitray finite number of parallel components is obvious by induction.

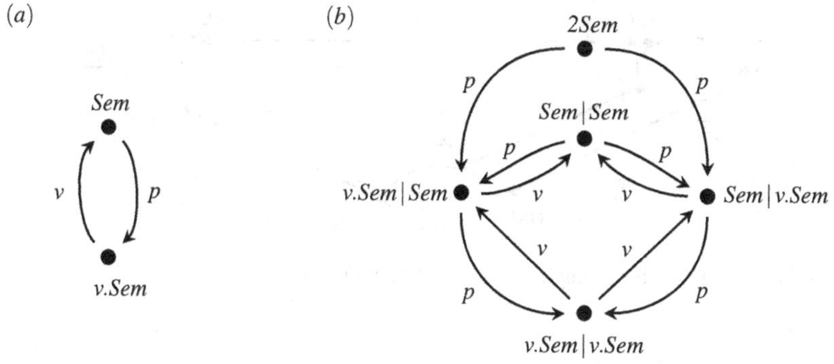

Fig. 3.8 One-position and two-position semaphores

$Sem \stackrel{def}{=} p.v.Sem$ — whose associated LTS is depicted in Figure 3.8(a) — where actions p and v are to be synchronized with the process requesting access to the shared resource.

As a simple, introductory example of a regular CCS process, let us consider a two-position semaphore, defined as follows: $2Sem \stackrel{def}{=} Sem\,|\,Sem$, whose associated LTS is in Figure 3.8(b). We may generalize this process to an n-position semaphore as follows: $nSem \stackrel{def}{=} \Pi_{1 \leq i \leq n} Sem$. The number of states of $nSem$ is $2^n + 1$ (by Exercise 3.22), which might make intractable the analysis of the LTS generated by $nSem$ for large n. In some cases, however, we can take advantage of the compositional definition of $nSem$. Suppose we want to demonstrate that $\mathscr{C}_{nSem}$ is deadlock-free. It is not difficult to observe (see Exercise 3.23) that if $\mathscr{C}_p$ or $\mathscr{C}_q$ is deadlock-free, then also $\mathscr{C}_{p|q}$ is deadlock-free. Hence, deadlock-freeness of $\mathscr{C}_{nSem}$ derives easily from deadlock-freeness of the LTS $\mathscr{C}_{Sem}$ for the one-position buffer Sem, so that we need not generate the huge LTS $\mathscr{C}_{nSem}$. □

Example 3.6. **(Simple communication protocol)** In order to show the advantage of the modular definition of a complex system, we describe the following simple communication protocol, originally outlined in [AILS07]. There are three components: a sender *Send* takes care of collecting a message from the environment on the port *acc* (for *accept*) and forwards it to a medium *Med*, which in turn forwards it to a receiver *Rec*, which will send it on the port $\overline{del}$ (for *deliver*) to the external environment and then send an ack message to the sender, so that the cycle can be repeated. The medium *Med* may be faulty: this is abstracted by an internal transition to some error state *Err*; as a consequence *Err* will ask *Send* to resend the message. The specification of the whole system is the regular CCS process

$$Protocol \stackrel{def}{=} (\nu send, error, trans, ack)(Send\,|\,Med\,|\,Rec),$$

where the three finite-state CCS components are specified as follows:

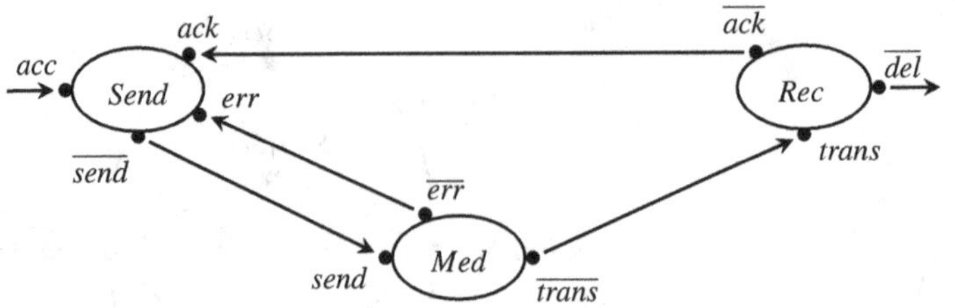

Fig. 3.9 The flow-graph for a simple communication protocol

$$\begin{array}{llll} Send \stackrel{def}{=} acc.Sending & & Med \stackrel{def}{=} send.Med' \\ Sending \stackrel{def}{=} \overline{send}.Wait & & Med' \stackrel{def}{=} \tau.Err + \overline{trans}.Med \\ Wait \stackrel{def}{=} ack.Send + error.Sending & & Err \stackrel{def}{=} \overline{error}.Med \end{array}$$

$$Rec \stackrel{def}{=} trans.Del$$
$$Del \stackrel{def}{=} \overline{del}.Ack$$
$$Ack \stackrel{def}{=} \overline{ack}.Rec$$

The flow-graph describing the interconnection architecture of the system is given in Figure 3.9, where, in order to help the reader, the names of the ports of the private communication channels have not been hidden (hence they are still visible).

The dynamic behavior of *Protocol* can be understood by observing the local behavior of each component. For instance, the sender *Send* first receives a message on port *acc*, then forwards it to *Med* along port $\overline{send}$ and, by doing so, enters the state *Wait* where it waits for either an *ack* message of successful delivery of the previous message from *Rec*, or for an *error* message from *Med*, asking for a retransmission of the previously lost message. The labeled transition system for *Protocol*, describing its dynamic behavior, is in Figure 3.10, where, for simplicity's sake, the restrictions have been omitted in the states and, to help the reader, the occurrences of label τ are indexed by the name of the synchronized action. Observe that three divergent states are present in the model, because of the possibility of having to retransmit the message some arbitrary number of times before the delivery is successful.

The observable behavior of this complex system can be described by the more abstract process

$$ProtSpec \stackrel{def}{=} acc.\overline{del}.ProtSpec,$$

called the *specification* of the communication protocol, whose associated LTS is very simple. This specification can be used to prove self-evidently that *acc* and $\overline{del}$ are strictly alternated, or that the system is deadlock-free. These properties are true also for the *implementation*, i.e., for *Protocol*, if we can prove that the implementation and the specification are bahaviorally equivalent w.r.t. a suitably discriminating equivalence which preserves such properties. As a matter of fact, it is easy to see that the two are weakly bisimilar, *ProtSpec* $\approx$ *Protocol*, as relation R below is a weak bisimulation, where, for simplicity's sake, we have omitted the restrictions in

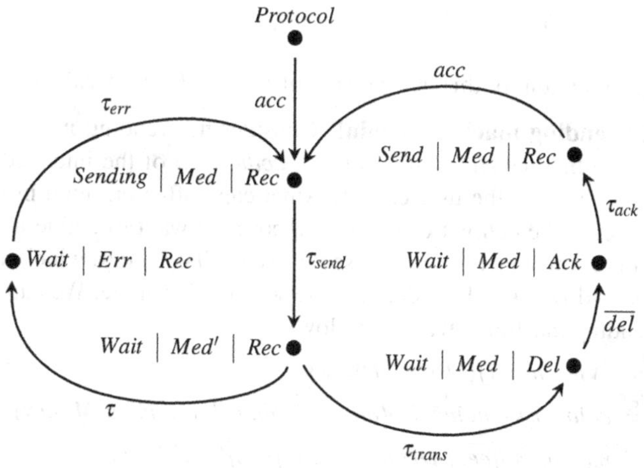

Fig. 3.10 The bahavior of the simple communication protocol

the state of the implementation, as done in Figure 3.10.

$$\begin{aligned}R = \{&(Protocol, ProtSpec), (Sending\,|\,Med\,|\,Rec, \overline{del}.ProtSpec),\\ &(Wait\,|\,Med'\,|\,Rec, \overline{del}.ProtSpec), (Wait\,|\,Err\,|\,Rec, \overline{del}.ProtSpec),\\ &(Wait\,|\,Med\,|\,Del, \overline{del}.ProtSpec), (Wait\,|\,Med\,|\,Ack, ProtSpec),\\ &(Send\,|\,Med\,|\,Rec, ProtSpec)\}\end{aligned}$$

Checking that R is a weak bisimulation is a very easy task; for instance, pair $(Wait\,|\,Med'\,|\,Rec,\ \overline{del}.ProtSpec)$ is a weak bisimulation pair because to transitions $Wait\,|\,Med'\,|\,Rec \xrightarrow{\tau} Wait\,|\,Err\,|\,Rec$ and $Wait\,|\,Med'\,|\,Rec \xrightarrow{\tau} Wait\,|\,Med\,|\,Del$, process $\overline{del}.ProtSpec$ can reply simply by idling, $\overline{del}.ProtSpec \stackrel{\varepsilon}{\Longrightarrow} \overline{del}.ProtSpec$, and the reached states are in R; symmetrically, transition $\overline{del}.ProtSpec \xrightarrow{\overline{del}} ProtSpec$ can be matched by $Wait\,|\,Med'\,|\,Rec \stackrel{\overline{del}}{\Longrightarrow} Wait\,|\,Med\,|\,Ack$ and the reached states are in R. (As a matter of fact, R is also a branching bisimulation.)

Observe that the specification is divergence-free; hence, by equating the specification and the implementation, weak bisimilarity (as well as branching bisimilarity) assumes fairness on the faulty bahavior of the medium, i.e., it assumes that the divergence in the LTS for the implementation will be eventually escaped (cf. the discussion about insensitivity of weak bisimilarity to divergence after Definition 2.21 in Section 2.4.2.) □

Exercise 3.43. Continuing Example 3.6, assume that now state Err is defined as

$$Err \stackrel{def}{=} \overline{error}.Med + \tau.Div \qquad\qquad Div \stackrel{def}{=} \tau.Div$$

This means that the faulty medium Med, when a message gets lost, can either ask for a retransmission of the lost message, or enter a livelock. Check if

$$ProtSpec' \stackrel{def}{=} acc.(\overline{del}.ProtSpec' + \tau.\mathbf{0})$$

is a correct specification for this new version of *Protocol* w.r.t. weak bisimilarity. □

Example 3.7. **(Vending machine, again!)** Consider the vending machine *AVM* of Example 3.4. We may consider this as the *specification* of the intended behavior: upon insertion of a coin, the user can get American coffee or, upon insertion of a supplementary coin, he can get espresso. Suppose we want to define a distributed implementation of this machine, composed of two units: *Money*, which handles the input of coins, and *Drink*, which delivers the selected beverage. We can model the distributed vending machine DVM_1 as follows:

$$\begin{aligned}
DVM_1 &\stackrel{def}{=} (\nu mc, me, d)(Money \,|\, Drink)\\
Money &\stackrel{def}{=} coin.(ask\text{-}am.\overline{mc}.d.Money + coin.ask\text{-}esp.\overline{me}.d.Money)\\
Drink &\stackrel{def}{=} mc.\overline{am\text{-}coffee}.\overline{d}.Drink + me.\overline{esp\text{-}coffee}.\overline{d}.Drink,
\end{aligned}$$

where action *mc* stands for *make-coffee*, action *me* stands for *make-espresso* and action *d* stands for *done*. After the insertion of a coin and the selection of the beverage, *Money* sends a request to *Drink* to prepare the selected beverage; when *Drink* has finished, it signals this to *Money*, so that the whole system can start again. An intermezzo:

Exercise 3.44. Draw the LTS for DVM_1. □

Given the LTS for DVM_1, it is not difficult to see that DVM_1 is weakly bisimilar to *AVM* ($DVM_1 \approx AVM$); it is crucial that, after coffee delivery, there is a synchronization between the two components on action *d*, so that a new coin is accepted by *Money* only after *Drink* has delivered the beverage. Hence, we can safely conclude that DVM_1 is a correct implementation of *AVM*. Second intermezzo:

Exercise 3.45. Build a weak bisimulation containing the pair (DVM_1, AVM). □

In the light of Exercise 3.40, one can define a bit more realistic model of the vending machine in that, if something goes wrong in the preparation of the coffee (e.g., water is unavailable), the coin(s) is (are) returned. A possible abstract specification may be

$$\begin{aligned}
AVM_2 \stackrel{def}{=} coin.(&coin.ask\text{-}esp.(\overline{esp\text{-}coffee}.AVM_2 + \tau.\overline{coin}.\overline{coin}.AVM_2)\\
&+ ask\text{-}am.(\overline{am\text{-}coffee}.AVM_2 + \tau.\overline{coin}.AVM_2)),
\end{aligned}$$

where, after the selection, the vending machine can either deliver the selected beverage or can internally choose to return the coin(s) and restart from the initial state. Similarly as above, a corresponding distributed vending machine DVM_2 may be defined as follows:

$$\begin{aligned}
DVM_2 &\stackrel{def}{=} (\nu mc, me, d, err)(Money' \,|\, Drink')\\
Money' &\stackrel{def}{=} coin.(ask\text{-}am.\overline{mc}.(d.Money' + err.\overline{coin}.Money')\\
&\qquad + coin.ask\text{-}esp.\overline{me}.(d.Money' + err.\overline{coin}.\overline{coin}.Money'))\\
Drink' &\stackrel{def}{=} mc.(\overline{am\text{-}coffee}.\overline{d}.Drink' + \overline{err}.Drink')\\
&\qquad + me.(\overline{esp\text{-}coffee}.\overline{d}.Drink' + \overline{err}.Drink'),
\end{aligned}$$

where action $\overline{err}$ signals that something goes wrong with the preparation of the beverage and so one or two coins are to be returned by $Money'$ to the user. It is not difficult to prove (do it!) that DVM_2 is weakly bisimilar to AVM_2. □

Exercise 3.46. (Two-position buffers) Draw the LTS associated to the following sequential, finite-state specification of a two-position buffer:

$$B_0 \stackrel{def}{=} in.B_1$$
$$B_1 \stackrel{def}{=} in.B_2 + \overline{out}.B_0$$
$$B_2 \stackrel{def}{=} \overline{out}.B_1$$

Consider then a one-position buffer:

$$B \stackrel{def}{=} in.B' \qquad B' \stackrel{def}{=} \overline{out}.B.$$

Draw the LTS for the regular CCS process $B\,|\,B$ and show that it is strongly bisimilar to B_0. Process $B\,|\,B$ may be considered the *parallel* (or *bag*) implementation of B_0, where the order of arrival of data in the buffer may be not respected in delivery.

Now, consider then the following *pipeline* buffer

$$Buf \stackrel{def}{=} (\nu d)(Buf_1\,|\,Buf_2) \quad Buf_1 \stackrel{def}{=} in.\overline{d}.Buf_1 \quad Buf_2 \stackrel{def}{=} d.\overline{out}.Buf_2.$$

Process *Buf* is such that the order of arrival of data in the buffer is respected in delivery. Show that B_0 is weakly bisimilar to *Buf*.

Hence, the parallel implementation and the pipeline one of a two-position buffer are weakly bisimilar. This result holds only because we are considering buffers that cannot handle values explicitly. (In Section 3.6 we will discuss an extension to CCS with explicit handling of data. See Exercise 3.78 for more concrete definitions of buffers.) □

Example 3.8. **(2-producers-consumer)** Another simple example of a regular CCS process is the system *2PC*, defined as follows:

$$2PC \stackrel{def}{=} (\nu send)((P_1\,|\,P_1)\,|\,C_1) \quad P_1 \stackrel{def}{=} produce.\overline{send}.P_1 \quad C_1 \stackrel{def}{=} send.consume.C_1,$$

where P_1 and C_1 represent a producer and a consumer, respectively. The whole system *2PC* is composed of the parallel composition of two producers and one consumer; the restriction on action *send* ensures that the consumer can only consume products coming from these two producers and, symmetrically, that the goods produced by the producers can only be consumed by this consumer. It is not difficult to see that *2PC* is deadlock-free, thanks to Exercise 3.23: indeed, both P_1 and C_1 are deadlock-free, hence also $((P_1\,|\,P_1)\,|\,C_1)$ is; finally, *2PC* is deadlock-free because for any state p such that $((P_1\,|\,P_1)\,|\,C_1) \longrightarrow^* p \stackrel{\alpha}{\longrightarrow}$ with $\alpha \in \{send, \overline{send}\}$, we have that $p \stackrel{\beta}{\longrightarrow}$ for some $\beta \in \{produce, \tau\}$. □

Exercise 3.47. Continuing Example 3.8, draw the finite-state LTS for *2PC* (it is composed of nine states and 18 transitions). How many occurrences of action *produce* can be performed before an occurrence of *consume* becomes mandatory? Generalize the process to three producers and two consumers; namely, consider:

$$3P2C \stackrel{def}{=} (\nu send)((P_1 \mid P_1 \mid P_1) \mid (C_1 \mid C_1)).$$

Can you draw easily its associated LTS? This latter exercise shows the necessity of tool support when considering examples that cannot be easily managed by hand. To this aim, a possible tool is the Concurrency Workbench [CWB]. □

Exercise 3.48. (Producer-consumer with one-position buffer) Another example of a regular CCS process is

$$PBC \stackrel{def}{=} (\nu in, out)((P_2 \mid B) \mid C_2) \quad P_2 \stackrel{def}{=} produce.\overline{in}.P_2 \quad C_2 \stackrel{def}{=} out.consume.C_2,$$

where B is the one-position buffer described in Exercise 3.46. (*i*) Draw the LTS for PBC. (*ii*) Prove that PBC and $2PC$ (see Example 3.8) are not (strong) trace equivalent. (*iii*) Show also that PBC and $2PC$ are weakly bisimilar, by providing a suitable weak bisimulation relation (possibly up to $\approx$), containing the pair $(PBC, 2PC)$. □

Exercise 3.49. (Producer-consumer with two-position buffer) One further example of a regular CCS process is

$$P2BC \stackrel{def}{=} (\nu in, out)((P_2 \mid B_0) \mid C_2),$$

where P_2 and C_2 are defined as in Exercise 3.48 and B_0 is the two-position buffer described in Exercise 3.46.

(*i*) Draw the LTS for $P2BC$. (*ii*) Show that $P2BC$ is not weakly trace equivalent to $2PC$ of Example 3.8 (*Hint:* How many occurrences of action *produce* can be performed before *consume* is mandatory?)

(*iii*) Consider also the variant system $P2BC' \stackrel{def}{=} (\nu in, out)((P_2 \mid (B \mid B)) \mid C_2)$ where B is the one-position buffer of Exercise 3.46. Show that $P2BC$ and $P2BC'$ are strongly bisimilar by building a suitable strong bisimulation relation.[11]

(*iv*) Finally, consider the variant $P2BC'' \stackrel{def}{=} (\nu in, out)((P_2 \mid Buf) \mid C_2)$ where Buf is the pipeline two-position buffer of Exercise 3.46. Show that $P2BC$ and $P2BC''$ are weakly bisimilar by building a suitable weak bisimulation relation.[12] □

---

[11] An alternative proof for $P2BC \sim P2BC'$ may be based on the fact that $\sim$ is a *congruence* for the CCS operators, as we will prove in Theorem 4.1; as a matter of fact, we have that $B_0 \sim B \mid B$ by Exercise 3.46, and such equivalence is preserved by the context $(\nu in, out)((P_2 \mid -)) \mid C_2)$, hence also $(\nu in, out)((P_2 \mid B_0)) \mid C_2) \sim (\nu in, out)((P_2 \mid (B \mid B)) \mid C_2)$. The thesis then follows by the following two facts: (*i*) $A \sim p$ whenever $A \stackrel{def}{=} p$, as observed in Remark 4.2; (*ii*) transitivity of $\sim$, as proved in Proposition 2.8.

[12] Even for this case, an alternative proof may be based on the fact that $\approx$ is a *congruence* for the CCS operators of parallel composition and restriction, as we will prove in Theorem 4.3; as a matter of fact, we have that $B_0 \approx Buf$ by Exercise 3.46, and such equivalence is preserved by the context $(\nu in, out)((P_2 \mid -)) \mid C_2)$, hence also $(\nu in, out)((P_2 \mid B_0)) \mid C_2) \approx (\nu in, out)((P_2 \mid Buf)) \mid C_2)$.

Example 3.9. (**n-position buffer**) For any natural n, an n-position buffer can be specified as follows:

$$B_0 \stackrel{def}{=} in.B_1$$
$$B_i \stackrel{def}{=} in.B_{i+1} + \overline{out}.B_{i-1} \quad \text{for } 0 < i < n$$
$$B_n \stackrel{def}{=} \overline{out}.B_{n-1}$$

A parallel implementation of this n-position buffer can be provided by means of multiple copies in parallel of the one-position buffer B of Exercise 3.46. We use B^k as a shorthand to denote the parallel process term $\Pi_{i=1}^k B$ (which is $\mathbf{0}$ when $k = 0$).

We prove that, for any n, the specification B_0 and the parallel implementation B^n are strongly bisimilar. Let us consider the relation

$$R = \{(B_k, (\Pi_{i=1}^k \overline{out}.B) \,|\, B^{n-k}) \;\mid\; 0 \le k \le n\}$$

It is not too difficult to show that R is a strong bisimulation up to $\sim$ (see Definition 2.15 and Proposition 2.10 for correctness of this proof principle), because we assume that parallel composition is associative[13] and commutative (with nil as neutral element), which indeed holds for strong bisimulation, as we will see in Section 4.1.1, Proposition 4.2.[14]

First, observe that for $k = 0$, the pair in R is $(B_0, \mathbf{0} \,|\, B^n)$. If R is a bisimulation up to $\sim$, then $B_0 \sim \mathbf{0} \,|\, B^n$. As $\mathbf{0} \,|\, B^n \sim B^n$, by transitivity we get $B_0 \sim B^n$. Now, we will prove that R is indeed a strong bisimulation up to $\sim$, i.e., we will prove that:

1. if $B_k \xrightarrow{\alpha} p$ for some action α and some process p, then there are some q and q' such that $(\Pi_{i=1}^k \overline{out}.B) \,|\, B^{n-k} \xrightarrow{\alpha} q$ with $q' \sim q$ and $(p, q') \in R$ (this is enough for application of the up-to technique, as $\sim$ is reflexive, hence $p \sim p$). And, symmetrically,
2. if $(\Pi_{i=1}^k \overline{out}.B) \,|\, B^{n-k} \xrightarrow{\alpha} q$ for some action α and some process q, then there are some processes q' and p such that $q \sim q'$, $B_k \xrightarrow{\alpha} p$ and $(p, q') \in R$.

(Case 1) If $k < n$, process B_k can do $B_k \xrightarrow{in} B_{k+1}$ and $(\Pi_{i=1}^k \overline{out}.B) \,|\, B^{n-k}$ can respond with *in* reaching $(\Pi_{i=1}^k \overline{out}.B) \,|\, \overline{out}.B \,|\, B^{n-(k+1)}$, which is strongly bisimilar to $(\Pi_{i=1}^{k+1} \overline{out}.B) \,|\, B^{n-(k+1)}$ and the pair $(B_{k+1}, (\Pi_{i=1}^{k+1} \overline{out}.B) \,|\, B^{n-(k+1)})$ is in R.

If $k > 0$, B_k can also do $B_k \xrightarrow{\overline{out}} B_{k-1}$ and $(\Pi_{i=1}^k \overline{out}.B) \,|\, B^{n-k}$ can respond by reaching $(\Pi_{i=1}^{k-1} \overline{out}.B \,|\, B) \,|\, B^{n-k}$, which is strongly bisimilar to $(\Pi_{i=1}^{k-1} \overline{out}.B) \,|\, B^{n-(k-1)}$ and the pair $(B_{k-1}, (\Pi_{i=1}^{k-1} \overline{out}.B) \,|\, B^{n-(k-1)})$ is in R.

(Case 2) Besides the *in*-labeled transition already considered in the previous case, $(\Pi_{i=1}^k \overline{out}.B) \,|\, B^{n-k}$ has other $n - (k-1)$ *in*-labeled transitions, all reaching states that are strongly bisimilar to $(\Pi_{i=1}^{k+1} \overline{out}.B) \,|\, B^{n-(k+1)}$ by associativity and commutativity of parallel composition. For instance, one of these states is

[13] Actually, when using the notation $\Pi_{i=1}^k B$, we are already assuming associativity of parallel composition implicitly.

[14] To be precise, we are also using the fact that $\sim$ is a congruence for parallel composition, i.e., if $p \sim q$, then $p \,|\, r \sim q \,|\, r$ for all r. This result is proved in Theorem 4.1.

$\Pi_{i=1}^{k}\overline{out}.B \,|\, B^j \,|\, \overline{out}.B \,|\, B^{n-(k+j+1)}$ for $1 \leq j,k \leq n$, $j+k < n$. To any of these transitions, B_k responds with $B_k \xrightarrow{in} B_{k+1}$ and $(B_{k+1}, (\Pi_{i=1}^{k+1}\overline{out}.B) \,|\, B^{n-(k+1)})$ is in R.

Similarly, if $k > 0$, besides the $\overline{out}$-labeled transition discussed above, process $(\Pi_{i=1}^{k}\overline{out}.B) \,|\, B^{n-k}$ has other $k-1$ $\overline{out}$-labeled transitions, all reaching states that are strongly bisimilar to $(\Pi_{i=1}^{k-1}\overline{out}.B) \,|\, B^{n-(k-1)}$. To any of these, B_k may respond with $B_k \xrightarrow{\overline{out}} B_{k-1}$ and $(B_{k-1}, (\Pi_{i=1}^{k-1}\overline{out}.B) \,|\, B^{n-(k-1)})$ is in R. And this completes the proof. □

Example 3.10. **(Linking operator)** By $p \frown q$ we denote the process obtained by *linking* process p with process q. In terms of flow-graph representation, the operation of linking p with q is similar to the parallel composition of p and q, but with the side effect that some non-complementary ports of the two are connected together. Its definition is not completely formal, as the actual choice of such ports depends on the definition of the two processes p and q. Assuming that the non-complementary ports to be connected are a in p and b in q, the linking (or *pipelining*) of p and q is defined as a derived operator in CCS as follows:

$$p \frown q = (\nu d)(p\{d/a\} \,|\, q\{d/b\}),$$

where d is a new name occurring neither in p nor in q, and the application of a substitution, say $\{d/a\}$, to process p (defined formally in Section 4.1.2, Definition 4.4) states that each occurrence of a ($\overline{a}$) in p is to be replaced by d ($\overline{d}$).

As an example, let us consider again process *Buf* of Exercise 3.46. We can define, equivalently, $Buf \stackrel{def}{=} B \frown B$, where $B \stackrel{def}{=} in.\overline{out}.B$ is the one-position buffer of Exercise 3.46 and the linking operator connects the *out* port of the left buffer to the *in* port the right buffer. As a matter of fact, if we choose d as the new action, $B \frown B$ is the same as $(\nu d)(B\{d/out\} \,|\, B\{d/in\})$, where the effect of applying the substitution, say $\{d/out\}$, to constant B is the definition of a new constant $B_{\{d/out\}}$ where the substitution is applied to its body: $B_{\{d/out\}} \stackrel{def}{=} in.\overline{d}.B_{\{d/out\}}$ and $B_{\{d/in\}} \stackrel{def}{=} d.\overline{out}.B_{\{d/in\}}$.

The definition of the linking operator can be generalized to vectors $(a_1, \ldots, a_n)$, $(b_1, \ldots, b_n)$ and $(d_1, \ldots, d_n)$ of actions as follows:
$p \frown q = (\nu d_1, \ldots, d_n)(p\{d_i/a_i\}_{1 \leq i \leq n} \,|\, q\{d_i/b_i\}_{1 \leq i \leq n})$. □

Example 3.11. **(Pipeline implementation of a n-position buffer)** As illustrated in Exercise 3.46, there are two natural implementations of a sequential specification of a buffer: the parallel implementation and the pipeline one. Example 3.9 discusses the parallel implementation of a n-position buffer. Here, a description of its pipeline implementation is given by means of the linking operator.

Some notation first. PB is used to denote either term B or term $\overline{out}.B$. Term $PB_{\pi}^{(n,k)}$ is used to denote the term

$$PB_1 \frown PB_2 \frown \ldots \frown PB_n$$

where the linking operator applied to two terms PB_i and PB_{i+1} (for $i = 1, \ldots n-1$) connects the *out* of PB_i to the *in* of PB_{i+1}, and where π is a binary vector of length n with only k elements set to 1: $\pi(i) = 1$ if PB_i is $\overline{out}.B$, $\pi(i) = 0$ if PB_i is B. If $\pi(i) = 1$ and $\pi(i+1) = 0$, for $0 \leq i < n$, we denote by $\pi[i+1/i]$ the vector $(\pi(1), \ldots, \pi(i-1), 0, 1, \pi(i+1), \ldots, \pi(n))$. Note that $PB_0^{(n,0)}$ — where the subscript 0 stands for a vector of 0's — is composed only of B components, and $PB_1^{(n,n)}$ — where the subscript 1 stands for a vector of 1's — is composed only of $\overline{out}.B$ components.

We would like to prove that, for any n, the sequential, finite-state specification B_0 of Example 3.9 and the regular, pipelined implementation $PB_0^{(n,0)}$ are weakly bisimilar. We will sketch a proof of this fact, where, for simplicity's sake, we implicitly assume associativity of the linking operator (up to strong bisimulation, formally proved in Example 4.4 of Section 4.1.2), so that we take the liberty of using our $PB_\pi^{(n,k)}$ notation, independently of the actual association of the n linking operators. Let us consider the relation

$$R = \{(B_k, PB_\pi^{(n,k)}) \mid 0 \leq k \leq n \text{ and } \pi \text{ is of length } n \text{ with } k \text{ elements set to } 1\}$$

It is not obvious that R is a weak bisimulation.[15] Indeed, we will prove that:

1. if $B_k \xrightarrow{\alpha} p$ for some action α and some process p, then there is some q such that $PB_\pi^{(n,k)} \stackrel{\alpha}{\Longrightarrow} q$ with $(p,q) \in R$. And, symmetrically,
2. if $PB_\pi^{(n,k)} \xrightarrow{\mu} q$ for some action μ and some process q, then there is some process p such that $B_k \xrightarrow{\mu} p$ and $(p,q) \in R$, or, if $\mu = \tau$, $(B_k, q) \in R$.

Before checking this, we need some auxiliary results.

Observe that $\overline{out}.B \frown B \xrightarrow{\tau} B \frown \overline{out}.B$, because the $\overline{out}$ transition of the left component is synchronized with the *in* transition of the right component. This means that, if $0 < k < n$,

$$PB_\pi^{(n,k)} \xrightarrow{\tau} PB_{\pi[i+1/i]}^{(n,k)}$$

assuming that the involved processes are of index i and $i+1$, respectively. This can be generalized in two ways:

(R_1) if $n \geq 1$, $PB_1^{(n,n)} \frown B \stackrel{\tau}{\Longrightarrow} B \frown PB_1^{(n,n)}$;

(R_2) similarly, if $n \geq 1$, $\overline{out}.B \frown PB_0^{(n,0)} \stackrel{\tau}{\Longrightarrow} PB_0^{(n,0)} \frown \overline{out}.B$.

We are now ready to check that R is a weak bisimulation. Consider $(B_k, PB_\pi^{(n,k)})$.

(Case 1) If $k < n$, process B_k can do $B_k \xrightarrow{in} B_{k+1}$. Let j be the least index such that $\pi(j) = 0$. If $j = 1$, then $PB_\pi^{(n,k)} \xrightarrow{in} PB_{\pi'}^{(n,k+1)}$ where π' differs from π only

[15] If we were not using the simplification about associativity of the linking operator, we would prove this relation R to be a weak bisimulation up to $\approx$ (see Definition 2.22 and Exercise 2.74 for the correctness of this proof technique). Moreover, in such a case, we would also need the fact that $\sim$ is a congruence for the linking operator, as discussed in Exercise 4.27.

for index 1, now set to 1. Otherwise (i.e., if $j > 1$), $PB_{\pi}^{(n,k)}$ can be represented as $PB_1^{(j-1,j-1)} \frown B \frown PB_{\pi_2}^{(n-j,k-(j-1))}$. By observation (R_1)

$$PB_1^{(j-1,j-1)} \frown B \frown PB_{\pi_2}^{(n-j,k-(j-1))} \stackrel{\tau}{\Longrightarrow} B \frown PB_1^{(j-1,j-1)} \frown PB_{\pi_2}^{(n-j,k-(j-1))} \stackrel{in}{\longrightarrow} PB_1^{(j,j)} \frown PB_{\pi_2}^{(n-j,k-(j-1))},$$

which can be represented as $PB_{\pi'}^{(n,k+1)}$, and the pair $(B_{k+1}, PB_{\pi'}^{(n,k+1)})$ is in R because the number of 1's in π' is one more than in π.
If $k > 0$, B_k can also do $B_k \xrightarrow{\overline{out}} B_{k-1}$. Let j be the greatest index such that $\pi(j) = 1$. If $j = n$, then $PB_{\pi}^{(n,k)} \xrightarrow{\overline{out}} PB_{\pi'}^{(n,k-1)}$, where π' differs from π only for index n, now set to 0. Otherwise, $PB_{\pi}^{(n,k)}$ can be seen as $PB_{\pi_1}^{(n-j,k-1)} \frown \overline{out}.B \frown PB_0^{(j-1,0)}$. By observation (R_2),

$$PB_{\pi_1}^{(n-j,k-1)} \frown \overline{out}.B \frown PB_0^{(j-1,0)} \stackrel{\tau}{\Longrightarrow} PB_{\pi_1}^{(n-j,k-1)} \frown PB_0^{(j-1,0)} \frown \overline{out}.B \xrightarrow{\overline{out}} PB_{\pi_1}^{(n-j,k-1)} \frown PB_0^{(j,0)},$$

which can be represented as $PB_{\pi'}^{(n,k-1)}$ and the pair $(B_{k-1}, PB_{\pi'}^{(n,k-1)})$ is in R because the number of 0's in π' is one more than in π.

(Case 2) If $k < n$ and $j = 1$ is the least index such that $\pi(j) = 0$, then $PB_{\pi}^{(n,k)}$ can do $PB_{\pi}^{(n,k)} \xrightarrow{in} PB_{\pi'}^{(n,k+1)}$ where π' differs from π only on the first index, now set to 1. B_k can respond with $B_k \xrightarrow{in} B_{k+1}$ and the pair $(B_{k+1}, PB_{\pi'}^{(n,k+1)})$ is in R.
If $k > 0$ and $j = n$ is the greatest index such that $\pi(j) = 1$, then we have that $PB_{\pi}^{(n,k)} \xrightarrow{\overline{out}} PB_{\pi'}^{(n,k-1)}$, where π' differs from π only on the last index, now set to 0. B_k can respond with $B_k \xrightarrow{\overline{out}} B_{k-1}$ and the pair $(B_{k-1}, PB_{\pi'}^{(n,k-1)})$ is in R.
Otherwise, let i be any index such that $\pi(i) = 1$ and $\pi(i+1) = 0$. In such a case, $PB_{\pi}^{(n,k)} \xrightarrow{\tau} PB_{\pi[i+i/i]}^{(n,k)}$. B_k can respond by idling and $(B_k, PB_{\pi[i+1/i]}^{(n,k)}) \in R$. This completes the proof, because no further transitions are possible from $PB_{\pi}^{(n,k)}$.

The parallel buffer B^n of Example 3.9 does not respect, in delivery, the order of arrival. However, it turns out that B^n and the pipeline buffer $PB_0^{(n,0)}$ are weakly bisimilar. In this simplified setting where the *in*/*out* actions do not carry values, we cannot observe any difference between the two implementations. However, we will see in Section 3.6, that the parallel (called also *bag*) buffer and the pipeline buffer are not equivalent in general, when data are explicitly handled. □

Exercise 3.50. Prove that the relation R of Example 3.11 is also a branching bisimulation. □

3.4.4 BPP: Basic Parallel Processes

The CCS subcalculus of *Basic Parallel Processes* (BPP for short) is generated by the following abstract syntax:

$$p ::= \mathbf{0} \mid \mu.q \mid p+p$$
$$q ::= p \mid q|q \mid C$$

where, as usual, we assume that process constants are always defined and guarded, that *Const(q)* is finite, and, additionally, that rule (Com) in Table 3.1 is not used (no synchronization). As the body of a constant is in syntactic category q, it may happen that parallel composition | occurs inside the body of recursively defined constants; hence, a BPP process may generate an infinite-state LTS, as illustrated below in Example 3.12. Under the assumptions listed in Remark 3.5, sometimes the syntax of BPP processes is more succinctly given as:

$$p ::= \Sigma_{j \in J} \mu_j.p_j \mid C \mid p|p$$

From a syntactical point of view, BPP is not a superclass of regular processes, because restriction is not allowed (as well as synchronization). However, from a semantical point of view, BPP is more general. On the one hand, for any regular process p there exists a BPP process q with associated LTS isomorphic to that of p, because a regular process generates a finite-state LTS and finite-state CCS processes are a subclass of BPP. On the other hand, the reverse of this implication does not hold: there are BPP processes that generate infinite-state LTSs. As a typical instance, consider the following example of a semi-counter [AILS07, San12], in turn inspired by the classic example of an unbounded bag buffer [BK84b, Fok00], described in the subsequent Exercise 3.55.

Example 3.12. (**Semi-counter**) Recalling Example 3.1, a semi-counter, i.e. a counter that cannot test for zero (see Section 3.4.6 for a real counter), can be represented by means of an unbounded number of constants $SCount_i$ for $i = 0, 1, \ldots$:

$$SCount_0 \stackrel{def}{=} inc.SCount_1$$
$$SCount_n \stackrel{def}{=} inc.SCount_{n+1} + dec.SCount_{n-1} \quad n > 0$$

The reader can easily check that the LTS for $SCount_0$ is isomorphic to the one in Figure 2.7(b), where process $SCount_i$ is mapped to state q_i, for any $i \in \mathbb{N}$. Observe that, for any $i \in \mathbb{N}$, the longest trace composed only of occurrences of action *dec* that $SCount_i$ can perform is of length i; such a trace is denoted by dec^i (where $dec^0 = \varepsilon$, $dec^{i+1} = dec\ dec^i$). Therefore, $SCount_i$ cannot be trace equivalent to any $SCount_j$ for $j \neq i$ because if, say, $j > i$ then trace dec^j can be executed by $SCount_j$, but not by $SCount_i$. This means that, for any $i \in \mathbb{N}$, $SCount_i$ cannot be trace equivalent to any other constant $SCount_j$ with a different index; hence, we can conclude that no finite-state CCS process q can be trace equivalent to $SCount_0$.

We want to show that there exists a simple BPP process, defined by means of a single constant

$$SC \stackrel{def}{=} inc.(SC \mid dec.\mathbf{0})$$

that is bisimulation equivalent to $SCount_0$, hence proving that the class of BPP is strictly more expressive than the class of finite-state CCS processes. First, an intermezzo:

Exercise 3.51. Draw the initial fragment of the infinite LTS for SC. Argue that the LTS for SC is not isomorphic to the LTS in Figure 2.7(b). Show that there are infinitely many pairs of bisimulation equivalent reachable states (e.g., SC and $SC \,|\, \mathbf{0}$ as well as $(SC \,|\, \mathbf{0}) \,|\, dec.\mathbf{0}$, $(SC \,|\, dec.\mathbf{0}) \,|\, \mathbf{0}$ and $SC \,|\, dec.\mathbf{0}$). Argue that the resulting LTS is not boundedly-branching, even if it is finitely-branching.[16] □

Now we prove that $SCount_0$ and SC are strongly bisimilar. Consider the relation

$$R = \{(SCount_n, SC \,|\, \Pi_{i=1}^{n} dec.\mathbf{0}) \mid n \geq 0\}.$$

It is not difficult to see that it is a strong bisimulation up to $\sim$, where we take advantage of the fact that parallel composition is associative,[17] commutative, with $\mathbf{0}$ as neutral element, with respect to strong bisimilarity $\sim$. This is proved in Section 4.1.1, Proposition 4.2. Moreover, we are also using the fact that $\sim$ is a congruence for parallel composition, i.e., if $p \sim q$, then $p \,|\, r \sim q \,|\, r$ for all r; this is proved in Theorem 4.1. (For concrete details, see Example 4.2.)

First, observe that for $n = 0$, the pair in R is $(SCount_0, SC \,|\, \mathbf{0})$. If R is a bisimulation up to $\sim$, then $SCount_0 \sim SC \,|\, \mathbf{0}$. As $SC \,|\, \mathbf{0} \sim SC$, by transitivity we get our expected result: $SCount_0 \sim SC$. Now, let us prove that R is indeed a strong bisimulation up to $\sim$.

Assume that $SCount_n \xrightarrow{\alpha} q$. Then *either* $\alpha = inc$ and $q = SCount_{n+1}$, *or* $n > 0$, $\alpha = dec$ and $q = SCount_{n-1}$. In the former case, the matching transition is

$$SC \,|\, \Pi_{i=1}^{n} dec.\mathbf{0} \xrightarrow{inc} (SC \,|\, dec.\mathbf{0}) \,|\, \Pi_{i=1}^{n} dec.\mathbf{0},$$

where the reached state is bisimilar to $SC \,|\, \Pi_{i=1}^{n+1} dec.\mathbf{0}$, and the pair $(SCount_{n+1}, SC \,|\, \Pi_{i=1}^{n+1} dec.\mathbf{0}) \in R$. In the latter case, (one of) the matching dec transition(s) starts from $SC \,|\, \Pi_{i=1}^{n} dec.\mathbf{0}$ and reaches $(SC \,|\, \Pi_{i=1}^{n-1} dec.\mathbf{0}) \,|\, \mathbf{0}$, which is strongly bisimilar to $SC \,|\, \Pi_{i=1}^{n-1} dec.\mathbf{0}$, and the pair $(SCount_{n-1}, SC \,|\, \Pi_{i=1}^{n-1} dec.\mathbf{0}) \in R$.

Assume now $SC \,|\, \Pi_{i=1}^{n} dec.\mathbf{0} \xrightarrow{\alpha} p$. Then, by inspecting the rules for parallel composition:

1. *Either* $SC \xrightarrow{inc} SC \,|\, dec.\mathbf{0}$ and thus $\alpha = inc$ and process $p = (SC \,|\, dec.\mathbf{0}) \,|\, \Pi_{i=1}^{n} dec.\mathbf{0}$ (which is bisimilar to $SC \,|\, \Pi_{i=1}^{n+1} dec.\mathbf{0}$). In such a case, the matching transition is $SCount_n \xrightarrow{inc} SCount_{n+1}$, and the pair $(SCount_{n+1}, SC \,|\, \Pi_{i=1}^{n+1} dec.\mathbf{0})$ is in R.

---

16 A BPP process p with infinitely many states is often unboundedly-branching; however, this is not always the case; for instance, $A \stackrel{def}{=} a.(A \,|\, \mathbf{0})$ is such that A has infinitely many states, but A is boundedly-branching.

17 To be precise, when using the notation $\Pi_{i=1}^{n} dec.\mathbf{0}$, we are already assuming associativity of parallel composition.

2. Or $n > 0$, $\alpha = dec$ and p is one of the following three terms: $(SC\,|\,\mathbf{0})\,|\,\Pi_{i=1}^{n-1} dec.\mathbf{0}$, $(SC\,|\,\Pi_{i=1}^{n-1} dec.\mathbf{0})\,|\,\mathbf{0}$ or $(((SC\,|\,\Pi_{i=1}^{n-k} dec.\mathbf{0})\,|\,\mathbf{0})\,|\,\Pi_{i=1}^{k-1} dec.\mathbf{0})$ for some $1 \leq k < n$. In any case, p is strongly bisimilar to $SC\,|\,\Pi_{i=1}^{n-1} dec.\mathbf{0}$. The matching transition is $SCount_n \xrightarrow{dec} SCount_{n-1}$, and the pair $(SCount_{n-1}, SC\,|\,\Pi_{i=1}^{n-1} dec.\mathbf{0})$ is in R.

And this completes the proof. So, we have shown that a semi-counter can be represented, up to $\sim$, by a simple BPP process. □

Exercise 3.52. Prove that a sequence $\sigma \in \{inc, dec\}^*$ is a trace in $Tr(SC)$ if and only if, in any prefix σ' of σ, the number of occurrences of dec is smaller than, or equal to, the number of occurrences of inc. If we denote by $\sharp(inc, \sigma)$ the number of occurrences of inc in σ, the requirement above can be formulated as follows: $\sharp(dec, \sigma') \leq \sharp(inc, \sigma')$ for all σ' such that there exist γ with $\sigma = \sigma'\gamma$. □

Exercise 3.53. (Half- and double-semi-counter) (i) Consider the BPP process HSC (half-semi-counter)

$$HSC \stackrel{def}{=} inc.inc.(HSC\,|\,dec.\mathbf{0})$$

Prove that a sequence $\sigma \in \{inc, dec\}^*$ is a trace in $Tr(HSC)$ if and only if, in any prefix σ' of σ, $\sharp(dec, \sigma') \leq \sharp(inc, \sigma')/2$.

(ii) Define a BPP process DSC (double-semi-counter) such that a sequence $\sigma \in \{inc, dec\}^*$ is a trace in $Tr(DSC)$ if and only if, in any prefix σ' of σ, $\sharp(dec, \sigma') \leq 2 \times \sharp(inc, \sigma')$. □

Exercise 3.54. (i) Define a BPP process ABC_1 such that a sequence $\sigma \in \{a, b, c\}^*$ is a trace in $Tr(ABC_1)$ if and only if, in any prefix σ' of σ, $\sharp(c, \sigma') \leq \sharp(b, \sigma') \leq \sharp(a, \sigma')$.

(ii) Define a BPP process ABC_2 such that a sequence $\sigma \in \{a, b, c\}^*$ is a trace in $Tr(ABC_2)$ iff, in any prefix σ' of σ, $\sharp(b, \sigma') \leq \sharp(a, \sigma')$ and $\sharp(c, \sigma') \leq \sharp(a, \sigma')$.

(iii) Define a BPP process ABC_3 such that a sequence $\sigma \in \{a, b, c\}^*$ is a trace in $Tr(ABC_3)$ if and only if, in any prefix σ' of σ, $\sharp(b, \sigma') + \sharp(c, \sigma') \leq \sharp(a, \sigma')$. □

Exercise 3.55. (Unbounded bag buffer) Consider the specification of the unbounded bag buffer UB_0:

$$UB_0 \stackrel{def}{=} in.UB_1$$
$$UB_i \stackrel{def}{=} in.UB_{i+1} + \overline{out}.UB_{i-1} \quad \text{for } 0 < i$$

Draw (the initial fragment of) the associated LTS. Consider the BPP process

$$UB \stackrel{def}{=} in.(UB\,|\,\overline{out}.\mathbf{0})$$

Following the steps in Example 3.12, show that $UB_0 \sim UB$. Process UB is a bag buffer because it does not respect in output the order of insertion of the elements in the buffer, even if in this setting, where channels do not carry values, this aspect is not observable. See Section 3.6 for more accurate descriptions of unbounded buffers; in particular, Example 3.23 and Example 3.25 for an unbounded *fifo* buffer (or queue). □

Definition 3.7. (BPP language) A language $L \subseteq (\mathscr{L} \cup \overline{\mathscr{L}})^*$ is a *BPP language* if there exists a BPP process p such that the set of its weak completed traces is L, i.e., $WCTr(p) = L$. □

As BPP is a superset of finite-state CCS, the class of regular languages is included in the class of BPP languages by Proposition 3.6. However, the class of BPP languages includes also non-regular languages, as the following exercise explains.

Exercise 3.56. (Completed traces may form a non-regular language) Consider the BPP process $A \stackrel{def}{=} a.(A \mid b.\mathbf{0}) + c.0$, discussed in [Ch93, BCMS01]. (i) Show that a sequence $\sigma \in \{a,b,c\}^*$ is a trace in $Tr(A)$ if and only if, in any prefix σ' of σ, $\sharp(b,\sigma') \leq \sharp(a,\sigma')$, $\sharp(c,\sigma') \leq 1$ and the only occurrence of c, if present, is always after the last occurrence of a. (ii) Show also that for any completed trace σ, additionally it holds that $\sharp(b,\sigma) = \sharp(a,\sigma)$. Argue that the set of completed traces $CTr(A)$ is not a *regular language* (see Section 1.3.2). (*Hint:* If $CTr(A)$ were regular, then $CTr(A) \cap a^*cb^*$ should be regular, because the intersection of two regular languages is a regular language [HMU01]. But the resulting set is $L = \{a^kcb^k \mid k \geq 0\}$, which is a typical example of a non-regular language.) □

Example 3.13. **(Completed traces may form a non-context-free language)** It is also possible to show that some BPP language is not a *context-free* language (i.e., it cannot be generated by a context-free grammar). For instance, consider the following process, originally introduced in [BCMS01]:

$$B \stackrel{def}{=} a.(B \mid b.\mathbf{0}) + c.(B \mid d.\mathbf{0}) + e.\mathbf{0}$$

If $CTr(B)$ were a context-free language, then also $CTr(B) \cap a^*c^*b^*d^*e$ would be context-free, as the intersection of a context-fee language with a regular language gives a context-free language [HMU01]. However, such an intersection is the set $\{a^kc^nb^kd^ne \mid k,n \geq 0\}$, which is a well-known example of a context-dependent language.

To complete the picture, there exist context-free languages not definable as the set of completed traces of any BPP process. For instance, in [Ch93, BCMS01] it is proved that the context-free language $L = \{a^kcb^k \mid k \geq 0\}$ is not a BPP language, i.e., there exists no BPP process p such that $WCTr(p) = L$. (See Example 3.15 and Exercise 3.60 for finite-net CCS definitions of variations of this language.) □

The problem of checking bisimulation equivalence over BPP processes is decidable [CHM93] and more recently it has been proved PSPACE-complete [Jan03]. Weak bisimilarity has been proved decidable in some restricted cases, e.g., when one of the two processes is finite-state [JKM01, KM02], but the problem in the general case is still open, even if a conjecture about its decidability has been recently proposed in [CHL11], based on the proof of decidability of branching bisimilarity for *normed* BPP processes[18] outlined there. On the contrary, trace equivalence over BPP is undecidable [Hir93]. For a gentle introduction to this topic, see [AIS12].

Exercise 3.57. Argue that trace equivalence is decidable for deterministic BPP processes, i.e., for those BPP processes generating deterministic LTSs. (*Hint:* Look at Exercise 2.44.) □

3.4.5 Finite-Net CCS

Finite-net CCS processes are generated by the abstract syntax

$$
\begin{aligned}
s &::= \mathbf{0} \mid \mu.t \mid s+s \\
t &::= s \mid t|t \mid C \\
p &::= t \mid (\nu a)p
\end{aligned}
$$

where, as usual, we assume that process constants are always defined and guarded and that $Const(p)$ is finite. It is easy to see that finite-net processes form a superclass of BPP (because they allow for a limited use of restriction and they model synchronization), and are syntactically incomparable with regular processes because, on the one hand, finite-net processes allow for the use of parallel composition inside the body of recursively defined constants, while, on the other hand, regular processes can mix the operators of restriction and parallel composition. However, semantically, regular processes are included in BPP, hence also in finite-net processes.

Under the assumptions listed in Remark 3.5, sometimes the syntax of finite-net CCS processes is more succinctly given as:

$$
\begin{aligned}
t &::= \Sigma_{j\in J}\mu_j.t_j \mid t|t \mid C \\
p &::= t \mid (\nu a)p
\end{aligned}
$$

Remark 3.13. **(Notational extension)** Note that the body of a constant C cannot contain occurrences of the restriction operator. However, for convenience, in order to give a name to every process, we will sometimes make use of constants that are non-recursively defined over the general class of finite-net processes. For instance, we consider acceptable the constant definition $A \stackrel{def}{=} (\nu a)(p\,|\,q)$ when A does not belong to $Const(p) \cup Const(q)$. This convention is extensively adopted in the following, e.g., for constant *HSC2* of Exercise 3.58 or constant *2PSC* of Example 3.15. □

This class is interesting because it has been shown in [Gor15, GV10] that its processes are in close connections with finite P/T Petri nets [Pet81, Rei85, DesRei98] with the distinguishing feature that net transitions have either one input arc, or two input arcs (but in such a case the transition is labeled τ). Indeed, in one direction it can be proved that a finite-net CCS process generates a finite P/T Petri net of that

[18] A BPP process is normed if each of its reachable states may terminate, i.e., may reach a deadlock.

form. Conversely, for any finite P/T net N of that form, we can find a finite-net CCS process p such that the net of p is isomorphic to N. Moreover, in [Gor15, GV10] a generalization of this result is presented: for any finite P/T net N, we can find a finite-net process p in a proper extension of CCS with multiway synchronization, called Multi-CCS (described in Chapter 6), such that the net of p is isomorphic to N. Hence, finite-net Multi-CCS processes are for Petri nets as fundamental as finite-state CCS processes are for LTSs.

Even if bisimulation equivalence is undecidable in general over finite-net CCS processes (as proved in Section 3.5.4), one might take advantage of the many decidable properties over finite P/T Petri nets that are hence also decidable for finite-net CCS processes, via the net semantics outlined in [Gor15, GV10]. Among these properties we mention the following:

- *Reachability*: given two finite-net processes p and q, we can decide if q is reachable from p [May81, May84, Kos82, Ler11]. This is important when checking some safety condition that is invalid at some erroneous state q.
- *Strong bisimilarity with a finite-state system*: given a finite-state process p and a finite-net process q, we can decide if the two are strongly bisimilar [JM95]. This problem is interesting because it enables us to perform equivalence-checking between the complex bahavior of an infinite-state implementation and its finite-state specification.
- *Strong regularity*: given a finite-net process q, we can decide if there exists a finite-state process p that is strongly bisimilar to q [JE96].

All these properties are difficult to check in practice, as they are at least exponential. The interested reader may find more details in [EM94, Esp98, Srba].

Exercise 3.58. (Half-semi-counter, again) Continuing Exercise 3.53, consider this new variant of the half-semi-counter *HSC2* defined as

$$\begin{aligned} HSC2 &\stackrel{def}{=} (\nu c)B \\ B &\stackrel{def}{=} inc.(B \,|\, (c.dec.\mathbf{0} + \overline{c}.\mathbf{0})) \end{aligned}$$

where, instead of two occurrences of *inc* being explicitly used, one restriction name is used to force a self-synchronization of two identical instances of $c.dec.\mathbf{0} + \overline{c}.\mathbf{0}$.

Show that the BPP process *HSC* and the finite-net process *HSC2* are weakly bisimilar, by checking that the relation

$$\begin{aligned} R = \{&(HSC, HSC2), (HSC, (\nu c)B)\} \cup \{(HSC \,|\, \Pi_{i=1}^{n} dec.\mathbf{0}, \\ &(\nu c)(B \,|\, \Pi_{i=1}^{2k}(c.dec.\mathbf{0} + \overline{c}.\mathbf{0}) \,|\, \Pi_{j=1}^{n-k} dec.\mathbf{0})) \;\big|\; 0 \le k \le n\} \cup \\ \{&(inc.(HSC \,|\, dec.\mathbf{0}) \,|\, \Pi_{i=1}^{n} dec.\mathbf{0}, \\ &(\nu c)(B \,|\, \Pi_{i=1}^{2k+1}(c.dec.\mathbf{0} + \overline{c}.\mathbf{0}) \,|\, \Pi_{j=1}^{n-k} dec.\mathbf{0})) \;\big|\; 0 \le k \le n\} \end{aligned}$$

is a weak bisimulation up to $\approx$. (*Hint:* Besides the laws for parallel composition holding for strong bisimulation equivalence $\sim$ — associativity, commutativity and nil as a neutral element — proved in Proposition 4.2, you also need to know that $\sim$ is a congruence for parallel composition and restriction, as proved in Theorem 4.1. For instance, as $B \,|\, \mathbf{0} \sim B$, it also holds that $(\nu c)(B \,|\, \mathbf{0}) \sim (\nu c)B$ by congruence.) □

Example 3.14. **(Unbounded producer-consumer)** We can generalize the producer-consumer system we have discussed in Example 3.8, by defining an unbounded, BPP producer $Pr = produce.(\overline{send}.\mathbf{0} \mid Pr)$, the usual consumer $C_1 = send.consume.C_1$ and the whole finite-net system $UPC \stackrel{def}{=} (\nu send)(Pr \mid C_1)$. □

Exercise 3.59. Consider process *UPC* of Example 3.14. (*i*) Draw the initial portion of its infinite-state LTS. (*ii*) Argue that *UPC* is not weakly bisimilar to *2PC* of Example 3.8. (*iii*) Consider the unbounded, BPP buffer *UB*, defined in Exercise 3.55. Then, taking the producer P_2 and the consumer C_2 as defined in Exercise 3.48, show that $PUBC \stackrel{def}{=} (\nu in, out)((P_2 \mid UB) \mid C_2)$ is weakly bisimilar to *UPC*. □

Definition 3.8. (Finite-net CCS language) A language $L \subseteq (\mathscr{L} \cup \overline{\mathscr{L}})^*$ is a *finite-net CCS language* if there exists a finite-net CCS process p such that the set of its weak completed traces is L, i.e., $WCTr(p) = L$. □

Example 3.15. **(Two-phase semi-counter)** A two-phase semi-counter is a process that can execute the increment action *inc* an unlimited number of times, followed by an equal number of occurrences of the decrement action *dec*. The two kinds of action cannot intertwine, as with the semi-counter *SC* of Example 3.12, and this explains why this semi-counter is called *two-phase*. The specification is as follows:

$$
\begin{aligned}
2PSC &\stackrel{def}{=} (\nu d)INC \\
INC &\stackrel{def}{=} inc.(INC \mid d.dec.\mathbf{0}) + \tau.DEC \\
DEC &\stackrel{def}{=} \overline{d}.DEC
\end{aligned}
$$

The set $WCTr(2PSC)$ of the weak completed traces of $2PSC$ is composed of all the traces of the form $inc^n dec^n$ for any $n \in \mathbb{N}$ (when $n = 0$, $inc^0 dec^0$ is meant to be the empty trace ε). This is a typical example of a *context-free* language. □

Exercise 3.60. Elaborate on the specification of the two-phase semi-counter above in order to produce a finite-net CCS process p whose associated language is $L = \{a^n c b^n \mid n \geq 0\}$. It has been proved in [Ch93, BCMS01] that no BPP process q is such that $WCTr(q) = WCTr(p)$, hence p (as well as $2PSC$ above) is a typical representative of the class of finite-net CCS processes. □

Exercise 3.61. Define a finite-net CCS process whose weak completed traces are of the form $a^n w$ for any $n \in \mathbb{N}$, where w is a sequence of length $2n$ composed of occurrences of b and c such that in any prefix of w the number of b's is greater or equal to the number of c's. □

The class of finite-net CCS languages includes the class of BBP languages, as BBP is a subset of finite-net CCS. However, it does not include all the context-free languages. It has been proved in [Pet81] that the typical context-free language $L = \{ww^R \mid w \in \{a,b\}^*\}$, where w^R means the reverse of w,[19] cannot be represented by means of a finite P/T Petri net. Since finite-net CCS processes may originate,

[19] The formal definition is as follows: $\varepsilon^R = \varepsilon$ and $(\alpha w)^R = w^R \alpha$.

via the semantics in [Gor15, GV10], only finite P/T Petri nets, then they cannot represent language L. (Exercise 3.64 describes a finitary CCS process whose set of weak completed traces is L.) Nonetheless, the class of finite-net CCS languages includes also context-dependent languages, as the following example shows.

Example 3.16. (**The language $a^n b^m c^m$ with $0 \le m \le n$**) A finite-net CCS process whose weak completed traces are of the form $a^n b^m c^m$, with $0 \le m \le n$, is

$$\begin{array}{ll} ABC & \stackrel{def}{=} (\nu d, e, f)A \\ A & \stackrel{def}{=} a.(A \,|\, d.b.\overline{e}.\mathbf{0}) + \tau.B \\ B & \stackrel{def}{=} \overline{d}.e.(B \,|\, f.c.\mathbf{0}) + \tau.C \\ C & \stackrel{def}{=} \overline{f}.C \end{array}$$

Observe that first a certain number of occurrences of action a are generated, as well as of subprocesses $d.b.\overline{e}.\mathbf{0}$. Then, when A performs the internal τ-labeled transition and becomes B, a certain number of activations of action b via a synchronization on d can be performed; these synchronizations cannot be more than the number of a's. When each occurrence of action b is performed, a *call-back* synchronization on e is executed, which activates a new instance of B, as well as of $f.c.\mathbf{0}$; hence, the number of processes $f.c.\mathbf{0}$ in parallel is equal to the number of b executed. Finally, when B internally moves to C, perhaps before having completed the activations of all the available occurrences of b, the same number of c can be performed after activation via synchronization on f. □

Exercise 3.62. (i) Specialize Example 3.16 in order to define a finite-net process such that its weak completed traces are of the form $a^n b^n$ with $n \in \mathbb{N}$.

(ii) Generalize Example 3.16 in order to define a finite-net process such that its weak completed traces are of the form $a^n b^m c^k d^k$ with $0 \le k \le m \le n$. □

To complete the picture, it has been proved [Pet81] that all the languages described by finite P/T Petri nets are context-dependent. Hence, also all the finite-net CCS languages are context-dependent. Figure 3.11 summarizes all the relationships among the different classes of languages, where the languages delineated in the figure are as follows:

- L_1 is the set $CTr(A)$ for $A \stackrel{def}{=} a.(A \,|\, b.\mathbf{0}) + c.\mathbf{0}$, as discussed in Exercise 3.56.
- $L_2 = CTr(B)$ for $B \stackrel{def}{=} a.(B \,|\, b.\mathbf{0}) + c.(B \,|\, d.\mathbf{0}) + e.\mathbf{0}$, as discussed in Example 3.13.
- $L_3 = \{a^n c b^n \mid n \ge 0\}$, as discussed in Exercise 3.60 (see also Example 3.15).
- L_4 is the language $\{ww^R \mid w \in \{a,b\}^*\}$, discussed after Exercise 3.61, and realized in finitary CCS in Exercise 3.64.
- L_5 is the context-dependent language $\{a^n b^m c^m \mid 0 \le m \le n\}$ discussed in Example 3.16.

In addition, we mention that Section 5.4.4 introduces a process algebra, called BPA, whose set of representable languages coincides with the class of context-free languages.

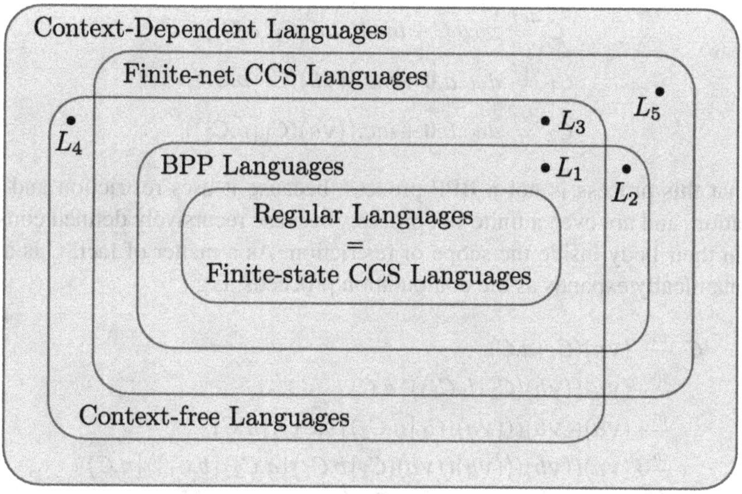

Fig. 3.11 Classification of language classes

3.4.6 Finitary CCS

The largest subclass of CCS processes is given by finitary CCS: the only constraint is that, for any p, $Const(p)$ is finite. From a syntactical point of view, this class is a superclass of finite-net processes, because restriction can occur in the body of recursively defined constants. Also, from a semantic point of view, there are finitary CCS processes that cannot be bisimilar to any finite-net process, as the following example shows.

Example 3.17. **(Counter)** Let us now consider a real counter, i.e., a semi-counter that can also test for zero:

$$Counter_0 \stackrel{def}{=} zero.Counter_0 + inc.Counter_1$$
$$Counter_n \stackrel{def}{=} inc.Counter_{n+1} + dec.Counter_{n-1} \quad n > 0$$

Its definition is not within finitary CCS because it uses infinitely many constants. Its LTS is very similar to the one for the semi-counter (see Figure 2.7(b)), with the only difference being that there is a self-loop transition on the first state q_0 labeled *zero*. Indeed, this process can *test for zero* — a crucial property in models of computation (as we will see in Section 3.5) — while the semi-counter process $SCount_0$ of Example 3.12 cannot. This process can be proved equivalent to a finitary CCS process using only three constants, as proposed in [Tau89]:

$$C \overset{def}{=} zero.C + inc.((\nu a)(C_1 \,|\, a.C))$$
$$C_1 \overset{def}{=} dec.\bar{a}.\mathbf{0} + inc.((\nu b)(C_2 \,|\, b.C_1))$$
$$C_2 \overset{def}{=} dec.\bar{b}.\mathbf{0} + inc.((\nu a)(C_1 \,|\, a.C_2))$$

Note that this process is not a BPP process, because it uses restriction and communication, and not even a finite-net process, because recursively defined constants occur in their body inside the scope of restriction. As a matter of fact, C is a term that dynamically expands as the computation proceeds. E.g.,

$$\begin{aligned}
C &\xrightarrow{inc} (\nu a)(C_1 \,|\, a.C) \\
&\xrightarrow{inc} (\nu a)((\nu b)(C_2 \,|\, b.C_1)) \,|\, a.C) \\
&\xrightarrow{inc} (\nu a)((\nu b)(((\nu a)(C_1 \,|\, a.C_2) \,|\, b.C_1)) \,|\, a.C) \\
&\xrightarrow{inc} (\nu a)((\nu b)(((\nu a)((\nu b)(C_2 \,|\, b.C_1) \,|\, a.C_2) \,|\, b.C_1)) \,|\, a.C) \\
&\xrightarrow{dec} (\nu a)((\nu b)(((\nu a)((\nu b)(\bar{b}.\mathbf{0} \,|\, b.C_1) \,|\, a.C_2) \,|\, b.C_1)) \,|\, a.C) \\
&\xrightarrow{\tau} (\nu a)((\nu b)(((\nu a)((\nu b)(\mathbf{0} \,|\, C_1) \,|\, a.C_2) \,|\, b.C_1)) \,|\, a.C) \\
&\xrightarrow{dec} (\nu a)((\nu b)(((\nu a)((\nu b)(\mathbf{0} \,|\, \bar{a}.\mathbf{0}) \,|\, a.C_2) \,|\, b.C_1)) \,|\, a.C) \\
&\xrightarrow{\tau} (\nu a)((\nu b)(((\nu a)((\nu b)(\mathbf{0} \,|\, \mathbf{0}) \,|\, C_2) \,|\, b.C_1)) \,|\, a.C) \\
&\xrightarrow{inc} (\nu a)((\nu b)(((\nu a)((\nu b)(\mathbf{0} \,|\, \mathbf{0}) \,|\, (\nu a)(C_1 \,|\, a.C_2) \,|\, b.C_1)) \,|\, a.C)
\end{aligned}$$

where the number of restrictions occuring in the term is determined by the number of executed actions *inc*, while the actual number represented by a term is determined by the number of active (alternating) restrictions. It is easy to see that $\mathscr{P}_C$ — the set of states reachable from C — is an infinite set. The initial portion of the LTS is depicted in Figure 3.12, where the dashed lines indicate the two connected states are weakly bisimilar.

We prove that $Counter_0$ and C are weakly bisimilar. Following [Tau89], first we need to define some auxiliary notation. Let $p_0 = C$ and $p_1 = (\nu a)(x \,|\, a.C)$, where x is a place holder for any term to be filled in. Define also $p_{2n} = p_{2n-1}[(\nu b)(x \,|\, b.C_1)/x]$ (for $n > 0$), where $p_n[q/x]$ stands for naive substitution of q for the (unique) occurrence of x in p_n. Similarly, define $p_{2n+1} = p_{2n}[(\nu a)(x \,|\, a.C_2)/x]$ for $n > 0$. Consider the relation R defined as

$$\begin{aligned}
R = \; & \{(C, Counter_0)\} \\
& \cup \{(p_{2n}[C_2/x],\, Counter_{2n}) \mid n > 0\} \\
& \cup \{(p_{2n+1}[C_1/x],\, Counter_{2n+1}) \mid n \geq 0\} \\
& \cup \{(p_{2n+1}[\bar{a}.\mathbf{0}/x],\, Counter_{2n}) \mid n \geq 0\} \\
& \cup \{(p_{2n}[\bar{b}.\mathbf{0}/x],\, Counter_{2n+1}) \mid n > 0\}.
\end{aligned}$$

We prove that R is a weak bisimulation up to $\approx$ (see Definition 2.22 and Exercise 2.74 for the correctness of this proof technique). In most cases, the bisimulation game is played as in the strong case, actually. In a couple of cases only, it needs to resort to weak transitions and the up-to condition.

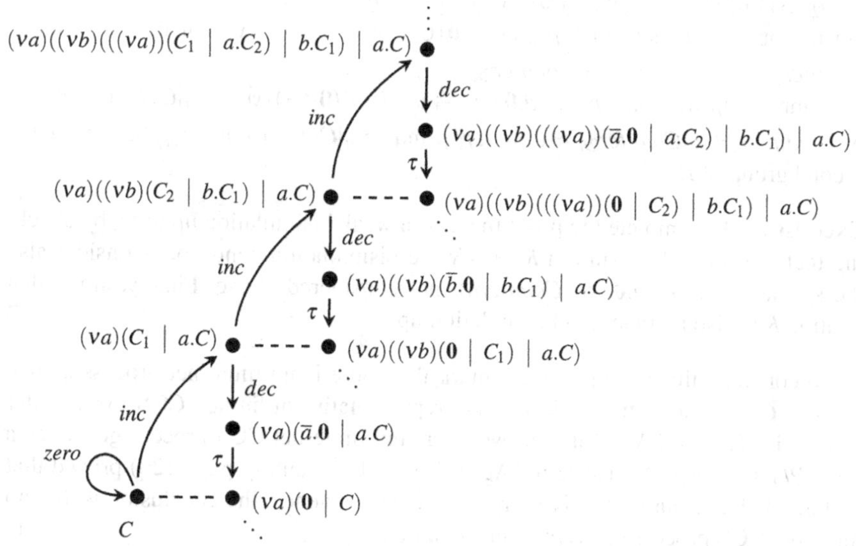

Fig. 3.12 The initial part of the infinite LTS of the counter C. (A dashed line connects weakly bisimilar states.)

First consider pair $(C, Counter_0)$. Transition $C \xrightarrow{zero} C$ is matched strongly (hence also weakly) by transition $Counter_0 \xrightarrow{zero} Counter_0$, and the pair of reached states $(C, Counter_0)$ is in R. Transition $C \xrightarrow{inc} (\nu a)(C_1 \,|\, a.C)$ is matched strongly by transition $Counter_0 \xrightarrow{inc} Counter_1$, and the pair of reached states $(p_1[C_1/x], Counter_1)$ belongs to the third group of R (when $n = 0$). No other transition is possible from C. Symmetrically, the two transitions from $Counter_0$ are the two listed above and are matched by the same transitions from C described above.

Now consider, more generally, a pair $(p_{2n+1}[C_1/x],\ Counter_{2n+1})$ for $n \geq 0$. Term $p_{2n+1}[C_1/x]$ can move only according to the inner occurrence of C_1, which is replacing the place-holder x. Hence we have transitions

$p_{2n+1}[C_1/x] \xrightarrow{inc} p_{2n+1}[(\nu b)(C_2 \,|\, b.C_1)/x] = p_{2n+2}[C_2/x]$ and

$p_{2n+1}[C_1/x] \xrightarrow{dec} p_{2n+1}[\bar{a}.\mathbf{0}/x]$.

In the former case, the matching transition is $Counter_{2n+1} \xrightarrow{inc} Counter_{2n+2}$, and $(p_{2n+2}[C_2/x], Counter_{2n+2})$ belongs to the second group in R. In the latter case, the matching transition is $Counter_{2n+1} \xrightarrow{dec} Counter_{2n}$, and $(p_{2n+1}[\bar{a}.\mathbf{0}/x], Counter_{2n})$ belongs to the fourth group of R. Symmetrically, if the first move comes from $Counter_{2n+1}$.

Consider now a pair of the fourth group $(p_{2n+1}[\bar{a}.\mathbf{0}/x],\ Counter_{2n})$ for $n > 0$. State $p_{2n+1}[\bar{a}.\mathbf{0}/x]$ is in fact $(p_{2n}[(\nu a)(x \,|\, a.C_2)/x])[\bar{a}.\mathbf{0}/x] = p_{2n}[(\nu a)(\bar{a}.\mathbf{0} \,|\, a.C_2)/x]$. From such a state there is only one possible move:

$p_{2n+1}[\bar{a}.\mathbf{0}/x] = p_{2n}[(\nu a)(\bar{a}.\mathbf{0} \,|\, a.C_2)/x] \xrightarrow{\tau} p_{2n}[(\nu a)(\mathbf{0} \,|\, C_2)/x]$.
On the one hand, note that $p_{2n}[(\nu a)(\mathbf{0} \,|\, C_2)/x] \sim p_{2n}[C_2/x]$. On the other hand, $Counter_{2n} \stackrel{\varepsilon}{\Longrightarrow} Counter_{2n} \approx Counter_{2n}$.

Summing up, transition $p_{2n+1}[\bar{a}.\mathbf{0}/x] \xrightarrow{\tau} p_{2n}[(\nu a)(\mathbf{0}|C_2)/x] \sim p_{2n}[C_2/x]$ is matched by $Counter_{2n} \stackrel{\varepsilon}{\Longrightarrow} Counter_{2n} \approx Counter_{2n}$, and $(p_{2n}[C_2/x], Counter_{2n})$ belongs to the second group of R.

Exercise 3.63. Complete the proof that R is a weak bisimulation up to $\approx$, by checking that also the other pairs in R satisfy the bisimulation condition. Consider also the special case $(p_1[\bar{a}.\mathbf{0}/x],\ Counter_0)$, not considered above. Finally, argue that relation R is also a branching bisimulation up to $\approx_{br}$. □

To conclude this example, we remark that there is no finite-net process q such that $q \approx C$, i.e., the counter C is a proper representative of finitary CCS. As a matter of fact, in [Gor15, GV10] it is proved that any finite-net CCS process generates a finite P/T Petri net. Agerwala in [Age75] (see also [Tau89], page 120) proved that no finite P/T net can faithfully represent a counter; hence, the conclusion is that no finite-net CCS process can represent a counter. □

Exercise 3.64. (The language ww^R) Consider $L = \{ww^R \mid w \in \{a,b\}^*\}$, where the reverse w^R of w is defined as follows: $\varepsilon^R = \varepsilon$ and $(\alpha w)^R = w^R \alpha$. We argued in Section 3.4.5 that L cannot be a finite-net CCS language, because [Pet81] shows that L is not a Petri net language. Argue that the finitary CCS process D

$$
\begin{aligned}
D &\stackrel{def}{=} a.((\nu d)(D_1 \,|\, d.a.\mathbf{0})) + b.((\nu d)(D_1 \,|\, d.b.\mathbf{0})) + \tau.\mathbf{0}\\
D_1 &\stackrel{def}{=} a.((\nu e)(D_2 \,|\, e.a.\bar{d}.\mathbf{0})) + b.((\nu e)(D_2 \,|\, e.b.\bar{d}.\mathbf{0})) + \bar{d}.\mathbf{0}\\
D_2 &\stackrel{def}{=} a.((\nu d)(D_1 \,|\, d.a.\bar{e}.\mathbf{0})) + b.((\nu d)(D_1 \,|\, d.b.\bar{e}.\mathbf{0})) + \bar{e}.\mathbf{0}
\end{aligned}
$$

represents language L, i.e., D is such that $WCTr(D) = L$. □

Example 3.18. **(The language $a^n b^n c^n$)** As a further example, consider the following process I: its weak completed traces are of the form $a^n b^n c^n$, for $n \in \mathbb{N}$.

$$
\begin{aligned}
I &\stackrel{def}{=} \tau.\mathbf{0} + a.((\nu l)(((\nu r)(End_r \,|\, B)) \,|\, I_1))\\
I_1 &\stackrel{def}{=} \tau.\bar{l}.End_l + a.((\nu r)(B \,|\, I_2))\\
I_2 &\stackrel{def}{=} \tau.\bar{r}.End_r + a.((\nu l)(B \,|\, I_1))\\
End_l &\stackrel{def}{=} l.\bar{l}.End_l \qquad B \stackrel{def}{=} l.b.\bar{r}.C + r.b.\bar{l}.C\\
End_r &\stackrel{def}{=} r.\bar{r}.End_r \qquad C \stackrel{def}{=} l.c.\bar{r}.D + r.c.\bar{l}.D \qquad D \stackrel{def}{=} \mathbf{0}
\end{aligned}
$$

An instance of what can be done by I is reported in Table 3.2. Note that after the first τ-labeled transition, performed by I_2 (second line), the computation is deterministic: the occurrence of B on the right must receive an input r from its right, and only after that be able to execute its b (fourth line). Then, it should output on l, by synchronizing with the first B on its left, while turning into C. This chain of left-activations of B to C, alternating synchronizations on r and l, goes on till the first

$$
\begin{aligned}
I \xrightarrow{a} ((\nu l)(((\nu r)(End_r \,|\, B)) \,|\, I_1)) \xrightarrow{a} & (\nu l)(((\nu r)(End_r \,|\, B)) \,|\, ((\nu r)(B \,|\, I_2)))) \\
\xrightarrow{\tau} & (\nu l)(((\nu r)(End_r \,|\, B)) \,|\, ((\nu r)(B \,|\, \bar{r}.End_r)))) \\
\xrightarrow{\tau_r} & (\nu l)(((\nu r)(End_r \,|\, B)) \,|\, ((\nu r)(b.\bar{l}.C \,|\, End_r)))) \\
\xrightarrow{b} & (\nu l)(((\nu r)(End_r \,|\, B)) \,|\, ((\nu r)(\bar{l}.C \,|\, End_r)))) \\
\xrightarrow{\tau_l} & (\nu l)(((\nu r)(End_r \,|\, b.\bar{r}.C)) \,|\, ((\nu r)(C \,|\, End_r)))) \\
\xrightarrow{b} & (\nu l)(((\nu r)(End_r \,|\, \bar{r}.C)) \,|\, ((\nu r)(C \,|\, End_r)))) \\
\xrightarrow{\tau_r} & (\nu l)(((\nu r)(\bar{r}.End_r \,|\, C)) \,|\, ((\nu r)(C \,|\, End_r)))) \\
\xRightarrow{\tau_r c} & (\nu l)(((\nu r)(End_r \,|\, \bar{l}.D)) \,|\, ((\nu r)(C \,|\, End_r)))) \\
\xRightarrow{\tau_l c} & (\nu l)(((\nu r)(End_r \,|\, D)) \,|\, ((\nu r)(\bar{r}.D \,|\, End_r))))
\end{aligned}
$$

Table 3.2 An execution of process I

B is encountered. Now a synchronization on r with the inner End_r takes place that enables c; after that, a synchronization on r with the adjacent C (the C that formerly was a B) on its right that now enables it; and so on, in a chain of right-activations of C to D. □

Exercise 3.65. Elaborate on the above example in order to define a process K such that its weak completed traces are of the form $a^n b^n c^n d^n$. □

Summing up, the six subclasses of CCS processes we have discussed in this section can be classified as in Figure 3.6 (Section 3.4). To show the precise inclusions of CCS subcalculi in the syntactic classification, it may be useful to think about the following examples.

Example 3.19. (*i*) Argue that $a.(a.0 \,|\, b.0)$ is a finite CCS process that is not a regular CCS process. (*ii*) Argue that $a.(((\nu b)b.0 \,|\, c.0)$ is a finite CCS process that is not a finite-net CCS process. (*iii*) Argue that $a.\mathbf{0} \,|\, ((\nu a)a.\mathbf{0})$ is a regular (and finite) CCS process that is not a finite-net CCS process. □

Example 3.20. **(Non-finitary CCS process)** As a possible representative of a full CCS process with no equivalent finitary CCS process, consider the family of process constants $\Omega_i = a_i.\Omega_{i+1}$ for $i \in \mathbb{N}$. Process Ω_0 uses infinitely many constants and, moreover, $sort(\Omega_0)$ (see Definition 2.5) is the infinite set $\{a_i \mid i \in \mathbb{N}\}$. In Section 4.1.2, Corollary 4.1 states that, for any finitary CCS process p, the set $sort(p)$ is finite; therefore, it is impossible to find a finitary CCS process equivalent to Ω_0. □

Exercise 3.66. (General representability theorem) Generalize the construction given in the proof of Theorem 3.2 for finite-state LTSs in order to show that any finitely-branching LTS with infinitely many states can be represented by a (full) CCS process term p using prefixing, summation and infinitely many constants. □

3.5 Turing-Completeness

In this section we show how to model faithfully in finitary CCS a well-known *Turing-complete* model of computation, namely *counter machines* (CMs, for short), so that we can conclude that also finitary CCS is Turing-complete. We then show that this has the expected consequence that all the behavioral equivalences we have discussed are undecidable for finitary CCS.[20] This is proved by showing that, if the bahavioral equivalences were decidable for finitary CCS, then we would solve the *halting problem* [Tur36, Dav58] for CMs, a well-known unsolvable problem for Turing-complete formalisms (see Section 1.3.5 for more detail).

3.5.1 Counter Machines

A counter machine (see, e.g., [ER64, Min67]) is a computational model composed of a finite set of *registers* (i.e., counters) that can hold arbitrarily large natural numbers, and of a program, which is a finite set of indexed instructions $\{(1 : I_1), \ldots, (m : I_m)\}$ such as operations on the contents of registers or conditional jumps. To perform a computation, the inputs are provided in registers $r_1, \ldots, r_n$; if no input is provided for a register, then it is supposed to contain the value 0 at the beginning of the computation. The execution of the program begins with the first instruction $(1 : I_1)$ and continues by executing the other instructions in sequence, unless a jump instruction is encountered. The execution stops when an instruction number higher than the length of the program is reached: this happens if the program is executing its last instruction and this instruction does not require a jump, or if the current instruction requires a jump to an instruction number not appearing in the program. If the program terminates, the result of the computation is the content of the registers specified as outputs (usually all the registers).

There are several different varieties of counter machines (CMs, for short), depending on the chosen instruction set. A very simple class, which in [Min67] is shown to be Turing-complete, has a very limited instruction set, composed only of the following two kinds:

- $(i : Inc(r_j))$: increment by 1 the content of the register r_j and execute the next instruction with index $i+1$;
- $(i : DecJump(r_j, s))$: if the content of the register r_j is not 0, then decrease it by 1 and execute the next instruction with index $i+1$, otherwise jump to instruction with index s.

Example 3.21. Assuming that register r_3 holds value 0, the following CM program computes the sum of the values stored in registers r_1 and r_2 — putting the result in register r_1 — by decrementing register r_2 and incrementing register r_1 at each

[20] However, remember that all the equivalences discussed in Chapter 2 are decidable for finite-state processes and that bisimulation equivalence is decidable even for BPP.

iteration, until register r_2 holds 0:

$$\{(1 : DecJump(r_2, 4)), (2 : Inc(r_1)), (3 : DecJump(r_3, 1))\}$$

Note that the third instruction corresponds to an unconditional jump, because register r_3 contains the value 0 at the beginning of the computation and its content is never modified by the program. The program halts when the content of r_2 is 0, so that the first instruction is a jump to instruction 4, not present in the program. □

Exercise 3.67. Continuing Example 3.21, what does CM compute when the value stored in register r_3 is not 0? (*Hint:* Be careful about the initial value of r_2.) □

Definition 3.9. (Counter machines) A *counter machine* (CM) M is defined as a pair $M = (I, n)$, where

$$I = \{(1 : I_1), \dots, (m : I_m)\}$$

is the set of indexed instructions of M, with $|I| = m$, n is the number of registers of M and each instruction I_i is of two possible kinds:

- $I_i = Inc(r_j) \quad (1 \leq j \leq n)$;
- $I_i = DecJump(r_j, s)) \quad (1 \leq j \leq n)$.

An *internal state* (or *configuration*) S of M is defined as a state vector

$$S = (i, v_1, \dots, v_n)$$

with i representing the *program counter* (i.e., the index of the next instruction to be executed) and $v_1, \dots, v_n \in \mathbb{N}$ the current values stored in the n registers of M.

The *initial* internal state (or *initial configuration*) is $(1, v_1, \dots, v_n)$, where 1 signals that the computation starts from the first instruction, and v_i is the input for register r_i, $i = 1, \dots, n$.

The (unlabeled) transition relation $\rightsquigarrow_M \subseteq \mathbf{S} \times \mathbf{S}$ is defined as follows:

$$(i, v_1, \dots, v_n) \rightsquigarrow_M (i', v'_1, \dots, v'_n)$$

whenever

- $I_i = Inc(r_j)$, so that $i' = i + 1$, $v'_j = v_j + 1$, $v'_k = v_k$ for any $k \neq j$; or
- $I_i = DecJump(r_j, s)$ and $v_j > 0$, so that $i' = i + 1$, $v'_j = v_j - 1$, $v'_k = v_k$ for any $k \neq j$; or
- $I_i = DecJump(r_j, s)$ and $v_j = 0$, so that $i' = s$, $v'_k = v_k$ for any $k = 1, \dots, n$.

As the three cases listed above are mutually exclusive, relation $\rightsquigarrow_M$ is *deterministic*, i.e., if $(i, v_1, \dots, v_n) \rightsquigarrow_M (i', v'_1, \dots, v'_n)$ and $(i, v_1, \dots, v_n) \rightsquigarrow_M (i'', v''_1, \dots, v''_n)$, then $i' = i''$ and $v'_j = v''_j$ for $j = 1, \dots n$.

We denote by $\rightsquigarrow^*_M$ the reflexive and transitive closure of relation $\rightsquigarrow_M$. Given the inputs $v_1, \dots, v_n$, a configuration $(i, v'_1, \dots, v'_n)$ is *terminal* if $(1, v_1, \dots, v_n) \rightsquigarrow^*_M (i, v'_1, \dots, v'_n)$ with the program counter $i > m$ (i.e., no further transition is possible from such a configuration); the values $v'_1, \dots, v'_n$ are the outputs of M. If there is no terminal configuration reachable from $(1, v_1, \ldots, v_n)$ (i.e., the computation diverges), we write $(1, v_1, \ldots, v_n) \Uparrow$.

The counter machine M computes the partial function f_M defined as follows: $f_M(v_1, \ldots, v_n) = (v'_1, \ldots, v'_n)$ if $(1, v_1, \ldots, v_n) \rightsquigarrow^*_M (i, v'_1, \ldots, v'_n)$ with the program counter $i > m$, otherwise $f_M(v_1, \ldots, v_n)$ is undefined. □

Exercise 3.68. Consider the CM M of Example 3.21. Compute the finite set of the configurations reachable from the initial one $(1, 3, 2, 0)$, i.e., when register r_1 holds value 3 and r_2 holds value 2 and r_3 is 0. Considering also the cases discussed in Exercise 3.67, what is the partial function $f_M(v_1, v_2, v_3)$ computed by M? □

Exercise 3.69. Consider the CM $M = (I, 1)$ defined by the following program I:

$$\{(1 : Inc(r_1)), (2 : DecJump(r_1, 1))), (3 : DecJump(r_1, 1))\}$$

Compute the finite set of the configurations reachable from the initial one $(1, 0)$. What is the partial function $f_M(v)$ computed by M? □

Exercise 3.70. Define a CM $M = (I, 3)$ such that the partial function $f_M(v_1, v_2, 0)$ it computes (assuming register r_3 holds 0 initially) is

$$f_M(v_1, v_2, 0) = \begin{cases} (v_1 - v_2, 0, 0) & \text{if } v_1 \geq v_2 \\ \text{undefined} & \text{otherwise} \end{cases}$$ □

Remark 3.14. **(Minimal Turing-complete class of counter machines)** In [Min67] (pp. 170-174) it is proved that the *three-counter machine* model (3CMs, for short), i.e., the class of counter machines that use three registers only, can simulate any Turing machine, hence, 3CMs can compute any partial recursive function of one variable. More precisely, a 3CM M starts with the input argument n in a counter, and it leaves the answer $f_M(n)$ in a possibly different counter, if it halts. The way inputs and outputs are treated is very important. As a matter of fact, [Min67] (pp. 255-258) proves that even the two-counter machine model (2CMs) can simulate any Turing machine, but only when the input argument n is appropriately encoded, and analogously for the output. In the same line, [Sch72] proves that a counter machine M can be simulated by a 2CM M', provided that one of the two counters of M' is initialized to the encoding, by Gödelization, of the values stored in the counters of M (and similarly for the outputs). □

3.5.2 Encoding Counter Machines into Finitary CCS

Now we want to define an encoding of any CM into some finitary CCS process. In the light of Remark 3.14, we may restrict our attention to a minimal Turing-complete class of counter machines. To avoid unnecessary complications, we assume to treat inputs and outputs plainly, and so in the following presentation we adopt three-counter machines (3CMs) as our Turing-complete model of computation.

Assuming that the tuple (v_1, v_2, v_3) denotes the initial inputs for the three registers of the CM $M = (I, 3)$, with $|I| = m$, the whole finitary CCS process $CM_{M(v_1,v_2,v_3)}$ is

$$CM_{M(v_1,v_2,v_3)} \stackrel{def}{=} (\nu L)(P_1 \mid \ldots \mid P_m \mid R_1 \mid R_2 \mid R_3 \mid B_{(v_1,v_2,v_3)}),$$

where each constant P_i defines the CM instruction of index i, each constant R_j defines the CM register r_j as a counter, and constant $B_{(v_1,v_2,v_3)}$ is used to perform the bootstrapping of the system by initializing the registers and by activating the first instruction. The set L is $\{p_i \mid \text{index } i \text{ occurs in some instruction}\} \cup \{inc_j, zero_j, dec_j \mid 1 \le j \le 3\}$, where such actions are used by an instruction P_i:

- to test the state of the register R_j, which can be zero (corresponding to an incoming answer on $zero_j$) or nonzero (answer on dec_j): in the latter case the register is decremented;
- to increment the register R_j (output on inc_j);
- to activate instruction i by executing the output $\overline{p}_i$.

Each instruction $(i : I_i)$ corresponds to the definition of a recursive process constant P_i, activated by an output on the channel p_i.

- An increment operation $(i : Inc(r_j))$ is modeled as:

$$P_i \stackrel{def}{=} p_i.P_i' \qquad P_i' \stackrel{def}{=} \overline{inc_j}.\overline{p}_{i+1}.P_i$$

  where p_i is the instruction activating input (by synchronizing with the output $\overline{p}_i$ executed by the currently active instruction); $\overline{inc_j}$ is the increment operation on register R_j; $\overline{p}_{i+1}$ is the output which activates instruction P_{i+1}, as it will synchronize with its initial input p_{i+1}. Note that the constant P_i is recursively defined as it must be always ready to process the required operation (increment of R_j) each time it is invoked.
- A "jump if zero/decrement" operation $(i : DecJump(r_j, s))$ is modeled as:

$$P_i \stackrel{def}{=} p_i.P_i' \qquad P_i' \stackrel{def}{=} \overline{zero_j}.\overline{p}_s.P_i + \overline{dec_j}.\overline{p}_{i+1}.P_i$$

  where the choice between $\overline{zero_j}$ and $\overline{dec_j}$ is driven the current status of register R_j: if R_j holds 0, then only the synchronization on $zero_j$ can take place (and the instruction of index s is activated), while if R_j is not 0, then only the synchronization on dec_j can take place (with the effect of decrementing the register and of activating instruction of index $i+1$).

Each register R_j is encoded as a counter (see Section 3.4.6 for details):

$$\begin{aligned} R_j &\stackrel{def}{=} zero_j.R_j + inc_j.((\nu a)(R_{j_1} \mid a.R_j)) \\ R_{j_1} &\stackrel{def}{=} dec_j.\bar{a}.\mathbf{0} + inc_j.((\nu b)(R_{j_2} \mid b.R_{j_1})) \\ R_{j_2} &\stackrel{def}{=} dec_j.\bar{b}.\mathbf{0} + inc_j.((\nu a)(R_{j_1} \mid a.R_{j_2})) \end{aligned}$$

Before starting the program I, we have first to introduce every input v_j in the corresponding register r_j. This is done by the following agent $B_{(v_1,v_2,v_3)}$ that performs the bootstrapping of the system by initializing the registers before emitting the initial program counter p_1.

$$B_{(v_1,v_2,v_3)} \stackrel{def}{=} \underbrace{\overline{inc_1}.\cdots.\overline{inc_1}}_{v_1 \text{ times}}.\underbrace{\overline{inc_2}.\cdots.\overline{inc_2}}_{v_2 \text{ times}}.\underbrace{\overline{inc_3}.\cdots.\overline{inc_3}}_{v_3 \text{ times}}.\overline{p}_1.\mathbf{0}$$

Hence, $CM_{M(v_1,v_2,v_3)} \longrightarrow^* (\nu L)(P'_1 \,|\, P_2 \,|\, \ldots \,|\, P_m \,|\, R'_1 \,|\, R'_2 \,|\, R'_3 \,|\, \mathbf{0})$, which represents the CCS process for the CM M ready to execute the first instruction: P'_1 derives from $P_1 \stackrel{def}{=} p_1.P'_1$, while R'_j denotes the register which the value v_j has been stored in.

Example 3.22. Consider the CM M of Example 3.21 and Exercise 3.68. The process $CM_{M(3,2,0)}$ is:

$$CM_{M(3,2,0)} \stackrel{def}{=} (\nu L)(P_1 \,|\, P_2 \,|\, P_3 \,|\, R_1 \,|\, R_2 \,|\, R_3 \,|\, B_{(3,2,0)})$$

where:

- $P_1 \stackrel{def}{=} p_1.P'_1 \qquad P'_1 \stackrel{def}{=} \overline{zero_2}.\overline{p}_4.P_1 + \overline{dec_2}.\overline{p}_2.P_1$
- $P_2 \stackrel{def}{=} p_2.P'_2 \qquad P'_2 \stackrel{def}{=} \overline{inc_1}.\overline{p}_3.P_2$
- $P_3 \stackrel{def}{=} p_3.P'_3 \qquad P'_3 \stackrel{def}{=} \overline{zero_3}.\overline{p}_1.P_3 + \overline{dec_3}.\overline{p}_4.P_3$
- $B_{(3,2,0)} \stackrel{def}{=} \overline{inc_1}.\overline{inc_1}.\overline{inc_1}.\overline{inc_2}.\overline{inc_2}.\overline{p}_1.\mathbf{0}$
- $L = \{inc_j, zero_j, dec_j \mid 1 \leq j \leq 3\} \cup \{p_i \mid 1 \leq i \leq 4\}$.

By performing the bootstrapping, $CM_{M(3,2,0)}$ reaches the state

$$(\nu L)(P'_1 \,|\, P_2 \,|\, P_3 \,|\, R'_1 \,|\, R'_2 \,|\, R_3 \,|\, \mathbf{0}),$$

which represents the CCS process for the CM M ready to execute the first instruction. R'_1 stands for $(\nu a)((\nu b)((\nu a)(R_{1_1} \,|\, a.R_{1_2}) \,|\, b.R_{1_1}) \,|\, a.R_1)$, while R'_2 stands for $(\nu a)((\nu b)(R_{2_2} \,|\, b.R_{2_1}) \,|\, a.R_2)$. □

For $i = 1, \ldots, m$, let $\langle CM_{(i,v_1,v_2,v_3)} \rangle$ be the set of all the terms of the form

$$(\nu L)(P_1 \,|\, \ldots \,|\, P_{i-1} \,|\, P'_i \,|\, P_{i+1} \,|\, \ldots \,|\, P_m \,|\, R'_1 \,|\, R'_2 \,|\, R'_3 \,|\, \mathbf{0})$$

where for $j = 1,2,3$, $R'_j \approx Counter_{v_j}$ and R'_j cannot perform τ initially, i.e., $R'_j \stackrel{\tau}{\not\longrightarrow}$. It is not difficult to see that if $Q, Q' \in \langle CM_{(i,v_1,v_2,v_3)} \rangle$ then $Q \sim Q'$.

The initial state $(1, v_1, v_2, v_3)$ of the CM M corresponds to a CCS process $Q \in \langle CM_{(1,v_1,v_2,v_3)} \rangle$. Each state change of M, e.g., $(i, v_1, v_2, v_3) \rightsquigarrow_M (i', v'_1, v'_2, v'_3)$, where the reached configuration (i', v'_1, v'_2, v'_3) is not terminal, determines a deterministic sequence of synchronizations[21] from any $Q \in \langle CM_{(i,v_1,v_2,v_3)} \rangle$ to some $Q' \in \langle CM_{(i',v'_1,v'_2,v'_3)} \rangle$.

[21] To be precise, the number of τs (synchronizations) is two when an increment instruction or a jump instruction is executed, while it is three when a decrement instruction is executed.

When $(i, v_1, v_2, v_3) \rightsquigarrow_M (i', v'_1, v'_2, v'_3)$ and the reached configuration is terminal (i.e., $i' > m$), from any $Q \in \langle CM_{(i,v_1,v_2,v_3)} \rangle$ it is possible to reach some Q' of the form

$$(\nu L)(P_1 \,|\, \ldots \,|\, P_{i-1} \,|\, \overline{p}_{i'}.P_i \,|\, P_{i+1} \,|\, \ldots \,|\, P_m \,|\, R'_1 \,|\, R'_2 \,|\, R'_3 \,|\, \mathbf{0})$$

for suitable R'_j ($j = 1,2,3$), where the ith instruction is stuck before the activation of the nonexisting instruction of index i'. For simplicity, we denote the class of terms of this form by $\langle CM_{(i',v'_1,v'_2,v'_3)} \rangle$ as well.

Observe that $CM_{M(v_1,v_2,v_3)}$ is deterministic, as in any reachable state only one synchronization is possible, reflecting the determinism of the CM M. Formally, one could prove the following proposition, which states that the encoding is correct.

Proposition 3.8. *Given a CM M with inputs v_1, v_2, v_3, let $CM_{M(v_1,v_2,v_3)}$ be the CCS process defined above, such that $CM_{M(v_1,v_2,v_3)} \longrightarrow^* Q \in \langle CM_{(1,v_1,v_2,v_3)} \rangle$. Then the following hold:*

- $(1, v_1, v_2, v_3) \rightsquigarrow^*_M (i, v'_1, v'_2, v'_3)$ *if and only if for all* $Q \in \langle CM_{(1,v_1,v_2,v_3)} \rangle$ *there exists some* $Q' \in \langle CM_{(i',v'_1,v'_2,v'_3)} \rangle$ *such that* $Q \longrightarrow^* Q'$;
- *if* $Q \in \langle CM_{(i,v'_1,v'_2,v'_3)} \rangle$ *and* $Q \longrightarrow^* Q'$, $Q' \longrightarrow Q_1$ *and* $Q' \longrightarrow Q_2$, *then* $Q_1 = Q_2$;
- *if* $Q \in \langle CM_{(i,v'_1,v'_2,v'_3)} \rangle$ *and* $Q \longrightarrow^* Q'$, *then there exists* $Q'' \in \langle CM_{(i',v''_1,v''_2,v''_3)} \rangle$ *such that* $Q' \longrightarrow^* Q''$, *for suitable* i', v''_1, v''_2, v''_3;
- $(1, v_1, v_2, v_3) \Uparrow$ *if and only if* $CM_{M(v_1,v_2,v_3)} \Uparrow$. □

As any 3CM can be modeled faithfully in finitary CCS, we have the following obvious consequence.

Corollary 3.3. *Finitary CCS is Turing-complete.* □

3.5.3 Undecidability of Behavioral Equivalences for Finitary CCS

We want to show that all the behavioral equivalences we have studied in Chapter 2 are undecidable over finitary CCS. To this aim, we elaborate a bit on the previous encoding of CMs in CCS in order to make termination observable, by performing a special termination action $\surd$. The introduction of this action is necessary because $CM_{M(v_1,v_2,v_3)}$ — be it terminating or not — is weakly bisimilar, as well as weak trace equivalent, to $\mathbf{0}$ in any case. Then we conclude that if the behavioral equivalence under scrutiny were decidable, we would be able to solve the *halting problem* [Tur36, Dav58], a well-known unsolvable problem for Turing-complete formalisms (see Section 1.3.5).

The halting problem for CMs can be formulated as follows. Given an enumeration of CMs, $M_1, M_2, M_3, \ldots$, function $halt(x,y)$ — where x is an index of a CM and y is an encoding of the inputs v_1, v_2, v_3 — returns 1 if M_x with inputs v_1, v_2, v_3 terminates, and 0 otherwise. The halting problem for CMs is solvable if and only if function *halt* is computable. As mentioned above, this problem is not solvable for Turing-complete formalisms, such as CMs.

Given a CM $M = (I, n)$, where $I = \{(1 : I_1), \ldots, (m : I_m)\}$ and v_1, v_2, v_3 are the inputs for r_1, r_2, r_3, we define the *Termination-observable* finitary CCS process

$$TCM_{M(v_1,v_2,v_3)} \stackrel{def}{=} (\nu L)(P_1 \,|\, \ldots \,|\, P_m \,|\, P_{m+1} \,|\, R_1 \,|\, R_2 \,|\, R_3 \,|\, B_{(v_1,v_2,v_3)})$$

where $L = \{inc_j, zero_j, dec_j \mid 1 \leq j \leq 3\} \cup \{p_i \mid 1 \leq i \leq m+1\}$, and where, w.r.t. $CM_{M(v_1,v_2,v_3)}$, we have included one further instruction of index $m+1$

$$P_{m+1} \stackrel{def}{=} p_{m+1}.\surd.\mathbf{0}$$

Action $\surd$ is a special *termination action*, as we require that any terminating computation will execute instruction of index $m+1$ as its last instruction. To achieve this, the encoding of a "jump if zero/decrement" operation $(i : DecJump(r_j, s))$, is now

$$P_i \stackrel{def}{=} p_i.P_i' \qquad P_i' \stackrel{def}{=} \begin{cases} \overline{zero}_j.\overline{p}_s.P_i + \overline{dec}_j.\overline{p}_{i+1}.P_i & \text{if } s \leq m, \\ \overline{zero}_j.\overline{p}_{m+1}.P_i + \overline{dec}_j.\overline{p}_{i+1}.P_i) & \text{otherwise,} \end{cases}$$

so that any terminating computation of the CM ends by activating the new instruction of index $m+1$. It is easy to see that the CM M with inputs v_1, v_2, v_3 terminates if and only if the CCS process $TCM_{M(v_1,v_2,v_3)}$ performs $\surd$. Hence, $TCM_{M(v_1,v_2,v_3)}$ is weakly bisimilar to $\surd.\mathbf{0}$ if and only if M with inputs v_1, v_2, v_3 terminates. Assuming that y is the encoding of v_1, v_2, v_3, the halting problem can be now restated as

$$halt(x,y) = \begin{cases} 1 & \text{if } TCM_{M_x(v_1,v_2,v_3)} \approx \surd.\mathbf{0} \\ 0 & \text{otherwise} \end{cases}$$

So, if weak bisimilarity $\approx$ were decidable, we could solve the halting problem: function $halt(x,y)$ can be computed by first extracting from x the code for M_x and from y the tuple v_1, v_2, v_3, then by generating the CCS process $TCM_{M_x(v_1,v_2,v_3)}$, and finally by checking if $TCM_{M_x(v_1,v_2,v_3)}$ is weakly bisimilar to $\surd.\mathbf{0}$. As we know that the halting problem is undecidable for Turing-complete formalisms, we conclude that also weak bisimilarity must be undecidable on finitary CCS, as all the other steps in the algorithm above are trivially computable.

The same argument above can be used in order to prove undecidability of weak trace equivalence $=_{wtr}$, weak simulation equivalence $\eqsim$, as well as of branching bisimilarity $\approx_{br}$. This argument can also be adapted slightly — it is enough to consider $\tau.\surd.\mathbf{0}$ instead of $\surd.\mathbf{0}$ — to prove undecidability of rooted weak bisimilarity $\approx^c$ as well as of rooted branching bisimilarity $\approx^c_{br}$.

We may wonder if undecidability holds also for strong bisimulation equivalence. Unfortunately, this is the case. We can compare the divergent process $Div \stackrel{def}{=} \tau.Div$ with $TCM_{M(v_1,v_2,v_3)}$: the CM M, with inputs v_1, v_2, v_3, diverges if and only if $TCM_{M(v_1,v_2,v_3)} \sim Div$. Hence, the halting problem can be now restated as follows:

$$halt(x,y) = \begin{cases} 0 & \text{if } TCM_{M_x(v_1,v_2,v_3)} \sim Div \\ 1 & \text{otherwise} \end{cases}$$

Therefore, we can conclude that also strong bisimulation equivalence is undecidable for finitary CCS. This argument can be used to prove undecidability of other strong equivalences because $TCM_{M_x(v_1,v_2,v_3)}$ and Div are also trace equivalent $=_{tr}$ as well as simulation equivalent $\simeq$.

Exercise 3.71. Argue that (weak as well as strong) bisimulation equivalence is not even *semidecidable* (see Section 1.3.5), while (weak and strong) bisimulation inequivalence is semidecidable. □

Exercise 3.72. Two CMs M_1 and M_2 are equivalent if, when given the same inputs (v_1, v_2, v_3), they end with the same values (v'_1, v'_2, v'_3) stored in the registers. Elaborate on the construction above in order to make observable not only the termination of the CM M, via action $\surd$, but also the actual contents of the registers, via actions $\overline{out}_1, \overline{out}_2, \overline{out}_3$ as follows: if the final value in r_j is v'_j for $j = 1, 2, 3$, then $OCM_{M(v_1,v_2,v_3)}$ produces a sequence of length v'_1 of occurrences of action $\overline{out}_1$, followed by a sequence of length v'_2 of occurrences of action $\overline{out}_2$, in turn followed by a sequence of length v'_3 of occurrences of action $\overline{out}_3$. With this construction, we can say that two CMs M_1 and M_2 are equivalent if and only if, for all possible input tuples (v_1, v_2, v_3), $OCM_{M_1(v_1,v_2,v_3)} \approx OCM_{M_2(v_1,v_2,v_3)}$. □

Remark 3.15. **(Set $sort(p)$ is not effectively decidable)** Given an enumeration of CCS processes $p_1, p_2, p_3, \ldots$, as well as an enumeration of actions $\mu_1, \mu_2, \mu_3, \ldots$, function

$$Srt(x,y) = \begin{cases} 1 & \text{if action } \mu_y \text{ belongs to } sort(p_x) \\ 0 & \text{otherwise} \end{cases}$$

cannot be computable. If *Srt* were computable, then we would solve the halting problem. In fact, in the construction above, action $\surd$ belongs to $sort(TCM_M)$ if and only if the CM M terminates. This observation has the consequence that, in general, for a finitary CCS process p the set *sort(p)* is *not effectively decidable*: even if set *sort(p)* is finite (hence decidable) by Corollary 4.1, it is not possible to give explicitly an algorithm that checks when a given action μ belongs to $sort(p)$, even if we know that such an algorithm must exist. As a matter of fact, if $sort(p_x)$ were effectively decidable for all x, then function *Srt* would be easily effectively computable.[22] □

Exercise 3.73. (Reachability is undecidable) With the same intuition as above, one can conclude that the *reachability problem* is undecidable for finitary CCS. This can be formalized by means of the following function $Reach : \mathscr{P} \times \mathscr{P} \to \{0, 1\}$:

[22] To be precise, *Srt* would be computable (but not effectively) even if, for finitely many x only, set $sort(p_x)$ is not effectively decidable. Hence, there are infinitely many x such that $sort(p_x)$ is not effectively decidable.

$$Reach(p,q) = \begin{cases} 1 & \text{if } p \longrightarrow^* q \\ 0 & \text{otherwise} \end{cases}$$

Argue that if *Reach* were computable, then we would solve the halting problem for CMs. (*Hint:* CM M with inputs (v_1, v_2, v_3) terminates if and only if $TCM_{M(v_1,v_2,v_3)}$ reaches a state where instruction of index $m+1$ has been activated, i.e., a state/term which contains $\surd.\mathbf{0}$) □

Exercise 3.74. Consider again Remark 3.8. We state that given a process p, a constant $C \in Const(p)$ is actually *unwound* by p if there exists a reachable process p' and a transition $p' \xrightarrow{\mu} p''$ whose proof makes use of rule (Cons) with C. Given an enumeration of CCS processes $p_1, p_2, p_3, \ldots$, as well as an enumeration of constants $C_1, C_2, C_3, \ldots$, argue that function

$$Unw(x,y) = \begin{cases} 1 & \text{if constant } C_y \text{ is unwound by } p_x \\ 0 & \text{otherwise} \end{cases}$$

is not computable, by showing that if Unw were computable, then also function *halt* becomes computable. (*Hint:* Consider the constant P_{m+1} in the construction of TCM_M above. When is it unwound?) □

3.5.4 Undecidability of Bisimilarity for Finite-Net CCS

Finite-net CCS is not a Turing-complete formalism. This can be proved by means of the Petri net semantics given to this sublanguage in [GV10]: a finite-net process generates a finite Petri net. For finite Petri nets, the *reachability problem* is decidable [May81, Kos82], hence function *Reach* of Exercise 3.73 is computable for finite-net processes. A formalism where the reachability problem is decidable cannot be Turing-complete because the halting problem is solvable: one can decide if a special termination state is reachable or, equivalently, if a special action $\surd$ is executable.

Despite this, strong bisimilarity is undecidable also for finite-net processes, because this is the case for finite Petri nets [Jan95]. Here we sketch, in the CCS setting, a process algebraic version of that proof.

Given a CM M, we define two finite-net CCS processes WR^1_M and WR^2_M with the property that if M, with inputs v_1, v_2, v_3, terminates, then $WR^1_{M(v_1,v_2,v_3)}$ and $WR^2_{M(v_1,v_2,v_3)}$ are not weakly bisimilar; conversely, if M, with inputs v_1, v_2, v_3, diverges, then $WR^1_{M(v_1,v_2,v_3)}$ and $WR^2_{M(v_1,v_2,v_3)}$ are weakly bisimilar. Hence, if we could decide weak bisimilarity on finite-net CCS, then we could solve the halting problem for CMs, which is impossible.

The finite-net CCS processes WR^1_M and WR^2_M are a sort of weak encoding of the CM M; such encodings are weak because, even if the CM M is deterministic, the resulting finite-net CCS processes are nondeterministic and exhibit also incorrect behavior.

The only components in the finitary CCS representation $TCM_{M(v_1,v_2,v_3)}$ of the CM M which are not finite-net CCS processes are the registers R_j, $j = 1,2,3$. We can replace such components with weak registers, WR_j, defined as BPP semi-counters (see Example 3.12):

$$WR_j \stackrel{def}{=} inc_j.(WR_j \,|\, dec_j.\mathbf{0}) \qquad j = 1,2,3$$

Note that this modification is important, as for weak registers the $zero_j$ action is unavailable, hence no *test-for-zero* is possible. As a consequence, the "jump if zero/decrement" operation $(i : DecJump(r_j, s))$ is to be modeled differently. A first attempt may be as follows:

$$P_i \stackrel{def}{=} p_i.P_i' \qquad P_i' \stackrel{def}{=} \tau.\overline{p}_s.P_i + \overline{dec}_j.\overline{p}_{i+1}.P_i$$

where the τ-transition corresponds to an unconditional jump (independently of the weak-register value), which makes nondeterministic the execution of the operation. Note that, if the weak register is unable to execute dec_j, then the unconditional jump correctly models this situation, so that the deterministic CM computation is also one of the computations of the weak CMs we are defining. However, when dec_j is possible, the τ-transition corresponds to a *cheat jump*.

Moreover, we need further elaboration. First, we need to make observable the various actions of increment and decrement of the weak counters to help in the weak bisimulation game. Hence, an increment operation $(i : Inc(r_j))$ is modeled as:

$$P_i \stackrel{def}{=} p_i.P_i' \qquad P_i' \stackrel{def}{=} \overline{inc}_j.up_j.\overline{p}_{i+1}.P_i$$

where action up_j is a visible action showing an increment on the weak register WR_j; moreover, the second attempt of modeling a "jump if zero/decrement" operation $(i : DecJump(r_j, s))$ is as follows:

$$P_i \stackrel{def}{=} p_i.P_i' \qquad P_i' \stackrel{def}{=} \tau.z_j.\overline{p}_s.P_i + \overline{dec}_j.down_j.\overline{p}_{i+1}.P_i$$

where z_j is an observable action denoting that a jump (assuming the weak counter holding 0) has been performed, and $down_j$ denotes that a decrement of weak register WR_j has been executed.

Then, we assume to have a trigger process $T_1 \stackrel{def}{=} \overline{t_1}.T_2$, with $T_2 \stackrel{def}{=} \overline{t_2}.T_1$, which has the duty of:

- activating the execution of the termination action $\surd$, performed by the last, additional instruction of index $m+1$, defined as follows:

$$P_{m+1} \stackrel{def}{=} p_{m+1}.t_1.\surd.\mathbf{0}$$

- or allowing for *explicit cheat* transitions in the "jump if zero/decrement" operation $(i : DecJump(r_j, s))$, which is actually modeled as follows:

$$P_i \stackrel{def}{=} p_i.P_i' \qquad P_i' \stackrel{def}{=} \begin{cases} \tau.z_j.\overline{p}_s.P_i + \overline{dec}_j.down_j.\overline{p}_{i+1}.P_i + \\ \overline{dec}_j.(t_1.\overline{inc}_j.z_j.\overline{p}_s.P_i + t_2.\overline{inc}_j.z_j.\overline{p}_s.P_i) & \text{if } s \leq m, \\ \tau.z_j.\overline{p}_{m+1}.P_i + \overline{dec}_j.down_j.\overline{p}_{i+1}.P_i + \\ \overline{dec}_j.(t_1.\overline{inc}_j.z_j.\overline{p}_{m+1}.P_i + t_2.\overline{inc}_j.z_j.\overline{p}_{m+1}.P_i) & \text{otherwise} \end{cases}$$

Note that, after the activation of the instruction (action p_i), an additional option is available: an explicit cheat jump, implemented by means of a sequence composed of the decrement action $\overline{dec}_j$ (hence, the weak register value is not 0), followed by transition t_1 or t_2 (depending on the current state of the trigger), then by a restore of the original weak-register value (action $\overline{inc}_j$), then by the observable jump action z_j, and finally by the activation of the instruction of index s (or $m+1$).

Given a CM $M = (I, 3)$, where $I = \{(1 : I_1), \ldots, (m : I_m)\}$ and v_1, v_2, v_3 are the inputs for r_1, r_2, r_3, we define the two *finite-net CCS* processes $WR^1_{M(v_1,v_2,v_3)}$ and $WR^2_{M(v_1,v_2,v_3)}$ as follows:

$$WR^1_{M(v_1,v_2,v_3)} \stackrel{def}{=} (\nu L)(P_1 \,|\, \ldots \,|\, P_m \,|\, P_{m+1} \,|\, T_1 \,|\, WR_1 \,|\, WR_2 \,|\, WR_3 \,|\, B_{(v_1,v_2,v_3)})$$
$$WR^2_{M(v_1,v_2,v_3)} \stackrel{def}{=} (\nu L)(P_1 \,|\, \ldots \,|\, P_m \,|\, P_{m+1} \,|\, T_2 \,|\, WR_1 \,|\, WR_2 \,|\, WR_3 \,|\, B_{(v_1,v_2,v_3)}),$$

where $L = \{inc_j, zero_j, dec_j \mid 1 \leq j \leq 3\} \cup \{p_i \mid 1 \leq i \leq m+1\} \cup \{t_1, t_2\}$. Note that the only difference between the two is that the trigger is T_1 for the former and T_2 for the latter.

Observe that, if we want to play a bisimulation game between the two, whenever $WR^1_{M(v_1,v_2,v_3)}$ executes an instruction, say P_i, the very same instruction is also executed by $WR^2_{M(v_1,v_2,v_3)}$ (and vice versa), because they both start by executing the same instruction P_1 and the following instruction is uniquely determined.

As a matter of fact, in case it is $(i : Inc(r_j))$, the next instruction must be P_{i+1} for both; in case it is $(i : DecJump(r_j, s))$, then we have to distinguish between two cases: if a decrement $down_j$ is executed, then, in order to match the same transition in the bisimulation game, also the other process will perform the same alternative (from among the three available) and the next instruction will be P_{i+1} for both; if a τ-jump is executed by also making observable action z_j, then the other process can execute either the same τ-jump or the explicit cheat sequence starting with a $\overline{dec}_j$, but in any case, the next state is P_s (or P_{m+1}) for both. Similarly, if an explicit cheat sequence starting with a $\overline{dec}_j$ is performed by also making observable action z_j, then the other process can execute either the same explicit cheat sequence or the τ-jump, but in any case, the next state is P_s (or P_{m+1}) for both. So, if we want to play a bisimulation game between the two, we are sure that the instructions they execute step by step are exactly the same at any step of the game.

It is easy to realize that when the CM M with inputs v_1, v_2, v_3 terminates, then the finite-net CCS process $WR^1_{M(v_1,v_2,v_3)}$ is able to mimic the same honest computation (as one of its computations — which never cheats — corresponds to the correct behavior of the CM M) by performing a suitable weak trace σ of actions, followed by $\surd$, because the trigger T_1 is available to activate the execution of the termination action $\surd$.

However, $WR^2_{M(v_1,v_2,v_3)}$ can reply to the honest computation of $WR^1_{M(v_1,v_2,v_3)}$, leading to the execution of $\surd$, only by executing the very same honest transitions (with observable weak trace σ): as the value stored in the weak register is 0 when an honest τ-jump is executed, the explicit cheat jump sequence is not possible, as it requires first to decrement the value of the weak register, and so its trigger T_2 is inactive during the execution of σ. Hence, $WR^2_{M(v_1,v_2,v_3)}$ cannot execute $\surd$ after σ, because the trigger T_2 cannot execute $\overline{t_1}$ initially and action t_1 is restricted. Hence, in such a case, the two finite-net processes are not bisimilar (actually, not even weak trace equivalent, as the trace $\sigma\surd$ is not a weak trace of $WR^2_{M(v_1,v_2,v_3)}$).

On the contrary, if the CM M with inputs v_1, v_2, v_3 diverges, then $WR^1_{M(v_1,v_2,v_3)}$ and $WR^2_{M(v_1,v_2,v_3)}$ are weakly bisimilar! Indeed, whatever honest computation is performed by $WR^1_{M(v_1,v_2,v_3)}$, the very same sequence of transitions can be performed also by $WR^2_{M(v_1,v_2,v_3)}$ and, conversely, whatever non-cheating computation is performed by $WR^2_{M(v_1,v_2,v_3)}$, the very same sequence of transitions can be performed also by $WR^1_{M(v_1,v_2,v_3)}$; hence, on honest transitions, the two finite-net CCS processes are (strongly) bisimilar.

Conversely, assume that at some point $WR^1_{M(v_1,v_2,v_3)}$ performs a τ-labeled cheat transition (wrong unconditional jump on a nonzero weak register, say WR_3),[23] on the ith instruction, i.e., it reaches a state of the form

$$(\nu L)(P_1 \mid \ldots \mid z_3.\overline{p}_s.P_i \mid \ldots \mid P_{m+1} \mid T_1 \mid WR'_1 \mid WR'_2 \mid WR'_3 \mid \mathbf{0})$$

In such a case, $WR^2_{M(v_1,v_2,v_n)}$, after having matched the previous transitions in a strict way, will respond to this τ-step by executing the explicit cheat transition sequence that includes t_2 (preceded by $\overline{dec_3}$ and followed by $\overline{inc_3}$ to restore the correct weak register value), which has the side effect of transforming its trigger T_2 into T_1, so that the state reached is

$$(\nu L)(P_1 \mid \ldots \mid z_3.\overline{p}_s.P_i \mid \ldots \mid P_{m+1} \mid T_1 \mid WR'_1 \mid WR'_2 \mid WR''_3 \mid \mathbf{0}),$$

which is almost the same state as the one reached by $WR^1_{M(v_1,v_2,v_n)}$: the only difference is that WR''_3 contains one additional $\mathbf{0}$ component w.r.t WR'_3. Hence, the two reached states are strongly bisimilar.

On the other hand, assume that, at some time, $WR^1_{M(v_1,v_2,v_3)}$ plays, as its ith instruction, the $\overline{dec_3}$ action as the first step of the explicit cheat-transition sequence, i.e., it reaches a state of the form

[23] Without loss of generality, in the following, the weak register involved in a cheat transition is alway WR_3.

$$Q = (\nu L)(P_1 \mid \ldots \mid t_1.\overline{inc_3}.z_3.\overline{p}_s.P_i + \ldots \mid \ldots \mid P_{m+1} \mid T_1 \mid WR'_1 \mid WR'_2 \mid WR'_3 \mid \mathbf{0}).$$

In such a case, $WR^2_{M(v_1,v_2,v_3)}$, after having matched the previous transitions in a strict way, will respond to this τ-step (as $\overline{dec_3}$ is synchronized with the corresponding, complementary action of the weak register WR_3) by executing the τ-labeled jump transition so that the reached state is

$$R = (\nu L)(P_1 \mid \ldots \mid z_3.\overline{p}_s.P_i \mid \ldots \mid P_{m+1} \mid T_2 \mid WR'_1 \mid WR'_2 \mid WR''_3 \mid \mathbf{0}),$$

where WR''_3 differs from WR'_3 because one of the $dec_3.\mathbf{0}$ components of the former is actually $\mathbf{0}$ in the latter, and the two reached states, Q and R, are weakly bisimilar. As a matter of fact, on the one hand, if Q executes the synchronization on t_1, reaching state

$$Q' = (\nu L)(P_1 \mid \ldots \mid \overline{inc_3}.z_3.\overline{p}_s.P_i \mid \ldots \mid P_{m+1} \mid T_2 \mid WR'_1 \mid WR'_2 \mid WR'_3 \mid \mathbf{0})$$

then R responds by idling, $R \stackrel{\varepsilon}{\Longrightarrow} R$, and Q' and R can be proved to be weakly bisimilar (check this! It is useful the following sentence). On the other hand, if R performs z_3, reaching

$$R' = (\nu L)(P_1 \mid \ldots \mid \overline{p}_s.P_i \mid \ldots \mid P_{m+1} \mid T_2 \mid WR'_1 \mid WR'_2 \mid WR''_3 \mid \mathbf{0})$$

then $Q \stackrel{z_3}{\Longrightarrow} Q''$, where

$$Q'' = (\nu L)(P_1 \mid \ldots \mid \overline{p}_s.P_i \mid \ldots \mid P_{m+1} \mid T_2 \mid WR'_1 \mid WR'_2 \mid WR'''_3 \mid \mathbf{0}).$$

Register WR'''_3 differs from WR''_3 only because of the additional presence of a $\mathbf{0}$ component in the former, so that R' and Q'' are strongly bisimilar.

Symmetrically, if $WR^2_{M(v_1,v_2,v_3)}$ plays a cheating τ-labeled jump transition, then $WR^1_{M(v_1,v_2,v_3)}$ will respond by executing the explicit cheat transition sequence, including t_1 (preceded by $\overline{dec_3}$ and followed by $\overline{inc_3}$ to restore the correct weak register value), which has the side effect of transforming its trigger T_1 into T_2, so that the two reached states are almost identical, and strongly bisimilar. Similarly, if $WR^2_{M(v_1,v_2,v_3)}$ plays the explicit cheat-transition sequence, including t_2, transforming the trigger T_2 into T_1, then $WR^1_{M(v_1,v_2,v_n)}$ will respond with the τ-labeled jump transition and the reached states are almost identical, and strongly bisimilar. Hence, in any case the two finite-net processes $WR^1_{M(v_1,v_2,v_3)}$ and $WR^2_{M(v_1,v_2,v_3)}$ are weakly bisimilar.

Summing up, assuming that x is an index of a CM and y is the encoding of the input tuple (v_1, v_2, v_3), the halting problem for the Turing-complete formalisms of CMs can be now restated as follows:

$$halt(x,y) = \begin{cases} 0 & \text{if } WR^1_{M_x(v_1,v_2,v_3)} \approx WR^2_{M_x(v_1,v_2,v_3)} \\ 1 & \text{otherwise} \end{cases}$$

So, if weak bisimilarity $\approx$ were decidable over finite-net CCS processes, we could solve the halting problem for a Turing-complete formalism, such as Counter Machines, which is clearly impossible.

Exercise 3.75. (Also strong bisimulation equivalence is undecidable) Modify the definition of constant P_i implementing the "jump if zero/decrement" operation $(i : DecJump(r_j, s))$ so that, when the CM M with inputs v_1, v_2, v_3 diverges, $WR^1_{M(v_1,v_2,v_3)}$ and $WR^2_{M(v_1,v_2,v_3)}$ are strongly bisimilar. This proves that also strong bisimilarity is undecidable for finite-net CCS. □

3.6 Value-Passing CCS

The calculus we have described so far, namely CCS, is such that no communication of data is possible, as the interaction between two processes is by pure synchronization. In practical cases, it may be useful to be able to model data exchange, even if, as we will see, from a theoretical point of view, we add nothing to the expressiveness of CCS by extending it with value-passing communication.

For simplicity, we assume that the set D of transmittable data is simply the set $\mathbb{N}$ of natural numbers, but any other set D of data may work as well.[24] On this set of data, we can build *arithmetic expressions*, ranged over by e (possibly indexed), by using arithmetic operators (e.g., $+$, $\times$, etc.), as well as *Boolean expressions*, ranged over by b (possibly indexed), by using relational arithmetic operators (such as $=, \leq$, etc.) or Boolean operators (e.g., *and*). The precise nature of these expressions is inessential for describing the calculus, hence these are not specified further.

Value-passing CCS is the calculus where inputs, as well as constants, may be parametrized by some variables on D. An example may help illustrate the features of the calculus. Assume we want to describe a simple one-position buffer B which inputs some natural number n, if empty, and then outputs n modulo 2 (i.e., the remainder of the division by 2), represented as $n\ mod\ 2$, becoming empty again. It may be defined as follows:

$$B \stackrel{def}{=} in(x).B'(x)$$

$$B'(x) \stackrel{def}{=} \overline{out}(x\ mod\ 2).B$$

The input prefix *in* carries the *variable* x as a parameter, whose scope is the process that is prefixed by that action (in this case $B'(x)$). The expected behavior of the

---

[24] It is assumed that D and *Act* are disjoint, so that a received value cannot be used later as a channel name. This is an important caution: if action names were transmittable values, the language would get new interesting features, so called *mobility* of channels, that are studied in the π-calculus [MPW92] (for an easy introduction, see [Mil99]; for a more advanced monograph, see [SW01]).

empty buffer B is that it is ready to accept in input any number n, it binds this value to the variable x, and in doing so the substitution $[n/x]$ is generated and applied to $B'(x)$, yielding $B'(n)$, which represents the full buffer holding the datum n. In order to see what $B'(n)$ can do, we have to match it with the left-hand side of the second defining equation, namely $B'(x) \stackrel{def}{=} \overline{out}(x\, mod\, 2).B$, hence binding x to n also in the body (in this case $\overline{out}(n\, mod\, 2).B$). Therefore, now $B'(n)$ can execute the output $\overline{out}(n\, mod\, 2)$ and then becomes B again. The intended meaning of the output $\overline{out}(n\, mod\, 2)$ is that the expression $n\, mod\, 2$ is first evaluated and then the result is given as output.

In general, a constant A may have many parameters: $A(x_1,\ldots,x_n) \stackrel{def}{=} p$, meaning that all the x_i's ($i = 1,\ldots,n$) are distinct data variables and that in p the only data variables that may occur are exactly the parameters of A. A process term p is a value-passing CCS process if, besides being fully defined and guarded, each occurrence of a variable x is bound by either an input prefix or a constant definition. For instance, the process constant $A(x) \stackrel{def}{=} \bar{a}(x+y).A(x)$ is not a process because variable y is not bound in A; similarly, $a(x).\bar{c}(x+y).\mathbf{0} + b(y).\bar{c}(x+y).\mathbf{0}$ is not a process.

Value-passing CCS offers a new construct: **if** b **then** p **else** q, where b is a Boolean expression. This construct is useful whenever some decison is to be taken, depending on the data tested in the Boolean expression b. For instance, consider a one-position buffer that outputs the factorial $n!$ of its input n:

$$\begin{aligned} Fact &\stackrel{def}{=} in(x).F(x,1) \\ F(x,y) &\stackrel{def}{=} \mathbf{if}\ x=0\ \mathbf{then}\ \overline{out}(y).Fact\ \mathbf{else}\ \tau.F(x-1, x\times y) \end{aligned}$$

where the number of the executed τ steps is equal to the value of the input.

Exercise 3.76. Check if the process Fib, defined as

$$\begin{aligned} Fib &\stackrel{def}{=} in(x).C(x,0,1,1) \\ C(x,y,w,z) &\stackrel{def}{=} \mathbf{if}\ x=y\ \mathbf{then}\ \overline{out}(w).Fib\ \mathbf{else}\ \tau.C(x,y+1,z,w+z), \end{aligned}$$

given in input a natural n, returns in output the nth Fibonacci number. □

Summarizing, the syntax of value-passing CCS is as follows:

$$\begin{aligned} p &::= \mathbf{0} \mid a(x).q \mid \bar{a}(e).q \mid \tau.q \mid p+p \mid \mathbf{if}\ b\ \mathbf{then}\ p\ \mathbf{else}\ p \\ q &::= p \mid q|q \mid (\nu a)q \mid C(x_1,\ldots,x_n) \end{aligned}$$

where a is any input, x is any variable, e is any arithmetic expression, b is any Boolean expression, the **if-then-else** construct composes sequential processes only, the number of parameters $x_1,\ldots,x_n$ of a constant C is any $n \geq 0$ and the x_i's are all distinct.

$$\text{(In)}\ \frac{}{a(x).p \xrightarrow{a(v)} p[v/x]}\ \text{for any } v \in D \qquad \text{(Tau)}\ \frac{}{\tau.p \xrightarrow{\tau} p}$$

$$\text{(Out)}\ \frac{}{\bar{a}(e).p \xrightarrow{\bar{a}(v)} p}\ \text{if } e \text{ has value } v$$

$$\text{(Cons)}\ \frac{p[v_1/x_1,\ldots,v_n/x_n] \xrightarrow{\mu} p'}{C(e_1,\ldots,e_n) \xrightarrow{\mu} p'}\ \text{if } C(x_1,\ldots,x_n) \stackrel{def}{=} p \text{ and each } e_i \text{ has value } v_i$$

$$\text{(Then)}\ \frac{p \xrightarrow{\mu} p'}{\textbf{if } b \textbf{ then } p \textbf{ else } q \xrightarrow{\mu} p'}\ \text{if } b \text{ has value } true$$

$$\text{(Else)}\ \frac{q \xrightarrow{\mu} q'}{\textbf{if } b \textbf{ then } p \textbf{ else } q \xrightarrow{\mu} q'}\ \text{if } b \text{ has value } false$$

Table 3.3 New inference rules for value-passing

We are now ready to define the SOS rules for value-passing CCS. They are described in Table 3.3, where rules (Sum$_1$), (Sum$_2$), (Par$_1$), (Par$_2$), (Com) and (Res) are omitted because they are identical to those in Table 3.1 for pure CCS. To be precise, rule (Com) needs a minor extension of the definition of complement of an action: $\overline{a(n)} = \bar{a}(n)$ and $\overline{\bar{a}(n)} = a(n)$. Similarly, the side condition of rules (Res) is that $\mu \neq a(n), \bar{a}(n)$ for any $n \in \mathbb{N}$.

Let us focus our attention on rule (In). It states that there are as many transitions from $a(x).p$ as the cardinality of the set D of data. For a particular choice of a value v in D, the label of the transition is $a(v)$ and the reached state is $p[v/x]$, i.e., p where each occurrence of the variable x in p is replaced by the value v. Rule (Out) states that the actual observed value v, sent along channel a, is the result of the evaluation of expression e; how such an evaluation is performed is not specified and depends on the actual choice of the set of expressions (which should include variables). The new rule (Cons) states that, given a constant definition $C(x_1,\ldots,x_n) \stackrel{def}{=} p$, a partially instantiated constant $C(e_1,\ldots,e_n)$ can do whatever p can do, where each occurrence of a variable x_i is replaced by the value v_i of expression e_i: notation $p[v_1/x_1,\ldots,v_n/x_n]$. Note that e_i can be a variable y_i, so that the substitution simply changes the name of the variable x_i to y_i. Rule (Then) and (Else) are straigthforward.

Let us consider again the value-passing CCS buffer B described at the beginning of this section. If we assume that the set of values that can be given in input to B is limited to $\{0,1\}$, then the resulting LTS is reported in Figure 3.13. This example may be used to show that there is a clear correspondence with *pure* CCS (i.e, CCS

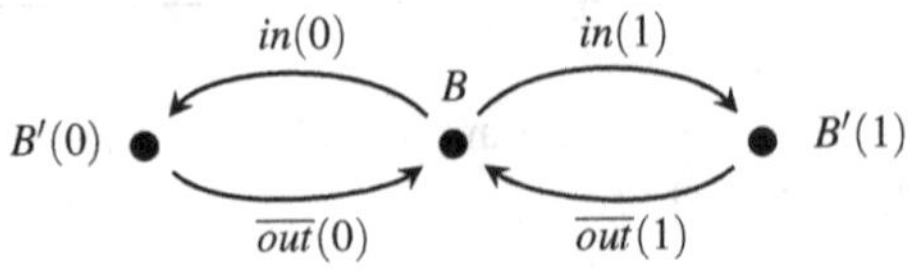

Fig. 3.13 The LTS model for the buffer B when $D = \{0,1\}$

with pure synchronization only): if we rename $in(0), in(1)$ with in_0, in_1, as well as $\overline{out}(0), \overline{out}(1)$ with $\overline{out}_0, \overline{out}_1$, and replace $B'(0), B'(1)$ with two process constants $B_0 \overset{def}{=} \overline{out}_0.PB$ and $B_1 \overset{def}{=} \overline{out}_1.PB$, respectively, then what we get is a pure CCS process: $PB \overset{def}{=} in_0.B_0 + in_1.B_1$. As a matter of fact, the LTS for PB generated by means of the rules of Table 3.1 is isomorphic (modulo the renaming of the actions above) to the LTS for B generated with the rules of Table 3.3, depicted in Figure 3.13.

Exercise 3.77. Consider again the value-passing CCS buffer B described at the beginning of this section. Draw the LTS in case $D = \{0,1,2,3\}$. Define a corresponding pure CCS process PB', generating an LTS isomorphic to the one for B. □

We can be more formal, and define a precise encoding of a value-passing CCS process p to a pure CCS process $[\![p]\!]$, as follows:

$$[\![\mathbf{0}]\!] = \mathbf{0} \qquad [\![\tau.p]\!] = \tau.[\![p]\!]$$

$$[\![a(x).p]\!] = \Sigma_{v \in D}\, a_v.[\![p[v/x]]\!] \qquad [\![\bar{a}(e).p]\!] = \overline{a_v}.[\![p]\!] \quad \text{if } e \text{ has value } v$$

$$[\![p_1 + p_2]\!] = [\![p_1]\!] + [\![p_2]\!] \qquad [\![p_1 \,|\, p_2]\!] = [\![p_1]\!] \,|\, [\![p_2]\!]$$

$$[\![(\nu a)p]\!] = (\nu\{a_v \,|\, v \in D\})[\![p]\!] \qquad [\![\mathbf{if}\ b\ \mathbf{then}\ p\ \mathbf{else}\ q]\!] = \begin{cases} [\![p]\!] \text{ if } b \text{ has value } true \\ [\![q]\!] \text{ if } b \text{ has value } false \end{cases}$$

In addition, we have $[\![A(e_1,\ldots,e_n)]\!] = A_{v_1,\ldots,v_n}$ if each e_i has value v_i. Furthermore, for each constant definition $A(x_1,\ldots,x_n) \overset{def}{=} p$, we have a family of definitions

$$A_{v_1,\ldots,v_n} \overset{def}{=} [\![p[v_1/x_1,\ldots,v_n/x_n]]\!],$$

one for each vector $(v_1,\ldots,v_n) \in D^n$.

Of course, the process $[\![p]\!]$ is really a pure *finitary* CCS process if the set D of data is finite; otherwise, we would need infinite summation in the encoding of $a(x).p$; moreover, we would have the unpleasant consequence that the encoding of any constant $A(x_1,\ldots,x_n)$ would generate infinitely many (pure) constants; finally, in the encoding of $(\nu a)p$ we would get a restriction over the infinite set $\{a_v \,|\, v \in D\}$.

Proposition 3.9. *Given a finite set D of values, the LTS for a value-passing CCS process p, generated with the rules of Table 3.3, is isomorphic (modulo renaming of $a(v)$ with a_v) to the LTS for the pure CCS process $[\![p]\!]$, generated with the rules of Table 3.1.*

Proof. *One has to prove that* $p \xrightarrow{a(v)} p'$ *implies* $[\![p]\!] \xrightarrow{a_v} [\![p']\!]$ *and, conversely, that* $[\![p]\!] \xrightarrow{a_v} q$ *implies there exists* p' *such that* $q = [\![p']\!]$ *and* $p \xrightarrow{a(v)} p'$. *(Analogously, for the other labels* $\overline{a}(v)/\overline{a}_v$ *and* τ.*) This can be proved by induction on the proof of the transitions and is left as an exercise to the reader. Then, the isomorphism bijection* f *maps a value-passing CCS process* p' *reachable from* p *to its encoding* $[\![p']\!]$. □

Exercise 3.78. (Two-position buffers, revisited) Continuing Exercise 3.46, we define two inequivalent, more concrete definitions of a buffer. The first one, B_{fifo}, defines a two-position pipeline buffer (or queue), according to the *first-in, first-out* (FIFO) discipline:

$$\begin{aligned}
B_{fifo} &\stackrel{def}{=} in(x).B_1(x)\\
B_1(x) &\stackrel{def}{=} in(y).B_2(y,x) + \overline{out}(x).B_{fifo}\\
B_2(x,y) &\stackrel{def}{=} \overline{out}(y).B_1(x)
\end{aligned}$$

The corresponding pure CCS process is defined by means of $|D|^2 + |D| + 1$ constants, where $|D|$ denotes the cardinality of the set D.

The second definition, B_{bag}, describes a two-position buffer as a bag of elements, so that the order of arrival of data in the buffer may be not respected on delivery:

$$\begin{aligned}
B_{bag} &\stackrel{def}{=} in(x).B'_1(x)\\
B'_1(x) &\stackrel{def}{=} in(y).B'_2(y,x) + \overline{out}(x).B_{bag}\\
B'_2(x,y) &\stackrel{def}{=} \overline{out}(x).B'_1(y) + \overline{out}(y).B'_1(x)
\end{aligned}$$

Show that B_{bag} is not trace equivalent to B_{fifo} (when $|D| > 1$, while B_{bag} is bisimulation equivalent to B_{fifo} if $|D| = 1$).

Consider then the one-position buffer $B \stackrel{def}{=} in(x).B'(x)$, $B'(x) \stackrel{def}{=} \overline{out}(x).B$. Draw the LTS for process $B\,|\,B$, when $D = \{0,1\}$. Check that relation

$$R = \{(B_{bag}, B\,|\,B)\} \cup \{(B'_1(v), B'(v)\,|\,B) \mid v \in D\} \cup \{(B'_2(u,v), B'(v)\,|\,B'(u)) \mid u,v \in D\}$$

is a strong bisimulation up to $\sim$, hence showing that $B_{bag} \sim B\,|\,B$. Note that the pure CCS process corresponding to $B\,|\,B$ can be described with only $|D| + 1$ constants.

Consider then the pipeline buffer *Buf*:

$$\begin{aligned}
&Buf \stackrel{def}{=} (\nu d)(Buf_1\,|\,Buf_2)\\
&Buf_1 \stackrel{def}{=} in(x).Buf'_1(x) \qquad Buf'_1(x) \stackrel{def}{=} \overline{d}(x).Buf_1\\
&Buf_2 \stackrel{def}{=} d(x).Buf'_2(x) \qquad Buf'_2(x) \stackrel{def}{=} \overline{out}(x).Buf_2
\end{aligned}$$

Show that B_{fifo} is weakly bisimilar to *Buf* by checking that relation

$$\begin{aligned}
S = \; &\{(B_{fifo}, Buf)\} \cup \{(B_{fifo}, (\nu d)(Buf_1\,|\,Buf_2))\} \cup\\
&\{(B_1(v), (\nu d)(Buf'_1(v)\,|\,Buf_2)) \mid v \in D\} \cup\\
&\{(B_1(v), (\nu d)(Buf_1\,|\,Buf'_2(v))) \mid v \in D\} \cup\\
&\{(B_2(u,v), (\nu d)(Buf'_1(u)\,|\,Buf'_2(v))) \mid u,v \in D\}
\end{aligned}$$

is a weak bisimulation. Note that the pure CCS process corresponding to Buf can be described with only $2 \times |D| + 3$ constants. □

Exercise 3.79. (Two-phase two-position buffer) Consider the following specification of a two-phase two-position buffer, i.e., a process that first fills the two positions of the buffer and then outputs the two data, respecting the arrival order:

$$B_{tp} \stackrel{def}{=} in(x).B'_{tp}(x) \qquad \begin{array}{r}B'_{tp}(x) \stackrel{def}{=} in(y).B''_{tp}(y,x)\\ B''_{tp}(x,y) \stackrel{def}{=} \overline{out}(y).\overline{out}(x).B_{tp}\end{array}$$

Observe that B_{tp} is trace equivalent neither to B_{bag} nor to B_{fifo}, not even if $|D| = 1$. Define a regular value-passing CCS process Buf_{tp} that implements (up to weak bisimilarity) the specification B_{tp} in a parallel way. □

Exercise 3.80. (Producer-consumer with a two-position buffer, revisited) Continuing Exercise 3.49, consider the more concrete version *VP2BC* of a producer/consumer with a two-position buffer:

$$VP2BC \stackrel{def}{=} (vin)(vout)((P \mid B_{bag}) \mid C)$$

where $P \stackrel{def}{=} produce(x).\overline{in}(x).P$ represents a producer, B_{bag} is the two-position bag-buffer described in Exercise 3.78 and $C \stackrel{def}{=} out(x).consume(x).C$ represents a consumer. Draw the LTS for *VP2BC*, assuming that $D = \{0,1\}$.

Consider the variant $VP2BC' \stackrel{def}{=} (vin)(vout)((P \mid (B \mid B)) \mid C)$, where B is the one-position bag-buffer described in Exercise 3.78. Show that *VP2BC* and *VP2BC'* are strongly bisimilar.

Consider also the variant system $VP2BC'' \stackrel{def}{=} (vin)(vout)((P \mid Buf) \mid C)$ where *Buf* is the pipeline two-position buffer of Exercise 3.78. Show that *VP2BC* and *VP2BC''* are not weak trace equivalent. (This is in contrast with the pure regular CCS variants of Exercise 3.49, where *P2BC* and *P2BC''* are weakly bisimilar.) □

Example 3.23. **(Unbounded bag buffer)** Continuing Exercise 3.55, a sequential specification of an unbounded *bag* buffer may be defined by means of a constant $UB(x_1, \ldots, x_n)$, parametrized over a tuple of values, as follows:

$$UB(\varepsilon) \stackrel{def}{=} in(x).UB(x)$$

$$UB(x_1, \ldots x_n) \stackrel{def}{=} \Sigma_{i=1}^{n} \overline{out}(x_i).UB(x_1, \ldots, x_{i-1}, x_{i+1}, \ldots x_n) + in(y).UB(y, x_1, \ldots x_n)$$

The corresponding pure CCS process would generate infinitely many constants $UB_{(v_1,\ldots,v_n)}$: even if the set D of data is finite, we get infinitely many tuples $(v_1, \ldots, v_n)$, if we do not impose an upper limit on the length of the buffer. Nonetheless, we can define a simple value-passing BPP process U, as follows:

$$U \stackrel{def}{=} in(x).(U \mid \overline{out}(x).\mathbf{0})$$

Note that the corresponding pure CCS process uses one single constant, even if the set D of data is infinite! □

Exercise 3.81. Prove that U is strongly bisimilar to $UB(\varepsilon)$, by adapting the proof given in Example 3.12 relating the semi-counters $SCounter_0$ and SC. (*Hint:* It may be useful to define a shorthand notation $Out(v_1,\ldots,v_n)$ to denote the parallel process $\overline{out}(v_1).\mathbf{0}\,|\ldots|\,\overline{out}(v_n).\mathbf{0}$, and then to show that, for any $1 \le k \le n$, $Out(v_1,\ldots,v_n) \xrightarrow{\overline{out}(v_k)} p \sim Out(v_1,\ldots,v_{k-1},v_{k+1},\ldots,v_n)$.) □

Example 3.24. **(Stack)** A simple description of a stack of elements, following the usual *last-in-first-out* (LIFO) discipline, can be easily given by means of a constant $Stack(\sigma)$, parametrized over a sequence of values σ:

$$\begin{aligned} Stack(\varepsilon) &\stackrel{def}{=} empty.Stack(\varepsilon) + push(x).Stack(x) \\ Stack(\sigma x) &\stackrel{def}{=} \overline{pop}(x).Stack(\sigma) + push(y).Stack(\sigma x y) \end{aligned}$$

The corresponding pure CCS process would generate infinitely many constants $Stack_\sigma$: even if the set D of data is finite, we get infinitely many sequences σ if we do not impose an upper limit on the length of the stack. Mimicking the counter process of Section 3.4.6, we can define a stack of elements as follows:

$$\begin{aligned} S &\stackrel{def}{=} empty.S + push(x).((\nu a)(S_1(x)\,|\,a.S)) \\ S_1(x) &\stackrel{def}{=} \overline{pop}(x).\bar{a}.\mathbf{0} + push(y).((\nu b)(S_2(y)\,|\,b.S_1(x))) \\ S_2(y) &\stackrel{def}{=} \overline{pop}(y).\bar{b}.\mathbf{0} + push(x).((\nu a)(S_1(x)\,|\,a.S_2(y))) \end{aligned}$$

Note that the corresponding pure CCS process uses $2 \times |D| + 1$ constants, hence finitely many if the set D of data is finite. □

Exercise 3.82. Rephrase the argument we used in Section 3.4.6 for proving that $Counter_0$ is weakly bisimilar to C, in order to prove that $Stack(\varepsilon)$ and S are weakly bisimilar too. □

Example 3.25. **(Queue or unbounded fifo buffer)** A much harder example is the finitary CCS representation of a queue of elements. Following the usual *first-in-first-out* (FIFO) discipline, a queue can be specified as follows:

$$\begin{aligned} Queue(\varepsilon) &\stackrel{def}{=} empty.Queue(\varepsilon) + in(x).Queue(x) \\ Queue(x\sigma) &\stackrel{def}{=} \overline{out}(x).Queue(\sigma) + in(y).Queue(x\sigma y) \end{aligned}$$

It is clear that, even if the elements to enqueue are chosen from a finite set D, such a specification is not finitary in pure CCS, unless we arbitrarily limit the length of the queue. Nevertheless it is possible to provide a finitary — although, obviously, not finite-state — implementation of $Queue(\varepsilon)$ with unbounded length by a clever usage of restricted names and parallel composition. This implementation consists of a list of processes which grows on the right when a new element is added, and shortens on the left when the last element is picked off. To a first approximation, after that some elements have been enqueued, the process would look like

$$\ldots L(v_1) \mid \ldots \mid M(v_2) \mid \ldots \mid M(v_3) \mid \ldots \mid H \ldots$$

where v_1 is the first element that has been enqueued (and the first to be dequeued), process L represents the "last" process of the list that is ready to send v_1 upon request, the processes $M(v_i)$'s are those in the middle of the queue, and H is the head process in charge of receiving the next element to be enqueued.

At each step of the computation, on the right of the list, the process H is ready to receive a new element over the channel *in* and then to spawn a new process M which keeps track of such an element on top of the queue; symmetrically, the last (not *nil*) process L is always ready to dequeue the oldest element of the queue over the channel *out* upon request.

Because of CCS semantics, the effect of dequeueing is not a real shortening of the list of these parallel processes, but the turning of the last non-*nil* process into *nil*. So, for example, after that v_1 has been dequeued and v_4 has been enqueued, the previous queue would become

$$\ldots \mathbf{0} \mid \ldots \mid L(v_2) \mid \ldots \mid M(v_3) \mid \ldots \mid M(v_4) \mid \ldots \mid H \ldots$$

As we are going to see in the following, in order to get a finitary implementation we need to use two sets of head, middle and last processes (H,M,L and H',M',L') which get alternated according to their even or odd position in the queue and are distinguished only by the alternate presence of the private names l and l'. These two names allow for the activation of the process next to the last one (for instance, in the previous example, this activation corresponds to the state change of $M(v_2)$ to $L(v_2)$); such an activation happens immediately before the last process (in the previous example, $L(v_1)$) becomes *nil*, so that the whole system is ready for the next dequeue event (from $L(v_2)$ in the example).

We are now going to fill the dots of the previous example by giving the full implementation of the finitary queue Q:

$$Q \stackrel{def}{=} empty.Q + in(x).((\nu l)(L(x) \mid H))$$

$$L(x) \stackrel{def}{=} \overline{out}(x).\bar{l} \qquad H \stackrel{def}{=} l.Q + in(x).((\nu l')(M'(x) \mid H')) \qquad M'(x) \stackrel{def}{=} l.L'(x)$$

$$L'(x) \stackrel{def}{=} \overline{out}(x).\bar{l}' \qquad H' \stackrel{def}{=} l'.Q + in(x).((\nu l)(M(x) \mid H)) \qquad M(x) \stackrel{def}{=} l'.L(x)$$

An example of execution of Q is in Table 3.4. N ote that the corresponding pure CCS process uses $4 \times |D| + 3$ constants.

Now we want to prove that the sequential specification $Queue(\varepsilon)$ and the parallel implementation Q are weakly bisimilar. To this aim, let us consider the following family of processes, where $v, u \in D$, x, y are place-holders for any term to be filled in, and $\sigma \in D^*$:

$$\begin{aligned} p_v &= (\nu l)(y \mid x) & p'_v &= (\nu l')(y \mid x) \\ q_{\sigma u} &= p_\sigma[(\nu l')(M'(u) \mid x)/x] & q'_{\sigma u} &= p'_\sigma[(\nu l)(M(u) \mid x)/x] \\ p_{\sigma u} &= q_\sigma[(\nu l)(M(u) \mid x)/x] & p'_{\sigma u} &= q'_\sigma[(\nu l')(M'(u) \mid x)/x] \end{aligned}$$

$$
\begin{array}{llr}
Q & \xrightarrow{in(v_1)} (\nu l)(L(v_1) \mid H) & \xrightarrow{in(v_2)} \\
 & (\nu l)(L(v_1) \mid (\nu l')(M'(v_2) \mid H')) & \xrightarrow{\overline{out}(v_1)} \xrightarrow{\tau} \\
 & (\nu l)(\mathbf{0} \mid (\nu l')(L'(v_2) \mid H')) & \xrightarrow{in(v_3)} \\
 & (\nu l)(\mathbf{0} \mid (\nu l')(L'(v_2) \mid (\nu l)(M(v_3) \mid H))) & \xrightarrow{\overline{out}(v_2)} \xrightarrow{\tau} \\
 & (\nu l)(\mathbf{0} \mid (\nu l')(\mathbf{0} \mid (\nu l)(L(v_3) \mid H))) & \xrightarrow{\overline{out}(v_3)} \xrightarrow{\tau} \\
 & (\nu l)(\mathbf{0} \mid (\nu l')(\mathbf{0} \mid (\nu l)(\mathbf{0} \mid Q))) &
\end{array}
$$

Table 3.4 An execution of process Q

The notation $p_\sigma[t/x]$ stands for naïve substitution of t for the (unique) occurrence of x in p_σ. Observe that any generated term p_σ (p'_σ) is such that the length of σ is odd, while any generated term q_σ (q'_σ) is such that the length of σ is even (and $\neq 0$). Then, let us consider the relation R, defined as follows:

$$
\begin{array}{rl}
R = & \{(Q, Queue(\varepsilon))\} \\
 & \cup\ \{(p_{v\sigma}[L(v)/y, H/x],\ Queue(v\sigma)) \mid v \in D, \sigma \in D^*\} \\
 & \cup\ \{(p'_{v\sigma}[L'(v)/y, H'/x],\ Queue(v\sigma)) \mid v \in D, \sigma \in D^*\} \\
 & \cup\ \{(p_{v\sigma}[\bar{l}/y, H/x],\ Queue(\sigma)) \mid v \in D, \sigma \in D^*\} \\
 & \cup\ \{(p'_{v\sigma}[\bar{l}'/y, H'/x],\ Queue(\sigma)) \mid v \in D, \sigma \in D^*\} \\
 & \cup\ \{(q_{v\sigma}[L(v)/y, H'/x],\ Queue(v\sigma)) \mid v \in D, \sigma \in D^+\} \\
 & \cup\ \{(q'_{v\sigma}[L'(v)/y, H/x],\ Queue(v\sigma)) \mid v \in D, \sigma \in D^+\} \\
 & \cup\ \{(q_{v\sigma}[\bar{l}/y, H'/x],\ Queue(\sigma)) \mid v \in D, \sigma \in D^+\} \\
 & \cup\ \{(q'_{v\sigma}[\bar{l}'/y, H/x],\ Queue(\sigma)) \mid v \in D, \sigma \in D^+\}
\end{array}
$$

Now we want to show that R is a weak bisimulation up to $\approx$ (see Definition 2.22 and Exercise 2.74 for the correctness of this proof technique).

First consider pair $(Q, Queue(\varepsilon))$. Transition $Q \xrightarrow{empty} Q$ is matched strongly (hence also weakly) by transition $Queue(\varepsilon) \xrightarrow{empty} Queue(\varepsilon)$ and the pair of reached states $(Q, Queue(\varepsilon))$ is in R. Transition $Q \xrightarrow{in(v)} (\nu l)(L(v) \mid H) = p_v[L(v)/y, H/x]$ is matched by transition $Queue(\varepsilon) \xrightarrow{in(v)} Queue(v)$ and the pair $(p_v[L(v)/y, H/x], Queue(v))$ belongs to the second group of R. No other transition is possible from Q. Symmetrically, the two transitions from $Queue(\varepsilon)$ are the two reported above and are matched by the same transitions from Q listed above.

Now consider the pair $(p_{v\sigma}[L(v)/y, H/x],\ Queue(v\sigma))$. Term $p_{v\sigma}[L(v)/y, H/x]$ can move according to the inner occurrence of H,

$$p_{v\sigma}[L(v)/y, H/x] \xrightarrow{in(u)} q_{v\sigma u}[L(v)/y, H'/x],$$

or according to the occurrence of $L(v)$ as follows:

$p_{v\sigma}[L(v)/y,H/x] \xrightarrow{\overline{out}(v)} p_{v\sigma}[\bar{l}/y,H/x]$

In the former case, $Queue(v\sigma) \xrightarrow{in(u)} Queue(v\sigma u)$ and the pair $(q_{v\sigma u}[L(v)/y,H'/x],$ $Queue(v\sigma u)) \in R$ (sixth group). In the latter case, $Queue(v\sigma) \xrightarrow{\overline{out}(v))} Queue(\sigma)$ and the pair of reached state $(p_{v\sigma}[\bar{l}/y,H/x], Queue(\sigma)) \in R$ (fourth group). Symmetrically, if $Queue(v\sigma)$ moves first.

Exercise 3.83. As done above, check the pair $(p'_{v\sigma}[L'(v)/y,H'/x],\ Queue(v\sigma))$. □

Now consider the pair $(p_{v\sigma}[\bar{l}/y,H/x], Queue(\sigma))$ with $\sigma = u\sigma'$ (and $\sigma' \neq \varepsilon$). Process $p_{v\sigma}[\bar{l}/y,H/x]$ can only do $p_{v\sigma}[\bar{l}/y,H/x] \xrightarrow{\tau} (\nu l)(\mathbf{0}\,|\,q'_{u\sigma'}[L'(u)/y,H/x]) \sim q'_{u\sigma'}[L'(u)/y,H/x]$. To this move, process $Queue(u\sigma')$ replies by idling: $Queue(u\sigma') \stackrel{\varepsilon}{\Longrightarrow} Queue(u\sigma')$ and $(q'_{u\sigma'}[L'(u)/y,H/x], Queue(u\sigma')) \in R$ (7th group). Conversely, if $Queue(u\sigma') \xrightarrow{in(z)} Queue(u\sigma' z)$, then

$$p_{v\sigma}[\bar{l}/y,H/x] \xrightarrow{\tau} (\nu l)(\mathbf{0}\,|\,q'_{u\sigma'}[L'(u)/y,H/x]) \xrightarrow{in(z)} (\nu l)(\mathbf{0}\,|\,p'_{u\sigma' z}[L'(u)/y,H'/x])$$

with $(\nu l)(\mathbf{0}\,|\,p'_{u\sigma' z}[L'(u)/y,H'/x]) \sim p'_{u\sigma' z}[L'(u)/y,H'/x]$ and the pair $(p'_{u\sigma' z}[L'(u)/y,H'/x],\ Queue(u\sigma' z)) \in R$ (third group).
Moreover, if $Queue(u\sigma') \xrightarrow{\overline{out}(u)} Queue(\sigma')$, then

$$p_{vu\sigma'}[\bar{l}/y,H/x] \xrightarrow{\tau} (\nu l)(\mathbf{0}\,|\,q'_{u\sigma'}[L'(u)/y,H/x]) \xrightarrow{\overline{out}(u)} (\nu l)(\mathbf{0}\,|\,q'_{u\sigma'}[\bar{l'}/y,H/x])$$

with $(\nu l)(\mathbf{0}\,|\,q'_{u\sigma'}[\bar{l'}/y,H/x]) \sim q'_{u\sigma'}[\bar{l'}/y,H/x]$ and the pair of states $(q'_{u\sigma'}[\bar{l'}/y,H/x],$ $Queue(\sigma')) \in R$ (ninth group).

Remark 3.16. Observe that in the special case of the above when $\sigma = \varepsilon$, process $p_v[\bar{l}/y,H/x]$ can only do $p_v[\bar{l}/y,H/x] \xrightarrow{\tau} (\nu l)(\mathbf{0}\,|\,Q) \sim Q$ and $(Q, Queue(\varepsilon)) \in R$.
Conversely, if $Queue(\sigma) \xrightarrow{empty} Queue(\varepsilon)$, then $p_v[\bar{l}/y,H/x] \stackrel{empty}{\Longrightarrow} (\nu l)(\mathbf{0}\,|\,Q) \sim Q$ and $(Q, Queue(\varepsilon)) \in R$. If $Queue(\varepsilon) \xrightarrow{in(v)} Queue(v)$, then

$$p_v[\bar{l}/y,H/x] \stackrel{in(v)}{\Longrightarrow} (\nu l)(\mathbf{0}\,|\,(\nu l)(L(x) \mid H)) \sim (\nu l)(L(x) \mid H)) = p_v[L(v)/y,H/x]$$

and $(p_v[L(v)/y,H/x], Queue(v)) \in R$ (second group). □

Exercise 3.84. As done above, check the pair $(p'_{v\sigma}[\bar{l'}/y,H'/x], Queue(\sigma))$ when $\sigma = u\sigma'$ (and $\sigma' \neq \varepsilon$), as well as when $\sigma = \varepsilon$. □

Now consider the pair $(q'_{v\sigma}[L'(v)/y,H/x],\ Queue(v\sigma))$. Term $q'_{v\sigma}[L'(v)/y,H/x]$ can move according to the inner occurrence of H,

$q'_{v\sigma}[L'(v)/y,H/x] \xrightarrow{in(u)} p'_{v\sigma u}[L'(v)/y,H'/x]$,

or according to the occurrence of $L'(v)$ as follows:

$q'_{v\sigma}[L'(v)/y,H/x] \xrightarrow{\overline{out}(v)} q'_{v\sigma}[\bar{l'}/y,H/x]$

In the former case, $Queue(v\sigma) \xrightarrow{in(u)} Queue(v\sigma u)$ and the pair $(p'_{v\sigma u}[L'(v)/y,H'/x],$ $Queue(v\sigma u)) \in R$ (third group). In the latter case, $Queue(v\sigma) \xrightarrow{\overline{out}(v))} Queue(\sigma)$ and the pair of reached state $(q'_{v\sigma}[\overline{l'}/y, H/x], Queue(\sigma)) \in R$ (ninth group). Symmetrically, if $Queue(v\sigma)$ moves first.

Exercise 3.85. As done above, check the pair $(q_{v\sigma}[L(v)/y, H'/x],\ Queue(v\sigma))$. $\square$

Now consider the pair $(q'_{v\sigma}[\overline{l'}/y, H/x], Queue(\sigma))$ with $\sigma = u\sigma'$.
Process $q'_{v\sigma}[\overline{l'}/y, H/x]$ can only do $q'_{v\sigma}[\overline{l'}/y, H/x] \xrightarrow{\tau} (\nu l')(\mathbf{0} \,|\, p_{u\sigma'}[L(u)/y, H/x]) \sim p_{u\sigma'}[L(u)/y, H/x]$. To this move, process $Queue(u\sigma')$ replies by idling:
$Queue(u\sigma') \stackrel{\varepsilon}{\Longrightarrow} Queue(u\sigma')$ and $(p_{u\sigma'}[L(u)/y, H/x], Queue(u\sigma')) \in R$ (2nd group).
Conversely, if $Queue(u\sigma') \xrightarrow{in(z)} Queue(u\sigma' z)$, then

$$q'_{v\sigma}[\overline{l'}/y, H/x] \xrightarrow{\tau} (\nu l')(\mathbf{0} \,|\, p_{u\sigma'}[L(u)/y, H/x]) \xrightarrow{in(z)} (\nu l')(\mathbf{0} \,|\, q_{u\sigma' z}[L(u)/y, H'/x])$$

with $(\nu l')(\mathbf{0} \,|\, q_{u\sigma' z}[L(u)/y, H'/x]) \sim q_{u\sigma' z}[L(u)/y, H'/x]$ and the pair $(q_{u\sigma' z}[L(u)/y, H'/x],\ Queue(u\sigma' z)) \in R$ (sixth group).
Moreover, if $Queue(u\sigma') \xrightarrow{\overline{out}(u)} Queue(\sigma')$, then

$$q'_{v\sigma}[\overline{l'}/y, H/x] \xrightarrow{\tau} (\nu l')(\mathbf{0} \,|\, p_{u\sigma'}[L(u)/y, H/x]) \xrightarrow{\overline{out}(u)} (\nu l')(\mathbf{0} \,|\, p_{u\sigma'}[\bar{l}/y, H/x])$$

with $(\nu l')(\mathbf{0} \,|\, p_{u\sigma'}[\bar{l}/y, H/x]) \sim p_{u\sigma'}[\bar{l}/y, H/x]$ and the pair of states $(p_{u\sigma'}[\bar{l}/y, H/x], Queue(\sigma')) \in R$ (fourth group).

Exercise 3.86. Check the pair $(q_{v\sigma}[\bar{l}/y, H'/x], Queue(\sigma))$ with $\sigma = u\sigma'$. $\square$

And this final exercise completes the proof that R is a weak bisimulation up to $\approx$, relating $Queue(\varepsilon)$ and Q. $\square$

Chapter 4
Algebraic Laws, Congruences and Axiomatizations

Abstract Behavioral equivalences, in particular those based on bisimulation, are shown to possess interesting algebraic laws. Moreover, we discuss which of them are congruences with respect to the CCS operators. Finally, behavioral congruences are axiomatized over finite CCS, and also finitely with the use of two auxiliary operators.

4.1 Some Algebraic Laws

In this section we examine the algebraic properties of the behavioral equivalences we have discussed in Chapter 2, with emphasis on bisimulation-based equivalences.

4.1.1 Laws for Strong Equivalences

Strong bisimulation equivalence $\sim$ (see Definition 2.14) enjoys some interesting properties. In particular, the choice operator $+$ satisfies the laws of a commutative monoid with $\mathbf{0}$ as neutral element, plus an idempotence law.

Proposition 4.1. (Laws of the choice operator for strong bisimilarity) *For any* $p, q, r \in \mathscr{P}$, *the following hold:*

$$\begin{array}{rcll} p+(q+r) & \sim & (p+q)+r & \textit{(associativity)} \\ p+q & \sim & q+p & \textit{(commutativity)} \\ p+\mathbf{0} & \sim & p & \textit{(identity)} \\ p+p & \sim & p & \textit{(idempotency)} \end{array}$$

Proof. The proof is easy. For each law, it is enough to exhibit a suitable bisimulation relation. For instance, for idempotency, take relation

$$R = \{(p+p,p) \mid p \in \mathscr{P}\} \cup \mathscr{I}$$

where $\mathscr{I} = \{(q,q) \mid q \in \mathscr{P}\}$ *is the identity relation. Any transition* $p+p \xrightarrow{\mu} p'$ *must be due to (by rule (Sum*$_1$*) or (Sum*$_2$*)) to a transition* $p \xrightarrow{\mu} p'$*, hence the thesis follows as* $(p',p') \in R$*. Symmetrically, any transition* $p \xrightarrow{\mu} p'$ *originates a transition* $p+p \xrightarrow{\mu} p'$ *(by, e.g., (Sum*$_1$*)) with* $(p',p') \in R$*. Therefore R is a bisimulation. As a further instance, for associativity, take relation*

$$R' = \{(p+(q+r),(p+q)+r) \mid p,q,r \in \mathscr{P}\} \cup \mathscr{I}$$

It is easy to observe that for any transition $p+(q+r) \xrightarrow{\mu} p'$ *there exists an analogous transition* $(p+q)+r \xrightarrow{\mu} p'$ *with* $(p',p') \in R'$*, and symmetrically if* $(p+q)+r$ *moves first. For instance, assume* $p+(q+r) \xrightarrow{\mu} p'$ *is due to application of rule (Sum*$_1$*) with premise* $p \xrightarrow{\mu} p'$*. Then, starting by the same premise, by double application of rule (Sum*$_1$*), also* $(p+q)+r \xrightarrow{\mu} p'$ *is derivable.* □

Exercise 4.1. Provide suitable bisimulation relations for the laws of commutativity and identity (i.e., nil as neutral element) of the choice operator $+$. □

These monoidal laws for the choice operator justify the notational convention, which we have sometimes adopted, of using an n-ary sum operator, e.g., $p_1 + p_2 + \ldots. + p_n$, shortened as $\Sigma_{1=i}^{n} p_i$.

Remark 4.1. As these laws of the choice operator hold for strong bisimulation, they also hold for any coarser behavioral equivalence, e.g., trace equivalence. This is an obvious instance of the following general fact: if $(p,q) \in R$ and $R \subseteq R'$, then $(p,q) \in R'$, where R, R' are behavioral relations. □

Exercise 4.2. Check whether the laws of summation also hold for LTS isomorphism. For instance, is $\mathscr{C}_{p+p}$ isomorphic to $\mathscr{C}_p$? □

Example 4.1. **(Distributivity of prefixing w.r.t. summation)** An important law that does not hold for bisimulation equivalence is the distributivity of prefixing w.r.t. summation:

$$\mu.(p+q) \not\sim \mu.p+\mu.q$$

As a specific instance, consider $a.(b+c)$ and $a.b+a.c$. This law does not even hold for simulation equivalence $\simeq$ (see Definition 2.12); nonetheless, it holds for trace equivalence (see Definition 2.9). □

Exercise 4.3. Check whether $\mu.(p+q) \sim \mu.(p+q)+\mu.q$. Verify that this law holds for simulation equivalence $\simeq$, hence also for trace equivalence $=_{tr}$. □

Parallel composition satisfies the laws of a commutative monoid, with **0** as neutral element.

Proposition 4.2. (Laws of the parallel operator for strong bisimilarity) *For any $p,q,r \in \mathscr{P}$, the following hold:*

$$\begin{array}{rcll} p\,|\,(q\,|\,r) & \sim & (p\,|\,q)\,|\,r & \textit{(associativity)} \\ p\,|\,q & \sim & q\,|\,p & \textit{(commutativity)} \\ p\,|\,\mathbf{0} & \sim & p & \textit{(identity)} \end{array}$$

Proof. The proof is easy. For each law, it is enough to exhibit a suitable bisimulation relation. For instance, for identity (i.e., nil absorption), take relation

$$R = \{(p\,|\,\mathbf{0}, p) \;\big|\; p \in \mathscr{P}\}$$

Note that any transition $p\,|\,\mathbf{0} \xrightarrow{\mu} q$ must be due (by rule (Par_1)) to a transition $p \xrightarrow{\mu} p'$ with $q = p'\,|\,\mathbf{0}$ and $(p'\,|\,\mathbf{0}, p') \in R$. Symmetrically, any transition $p \xrightarrow{\mu} p'$ originates $p\,|\,\mathbf{0} \xrightarrow{\mu} p'\,|\,\mathbf{0}$ with $(p'\,|\,\mathbf{0}, p') \in R$. Hence, R is a bisimulation. □

Exercise 4.4. Check that relation $S = \{(p\,|\,q, q\,|\,p) \;\big|\; p,q \in \mathscr{P}\}$ is a bisimulation, hence proving commutativity of parallel composition w.r.t. $\sim$. Provide a suitable bisimulation relation for the associativity law of parallel composition. □

The following example and subsequent exercise attempt to show the use of these algebraic laws in the technique of bisimulation up to. They are rather technical and can be skipped on a first reading.

Example 4.2. In Example 3.12 (Section 3.4.4), the strong bisimulation R (up to $\sim$) was using states of the form $SC\,|\,\Pi_{i=1}^{n} dec.\mathbf{0}$ for $n \geq 0$. For simplicity's sake, let us assume that $n = 3$; hence, process $\Pi_{i=1}^{3} dec.\mathbf{0}$ is a shorthand notation for process $((dec.\mathbf{0}\,|\,dec.\mathbf{0})\,|\,dec.\mathbf{0})$.

We want to show that for any transition $SC\,|\,\Pi_{i=1}^{3} dec.\mathbf{0} \xrightarrow{\mu} p$ (for suitable μ and p), the reached state p is strongly bisimilar to a process $SC\,|\,\Pi_{i=1}^{k} dec.\mathbf{0}$ for $k = 2$ or $k = 4$ in the form required by relation R, by using the laws of parallel composition. First, consider $SC\,|\,\Pi_{i=1}^{3} dec.\mathbf{0} \xrightarrow{inc} (SC\,|\,dec.\mathbf{0})\,|\,\Pi_{i=1}^{3} dec.\mathbf{0}$. By using the associativity law, we get:

$(SC\,|\,dec.\mathbf{0})\,|\,\Pi_{i=1}^{3} dec.\mathbf{0} =$
$(SC\,|\,dec.\mathbf{0})\,|\,((dec.\mathbf{0}\,|\,dec.\mathbf{0})\,|\,dec.\mathbf{0}) \sim SC\,|\,(dec.\mathbf{0}\,|\,((dec.\mathbf{0}\,|\,dec.\mathbf{0})\,|\,dec.\mathbf{0}))$

Since $((dec.\mathbf{0}\,|\,dec.\mathbf{0})\,|\,dec.\mathbf{0}) \sim (dec.\mathbf{0}\,|\,(dec.\mathbf{0}\,|\,dec.\mathbf{0}))$, by congruence of $\sim$ w.r.t. parallel composition (see Proposition 4.1), we get:

$SC\,|\,(dec.\mathbf{0}\,|\,((dec.\mathbf{0}\,|\,dec.\mathbf{0})\,|\,dec.\mathbf{0})) \sim SC\,|\,(dec.\mathbf{0}\,|\,(dec.\mathbf{0}\,|\,(dec.\mathbf{0}\,|\,dec.\mathbf{0})))$

Since $(dec.\mathbf{0}\,|\,(dec.\mathbf{0}\,|\,(dec.\mathbf{0}\,|\,dec.\mathbf{0}))) \sim ((dec.\mathbf{0}\,|\,dec.\mathbf{0})\,|\,(dec.\mathbf{0}\,|\,dec.\mathbf{0}))$, by congruence we get:

$SC\,|\,(dec.\mathbf{0}\,|\,(dec.\mathbf{0}\,|\,(dec.\mathbf{0}\,|\,dec.\mathbf{0}))) \sim SC\,|\,((dec.\mathbf{0}\,|\,dec.\mathbf{0})\,|\,(dec.\mathbf{0}\,|\,dec.\mathbf{0}))$

Since $(((dec.\mathbf{0}\,|\,dec.\mathbf{0})\,|\,(dec.\mathbf{0}\,|\,dec.\mathbf{0})) \sim (((dec.\mathbf{0}\,|\,dec.\mathbf{0})\,|\,dec.\mathbf{0})\,|\,dec.\mathbf{0})$, by congruence we get:

$SC\,|\,((dec.\mathbf{0}\,|\,dec.\mathbf{0})\,|\,(dec.\mathbf{0}\,|\,dec.\mathbf{0})) \sim SC\,|\,(((dec.\mathbf{0}\,|\,dec.\mathbf{0})\,|\,dec.\mathbf{0})\,|\,dec.\mathbf{0})$
$= SC\,|\,\Pi_{i=1}^{4} dec.\mathbf{0}$

Summing up, $SC \,|\, \Pi_{i=1}^{3} dec.\mathbf{0} \xrightarrow{inc} (SC \,|\, dec.\mathbf{0}) \,|\, \Pi_{i=1}^{3} dec.\mathbf{0} \sim SC \,|\, \Pi_{i=1}^{4} dec.\mathbf{0}$.

Second, $SC \,|\, \Pi_{i=1}^{3} dec.\mathbf{0} = SC \,|\, ((dec.\mathbf{0} \,|\, dec.\mathbf{0}) \,|\, dec.\mathbf{0}) \xrightarrow{dec} SC \,|\, ((\mathbf{0} \,|\, dec.\mathbf{0}) \,|\, dec.\mathbf{0})$.
Since $\mathbf{0} \,|\, dec.\mathbf{0} \sim dec.\mathbf{0} \,|\, \mathbf{0}$, by congruence, we get:
$SC \,|\, ((\mathbf{0} \,|\, dec.\mathbf{0}) \,|\, dec.\mathbf{0}) \sim SC \,|\, ((dec.\mathbf{0} \,|\, \mathbf{0}) \,|\, dec.\mathbf{0})$
Since $dec.\mathbf{0} \,|\, \mathbf{0} \sim dec.\mathbf{0}$, by congruence, we get:
$SC \,|\, ((dec.\mathbf{0} \,|\, \mathbf{0}) \,|\, dec.\mathbf{0}) \sim SC \,|\, (dec.\mathbf{0} \,|\, dec.\mathbf{0}) = SC \,|\, \Pi_{i=1}^{2} dec.\mathbf{0}$
Summing up, $SC \,|\, \Pi_{i=1}^{3} dec.\mathbf{0} \xrightarrow{dec} SC \,|\, ((\mathbf{0} \,|\, dec.\mathbf{0}) \,|\, dec.\mathbf{0}) \sim SC \,|\, \Pi_{i=1}^{2} dec.\mathbf{0}$.

The case for transition $SC \,|\, \Pi_{i=1}^{3} dec.\mathbf{0} \xrightarrow{dec} SC \,|\, ((dec.\mathbf{0} \,|\, \mathbf{0}) \,|\, dec.\mathbf{0})$, as well as for transition $SC \,|\, \Pi_{i=1}^{3} dec.\mathbf{0} \xrightarrow{dec} SC \,|\, ((dec.\mathbf{0} \,|\, dec.\mathbf{0}) \,|\, \mathbf{0})$, is similar, hence left as an exercise. □

Exercise 4.5. In Example 3.9 (Section 3.4.3), the strong bisimulation R (up to $\sim$) was using state of the form $(\Pi_{i=1}^{k} \overline{out}.B) \,|\, B^{n-k}$, for $0 \leq k \leq n$. For simplicity's sake, let us assume that $n = 4$. Show that for any transition $((\overline{out}.B \,|\, \overline{out}.B) \,|\, \overline{out}.B) \,|\, B \xrightarrow{\mu} p$ (for suitable μ and p), the reached state p is strongly bisimilar to a state of the form $(\Pi_{i=1}^{k} \overline{out}.B) \,|\, B^{4-k}$ for $k = 2$ or $k = 4$, by using the parallel composition laws and the congruence property for parallel composition. □

The idempotence law $p \,|\, p \sim p$ does not hold for bisimulation equivalence (and not even for trace equivalence). For instance, $a \,|\, a \not\sim a$. In some cases, it may hold, e.g., for $A \,|\, A \sim A$ with $A \stackrel{def}{=} a.A$.

These monoidal laws of $|$ justify the notational convention, which we have sometimes adopted, of using an n-ary parallel operator, e.g., $p_1 \,|\, p_2 \,|\, \ldots. \,|\, p_n$, shortened as $\Pi_{1=i}^{n} p_i$.

Exercise 4.6. Argue that the laws of the parallel operator of Proposition 4.2 hold also for LTS isomorphism. □

Remark 4.2. **(Distingusihing bisimulation equivalence from LTS isomorphism)** As an instance of a law that holds for bisimulation equivalence, but does not hold for LTS isomorphism, consider the following:

$$C \sim p \quad \text{if } C \stackrel{def}{=} p$$

Indeed, this holds for bisimulation equivalence (check this by providing a suitable bisimulation relation), but the LTS for C is, in general, not isomorphic to that of p. For instance, for $A \stackrel{def}{=} a.A$, we have that the LTS for A is reported in Figure 3.2(a), while the LTS for $a.A$ is isomorphic to the one in Figure 3.2(b). □

As anticipated by Remark 3.11 (Section 3.4.1), an interesting law that holds for bisimulation equivalence relates parallel composition and summation; it is called the *expansion law* (sometimes also called the *interleaving law*), because it explains how to expand the parallel composition of two summations into the summation of sequential processes. This law is sound for LTS isomorphism, too.

Proposition 4.3. (Expansion Law) *Let* $p = \Sigma_{i=1}^{n} \mu_i.p_i$ *and* $q = \Sigma_{j=1}^{m} \mu'_j.q_j$*. Then*

$$p\,|\,q \sim \Sigma_{i=1}^{n} \mu_i.(p_i\,|\,q) + \Sigma_{j=1}^{m} \mu'_j.(p\,|\,q_j) + \Sigma_{i,j:\overline{\mu_i}=\mu'_j} \tau.(p_i\,|\,q_j)$$

Proof. *It follows trivially from the following observation: let* T_r *be the (finite) set of all transitions outgoing from* r*, i.e.* $\{(r,\mu_k,r_k) \in \rightarrow \;\mid\; \mu_k \in Act, r_k \in \mathscr{P}\}$*; then* $r \sim \Sigma_{(r,\mu_k,r_k)\in T_r} \mu_k.r_k$. □

Exercise 4.7. (Distributivity of parallel composition w.r.t. summation) Show that the law $(p+q)\,|\,r \sim p\,|\,r + q\,|\,r$ is not valid for bisimulation equivalence. Is it valid for simulation equivalence? And for trace equivalence? □

Also, for restriction we have some interesting laws. More in the next subsection.

Proposition 4.4. *For any process* p*, the following hold:*

$$\begin{array}{ll}
(i) & (\nu a)\mathbf{0} \sim \mathbf{0} \\
(ii) & (\nu a)((\nu b)p) \sim (\nu b)((\nu a)p) \;\; \text{if } a \neq b \\
(iii) & (\nu a)((\nu a)p) \sim (\nu a)p \\
(iv) & (\nu a)(\mu.p) \sim \begin{cases} \mathbf{0} & \text{if } \mu = a \text{ or } \mu = \overline{a} \\ \mu.(\nu a)p & \text{otherwise} \end{cases}
\end{array}$$

Proof. *The proofs of the first three laws are left as an exercise for the reader. For the fourth law, it is easy to check that relation*

$$R = \{((\nu a)(\mu.p), \mu.(\nu a)p) \;\mid\; \mu \neq a, \overline{a}\} \cup \{((\nu a)(\mu.p), \mathbf{0}) \;\mid\; \mu = a \vee \mu = \overline{a}\} \cup \mathscr{I}$$

is a strong bisimulation. □

The second law — which states that the actual order of restricted names is inessential — together with third law — which states that restricting twice by the same action has the same effect as restricting once — justifies the notational convention, that we often adopt, of using the restriction operator over a *set* of names; e.g., for $(\nu a)((\nu b)((\nu a)p))$, we can use the notation $(\nu a,b)p$.

Remark 4.3. **(Distributivity of restriction w.r.t. summation)** Another law for restriction is usually listed, namely:

$$(\nu a)(p+q) \;\sim\; (\nu a)p + (\nu a)q$$

Strictly speaking, the term $(\nu a)p + (\nu a)q$ is not a CCS process. However, if we consider a more generous syntax for CCS — as is mostly done in other books, e.g., [Mil89] (but not in [Mil99]), and as we will do in Section 4.3 — it is easy to see that this law holds; it is enough to check that relation R below is a strong bisimulation:

$$R = \{((\nu a)(p+q), (\nu a)p + (\nu a)q) \;\mid\; p,q \in \mathscr{P}\} \cup \mathscr{I}$$ □

4.1.2 Syntactic Substitution and Alpha-Conversion

Other interesting laws hold for bisimulation equivalence. In particular the so-called law of *alpha-conversion* that allows for the replacement of a restricted name with another one not in use, hence expressing that the actual name used in a restriction is inessential:

$$(\nu a)p \sim (\nu b)(p\{b/a\}) \quad \text{if } b \notin fn(p) \cup bn(p)$$

In this law we use some auxiliary notations: $fn(p)$ and $bn(p)$ are the sets of *free* and *bound* names of p, respectively, while $p\{b/a\}$ stands for the term obtained by replacing the free occurrences of a in p by b. These definitions are formally given as follows.

Definition 4.1. (Free names) The *free names* of a process p, denoted $fn(p)$, are defined as the set $F(p,\emptyset)$, where $F(p,I)$, with I a set of process constants, is defined as follows:

$$\begin{aligned}
F(\mathbf{0},I) &= \emptyset \\
F(a.p,I) &= F(\bar{a}.p,I) \qquad = F(p,I) \cup \{a\} \\
F(\tau.p,I) &= F(p,I) \\
F(p+q,I) &= F(p\,|\,q,I) \qquad = F(p,I) \cup F(q,I) \\
F((\nu a)p,I) &= F(p,I) \setminus \{a\} \\
F(C,I) &= \begin{cases} F(q,I \cup \{C\}) & \text{if } C \stackrel{def}{=} q \text{ and } C \notin I \\ \emptyset & \text{if } C \in I \end{cases}
\end{aligned}$$

□

According to this definition, $fn(p)$ is effectively computable only for those processes p where the set *Const(p)* of the constants used in p is finite, i.e., for processes in finitary CCS.

Exercise 4.8. Compute the set of free names of the following CCS processes:

$$(\nu a)(a.b\,|\,\bar{a})\,|\,a \qquad b.d\,|\,(\nu a)(a.c) \qquad (\nu a)(b\,|\,(\nu a)(b.\bar{a}))$$

Compute also the set of free names for the counter process C of Example 3.17. □

Observe that c belongs to $fn(b.d\,|\,(\nu a)(a.c))$, while $c \notin sort(b.d\,|\,(\nu a)(a.c))$, because c cannot be executed. (For the definition of $sort(p)$, see Definition 2.5.)

Exercise 4.9. (Size of a finitary CCS process) Continuing Exercise 3.29, extend the definition of function *size*(p), computing the total number of prefixes occurring in p, by mimicking the definition of function $F(p,I)$ in Definition 4.1. □

Proposition 4.5. *For any finitary CCS process p, the set $fn(p)$ is finite.*

Proof. *The size of $fn(p)$ is less than the size of the set of all prefixes occurring in p and in the body of any constant $A \in$ Const(p); since Const(p) is finite, $fn(p)$ is finite as well.* □

Proposition 4.6. *For any p, if $p \xrightarrow{\mu} p'$, then $fn(p') \subseteq fn(p)$, and if $\mu \in \{a, \overline{a}\}$, then $a \in fn(p)$.*

Proof. *By induction on the proof of transition $p \xrightarrow{\mu} p'$.* □

Let us denote with $\overline{fn(p)}$ the set $\{\overline{a} \mid a \in fn(p)\}$. Hence, the proposition above ensures that any executable action μ must belong to set $fn(p) \cup \overline{fn(p)} \cup \{\tau\}$. This means that, for any p in finitary CCS, $fn(p) \cup \overline{fn(p)} \cup \{\tau\}$ is an effectively decidable, finite superset of the finite set $sort(p)$ of the actions that are actually executable: $sort(p) \subseteq fn(p) \cup \overline{fn(p)} \cup \{\tau\}$. (See also Remark 3.15, stating that $sort(p)$ is, in general, not effectively decidable.)

Corollary 4.1. *For any finitary CCS process p, the set $sort(p)$ is finite.*

Proof. *By Proposition 4.6, we know that any action $\mu \in sort(p)$ is also an action in $fn(p) \cup \overline{fn(p)} \cup \{\tau\}$. The thesis follows by Proposition 4.5, which ensures that $fn(p)$ is finite for any finitary CCS process p.* □

As an instance of a (non-finitary) CCS process such that $sort(p)$ is infinite, consider Example 3.20.

Exercise 4.10. Argue that if $p \sim q$, then $sort(p) = sort(q)$. Show two finite CCS processes p and q such that $p \sim q$ and $fn(p) \neq fn(q)$. □

Definition 4.2. (Bound names) The *bound names* of a process p, denoted $bn(p)$, are defined as the set $B(p, \emptyset)$, where $B(p, I)$, with I a set of process constants, is defined as follows:

$$\begin{aligned}
B(\mathbf{0}, I) &= \emptyset \\
B(\mu.p, I) &= B(p, I) \\
B(p+q, I) &= B(p \mid q, I) \quad = B(p, I) \cup B(q, I) \\
B((\nu a)p, I) &= B(p, I) \cup \{a\} \\
B(C, I) &= \begin{cases} B(q, I \cup \{C\}) & \text{if } C \stackrel{def}{=} q \text{ and } C \notin I \\ \emptyset & \text{if } C \in I \end{cases}
\end{aligned}$$

Exercise 4.11. Compute the sets of bound and free names of the following CCS processes:

$$(\nu a)(a.b \mid \overline{a}) \mid a \qquad b.d \mid (\nu a)(a.c) \qquad (\nu a)(b \mid (\nu a)(b.\overline{a}))$$

Compute also the set of bound names for the counter process C of Example 3.17. □

Remark 4.4. **(Free occurrence of a name)** By definition, $fn(p) \subseteq \mathscr{L}$, i.e., a free name is not a co-name. However, for convenience, we say α *occurs free* in p to mean that α is a name (or a co-name) syntactically occurring in p that is not bound (i.e., not occurring in the scope of a restriction operator for that name). For instance, if $p = (\nu c)(\overline{a}.c.\mathbf{0} \mid (\nu a)(b.a.\mathbf{0}))$, we have that $fn(p) = \{a, b\}$ and $bn(p) = \{a, c\}$; nonetheless, $\overline{a}$ occurs free in p while a does not occur free in p. □

Note that for any p, if $p \xrightarrow{\mu} p'$, then $bn(p') = bn(p)$, because restriction is a static operator, hence it never disappears during the computation.

Exercise 4.12. Show that if $p \sim q$, then it is not necessary that $bn(p) = bn(q)$. □

Proposition 4.7. *For any finitary CCS process p, the set $bn(p)$ is finite.*

Proof. *Set $bn(p)$ is formed by all the actions occurring in any restriction operator inside p and inside the body of any constant $A \in Const(p)$; since Const(p) is finite, $bn(p)$ is finite as well.* □

Remark 4.5. (**Semantic bound names**) Even if, from a syntactic point of view, the set of bound names of any finitary CCS process is finite, the number of different names that a finitary process may use *semantically* may be unbounded. Consider again the counter process C of Example 3.17: clearly, $bn(C) = \{a, b\}$, but the use of nested, alternated occurrences of the restriction operator allows for an unbounded generation of *new* names. To explain the issue, consider the reachable process $p = (\nu a)((\nu b)(((\nu a)((\nu b)(C_2 \,|\, b.C_1) \,|\, a.C_2) \,|\, b.C_1)) \,|\, a.C)$. From a semantical point of view, p is using four different restricted names: the rightmost occurrence of a prefix a (in $a.C$) is subject to the leftmost occurrence of the operator (νa), while the leftmost occurrence of a (in $a.C_2$) is subject to the rightmost occurrence of (νa), hence the two occurrences of a are denoting two *different* semantic actions, even if they are both called a. And similarly for the occurrences of prefixes b and operators (νb). Indeed, from a semantical point of view, C may generate unboundedly many different *semantic* bound names, but these are obtained by means of two syntactic bound names only, namely a and b, by a clever alternation of their use. □

Definition 4.3. (Substitutions) A substitution is a set $\{b_i/a_i\}_{i \in I}$ of associations of the form b_i/a_i for $i \in I$, meaning that action $a_i \in \mathscr{L}$ is to be replaced by action $b_i \in \mathscr{L}$, when applied to some term. A substitution $\{b_i/a_i\}_{i \in I}$ is *admissible* when $a_i \neq a_j$ for all $i \neq j$ and $b_i \neq a_j$ for all i, j. Hence, for instance, the following are not admissible: $\{a/a\}$, $\{b/a, c/a\}$ and $\{b/a, a/b\}$. On the contrary, $\{b/a\}$, $\{b/a, b/c\}$ and $\{b/a, d/c\}$ are admissible. We use θ to range over the set of admissible substitutions.

A substitution $\{b_i/a_i\}_{i \in I}$ is *empty*, denoted by ε, if $|I| = 0$, i.e., there is no association in the set. Of course, the empty substitution ε is admissible. A substitution $\{b_i/a_i\}_{i \in I}$ is *unary* if $|I| = 1$, i.e., it is of the form $\{b/a\}$. A unary substitution $\{b/a\}$ is admissible provided that $a \neq b$.

The composition of an admissible substitution $\theta = \{b_1/a_1, \ldots, b_n/a_n\}$ with a unary admissible substitution $\{b/a\}$, denoted by $\theta \circ \{b/a\}$, is defined — provided that $b \neq a_j$ for all $j = 1, \ldots, n$ — as follows:

- $\{b_1/a_1, \ldots, b_n/a_n\}$ if there exists an index j such that $a = a_j$;
- $\{b'_1/a_1, \ldots, b'_n/a_n, b/a\}$ if $a \neq a_j$ for all $j = 1, \ldots, n$, where $b'_i = b_i$ if $a \neq b_i$, otherwise $b'_i = b$, for $i = 1, \ldots, n$.

Note that $\theta \circ \{b/a\}$, if defined, is an admissible substitution. Examples of substitution composition are the following: $\varepsilon \circ \{b/a\} = \{b/a\}$, $\{b/a\} \circ \{b/a\} = \{b/a\}$ and $\{b/a, d/c\} \circ \{d/b\} = \{d/a, d/c, d/b\}$.

A nonemtpy admissible substitution $\theta = \{b_1/a_1, \dots, b_n/a_n\}$ can be represented as the composition of the unary admissible substitution $\{b_1/a_1\}$ and of the admissible substitution $\theta' = \{b_2/a_2, \dots, b_n/a_n\}$, i.e., $\theta = \{b_1/a_1\} \circ \theta'$. □

Remark 4.6. **(Parametrized constants)** In the following definition of syntactic substitution, we are applying a unary substitution $\{b/a\}$ also to any constant A, resulting in the new constant $A_{\{b/a\}}$. In order to get a completely satisfactory definition, we have to assume that constants are parametrized by admissible substitutions, with the proviso that a normal constant A is simply parametrized by the empty substitution A_ε. Then, applying a unary substitution $\{b/a\}$ to A_θ, $A_\theta\{b/a\}$ would generate a new constant with index $\theta \circ \{b/a\}$, i.e., $A_{\theta \circ \{b/a\}}$.

As we usually apply unary substitutions of the form $\{b/a\}$ to a process p when b is a fresh name (neither free nor bound in p), we are sure that if in p there is an occurrence of A_θ, then $\theta \circ \{b/a\}$ is definable and gives rise to another admissible substitution. □

Definition 4.4. (Syntactic Substitution) The syntactic substitution $p\{b/a\}$ of action b for a different action a inside a CCS process p is defined as follows:

$$
\begin{aligned}
\mathbf{0}\{b/a\} &= \mathbf{0}\\
(a.p)\{b/a\} &= b.(p\{b/a\})\\
(\overline{a}.p)\{b/a\} &= \overline{b}.(p\{b/a\})\\
(\mu.p)\{b/a\} &= \mu.(p\{b/a\}) \qquad \text{if } \mu \neq a, \overline{a}\\
(p+q)\{b/a\} &= p\{b/a\} + q\{b/a\}\\
(p\,|\,q)\{b/a\} &= p\{b/a\} \,|\, q\{b/a\}\\
((\nu c)p)\{b/a\} &= (\nu c)(p\{b/a\}) \qquad \text{if } c \neq a, b\\
((\nu a)p)\{b/a\} &= (\nu a)p\\
((\nu b)p)\{b/a\} &= \begin{cases} (\nu b)p & \text{if } a \notin fn(p)\\ (\nu c)((p\{c/b\})\{b/a\}) & \text{otherwise, with } c \notin fn(p) \cup bn(p) \end{cases}\\
C_\theta\{b/a\} &= \begin{cases} C_\theta & \text{if } a \notin fn(C_\theta)\\ C_{\theta \circ \{b/a\}} & \text{otherwise, with } C_{\theta \circ \{b/a\}} \stackrel{def}{=} q\{b/a\} \text{ if } C_\theta \stackrel{def}{=} q \end{cases}
\end{aligned}
$$

The application of an admissible substitution $\theta = \{b/a\} \circ \theta'$ to a process p can be computed as follows: $p\theta = (p\{b/a\})\theta'$, with the proviso that $p\varepsilon = p$. □

Example 4.3. Let us consider again Example 3.10, where we have defined a pipelined buffer $Buf \stackrel{def}{=} B \frown B$, where $B \stackrel{def}{=} in.\overline{out}.B$ is the one-position buffer of Exercise 3.46 and the linking connects the *out* port of the left buffer to the *in* port of the right buffer. As a matter of fact, the explicit definition of *Buf* is

$$Buf \stackrel{def}{=} (\nu d)(B\{d/out\} \,|\, B\{d/in\})$$

and the effect of applying the substitution to B is the definition of a new constant where the substitution is applied to its body:

$$B_{\{d/out\}} \stackrel{def}{=} (in.\overline{out}.B)\{d/out\} = in.\overline{d}.B_{\{d/out\}}$$
$$B_{\{d/in\}} \stackrel{def}{=} (in.\overline{out}.B)\{d/in\} = d.\overline{out}.B_{\{d/in\}}$$
□

Exercise 4.13. Let us consider constants $A \stackrel{def}{=} a.A + b.B$ and $B \stackrel{def}{=} c.d.A + a.B$. Suppose we want to replace action a with action c. Compute the result of $A\{c/a\}$. □

Exercise 4.14. Consider $A \stackrel{def}{=} (\nu a)(a.b.A \,|\, \overline{a}.c.A)$. Compute $A\{b/a\}$ and $A\{a/b\}$. □

Proposition 4.8. *For any p in finitary CCS and for any $a, b \in \mathscr{L}, a \neq b$, $p\{b/a\}$ is a finitary CCS process.*

Proof. *The proof is by case analysis, following the definition of syntactic substitution. As Const(p) is finite, the recursive application of the syntactic substitution to the body of each constant will eventually terminate.* □

Note that if p is not in finitary CCS, then the procedure above for computing $p\{b/a\}$ may not terminate.

Proposition 4.9. *For any process $p \in \mathscr{P}$, the following hold:*

$$\begin{array}{lll}
(i) & a \notin fn(p\{b/a\}) & \\
(ii) & p\{b/a\} = p & \text{if } a \notin fn(p) \\
(iii) & p\{b/a\}\{d/a\} = p\{b/a\} &
\end{array}$$

Proof. *The proofs of (i) and (ii) are left as an exercise. Item (iii) holds because $a \neq b$ by Definition 4.4 and also because of (ii) as $a \notin fn(p\{b/a\})$ by (i).* □

Proposition 4.10. *For any process $p \in \mathscr{P}$, let $b \notin fn(p)$. Then, the following hold:*

$$\begin{array}{llll}
(i) & p\{b/a\} \xrightarrow{b} q & \text{iff} & \exists p'. q = p'\{b/a\} \text{ and } p \xrightarrow{a} p' \\
(ii) & p\{b/a\} \xrightarrow{\overline{b}} q & \text{iff} & \exists p'. q = p'\{b/a\} \text{ and } p \xrightarrow{\overline{a}} p' \\
(iii) & p\{b/a\} \xrightarrow{\mu} q & \text{iff} & \exists p'. q = p'\{b/a\} \text{ and } p \xrightarrow{\mu} p' \;\; \text{if } \mu \notin \{a, \overline{a}, b, \overline{b}\} \\
(iv) & p\{b/a\} \xrightarrow{\mu} q & \text{implies} & \mu \neq a, \overline{a}
\end{array}$$

Proof. *The proof is by induction on the proof of the involved transitions and is left as an exercise for the reader.* □

Exercise 4.15. Prove that, for any process $p \in \mathscr{P}$, and for any four actions a, b, c, d — where only b and d can be the same action — the following holds:
$p\{b/a\}\{d/c\} \sim p\{d/c\}\{b/a\}$. □

Proposition 4.11. (Alpha-conversion and scope-enlargement) *For any $p, q \in \mathscr{P}$, the following additional laws of restriction hold:*

$$\begin{array}{llll}
(i) & (\nu a)p \sim p & \text{if } a \notin fn(p) & \\
(ii) & (\nu a)p \,|\, q \sim (\nu a)(p \,|\, q) & \text{if } a \notin fn(q) & (\text{scope-enlargement}_1) \\
(iii) & p \,|\, (\nu a)q \sim (\nu a)(p \,|\, q) & \text{if } a \notin fn(p) & (\text{scope-enlargement}_2) \\
(iv) & (\nu a)p \sim (\nu b)(p\{b/a\}) & \text{if } b \notin fn(p) \cup bn(p) & (\text{alpha-conversion})
\end{array}$$

Proof. The proofs of (i)–(iii) are left as an exercise for the reader. To prove (iv), consider the relation

$$R = \{((\nu a)p, (\nu b)(p\{b/a\}))) \mid p \in \mathscr{P} \wedge b \notin fn(p) \cup bn(p)\}.$$

If $(\nu a)p \xrightarrow{\mu} q$, *then by rule (Res) this is due to* $p \xrightarrow{\mu} p'$ *with* $\mu \neq a, \overline{a}$, *so that* $q = (\nu a)p'$. *Since* $b \notin fn(p)$, *then* $\mu \neq b, \overline{b}$; *hence, by Proposition 4.10(iii), transition* $p\{b/a\} \xrightarrow{\mu} p'\{b/a\}$ *is derivable, and also* $(\nu b)(p\{b/a\}) \xrightarrow{\mu} (\nu b)(p'\{b/a\})$ *is derivable by (Res), with* $((\nu a)p', (\nu b)(p'\{b/a\}) \in R$.

Symmetrically, if $(\nu b)(p\{b/a\}) \xrightarrow{\mu} q$, *then by (Res) this is due to* $p\{b/a\} \xrightarrow{\mu} q'$, *with* $\mu \neq b, \overline{b}$, *so that* $q = (\nu b)q'$. *Since* $a \notin fn(p\{b/a\})$ *by Proposition 4.9(i), then* $\mu \neq a, \overline{a}$. *Hence, by Proposition 4.10(iii),* $q' = p'\{b/a\}$ *and transition* $p \xrightarrow{\mu} p'$ *is derivable; by rule (Res), also transition* $(\nu a)p \xrightarrow{\mu} (\nu a)p'$ *is derivable, with* $((\nu a)p', (\nu b)(p'\{b/a\}) \in R$. □

Exercise 4.16. Pinpoint the conditions under which $(\nu a)(p \mid q) \sim (\nu a)p \mid (\nu a)q$. Show an example for which this law does not hold. □

Exercise 4.17. By using scope-enlargement and alpha-conversion, show that

$$\begin{aligned}
a.(\nu a)b \mid (\nu a)a.c &\sim (\nu a')(a.(\nu a)b \mid a'.c)\\
(\nu a)(\overline{a}.b + c) \mid (\nu a)(a.b + d) &\sim (\nu a)((\nu a')((\overline{a}.b + c) \mid (a'.b + d)))\\
(\nu a)((((\nu b)(\overline{a}.b.c) \mid a.d) \mid (\nu a)(a.\overline{b}.c)) &\sim (\nu a, b', a')((\overline{a}.b'.c \mid a.d) \mid a'.\overline{b}.c))
\end{aligned}$$

where $(\nu a, b', a')p$ stands for $(\nu a)((\nu b')((\nu a')p))$. □

The exercise above shows that, by means of scope-enlargement and alpha-conversion, it is possible, in some cases, to put all the restrictions at the top level, by giving a different syntactic name to each (semantic) bound name.

Definition 4.5. (Prenex form) A CCS process p is *nonrestricted* if there is no occurrence of any restriction operator in p. Process p is in *weak prenex form* if *either* p is nonrestricted, *or* $p = (\nu a)q$ with q in weak prenex form.

A CCS process p is *restriction-free* if $bn(p) = \emptyset$. Process p is in *strong prenex form* if *either* p is restriction-free, *or* $p = (\nu a)q$ with q in strong prenex form. □

For instance, process $(\nu a)((\nu a')((\overline{a}.b + c) \mid (a'.b + d)))$ is in strong prenex form, while process $(\nu a')(a.(\nu a)b \mid a'.c)$ is not even in weak prenex form. Moreover, given the constant definition $A \stackrel{def}{=} (\nu b)(a.A)$, process $c.A$ is in weak prenex form (as it is nonrestricted), but not in strong prenex form (because $bn(c.A) = \{b\}$). As a shorthand notation, a process in weak (strong) prenex form is denoted by $(\nu L)p$, where L is a set of actions and p is a non-restricted (restriction-free) process.

Note that, by syntactic definition, a finite-net CCS process is a strong prenex form, because the possible occurrences of the restriction operator can only appear at the top level. In general, it is not always possible to turn a finitary CCS process into an equivalent strong prenex form because of the occurrence of restriction inside the body of recursively defined constants. However, with the help of the additional scope-enlargment laws for the dynamic operators, listed below, it is always possible to turn a finitary CCS process into a bisimulation equivalent weak prenex form, as well as a finite CCS process into a bisimulation equivalent strong prenex form.

Proposition 4.12. (Scope-enlargement for dynamic operators) *For any $p, q \in \mathscr{P}$, the following additional laws of restriction hold:*

$$\begin{array}{ll} (i) \ \alpha.(\nu a)p \sim (\nu b)(\alpha.(p\{b/a\})) & \\ \qquad \text{if } \alpha = a \vee \alpha = \overline{a} \text{ and } b \notin fn(p) \cup bn(p) & (\textit{prefix-scope-enlargement}) \\ (ii) \ (\nu a)p + q \sim (\nu a)(p+q) \quad \text{if } a \notin fn(q) & (\textit{choice-scope-enlargement}_1) \\ (iii) \ p + (\nu a)q \sim (\nu a)(p+q) \quad \text{if } a \notin fn(p) & (\textit{choice-scope-enlargement}_2) \end{array}$$

Proof. (i) *Consider* $S = \{((\alpha.(\nu a)p, (\nu b)(\alpha.(p\{b/a\}))) \mid \alpha = a \vee \alpha = \overline{a} \wedge p \in \mathscr{P} \wedge b \notin fn(p) \cup bn(p)\} \cup R$, *where* R *is the bisimulation relation reported in the proof of Proposition 4.11. It is easy to check that* S *is a bisimulation, proving the prefix-scope-enlargement law.* (ii) *and* (iii) *are left as an exercise for the reader.* □

Note that the complementary law of prefix-scope-enlargement is

$$\mu.(\nu a)p \sim (\nu a)(\mu.p) \quad \text{if } \mu \neq a, \overline{a}$$

which was illustrated in Proposition 4.4. Note also that in the law of choice-scope-enlargement, process $(\nu a)p + q$ is not a CCS process, as discussed in Remark 4.7 about distributivity of restriction w.r.t. summation.

Exercise 4.18. By using the scope-enlargement laws and the alpha-conversion law, show that the following hold:

$$\begin{array}{r} a.(\nu a)(b.a.c) \mid (\nu a)(a.c) \sim (\nu a_1, a_2)(a.b.a_1.c \mid a_2.c) \\ b.(\nu a)(b.a.A) + a.B \sim (\nu a_1)(b.b.a_1.A_{\{a_1/a\}} + a.B) \end{array}$$

□

Exercise 4.19. (Prenex form for finitary CCS processes) By using the restriction laws (in particular, the scope-enlargement laws for parallel composition and for the dynamic operators), prove that, for any finitary CCS process p there exists a finitary CCS process q in weak prenex form such that $p \sim q$.

(*Hint:* By double induction, first by structural induction (base cases: $\mathbf{0}$ and constant A), and then by induction on the *b-size* (counting the number of bound names occurring in a process) of the involved processes, for the cases of the choice operator and the parallel one. Function *b-size*(p) is defined as follows:

$$\begin{array}{rlrl} b\text{-}size(\mathbf{0}) &= 0 & b\text{-}size(\mu.p) &= b\text{-}size(p) \\ b\text{-}size(p_1 + p_2) &= b\text{-}size(p_1) + b\text{-}size(p_2) & b\text{-}size((\nu a)p) &= 1 + b\text{-}size(p) \\ b\text{-}size(p_1 \mid p_2) &= b\text{-}size(p_1) + b\text{-}size(p_2) & b\text{-}size(C) &= 0 \end{array}$$

You need also the congruence property of $\sim$ w.r.t. CCS operators, proved in Theorem 4.1.)

This proof can be used to show that for any finite CCS process, there is a bisimulation equivalent finite CCS process in strong prenex form. □

Example 4.4. **(Linking is associative)** In Example 3.10 we have introduced the linking operator. We want to prove that the linking operator is associative, up to strong bisimulation $\sim$, i.e.,

$$p^\frown(q^\frown r) \sim (p^\frown q)^\frown r$$

by exploiting the fact that parallel composition is associative (Proposition 4.2), the laws in Proposition 4.4 and Proposition 4.11, the definition of syntactic substitution in Definition 4.4 together with its properties in Proposition 4.9 and Exercise 4.15 and, finally, the fact that $\sim$ is a congruence for parallel composition and restriction, as we will see in Theorem 4.1, Section 4.2.

As a matter of fact, by taking new names d_1 and d_2, and assuming that action a of p is linked to b of q, and that action a of q is linked to action b of r, we have:

$$\begin{aligned}
p^\frown(q^\frown r) &= (\nu d_1)(p\{d_1/a\} \mid (q^\frown r)\{d_1/b\}) \\
&= (\nu d_1)(p\{d_1/a\} \mid ((\nu d_2)(q\{d_2/a\} \mid r\{d_2/b\}))\{d_1/b\}) \\
&= (\nu d_1)(p\{d_1/a\} \mid (\nu d_2)((q\{d_2/a\} \mid r\{d_2/b\})\{d_1/b\})) \\
&= (\nu d_1)(p\{d_1/a\} \mid (\nu d_2)(q\{d_2/a\}\{d_1/b\} \mid r\{d_2/b\}\{d_1/b\})) \\
&= (\nu d_1)(p\{d_1/a\} \mid (\nu d_2)(q\{d_2/a\}\{d_1/b\} \mid r\{d_2/b\})) \\
&\sim (\nu d_1)(p\{d_1/a\} \mid (\nu d_2)(q\{d_1/b\}\{d_2/a\} \mid r\{d_2/b\})) \\
&\sim (\nu d_1)(\nu d_2)(p\{d_1/a\} \mid (q\{d_1/b\}\{d_2/a\} \mid r\{d_2/b\})) \\
&\sim (\nu d_1)(\nu d_2)((p\{d_1/a\} \mid q\{d_1/b\}\{d_2/a\}) \mid r\{d_2/b\}) \\
&\sim (\nu d_2)(\nu d_1)((p\{d_1/a\} \mid q\{d_1/b\}\{d_2/a\}) \mid r\{d_2/b\}) \\
&\sim (\nu d_2)((\nu d_1)(p\{d_1/a\} \mid q\{d_1/b\}\{d_2/a\}) \mid r\{d_2/b\}) \\
&= (\nu d_2)((\nu d_1)(p\{d_1/a\}\{d_2/a\} \mid q\{d_1/b\}\{d_2/a\}) \mid r\{d_2/b\}) \\
&= (\nu d_2)((\nu d_1)((p\{d_1/a\} \mid q\{d_1/b\})\{d_2/a\}) \mid r\{d_2/b\}) \\
&= (\nu d_2)(((\nu d_1)(p\{d_1/a\} \mid q\{d_1/b\}))\{d_2/a\} \mid r\{d_2/b\}) \\
&= (\nu d_2)((p^\frown q)\{d_2/a\} \mid r\{d_2/b\}) \;=\; (p^\frown q)^\frown r
\end{aligned}$$

□

4.1.3 Laws for Weak Equivalences

For weak bisimulation equivalence $\approx$ (see Definition 2.20) a very simple law holds:

$$\tau.p \approx p.$$

This can be easily proved by observing that (i) if $\tau.p \xrightarrow{\tau} p$, then $p \stackrel{\varepsilon}{\Longrightarrow} p$ and $p \approx p$, and (ii) if $p \xrightarrow{\mu} p'$, then $\tau.p \stackrel{\mu}{\Longrightarrow} p'$ and $p' \approx p'$. However, such a law does not hold for rooted weak bisimilarity $\approx^c$ (see Definition 2.23), because if $\tau.p \xrightarrow{\tau} p$, then p may be unable to perform $\stackrel{\tau}{\Longrightarrow}$ reaching a state weakly bisimilar to p itself; e.g., consider $\tau.a.\mathbf{0}$ and $a.\mathbf{0}$. This law holds also for weak trace equivalence, namely $\tau.p =_{wtr} p$, and for all the weak equivalences in between these two, according to the classification in Figure 2.33. As we will see, it also holds for branching bisimilarity.

Exercise 4.20. Prove that $p \approx q$ if and only if $\mu.p \approx^c \mu.q$, for any $\mu \in Act$. (*Hint:* For the implication from right to left, it may be useful to generalize Exercise 2.78 as follows: if $q \stackrel{\tau}{\Longrightarrow} q'$ with $q' \approx p$ and $p \stackrel{\tau}{\Longrightarrow} p'$ with $p' \approx q$, then $p \approx^c q$.) □

The following three τ-laws hold for $\approx^c$. Hence, they also hold for weak bisimilarity $\approx$ and for any coarser weak equivalence (e.g., weak trace equivalence).

Proposition 4.13. (τ-laws for rooted weak bisimilarity) *For any processes p,q and for any $\mu \in Act$, the following hold:*

$$\begin{array}{ll} (i) & \mu.\tau.p \approx^c \mu.p \\ (ii) & p+\tau.p \approx^c \tau.p \\ (iii) & \mu.(p+\tau.q) \approx^c \mu.(p+\tau.q)+\mu.q \end{array}$$

Proof. *The first law follows directly by the fact that $\tau.p \approx p$ and by Exercise 4.20.*

For the second law, observe that the only move from $\tau.p$, namely $\tau.p \stackrel{\tau}{\longrightarrow} p$, can be easily matched by $p+\tau.p$, with $p+\tau.p \stackrel{\tau}{\longrightarrow} p$ and $p \approx p$. Conversely, if $p+\tau.p \stackrel{\mu}{\longrightarrow} p'$, then $\tau.p \stackrel{\mu}{\Longrightarrow} p'$ with $p' \approx p'$. Indeed, $p+\tau.p \stackrel{\mu}{\longrightarrow} p'$ is derivable if either $p \stackrel{\mu}{\longrightarrow} p'$ or $\tau.p \stackrel{\tau}{\longrightarrow} p$ (and $\mu = \tau$ and $p' = p$). In the former case $\tau.p \stackrel{\tau}{\longrightarrow} p \stackrel{\mu}{\longrightarrow} p'$ and $p' \approx p'$; in the latter case, $\tau.p \stackrel{\tau}{\longrightarrow} p$ and $p \approx p$.

For the third law, the only nontrivial case is when $\mu.(p+\tau.q)+\mu.q \stackrel{\mu}{\longrightarrow} q$; in such a case $\mu.(p+\tau.q) \stackrel{\mu}{\Longrightarrow} q$ with $q \approx q$. □

Exercise 4.21. (i) Argue that $\mu.(p+\tau.q) \approx^c \mu.(p+q)+\mu.q$ is not valid. Find a suitable weak equivalence for which it holds. (ii) Argue that also $\mu.(p+q) \approx^c \mu.(\tau.p+\tau.q)$ is not valid as well. □

Lemma 4.1. (Hennessy Lemma) *[Mil89]*
For any processes p and q, $p \approx q$ if and only if ($p \approx^c q$ or $p \approx^c \tau.q$ or $\tau.p \approx^c q$).

Proof. $\Leftarrow$) *We have three cases: (i) If $p \approx^c q$, then $p \approx q$ by Exercise 2.76. (ii) If $p \approx^c \tau.q$, then $p \approx \tau.q \approx q$. ($iii$) Symmetric to the previous one.*

$\Rightarrow$) *We assume that $p \approx q$. We consider three cases: (i) If there exists p' such that $p \stackrel{\tau}{\longrightarrow} p' \approx q$, then it is easy to observe that $p \approx^c \tau.q$. As a matter of fact, in one direction, if $\tau.q \stackrel{\tau}{\longrightarrow} q$, then $p \stackrel{\tau}{\longrightarrow} p'$ with $p' \approx q$, as required. In the other direction, if $p \stackrel{\tau}{\longrightarrow} p'$ with $p' \approx q$, then $\tau.q \stackrel{\tau}{\longrightarrow} q$, as required; if $p \stackrel{\tau}{\longrightarrow} p''$ with $p'' \not\approx q$, then $\tau.q \stackrel{\tau}{\longrightarrow} q \stackrel{\varepsilon}{\Longrightarrow} q''$ with $p'' \approx q''$, because $p \approx q$; finally, if $p \stackrel{\alpha}{\longrightarrow} p''$, then $\tau.q \stackrel{\tau}{\longrightarrow} q \stackrel{\alpha}{\Longrightarrow} q''$ with $p'' \approx q''$, because $p \approx q$; hence the rooted weak bisimulation condition is respected also in this direction. (ii) Symmetrically to the above case, if $q \stackrel{\tau}{\longrightarrow} q' \approx p$, then $\tau.p \approx^c q$. (iii) If neither of the above two holds, then we can show that $p \approx^c q$ as follows. If $p \stackrel{\alpha}{\longrightarrow} p'$, then $q \stackrel{\alpha}{\Longrightarrow} q'$, with $p' \approx q'$, and the definition of rooted weak bisimilarity $\approx^c$ is respected. If $p \stackrel{\tau}{\longrightarrow} p'$, then $q \stackrel{\varepsilon}{\Longrightarrow} q'$, with $p' \approx q'$; note that q' cannot be q itself, otherwise we would be in case (i) above, This means that $q \stackrel{\tau}{\Longrightarrow} q'$ and the definition of rooted weak bisimilarity $\approx^c$ is respected. Symmetrically, if q moves first.* □

Exercise 4.22. (i) Prove that $p \mid \tau.q \approx p \mid q$ for any p,q. (ii) Show that $p \mid \tau.q \not\approx^c p \mid q$ for some p,q. (iii) Prove also that $p \mid \tau.q \approx^c \tau.(p \mid q)$ for any p,q. □

For branching bisimulation equivalence $\approx_{br}$ (see Definition 2.24), we have that $\tau.p \approx_{br} p$. Indeed, relation $R = \{(\tau.p,p)\} \cup \mathscr{I}$ is a branching bisimulation, where $\mathscr{I} = \{(p,p) \mid p \in \mathscr{P}\}$: to transition $\tau.p \xrightarrow{\tau} p$, p can reply with $p \overset{\varepsilon}{\Longrightarrow} p$ and the branching conditions $(p,p) \in R$ and $(\tau.p,p) \in R$ are satisfied; symmetrically, if $p \xrightarrow{\mu} p'$, then $\tau.p \xrightarrow{\tau} p \xrightarrow{\mu} p'$ and the branching conditions $(p,p) \in R$ and $(p',p') \in R$ are satisfied.

This law, however, does not hold for rooted branching bisimulation equivalence $\approx_{br}^c$ (see Definition 2.26). As an example, consider $\tau.a$ and a: transition $\tau.a \xrightarrow{\tau} a$ cannot be matched by a in a strong way, as required by the definition of rooted branching bisimilarity.

Exercise 4.23. Prove that $p \approx_{br} q$ if and only if $\mu.p \approx_{br}^c \mu.q$, for any $\mu \in Act$. □

As rooted branching bisimilarity $\approx_{br}^c$ is strictly finer than rooted weak bisimulation equivalence $\approx^c$, it may be not a surprise that some of the τ-laws in Proposition 4.13 do not hold for it. Even if the first τ-law $\mu.\tau.p \approx_{br}^c \mu.p$ is valid (because $\tau.p \approx_{br} p$ holds and by Exercise 4.23), the other two τ-laws are invalid.

In particular, to get convinced that the third τ-law $\mu.(p+\tau.q) \approx_{br}^c \mu.(p+\tau.q) + \mu.q$ is invalid, assume that $\mu = a$, $p = c.\mathbf{0}$ and $q = b.\mathbf{0}$ (this is illustrated in Figure 2.29). Then, move $a.(c.\mathbf{0}+\tau.b.\mathbf{0}) + a.b.\mathbf{0} \xrightarrow{a} b.\mathbf{0}$ is matched (strongly, as required by the definition of rooted branching bisimilarity) by $a.(c.\mathbf{0}+\tau.b.\mathbf{0}) \xrightarrow{a} c.\mathbf{0}+\tau.b.\mathbf{0}$, but of course $b.\mathbf{0} \not\approx_{br} c.\mathbf{0}+\tau.b.\mathbf{0}$.

Similarly, the second τ-law $p+\tau.p \approx_{br}^c \tau.p$ is invalid. E.g., take $p = \tau.a.\mathbf{0}+b.\mathbf{0}$; then, to transition $p+\tau.p \xrightarrow{\tau} a.\mathbf{0}$ (which is due to transition $p \xrightarrow{\tau} a.\mathbf{0}$), $\tau.p$ can react only with $\tau.p \xrightarrow{\tau} p$, but $a.\mathbf{0} \not\approx_{br} p$. Hence, also the second τ-law is invalid.

The typical law for rooted branching bisimulation is described below. A specific instance of this law is depicted in Figure 2.32(a).

Proposition 4.14. **(τ-law for rooted branching bisimilarity)** *For any processes p,q and for any $\mu \in Act$, the following holds:*

$$\mu.(\tau.(p+q)+p) \approx_{br}^c \mu.(p+q)$$

Proof. *Observe that the two terms match their initial μ-labeled transition as in strong bisimulation, as required, and so it remains to prove that $\tau.(p+q)+p \approx_{br} p+q$. To achieve this, we will prove that the relation*

$$R = \{(\tau.(p+q)+p, p+q)\} \cup \mathscr{I}$$

is a branching bisimulation.

If $\tau.(p+q)+p \xrightarrow{\tau} p+q$, then $p+q \overset{\varepsilon}{\Longrightarrow} p+q$ and $(\tau.(p+q)+p, p+q) \in R$ as well as $(p+q,p+q) \in R$. Instead, if $\tau.(p+q)+p \xrightarrow{\mu} p'$, because $p \xrightarrow{\mu} p'$, then $p+q \overset{\varepsilon}{\Longrightarrow} p+q \xrightarrow{\mu} p'$ and $(p',p') \in R$ as well as $(\tau.(p+q)+p, p+q) \in R$.

Conversely, if $p+q$ *moves first, we have the following two cases:* (*i*) $p+q \xrightarrow{\mu} q'$ *because* $q \xrightarrow{\mu} q'$*; in such a case,* $\tau.(p+q)+p \xrightarrow{\tau} p+q \xrightarrow{\mu} q'$*, with* $(q',q') \in R$ *and* $(p+q,p+q) \in R$*;* (*ii*) $p+q \xrightarrow{\mu} p'$ *because* $p \xrightarrow{\mu} p'$*; in such a case, we have that* $\tau.(p+q)+p \xrightarrow{\tau} p+q \xrightarrow{\mu} p'$*, with* $(p+q,p+q) \in R$ *and* $(p',p') \in R$*.*

In any case, the branching bisimulation conditions are respected, hence R *is a branching bisimulation proving that* $\tau.(p+q)+p \approx_{br} p+q$*, and consequently,* $\mu.(\tau.(p+q)+p) \approx_{br}^{c} \mu.(p+q)$ *holds.* □

Exercise 4.24. Prove that the Hennessy Lemma (see Lemma 4.1) does not hold for branching bisimilarity $\approx_{br}$, i.e., it is not true that $p \approx_{br} q$ if and only if ($p \approx_{br}^{c} q$ or $p \approx_{br}^{c} \tau.q$ or $\tau.p \approx_{br}^{c} q$). (*Hint:* Consider the processes $\tau.(a+b)+a$ and $a+b$.) □

We prove that the following weakening of the Hennessy Lemma, called the *Deng Lemma* [Den07], holds for branching bisimulation.

Lemma 4.2. (Deng Lemma)
If $p \approx_{br} q$*, then one of the following three cases holds:*

(*i*) $\exists p'$ *such that* $p \xrightarrow{\tau} p'$ *and* $p' \approx_{br} q$*, or*
(*ii*) $\exists q'$ *such that* $q \xrightarrow{\tau} q'$ *and* $p \approx_{br} q'$*, or*
(*iii*) $p \approx_{br}^{c} q$*.*

Proof. Let us examine how p *and* q *match each other's transitions. We distinguish four possible cases.*

1. $p \xrightarrow{\tau} p'$ *and the first option of the branching bisimulation game is satisfied, namely there exists* q' *such that* $q \stackrel{\varepsilon}{\Longrightarrow} q'$ *with* $p \approx_{br} q'$ *and* $p' \approx_{br} q'$*; by transitivity,* $p' \approx_{br} q$*. This is exactly what Clause* (*i*) *says.*
2. $p \xrightarrow{\mu} p'$*, and there exist* q',q'',q''' *such that* $q \xrightarrow{\tau} q' \stackrel{\varepsilon}{\Longrightarrow} q'' \xrightarrow{\mu} q'''$ *with* $p \approx_{br} q''$*,* $p' \approx_{br} q'''$*. Hence, by transitivity,* $q \approx_{br} q''$*, and so by Exercise 2.82 (Stuttering Lemma), we derive* $q \approx_{br} q'$*, and also that* $p \approx_{br} q'$ *by transitivity. Hence Clause* (*ii*) *above is satisfied.*
3. *The symmetric cases of 1 and 2 by exchanging the roles of* p *and* q*.*
4. *None of the first three cases holds, i.e., either both* p *and* q *have no transitions at all or each strong transition* $p \xrightarrow{\mu} p'$ *is matched by a strong transition* $q \xrightarrow{\mu} q'$ *from* q *with* $p' \approx_{br} q'$ *and vice versa. In this case it holds that* $p \approx_{br}^{c} q$*.* □

Exercise 4.25. Prove that the Deng Lemma holds also for weak bisimilarity, i.e, if $p \approx q$, then one of the following three cases holds:

(i) $\exists p'$ such that $p \xrightarrow{\tau} p'$ and $p' \approx q$, or
(ii) $\exists q'$ such that $q \xrightarrow{\tau} q'$ and $p \approx q'$, or
(iii) $p \approx^{c} q$. □

Exercise 4.26. Prove that if $\tau.p+q \approx_{br} p$ then $p+q \approx_{br} p$. Does this hold for weak bisimilarity? (*Hint:* Show that relation $R = \{(p+q,p) \mid \tau.p+q \approx_{br} p\} \cup \{(p,p) \mid p \in \mathscr{P}\}$ is a branching bisimulation up to $\approx_{br}$.) □

4.2 Congruence

When we compare the behavior of different systems, we expect to use a behavioral relation that is not only an equivalence, but rather a *congruence*. An equivalence relation is a congruence if it is preserved by the operators of the process calculus. In our setting, we expect to be able to prove that if $p \sim q$, then $\mu.p \sim \mu.q$ for any $\mu \in Act$, $p \,|\, r \sim q \,|\, r$ for any process r, and similarly for the other operators. The reason why we require that our chosen behavioral equivalence be a congruence is that it will support *substitution of equals for equals in any context*: if a subcomponent p in a complex system $\mathscr{C}[p]$ becomes faulty, we can safely replace it with a congruent process q, so that the whole system $\mathscr{C}[q]$ behaves as $\mathscr{C}[p]$ (where a *context* $\mathscr{C}[-]$ is a process expression with a single occurrence of a hole $[-]$ in it, as a subexpression).

This idea is at the base of *compositional reasoning*: if we want to check if $p_1 \,|\, p_2$ is equivalent to $q_1 \,|\, q_2$, we can decompose the problem into two smaller subproblems, namely checking if p_i is equivalent to q_i, for $i = 1,2$; if this is the case, then we can conclude — by congruence — that $p_1 \,|\, p_2$ is equivalent to $q_1 \,|\, q_2$. When a compositional reasoning is not allowed, we may anyway take advantage of the behavioral congruence for building smaller state spaces. For instance, if we want to analyze, with the help of a verification tool (e.g., the Concurrency Workbench [CWB]), the large state space of $p_1 \,|\, p_2 \,|\, \dots \,|\, p_n$ — so large that it cannot be generated by the tool — we can first minimize the state space of the n processes p_i w.r.t. the chosen behavioral congruence, obtaining the smaller state spaces for p_i', and then generate the state space for $p_1' \,|\, p_2' \,|\, \dots \,|\, p_n'$ (possibly minimizing it too), which is often considerably smaller than the state space for $p_1 \,|\, p_2 \,|\, \dots \,|\, p_n$; such a procedure is sound if the chosen behavioral equivalence is a congruence for parallel composition.

Not all the equivalences we have discussed in Chapter 2 are congruences. For instance, it is not difficult to see that LTS ismorphism $\cong$ (see Definition 2.8) is not a congruence for $+$. As an instance, consider $a.\mathbf{0}$ and $a.(\mathbf{0}+\mathbf{0})$, which generate isomorphic LTSs; however, when put in a context $\mathscr{C}[-] = - + a.\mathbf{0}$, we get the two processes $\mathscr{C}[a.\mathbf{0}] = a.\mathbf{0} + a.\mathbf{0}$ and $\mathscr{C}[a.(\mathbf{0}+\mathbf{0})] = a.(\mathbf{0}+\mathbf{0}) + a.\mathbf{0}$ that generate non-isomorphic LTSs: the former is composed of two states only, while the latter of three. Similarly, it is not difficult to prove that completed trace equivalence (see Definition 2.11) is not a congruence for restriction: take processes $p = a.(b+c)$ and $q = a.b + a.c$, which are completed trace equivalent; however, $(\nu b)p$ and $(\nu b)q$ are not completed trace equivalent, as trace a is completed only for the latter. Nonetheless, the most fundamental behavioral equivalences for our aims are actually congruences, as the following sections show.

4.2.1 Strong Bisimulation Equivalence Is a Congruence

Theorem 4.1. *If $p \sim q$, then the following hold:*

1) $\mu.p \sim \mu.q$ *for all* $\mu \in Act$,
2) $p+r \sim q+r$ *for all* $r \in \mathscr{P}$,
3) $p \mid r \sim q \mid r$ *for all* $r \in \mathscr{P}$,
4) $(\nu a)p \sim (\nu a)q$ *for all* $a \in \mathscr{L}$.

Proof. *Assume* R *is a bisimulation such that* $(p,q) \in R$.

For case 1, consider relation $R_1 = \{(\mu.p, \mu.q) \mid \mu \in Act\} \cup R$. *It is very easy to check that* R_1 *is a bisimulation.*

For case 2, we show that $R_2 = \{(p+r, q+r) \mid r \in \mathscr{P}\} \cup R \cup \mathscr{I}$ *is a bisimulation, where* $\mathscr{I} = \{(r,r) \mid r \in \mathscr{P}\}$. *If* $p+r \xrightarrow{\mu} p'$, *this may be due to either* $p \xrightarrow{\mu} p'$ *(due to rule (Sum*$_1$*)) or* $r \xrightarrow{\mu} p'$ *(due to rule (Sum*$_2$*)). In the first case, as* $(p,q) \in R$, *there exists a transition* $q \xrightarrow{\mu} q'$ *with* $(p',q') \in R$; *by rule (Sum*$_1$*) also transition* $q+r \xrightarrow{\mu} q'$ *is derivable with* $(p',q') \in R_2$. *In the other case,* $q+r$ *can do exactly the same transition from* r: $q+r \xrightarrow{\mu} p'$ *with* $(p',p') \in R_2$. *The symmetric case when* $q+r$ *moves first is analogous, hence omitted.*

For case 3, we show that relation $R_3 = \{(p' \mid r', q' \mid r') \mid (p',q') \in R, r' \in \mathscr{P}\}$ *is a bisimulation. As* $(p,q) \in R$, *the thesis follows. Let us consider* $(p' \mid r', q' \mid r') \in R_3$, *so that* $(p',q') \in R$. *If* $p' \mid r' \xrightarrow{\mu} s$, *then this is due to either* $p' \xrightarrow{\mu} p''$ *and* $s = p'' \mid r'$ *(rule (Par*$_1$*)), or* $r' \xrightarrow{\mu} r''$ *and* $s = p' \mid r''$ *(rule (Par*$_2$*)), or* $p' \xrightarrow{\alpha} p''$, $r' \xrightarrow{\overline{\alpha}} r''$ *and* $s = p'' \mid r''$ *(rule (Com)). In the first case, since* $(p',q') \in R$, *we have that* $q' \xrightarrow{\mu} q''$ *with* $(p'',q'') \in R$, *and by rule (Par*$_1$*) also that* $q' \mid r' \xrightarrow{\mu} q'' \mid r'$ *with* $(p'' \mid r', q'' \mid r') \in R_3$. *In the second case, by rule (Par*$_2$*) we have that* $q' \mid r' \xrightarrow{\mu} q' \mid r''$ *with* $(p' \mid r'', q' \mid r'') \in R_3$. *In the third case, since* $(p',q') \in R$, *we have* $q' \xrightarrow{\alpha} q''$ *with* $(p'',q'') \in R$. *By rule (Com), also transition* $q' \mid r' \xrightarrow{\tau} q'' \mid r''$ *with* $(p'' \mid r'', q'' \mid r'') \in R_3$. *The symmetric case when* $q' \mid r'$ *moves first is analogous, and hence omitted.*

For case 4, the reader may show that $R_4 = \{((\nu a)p', (\nu a)q') \mid a \in \mathscr{L}$ *and* $(p',q') \in R\}$ *is a strong bisimulation.* □

Note that the symmetric cases $r+p \sim r+q$ and $r \mid p \sim r \mid q$ are implied by the fact that the operators of choice and parallelism are commutative w.r.t. $\sim$ (see Proposition 4.1 and 4.2).

Exercise 4.27. (Linking, again) Continuing Example 3.10, show that strong bisimulation equivalence is a congruence also for the linking operator, i.e., if $p \sim q$, then $p \frown r \sim q \frown r$ and $r \frown p \sim r \frown q$ for all $r \in \mathscr{P}$.[1] □

Exercise 4.28. Prove that trace equivalence (see Definition 2.9) is a congruence for the CCS operators: if $Tr(p) = Tr(q)$, then

1) $Tr(\mu.p) = Tr(\mu.q)$ for any $\mu \in Act$,
2) $Tr(p+r) = Tr(q+r)$ for any $r \in \mathscr{P}$,
3) $Tr(p \mid r) = Tr(q \mid r)$ for any $r \in \mathscr{P}$,
4) $Tr((\nu a)p) = Tr((\nu a)q)$ for any $a \in \mathscr{L}$.

[1] Both statements are to be proved because the linking operator is not commutative, i.e., $p \frown q$ is not bisimilar to $q \frown p$, in general.

(*Hint:* First define auxiliary operators on sets of traces: $\mu.L = \{\mu\sigma \mid \sigma \in L\}$; $L_1 \otimes L_2$ as the set of all the possible interleavings among each trace from L_1 and each trace from L_2; and $L \setminus a$ as the set composed of the traces in L with no occurrence of a or $\overline{a}$. Then, show that $Tr(\mu.p) = \{\varepsilon\} \cup \mu.Tr(p)$, $Tr(p+r) = Tr(p) \cup Tr(r)$, $Tr(p \mid r) = Tr(p) \otimes Tr(r)$, and, finally, that $Tr((\nu a)p) = Tr(p) \setminus a$. For simplicity's sake, you may restrict yourself to finite CCS only.) □

Exercise 4.29. Is simulation equivalence (see Definition 2.12) a congruence for CCS operators? And what about ready simulation equivalence (see Exercise 2.33)? □

Exercise 4.30. We have already shown that completed trace equivalence is not a congruence for restriction. Show that completed simulation equivalence (see Definition 2.13) is not a congruence for restriction as well. □

4.2.2 Recursion

In the previous section, we have proved that $\sim$ is a congruence for all the static and dynamic CCS operators. Still there is one construct missing, namely recursion. Here we simply sketch the issue, following [Mil89].

Let us consider an extension of CCS which includes variables $X, Y, \ldots$ from a set *Var*, as sketched also in Section 3.3.2. Formally, *open* CCS terms are defined by means of the abstract syntax

$$\begin{aligned} p &::= \mathbf{0} \mid X \mid \mu.q \mid p+p \\ q &::= p \mid X \mid q \mid q \mid (\nu a)q \mid C \end{aligned}$$

where $X \in Var$. For instance, $p_1 = a.(b.\mathbf{0} \mid X) + c.\mathbf{0}$, $p_2 = c.\mathbf{0} + a.(Y \mid b.\mathbf{0})$ and $p_3 = a.X + b.(c.Y \mid X) + Y$ are open CCS terms, while $a.X.\mathbf{0}$ is not. We often write $p(X_1, \ldots, X_n)$ to denote an open term p such that all the variables occurring in it are in the set $\{X_1, \ldots, X_n\}$; for instance, the three terms above can be denoted by $p_1(X), p_2(Y), p_3(X,Y)$. An open term p with no occurrence of any variable is called *closed* and corresponds to a CCS term, as defined in Section 3.1.2.

An open term $p(X_1, \ldots, X_n)$ can be *closed* by means of a substitution as follows:

$$p(X_1, \ldots, X_n)\{r_1/X_1, \ldots, r_n/X_n\}$$

with the effect that each occurrence of variable X_i is replaced by the *closed* CCS process r_i, for $i = 1, \ldots, n$. For instance, $p_1(X)\{d.\mathbf{0}/X\} = a.(b.\mathbf{0} \mid d.\mathbf{0}) + c.\mathbf{0}$.

A natural extension of bisimulation equivalence $\sim$ over open terms is as follows:

$p(X_1, \ldots, X_n) \sim q(Y_1, \ldots, Y_n)$ if for all tuples of (closed) CCS processes $(r_1, \ldots, r_n)$,
$p(X_1, \ldots, X_n)\{r_1/X_1, \ldots, r_n/X_n\} \sim q(Y_1, \ldots, Y_n)\{r_1/Y_1, \ldots, r_n/Y_n\}$.

For instance, it is easy to see that $p_1(X) \sim p_2(Y)$. As a matter of fact, for all r, $p_1(X)\{r/X\} = a.(b.\mathbf{0} \mid r) + c.\mathbf{0} \sim c.\mathbf{0} + a.(r \mid b.\mathbf{0}) = p_2(Y)\{r/Y\}$, which can be easily proved by means of the algebraic properties (and the congruence ones) of $\sim$ for the operators of parallel composition and choice.

For simplicity's sake, let us now restrict our attention to open terms using a single variable.

We can *recursively close* an open term $p(X)$ by means of a recursively defined constant. For instance, $A \stackrel{def}{=} p(X)\{A/X\}$. The resulting process constant A is a CCS process. By saying that strong bisimilarity is a congruence for recursion we mean the following: If $p(X) \sim q(Y)$ and $A \stackrel{def}{=} p(X)\{A/X\}$ and $B \stackrel{def}{=} q(Y)\{B/Y\}$, then $A \sim B$. For instance, if $C_1 \stackrel{def}{=} p_1(X)\{C_1/X\}$ and $C_2 \stackrel{def}{=} p_2(Y)\{C_2/Y\}$, then $C_1 \sim C_2$ because $p_1(X) \sim p_2(Y)$. The following theorem proves this fact for terms using at most one variable.

Theorem 4.2. *Let p and q be two open terms containing variable X at most. Let $A \stackrel{def}{=} p\{A/X\}$, $B \stackrel{def}{=} q\{B/X\}$ and $p \sim q$. Then $A \sim B$.*

Proof. *Consider relation*

$$R = \{(G\{A/X\}, G\{B/X\}) \mid G \text{ contains at most variable } X\}.$$

Note that when G is X, we get $(A,B) \in R$. If we prove that R is a strong bisimulation up to $\sim$, then we get the required thesis $A \sim B$ by Proposition 2.10 (stating the correctness of this proof principle).

By symmetry, it is enough to prove that if $G\{A/X\} \xrightarrow{\mu} p'$, then $G\{B/X\} \xrightarrow{\mu} q'$ with $p' \sim R \sim q'$. The proof proceeds by induction of the proof of $G\{A/X\} \xrightarrow{\mu} p'$. We examine the possible shapes of G.

- *$G = X$. Then, we have $G\{A/X\} = A$ and $G\{B/X\} = B$. We want to prove that for any transition from A, e.g., $A \xrightarrow{\mu} p'$, there exists a process q' such that $B \xrightarrow{\mu} q'$ with $p' \sim R \sim q'$. By hypothesis, $A \stackrel{def}{=} p\{A/X\}$, hence by rule (Cons), also $p\{A/X\} \xrightarrow{\mu} p'$ (by a shorter inference); hence, induction can be applied to conclude that $p\{B/X\} \xrightarrow{\mu} q'''$ with $p' \sim p'' R q'' \sim q'''$ for some suitable p'', q''. But we also have that $p \sim q$, so $q\{B/X\} \xrightarrow{\mu} q'$ with $q''' \sim q'$. Since $B \stackrel{def}{=} q\{B/X\}$, by rule (Cons), also $B \xrightarrow{\mu} q'$ with $p' \sim p'' R q'' \sim q'$, as required.*
- *$G = \mu.G'$. In this case, $G\{A/X\} = \mu.G'\{A/X\} \xrightarrow{\mu} G'\{A/X\}$. Similarly, $G\{B/X\} = \mu.G'\{B/X\} \xrightarrow{\mu} G'\{B/X\}$, and $(G'\{A/X\}, G'\{B/X\}) \in R$.*
- *$G = G_1 + G_2$. In this case, $G\{A/X\} = G_1\{A/X\} + G_2\{A/X\}$. A transition from $G\{A/X\}$, e.g., $G_1\{A/X\} + G_2\{A/X\} \xrightarrow{\mu} p'$, is derivable only if $G_i\{A/X\} \xrightarrow{\mu} p'$ for some $i = 1, 2$. Without loss of generality, assume the transition is due to $G_1\{A/X\} \xrightarrow{\mu} p'$. We can apply induction (as we are shortening the proof tree) and conclude that $G_1\{B/X\} \xrightarrow{\mu} q'$ with $p' \sim R \sim q'$. By rule (Sum$_1$), transition*

$G_1\{B/X\} \xrightarrow{\mu} q'$ *ensures that also* $G\{B/X\} = G_1\{B/X\} + G_2\{B/X\} \xrightarrow{\mu} q'$, *with* $p' \sim R \sim q'$, *as required.*

- $G = G_1 \mid G_2$. *In this case,* $G\{A/X\} = G_1\{A/X\} \mid G_2\{A/X\}$. *A transition from* $G\{A/X\}$, *e.g.,* $G_1\{A/X\} \mid G_2\{A/X\} \xrightarrow{\mu} p'$, *is derivable if* $G_1\{A/X\} \xrightarrow{\mu} p'_1$ *with* $p' = p'_1 \mid G_2\{A/X\}$; or $G_2\{A/X\} \xrightarrow{\mu} p'_2$ *with* $p' = G_1\{A/X\} \mid p'_2$; or $\mu = \tau$, $G_1\{A/X\} \xrightarrow{\alpha} p'_1$, $G_2\{A/X\} \xrightarrow{\overline{\alpha}} p'_2$ *with* $p' = p'_1 \mid p'_2$. *The first two cases are trivial, hence omitted. We shall treat only the last case. By induction, as both transitions* $G_1\{A/X\} \xrightarrow{\alpha} p'_1$ *and* $G_2\{A/X\} \xrightarrow{\overline{\alpha}} p'_2$ *have a shorter proof, we have* $G_1\{B/X\} \xrightarrow{\alpha} q'_1$ *with* $p'_1 \sim p''_1 R q''_1 \sim q'_1$ *for suitable* p''_1 *and* q''_1, *as well as* $G_2\{B/X\} \xrightarrow{\overline{\alpha}} q'_2$ *with* $p'_2 \sim p''_2 R q''_2 \sim q'_2$ *for suitable* p''_2 *and* q''_2. *Hence, setting* $q' = q'_1 \mid q'_2$, *we have by rule (Com) that* $G\{B/X\} = G_1\{B/X\} \mid G_2\{B/X\} \xrightarrow{\tau} q'$. *So it remains to prove that* $p' \sim R \sim q'$. *We know that* $p'_1 \sim p''_1 R q''_1 \sim q'_1$ *as well as* $p'_2 \sim p''_2 R q''_2 \sim q'_2$. *By congruence, we have* $p' = p'_1 \mid p'_2 \sim p''_1 \mid p''_2 = p''$ *and* $q'' = q''_1 \mid q''_2 \sim q'_1 \mid q'_2 = q'$. *So, it remains to prove that* $p'' R q''$. *Since* $p''_1 R q''_1$, *there exists a term* H_1 *such that* $p''_1 = H_1\{A/X\}$ *and* $q''_1 = H_1\{B/X\}$. *Similarly, since* $p''_2 R q''_2$, *there exists a term* H_2 *such that* $p''_2 = H_2\{A/X\}$ *and* $q''_2 = H_2\{B/X\}$. *Therefore, if we set* $H = H_1 \mid H_2$, *we have that* $p'' = p''_1 \mid p''_2 = H_1\{A/X\} \mid H_2\{A/X\} = H\{A/X\}$ *as well as* $q'' = q''_1 \mid q''_2 = H_1\{B/X\} \mid H_2\{B/X\} = H\{B/X\}$; *summing up,* $p'' R q''$, *as required.*
- $G = (\nu a)G'$, *left as an exercise for the reader.*
- $G = C$. *Since no instance of* X *occurs in* G, $G\{A/X\} = C \xrightarrow{\mu} p'$, *which is possible only if* $C \stackrel{def}{=} p_1$ *and* $p_1 \xrightarrow{\mu} p'$. *Clearly, also* $G\{B/X\} = C \xrightarrow{\mu} p'$ *and* $(p', p') = (p'\{A/X\}, p'\{B/X\}) \in R$. □

The generalization to open terms over a set $\{X_1, \ldots, X_n\}$ of variables is more complex, but obvious. Consider a set of pairs of equivalent open terms of this form:

$$
\begin{array}{l}
p_1(X_1,\ldots,X_n) \sim q_1(X_1,\ldots,X_n)\\
p_2(X_1,\ldots,X_n) \sim q_2(X_1,\ldots,X_n)\\
\ldots\\
p_n(X_1,\ldots,X_n) \sim q_n(X_1,\ldots,X_n)
\end{array}
$$

Then, we can recursively close these terms as follows:

$$
\begin{array}{llll}
A_1 \stackrel{def}{=} p_1\{A_1/X_1,\ldots,A_n/X_n\} & & B_1 \stackrel{def}{=} q_1\{B_1/X_1,\ldots,B_n/X_n\}\\
A_2 \stackrel{def}{=} p_2\{A_1/X_1,\ldots,A_n/X_n\} & & B_2 \stackrel{def}{=} q_2\{B_1/X_1,\ldots,B_n/X_n\}\\
\ldots & & \\
A_n \stackrel{def}{=} p_n\{A_1/X_1,\ldots,A_n/X_n\} & & B_n \stackrel{def}{=} q_n\{B_1/X_1,\ldots,B_n/X_n\}
\end{array}
$$

Then, the thesis is that $A_i \sim B_i$ for all $i = 1, 2, \ldots, n$.

4.2.3 Weak Equivalences Are Congruences

Weak bisimulation equivalence $\approx$ (see Definition 2.20) is a congruence for the operators of action prefixing, parallelism and restriction.

Theorem 4.3. *If $p \approx q$, then the following hold:*

1) $\mu.p \approx \mu.q$ *for all* $\mu \in Act$,
2) $p \,|\, r \approx q \,|\, r$ *for all* $r \in \mathscr{P}$,
3) $(\nu a)p \approx (\nu a)q$ *for all* $a \in \mathscr{L}$.

Proof. *Assume R is a weak bisimulation such that $(p,q) \in R$.*

For case 1, consider relation $R_1 = \{(\mu.p,\mu.q) \mid \mu \in Act\} \cup R$. It is very easy to check that R_1 is a weak bisimulation.

For case 2, we show that relation $R_2 = \{(p'\,|\,r', q'\,|\,r') \mid (p',q') \in R, r' \in \mathscr{P}\}$ is a weak bisimulation. As $(p,q) \in R$, the thesis follows. Let us consider pair $(p'\,|\,r', q'\,|\,r') \in R_2$, hence with $(p',q') \in R$. If $p'\,|\,r' \xrightarrow{\mu} s$, then this is due to one of the following three cases: $p' \xrightarrow{\mu} p''$ and $s = p''\,|\,r'$ (rule (Par$_1$)), or $r' \xrightarrow{\mu} r''$ and $s = p'\,|\,r''$ (rule (Par$_2$)), or $p' \xrightarrow{\alpha} p''$, $r' \xrightarrow{\overline{\alpha}} r''$, with $\mu = \tau$ and $s = p''\,|\,r''$ (rule (Com)).

In the first case, since $(p',q') \in R$, we have that either $q' \stackrel{\alpha}{\Longrightarrow} q''$ (if $\mu = \alpha$) or $q' \stackrel{\varepsilon}{\Longrightarrow} q''$ (if $\mu = \tau$) with $(p'',q'') \in R$. By (possibly repeated applications of) rule (Par$_1$) also $q'\,|\,r' \stackrel{\alpha}{\Longrightarrow} q''\,|\,r'$ or $q'\,|\,r' \stackrel{\varepsilon}{\Longrightarrow} q''\,|\,r'$ with $(p''\,|\,r', q''\,|\,r') \in R_2$.

In the second case, by rule (Par$_2$) we have that $q'\,|\,r' \xrightarrow{\mu} q'\,|\,r''$, hence also $q'\,|\,r' \stackrel{\mu}{\Longrightarrow} q'\,|\,r''$, with $(p'\,|\,r'', q'\,|\,r'') \in R_2$.

In the third case, since $(p',q') \in R$, we have that $q' \stackrel{\alpha}{\Longrightarrow} q''$ with $(p'',q'') \in R$. By possibly repeated applications of rule (Par$_1$) and one application of rule (Com), $q'\,|\,r' \stackrel{\tau}{\Longrightarrow} q''\,|\,r''$ with $(p''\,|\,r'', q''\,|\,r'') \in R_2$ is also derivable.

The symmetric cases when $q'\,|\,r'$ moves first are analogous, hence omitted.

For case 3, the reader may show that $R_3 = \{((\nu a)p', (\nu a)q') \mid (p',q') \in R$ and $a \in \mathscr{L}\}$ is a weak bisimulation. □

Unfortunately, $\approx$ is not a congruence for $+$. As a matter of fact, $\tau.a \approx a$, but $\tau.a + b \not\approx a + b$, because $\tau.a + b \xrightarrow{\tau} a$ and the only silent step that can be taken from $a+b$ is $a+b \stackrel{\varepsilon}{\Longrightarrow} a+b$, but a and $a+b$ are not weakly bisimilar.

Exercise 4.31. Find suitable constraints on CCS syntax so that $\approx$ is a congruence also w.r.t. to the $+$ operator. (*Hint:* Avoid τ prefixes as guards of summands.) □

However, rooted weak bisimulation equivalence $\approx^c$ (see Definition 2.23) is a congruence for all the CCS operators.

Theorem 4.4. *If $p \approx^c q$, then the following hold:*

1) $\mu.p \approx^c \mu.q$ *for all* $\mu \in Act$,
2) $p + r \approx^c q + r$ *for all* $r \in \mathscr{P}$,
3) $p \,|\, r \approx^c q \,|\, r$ *for all* $r \in \mathscr{P}$,
4) $(\nu a)p \approx^c (\nu a)q$ *for all* $a \in \mathscr{L}$.

Proof. *Case 1 follows by direct application of Exercises 2.76 and 4.20.*

For case 2, if $p+r \xrightarrow{\mu} s$, *this can be due to either* $p \xrightarrow{\mu} s$ *or* $r \xrightarrow{\mu} s$. *In the former case, since* $p \approx^c q$, *we have that* $q \stackrel{\mu}{\Longrightarrow} q'$, *with* $s \approx q'$. *Hence, by induction on the length of the path associated to* $q \stackrel{\mu}{\Longrightarrow} q'$ *and by applying rule (*Sum_1*) on the first transition on that path, we can derive* $q+r \stackrel{\mu}{\Longrightarrow} q'$ *with* $s \approx q'$. *In the latter case, by rule (*Sum_2*) we can derive* $q+r \xrightarrow{\mu} s$, *hence* $q+r \stackrel{\mu}{\Longrightarrow} s$ *with* $s \approx s$. *The symmetric case when* $q+r$ *moves first is omitted.*

For case 3, if $p \mid r \xrightarrow{\mu} s$, *then this is due to one of the following three cases:* $p \xrightarrow{\mu} p'$ *and* $s = p' \mid r$ *(rule (*Par_1*)), or* $r \xrightarrow{\mu} r'$ *and* $s = p \mid r'$ *(rule (*Par_2*)), or* $p \xrightarrow{\alpha} p'$, $r \xrightarrow{\overline{\alpha}} r'$, *with* $\mu = \tau$ *and* $s = p' \mid r'$ *(rule (Com)). In the first case, as* $p \approx^c q$, *we have that* $q \stackrel{\mu}{\Longrightarrow} q'$ *with* $p' \approx q'$. *By (possibly repeated applications of) rule (*Par_1*) also* $q \mid r \stackrel{\mu}{\Longrightarrow} q' \mid r$. *As* $p' \approx q'$ *and* $\approx$ *is a congruence for parallel composition (as proved in Theorem 4.3(2)), we have that* $p' \mid r \approx q' \mid r$, *as required. In the second case, by rule (*Par_2*) we have that* $q \mid r \xrightarrow{\mu} q \mid r'$, *hence also* $q \mid r \stackrel{\mu}{\Longrightarrow} q \mid r'$, *with* $p \mid r' \approx q \mid r'$, *as required, because* $p \approx^c q$, *hence* $p \approx q$, *and* $\approx$ *is a congruence for parallel composition. In the third case, since* $p \approx^c q$, *we have* $q \stackrel{\alpha}{\Longrightarrow} q'$ *with* $p' \approx q'$. *By possibly repeated applications of rule (*Par_1*) and one application of rule (Com),* $q \mid r \stackrel{\tau}{\Longrightarrow} q' \mid r'$ *is also derivable, with* $p' \mid r' \approx q' \mid r'$ *as required, because* $\approx$ *is a congruence for parallel composition. The symmetric case when* $q \mid r$ *moves first is omitted.*

Case 4 is left as an exercise for the reader. (Hint: One has to take advantage of the fact that $\approx$ *is a congruence for restriction, as proved in Theorem 4.3(3).)* □

Actually, one can prove that $\approx^c$ is the *coarsest congruence* (i.e., the congruence that identifies as much as possible) contained in weak bisimulation equivalence. This is made precise by the following theorem that justifies rooted weak bisimilarity as the compositional refinement of weak bisimilarity.

Theorem 4.5. (Coarsest congruence contained in $\approx$**)** *Assume that* $fn(p) \cup fn(q) \neq \mathscr{L}$. *Then* $p \approx^c q$ *if and only if* $p+r \approx q+r$ *for all* $r \in \mathscr{P}$.

Proof. *The implication from left to right follows by Theorem 4.4(2) and Exercise 2.76. For the implication from right to left, suppose that* $p+r \approx q+r$ *for all* $r \in \mathscr{P}$. *Take any action* $a \in \mathscr{L}$ *such that* $a \notin fn(p) \cup fn(q)$[2] *and assume* $p \xrightarrow{\mu} p'$. *Then also* $p+a \xrightarrow{\mu} p'$ *(by rule (*Sum_1*)). As* $p+a \approx q+a$, *then also* $q+a$ *must respond to this transition. We have to examine two different cases: either* $\mu = \tau$ *and* $q+a \stackrel{\varepsilon}{\Longrightarrow} q+a$, *or the transition truly originates from* q, *i.e.,* $q+a \stackrel{\mu}{\Longrightarrow} q'$ *(*μ *can be* τ*). The former case is impossible:* p' *cannot be weakly bisimilar to* $q+a$, *as* p' *cannot execute* a. *Hence, the second case must be true; but this is indeed what is requested by rooted*

[2] The assumption that $\mathscr{L}$ is not covered by the free names of p and q is not strictly necessary [vGl05], but makes the proof easier. Such an assumption is satisfied when p and q are finitary CCS processes (see Proposition 4.5).

weak bisimulation: if $p \xrightarrow{\mu} p'$*, then* $q \stackrel{\mu}{\Longrightarrow} q'$ *with* $p' \approx q'$*. The symmetric case when* q *moves first is omitted.* □

Also branching bisimilarity $\approx_{br}$ (see Definition 2.24) is a congruence for action prefixing, parallel composition and restriction, but not for $+$: it is easy to see that $\tau.a \approx_{br} a$ but $\tau.a + b \not\approx_{br} a + b$.

Theorem 4.6. *If* $p \approx_{br} q$*, then the following hold:*

1) $\mu.p \approx_{br} \mu.q$ *for all* $\mu \in Act$*,*
2) $p\,|\,r \approx_{br} q\,|\,r$ *for all* $r \in \mathscr{P}$*,*
3) $(\nu a)p \approx_{br} (\nu a)q$ *for all* $a \in \mathscr{L}$*.*

Proof. *Assume* R *is a branching bisimulation such that* $(p,q) \in R$*.*

For case 1, consider the relation $R_1 = \{(\mu.p, \mu.q) \mid \mu \in Act\} \cup R$*. It is very easy to check that* R_1 *is a branching bisimulation.*

For case 2, we show that relation $R_2 = \{(p'\,|\,r', q'\,|\,r') \mid (p',q') \in R, r' \in \mathscr{P}\}$ *is a branching bisimulation. As* $(p,q) \in R$*, the thesis follows. Let us consider pair* $(p'\,|\,r', q'\,|\,r') \in R_2$*, hence with* $(p',q') \in R$*. If* $p'\,|\,r' \xrightarrow{\mu} s$*, then this is due to one of the following three cases:* $p' \xrightarrow{\mu} p''$ *and* $s = p''\,|\,r'$ *(rule (Par*$_1$*)), or* $r' \xrightarrow{\mu} r''$ *and* $s = p'\,|\,r''$ *(rule (Par*$_2$*)), or* $p' \xrightarrow{\alpha} p''$*,* $r' \xrightarrow{\overline{\alpha}} r''$ *and* $s = p''\,|\,r''$ *(rule (Com)).*

In the first case, since $(p',q') \in R$*, if* $\mu = \alpha$ *we have that* $q' \stackrel{\varepsilon}{\Longrightarrow} q_0 \xrightarrow{\alpha} q''$ *with* $(p',q_0) \in R$ *as well as* $(p'',q'') \in R$*. By (possibly repeated applications of) rule (Par*$_1$*) also* $q'\,|\,r' \stackrel{\varepsilon}{\Longrightarrow} q_0\,|\,r' \xrightarrow{\alpha} q''\,|\,r'$ *with* $(p'\,|\,r', q_0\,|\,r') \in R_2$ *and* $(p''\,|\,r', q''\,|\,r') \in R_2$*. Instead, if* $\mu = \tau$*, we have two subcases:* either $q' \stackrel{\varepsilon}{\Longrightarrow} q_0 \xrightarrow{\tau} q''$ *with* $(p',q_0) \in R$ *as well as* $(p'',q'') \in R$*,* or $q' \stackrel{\varepsilon}{\Longrightarrow} q''$ *with* $(p',q'') \in R$ *as well as* $(p'',q'') \in R$*. The former subcase is analogous to the case above for* $\mu = \alpha$*, hence omitted. For the latter subcase, by (possibly repeated applications of) rule (Par*$_1$*) also* $q'\,|\,r' \stackrel{\varepsilon}{\Longrightarrow} q''\,|\,r'$ *with* $(p'\,|\,r', q''\,|\,r') \in R_2$ *as well as* $(p''\,|\,r', q''\,|\,r') \in R_2$*.*

In the second case, i.e., when $r' \xrightarrow{\mu} r''$*, by rule (Par*$_2$*) we have that* $q'\,|\,r' \xrightarrow{\mu} q'\,|\,r''$*, with* $(p'\,|\,r'', q'\,|\,r'') \in R_2$*, hence the branching bisimulation condition is satisfied.*

In the third case, since $(p',q') \in R$*, we have that* $q' \stackrel{\varepsilon}{\Longrightarrow} q_0 \xrightarrow{\alpha} q''$ *with* $(p',q_0) \in R$ *as well as* $(p'',q'') \in R$*. By possibly repeated applications of rule (Par*$_1$*) and one final application of rule (Com), we can derive transition* $q'\,|\,r' \stackrel{\varepsilon}{\Longrightarrow} q_o\,|\,r' \xrightarrow{\tau} q''\,|\,r''$ *with* $(p'\,|\,r', q_0\,|\,r') \in R_2$ *as well as* $(p''\,|\,r'', q''\,|\,r'') \in R_2$*.*

The symmetric cases when $q'\,|\,r'$ *moves first are analogous, hence omitted.*

For case 3, the reader may show that $R_3 = \{((\nu a)p', (\nu a)q') \mid (p',q') \in R$ *and* $a \in \mathscr{L}\}$ *is a branching bisimulation.* □

Similarly to the case of rooted weak bisimilarity, the rooted version of branching bisimilarity $\approx_{br}^{c}$ (Definition 2.26) is a congruence for all CCS operators.

Theorem 4.7. *If* $p \approx_{br}^{c} q$*, then the following hold:*

1) $\mu.p \approx_{br}^{c} \mu.q$ *for all* $\mu \in Act$*,*
2) $p + r \approx_{br}^{c} q + r$ *for all* $r \in \mathscr{P}$*,*
3) $p\,|\,r \approx_{br}^{c} q\,|\,r$ *for all* $r \in \mathscr{P}$*,*
4) $(\nu a)p \approx_{br}^{c} (\nu a)q$ *for all* $a \in \mathscr{L}$*.*

Proof. *For case 1, transition* $\mu.p \xrightarrow{\mu} p$ *is matched by transition* $\mu.q \xrightarrow{\mu} q$, *with* $p \approx_{br} q$ *(due to Exercise 2.84), as required.*

For case 2, if $p+r \xrightarrow{\mu} s$, *this can be due to either* $p \xrightarrow{\mu} s$ *or* $r \xrightarrow{\mu} s$. *In the former case, since* $p \approx_{br}^{c} q$, *we have that* $q \xrightarrow{\mu} q'$, *with* $s \approx_{br} q'$. *Hence, by rule (Sum*$_1$*), we can derive* $q+r \xrightarrow{\mu} q'$ *with* $s \approx_{br} q'$, *as required. In the latter case, by rule (Sum*$_2$*) we can derive* $q+r \xrightarrow{\mu} s$, *with* $s \approx_{br} s$. *The symmetric case when* $q+r$ *moves first is omitted.*

For case 3, if $p \mid r \xrightarrow{\mu} s$, *this can be due to one of the following three cases:* $p \xrightarrow{\mu} p'$ *and* $s = p' \mid r$, or $r \xrightarrow{\mu} r'$ *and* $s = p \mid r'$, or $p \xrightarrow{\alpha} p'$, $r \xrightarrow{\overline{\alpha}} r'$ *and* $s = p' \mid r'$.

In the first case, since $p \approx_{br}^{c} q$, *we have* $q \xrightarrow{\mu} q'$, *with* $p' \approx_{br} q'$. *By rule (Par*$_1$*), we can derive* $q \mid r \xrightarrow{\mu} q' \mid r$, *with* $p' \mid r \approx_{br} q' \mid r$, *by Theorem 4.6(2), as required.*

In the second case, by rule (Par$_2$*),* $q \mid r \xrightarrow{\mu} q \mid r'$. *As* $p \approx_{br} q$ *by Exercise 2.84, by Theorem 4.6(2), we have that* $p \mid r' \approx_{br} q \mid r'$, *as required.*

In the third case, since $p \approx_{br}^{c} q$, *we have that* $q \xrightarrow{\alpha} q'$, *with* $p' \approx_{br} q'$. *By rule (Com),* $q \mid r \xrightarrow{\tau} q' \mid r'$, *with* $p' \mid r' \approx_{br} q' \mid r'$ *by Theorem 4.6(2), as required.*

Case 4 is left as an exercise for the reader. □

Exercise 4.32. Prove that rooted branching bisimilarity $\approx_{br}^{c}$ is the coarsest congruence contained in branching bisimulation equivalence $\approx_{br}$. (*Hint:* Follow the steps of the proof of Theorem 4.5.) □

Example 4.5. **(Predictive operator)** We may wonder if there exists an operator that preserves (rooted) branching bisimilarity but not rooted weak bisimilarity, hence supporting the need for (rooted) branching bisimilarity. The answer is positive. One of these operators is the predictive operator, originally introduced in [vGW96]. Its informal definition is as follows: $pred(p)$ can do whatever p can do, but it can also predict initial observable moves of p without actually executing them and without changing state. Formally, the operational rules are as follows:

$$(\mathrm{P}_1)\ \frac{p \xrightarrow{\mu} p'}{pred(p) \xrightarrow{\mu} pred(p')} \qquad (\mathrm{P}_2)\ \frac{p \xrightarrow{\alpha} p'}{pred(p) \xrightarrow{can\ do\ \alpha} pred(p)}$$

It is easy to see that $b+\tau.a \approx^{c} b+\tau.a+a$ (see also Figure 2.31 in Section 2.4.3, where we discuss that these two systems are not branching bisimilar). However, $pred(b+\tau.a)$ is not rooted weak bisimilar to $pred(b+\tau.a+a)$, because the latter can do $pred(b+\tau.a+a) \xrightarrow{can\ do\ a} pred(b+\tau.a+a)$ and the former can only reply with $pred(b+\tau.a) \xRightarrow{can\ do\ a} pred(a)$ and the reached states are clearly not equivalent as only $pred(b+\tau.a+a)$ can execute action b.

Now we prove that branching bisimilarity $\approx_{br}$ is preserved by the predictive operator, i.e., if $p \approx_{br} q$ then $pred(p) \approx_{br} pred(q)$. Let R be a branching bisimulation such that $(p,q) \in R$. Then we prove that $S = \{(pred(p'), pred(q')) \mid (p',q') \in R\}$ is a branching bisimulation.

If $pred(p') \xrightarrow{can\ do\ \alpha} pred(p')$, then $p' \xrightarrow{\alpha} p''$ for some p''. Since $(p',q') \in R$, then q' can reply with $q' \xRightarrow{\varepsilon} q'' \xrightarrow{\alpha} q'''$, with $(p',q'') \in R$ and $(p'',q''') \in R$. Hence, by means of the operational rules above, $pred(q') \stackrel{\varepsilon}{\Longrightarrow} pred(q'') \stackrel{can\ do\ \alpha}{\longrightarrow} pred(q'')$ with $(pred(p'), pred(q'')) \in S$, as required.

If $pred(p') \stackrel{\alpha}{\longrightarrow} pred(p'')$, then $p' \stackrel{\alpha}{\longrightarrow} p''$ for some p''. Since $(p',q') \in R$ then q' can reply with $q' \stackrel{\varepsilon}{\Longrightarrow} q'' \stackrel{\alpha}{\longrightarrow} q'''$ with $(p',q'') \in R$ and $(p'',q''') \in R$. Hence, by means of the operational rules above, $pred(q') \stackrel{\varepsilon}{\Longrightarrow} pred(q'') \stackrel{\alpha}{\longrightarrow} pred(q''')$ with $(pred(p'), pred(q'')) \in S$, as well as $(pred(p''), pred(q''')) \in S$, as required.

If $pred(p') \stackrel{\tau}{\longrightarrow} pred(p'')$, then $p' \stackrel{\tau}{\longrightarrow} p''$ for some p''. Since $(p',q') \in R$ then q' can reply either with $q' \stackrel{\varepsilon}{\Longrightarrow} q''$, such that $(p',q'') \in R$, as well as $(p'',q'') \in R$, or with $q' \stackrel{\varepsilon}{\Longrightarrow} q'' \stackrel{\tau}{\longrightarrow} q'''$, with $(p',q'') \in R$ and $(p'',q''') \in R$. Hence, by means of the operational rules above, we can derive: in the former case, $pred(q') \stackrel{\varepsilon}{\Longrightarrow} pred(q'')$ with $(pred(p'), pred(q'')) \in S$, as well as $(pred(p''), pred(q'')) \in S$, as required; in the latter case, $pred(q') \stackrel{\varepsilon}{\Longrightarrow} pred(q'') \stackrel{\tau}{\longrightarrow} pred(q''')$ with $(pred(p'), pred(q'')) \in S$ as well as $(pred(p''), pred(q''')) \in S$, as required.

The symmetric case when $pred(q')$ moves first is analogous, hence omitted. Summing up, S is a branching bisimulation and so $pred(p) \approx_{br} pred(q)$. □

Exercise 4.33. Continuing Exercise 4.28, prove that also weak trace equivalence (see Definition 2.17) is a congruence for the CCS operators. □

Exercise 4.34. Is weak simulation equivalence (see Definition 2.19) a congruence for the CCS operators? □

Exercise 4.35. Continuing Exercise 4.30, show that weak completed trace equivalence (see Definition 2.18), as well as weak completed simulation (see Exercise 2.60), is not a congruence not only for restriction, but also for choice. (*Hint:* Consider $\tau.\mathbf{0}$ and $\mathbf{0}$, which are weakly completed trace (as well as weakly completed simulation) equivalent.) □

The recursion congruence proof in Section 4.2.2 can be extended also to the case of weak bisimilarity and branching bisimilarity.

Exercise 4.36. (Weak bisimilarity is a congruence for recursion) Let p and q be two open terms containing variable X at most. We say that $p \approx q$ if for every (closed) CCS process r, $p\{r/X\} \approx q\{r/X\}$. Let $A \stackrel{def}{=} p\{A/X\}$, $B \stackrel{def}{=} q\{B/X\}$ and $p \approx q$. Then prove that $A \approx B$ by adapting the proof of Theorem 4.2. Redo the same for $\approx_{br}$. □

4.3 Axiomatization of Finite Processes

Behavioral equivalences for full CCS are undecidable, as we have discussed in Section 3.5.3. However, for certain subclasses decidability holds. In these cases it may be possible to give algebraic characterizations of bisimilarity, in the form of axiomatizations. An axiomatization of a behavioral congruence is an elegant way of pinpointing what are the basic features of such a behavioral congruence: different congruences are characterized by different axioms. For instance, the basic difference between bisimulation congruence and trace congruence is the axiom of distributivity of action prefixing w.r.t. summation $\mu.(x+y) = \mu.x+\mu.y$, which is an axiom for the latter but not for the former. Moreover, an axiomatization may offer an alternative proof method which, in some cases, is more efficient than checking the behavioral congruence on the associated LTSs of the two processes: once we have proved that the axiomatization is sound and complete w.r.t. the behavioral congruence, we can ignore the latter (i.e., we do not need to generate the LTSs associated to the two processes and check that the two are congruent) and we can reason on the former in purely algebraic terms. For instance, proving that $(p+q)+r$ is bisimilar to $p+(q+r)$ (for processes p,q,r) can be easily achieved by instantiation of the axiom of associativity of the choice operator: $(x+y)+z = x+(y+z)$.

For the sake of simplicity, we restrict our attention to $\mathscr{P}_{fin}$, the class of finite CCS processes (no occurrence of constants). See [Mil89] for an axiomatization of strong and rooted weak bisimulation equivalences over the class of finite-state processes (originally presented in [Mil84, Mil89b]) and [BB08] for an axiomatization of such equivalences over regular processes.

In the following, we briefly recall the basic definitions about axiomatizations (usually called *equational theories*) and equational deduction, then we present a couple of axiomatizations for strong equivalences, and finally some axiomatizations for weak equivalences. A more advanced presentation of these topics can be found in [BBR10], where a variety of process algebras are axiomatized.

4.3.1 Equational Deduction

In general, given a (one-sorted) *signature* Σ (i.e., a set of constants and function symbols with their arities), we can define the set of terms over Σ and over a set of variables V (notation $\mathscr{T}(\Sigma,V)$) as the smallest set that satisfies the following:

(i) each variable x in V is a term in $\mathscr{T}(\Sigma,V)$
(ii) each constant c in Σ is a term of $\mathscr{T}(\Sigma,V)$
(iii) if f is a n-ary function symbol and $t_1,\dots,t_n$ are terms in $\mathscr{T}(\Sigma,V)$, then also $f(t_1,\dots,t_n)$ is a term in $\mathscr{T}(\Sigma,V)$

Usually, we write $\mathscr{T}(\Sigma)$ to denote the set of *closed* (or *ground*) terms (i.e., terms with no occurrence of variables).

Example 4.6. **(Signature for finite CCS)** We can define a signature Σ_{CCS} for finite *CCS* as follows: we have one constant $\mathbf{0}$, a set of unary operators of prefixing $\mu.\_$ (one for each $\mu \in Act$, which is assumed to be finite), another set of unary operators of restriction $(\nu a)\_$ (one for each $a \in \mathscr{L}$, assumed finite as well) and the binary operators of choice $\_+\_$ and parallelism $\_|\_$, which are used in infix form (while f above was defined in prefix form).

The set $\mathscr{T}(\Sigma_{CCS},V)$ includes, for instance, the (open) term $(a.x+\mathbf{0})\,|\,b.\mathbf{0}$, where x is a variable. The set $\mathscr{T}(\Sigma_{CCS})$ of closed terms is slightly more general than the set

1. Reflexivity	$\dfrac{}{t = t}$	
2. Symmetry	$\dfrac{t_1 = t_2}{t_2 = t_1}$	
3. Transitivity	$\dfrac{t_1 = t_2 \quad t_2 = t_3}{t_1 = t_3}$	
4. Substitutivity	$\dfrac{t_i = t_i'}{f(t_1, \ldots t_i, \ldots t_k) = f(t_1, \ldots t_i', \ldots t_k)}$	for any f and $1 \leq i \leq n$
5. Instantiation	$\dfrac{t_1 = t_2}{t_1[\rho] = t_2[\rho]}$	for any substitution ρ
6. Axioms	$\dfrac{}{t_1 = t_2}$	for all axioms $t_1 = t_2$ in E

Table 4.1 Rules of equational deduction

$\mathscr{P}_{fin}$ of Section 3.4.1 because now we admit unguarded sums. As a matter of fact, $\mathscr{P}_{fin}$ is a two-sorted signature (one sort p for sequential processes and one sort q for general processes), but in this section, for simplicity's sake, we prefer to consider only one-sorted signatures. □

A *substitution* ρ is a mapping from V to $\mathscr{T}(\Sigma, V)$. For any term t, by $t[\rho]$ we denote the term obtained as follows:

(i) for each variable x in V, $x[\rho] = \rho(x)$,
(ii) for each constant c in Σ, $c[\rho] = c$,
(iii) for any n-ary function symbol f and terms $t_1, \ldots, t_n$, $f(t_1, \ldots, t_n)[\rho]$ is the term $f(t_1[\rho], \ldots, t_n[\rho])$.

A substitution ρ is *closed* if $\rho(x) \in \mathscr{T}(\Sigma)$ for each variable $x \in V$. Hence, the application of a closed substitution to an open term yields a closed term.

An *equational theory* is a tuple (Σ, E), where Σ is a signature and E is a set of equations of the form $t_1 = t_2$, where t_1 and t_2 are terms in $\mathscr{T}(\Sigma, V)$. The equations in E are usually called *axioms* and the equational theory (Σ, E) is often called an *axiomatization.*

For any equational theory (Σ, E), there is a standard set of proof rules for deriving equalities on $\mathscr{T}(\Sigma, V)$. This set of proof rules forms an *equational deductive proof system* $D(E)$, composed of the rules in Table 4.1.

A *proof* is a finite sequence of equalities $t_1 = t_1'$, $t_2 = t_2'$, $\ldots t_k = t_k'$ such that each $t_i = t_i'$ is either an axiom (rule 1 or 6) or can be derived by using one of the rules

A1	Associativity	$x+(y+z) = (x+y)+z$
A2	Commutativity	$x+y = y+x$
A3	Identity	$x+\mathbf{0} = x$
A4	Idempotence	$x+x = x$
T	Distributivity	$\mu.(x+y) = \mu.x+\mu.y$

Table 4.2 Axioms for choice

$2-5$ with premises some of the previous equalities $t_1 = t_1'$, $t_2 = t_2', \ldots t_{i-1} = t_{i-1}'$. This linear representation of a proof can be more conveniently expressed in the form of a tree (proof tree).

We use the notation $E \vdash t_1 = t_2$ if there exists a proof with $t_1 = t_2$ as its last equality, using only the axioms in E (plus rule 1). This directly determines a congruence over the set of terms: $t_1 =_E t_2$ if and only if $E \vdash t_1 = t_2$. Indeed, note that rules $1-3$ ensure that the relation $=_E$ that $D(E)$ is inducing on $\mathscr{T}(\Sigma, V)$ is an equivalence relation; similarly, rule 4 ensures substitutivity of equals for equals in any context, i.e., that the relation $=_E$ that $D(E)$ is inducing is a congruence.

Example 4.7. Let us consider the axioms for the choice operator in Table 4.2. The proof that $\{\mathbf{A1}, \mathbf{A2}, \mathbf{A4}\} \vdash a.\mathbf{0} + (b.w + a.\mathbf{0}) = a.\mathbf{0} + b.w$ is as follows:

1. $x+y=y+x$	Axiom **A2**
2. $b.w+a.\mathbf{0}=a.\mathbf{0}+b.w$	Rule 5: Instantiation of line 1
3. $a.\mathbf{0}+(b.w+a.\mathbf{0})=a.\mathbf{0}+(a.\mathbf{0}+b.w)$	Rule 4: Substitutivity on line 2
4. $x+(y+z)=(x+y)+z$	Axiom **A1**
5. $a.\mathbf{0}+(a.\mathbf{0}+b.w)=(a.\mathbf{0}+a.\mathbf{0})+b.w$	Rule 5: Instantiation of line 4
6. $x+x=x$	Axiom **A4**
7. $a.\mathbf{0}+a.\mathbf{0}=a.\mathbf{0}$	Rule 5: Instantiation of line 6
8. $(a.\mathbf{0}+a.\mathbf{0})+b.w=a.\mathbf{0}+b.w$	Rule 4: Substitutivity on line 7
9. $a.\mathbf{0}+(b.w+a.\mathbf{0})=(a.\mathbf{0}+a.\mathbf{0})+b.w$	Rule 3: Transitivity on lines 3 and 5
10. $a.\mathbf{0}+(b.w+a.\mathbf{0})=a.\mathbf{0}+b.w$	Rule 3: Transitivity on lines 9 and 8

This linear representation of the proof may be more conveniently represented as the proof tree

$$\dfrac{\dfrac{\dfrac{\dfrac{\dfrac{}{x+y=y+x}}{b.w+a.\mathbf{0}=a.\mathbf{0}+b.w}}{a.\mathbf{0}+(b.w+a.\mathbf{0})=a.\mathbf{0}+(a.\mathbf{0}+b.w)} \qquad \dfrac{\dfrac{}{x+(y+z)=(x+y)+z}}{a.\mathbf{0}+(a.\mathbf{0}+b.w)=(a.\mathbf{0}+a.\mathbf{0})+b.w}}{a.\mathbf{0}+(b.w+a.\mathbf{0})=(a.\mathbf{0}+a.\mathbf{0})+b.w} \qquad \dfrac{\dfrac{\dfrac{}{x+x=x}}{a.\mathbf{0}+a.\mathbf{0}=a.\mathbf{0}}}{(a.\mathbf{0}+a.\mathbf{0})+b.w=a.\mathbf{0}+b.w}}{a.\mathbf{0}+(b.w+a.\mathbf{0})=a.\mathbf{0}+b.w}$$

R1		$(\nu a)\mathbf{0} = \mathbf{0}$
R2	if $\mu \notin \{a,\overline{a}\}$	$(\nu a)\mu.x = \mu.(\nu a)x$
R3	if $\mu \in \{a,\overline{a}\}$	$(\nu a)\mu.x = \mathbf{0}$
R4		$(\nu a)(x+y) = (\nu a)x + (\nu a)y$
Exp	if $x = \Sigma_{i=1}^{n} \mu_i.x_i$ and $y = \Sigma_{j=1}^{m} \mu'_j.y_j$	$x \mid y = \Sigma_i \mu_i.(x_i \mid y) + \Sigma_j \mu'_j.(x \mid y_j) + \Sigma_{i,j:\overline{\mu_i}=\mu'_j} \tau.(x_i \mid y_j)$

Table 4.3 Axioms for restriction and expansion law

where, for lack of space, we have omitted the names of the axioms and rules we have used; note that the leaves of the tree are the three axioms and that the root is the equality being proved. □

Let $R \subseteq \mathscr{T}(\Sigma) \times \mathscr{T}(\Sigma)$ be a relation on closed terms. The equational deductive proof system $D(E)$ for the equational theory (Σ, E) is (ground) *sound* w.r.t. R if

$$E \vdash t_1 = t_2 \text{ implies } (t_1, t_2) \in R$$

If R is a behavioral congruence, this means that the axiomatization is sound whenever $D(E)$ proves equalities only between two behaviorally congruent terms, i.e., it never equates two terms that are not behaviorally congruent.

$D(E)$ is (ground) *complete* w.r.t. R if

$$(t_1, t_2) \in R \text{ implies } E \vdash t_1 = t_2$$

So, $D(E)$ is complete if it can prove that each pair of behaviorally congruent terms is equated by the axioms, i.e., it never misses to prove equality of two behaviorally congruent terms.

In other words, a sound and complete axiomatization (Σ, E) of a congruence R on $\mathscr{T}(\Sigma)$ is such that the congruence $t_1 =_E t_2$ induced by $D(E)$ is exactly the same congruence defined by R.

4.3.2 Axiomatizing Strong Equivalences

As our first behavioral congruence over *finite* CCS, we consider strong bisimulation equivalence $\sim$. Let $\mathscr{SB}$ be the set of axioms **A1–A4** of Table 4.2, and **R1–R4** and **Exp** of Table 4.3. We want to show that the equational theory $(\Sigma_{CCS}, \mathscr{SB})$ is a sound and complete axiomatization of $\sim$.

Remark 4.7. **(Axiom vs axiom schematon)** To be precise, the four axioms **R1–R4** are not axioms, but rather *axiom schemata*, i.e., there is an instance of these axioms for any possible action $a \in \mathscr{L}$ and prefix $\mu \in Act$. Hence, under the assumption that

Act (hence also $\mathscr{L}$) is finite, we still have a finite number of axioms for restriction. Similarly, **Exp** is not an axiom, but an axiom schematon, which stands for an infinite collection of axioms, each one for the fixed choice of n, m and for the choice of the initial prefixes μ_i and μ'_j. This means that $\mathscr{S}\mathscr{B}$ is not a finite axiomatization of $\sim$ over finite CCS. A finite axiomatization can be obtained with the use of the auxiliary operators of *left merge* and *communication merge* of ACP [BK84a, BK85], as we will show in Section 4.3.4. Moller in [Mol90a, Mol90b] proved that $\sim$ is not finitely axiomatizable without introducing some auxiliary operators. □

For *soundness*, we have to prove that for any pair of closed terms p and q, if $\mathscr{S}\mathscr{B} \vdash p = q$, then $p \sim q$. This must be done by induction on the proof tree for $\mathscr{S}\mathscr{B} \vdash p = q$. The base cases are the two axioms: rule 1 and rule 6 in Table 4.1. Reflexivity (rule 1) is sound because $p \sim p$ for any $p \in \mathscr{P}$. Rule 6 (Axioms) is sound if for each axiom in E, each ground instantiation (rule 5) of the axiom holds when $=$ is replaced by $\sim$. For instance, **A2** is sound if for all possible finite CCS processes p and q, we have that $p+q \sim q+p$. For the axioms for choice, this has already been proved in Proposition 4.1. Axiom **Exp** is sound because of Proposition 4.3, while the axioms of restriction are sound for Proposition 4.4 (and Remark 4.3). The other rules in Table 4.1 (rules 2–4) are obviously sound, because assuming that the thesis holds for the premise of the rule, it trivially holds also for the conclusion of the premise, as $\sim$ is an equivalence relation, actually a congruence. Summing up, we have the following:

Theorem 4.8. (Soundness) *For any pair of closed terms p and q, if $\mathscr{S}\mathscr{B} \vdash p = q$, then $p \sim q$.*

Proof. It follows by Proposition 4.1 (for the axioms for +), by Proposition 4.3 (for the expansion law), and by Proposition 4.4 (for the axioms for restriction). □

For *completeness*, we have to resort to some *normal form* and to prove first the completeness result for normal forms only. Then, we extend the result to all processes by proving that any process can be reduced to a normal form.

Definition 4.6. A process of the form $\Sigma_{i\in I}\mu_i.p_i$ is in *head standard form.* A process p is in *normal form* if p and all its subterms are in head standard form. □

In other words, a normal form is a term built only with $\mathbf{0}$, action prefixing and (n-ary) summation, where the order and grouping of summands may be ignored in virtue of axioms **A1** of associativity, **A2** of commutativity and **A3** of identity. Note that $\mathbf{0}$ is a normal form $\Sigma_{i\in I}\mu_i.P_i$ when $I = \emptyset$. The proofs we will present are mainly based on induction on a suitable measure of a normal form, called *depth*. The depth of a normal form is the maximal number of nested prefixes in its syntax; formally, it is defined as

$$depth(\Sigma_{i\in I}\mu_i.p_i) = max\{depth(\mu_i.p_i) \mid i \in I\}$$
$$depth(\mu.p) = 1 + depth(p)$$

with the proviso that $max\,\emptyset = 0$; hence $depth(\Sigma_{i\in I}\mu_i.p_i) = 0$ if and only if $I = \emptyset$. The following obvious fact holds for normal forms.

Proposition 4.15. *For any normal form p, if $p \xrightarrow{\mu} p'$, then $\mu.p'$ is a summand of p, p' is a normal form and $depth(p') < depth(p)$.* □

Exercise 4.37. Show that, for p and q normal forms, if $p \sim q$, then $depth(p) = depth(q)$. □

Proposition 4.16. (Completeness for normal forms) *If p and q are normal forms such that $p \sim q$, then $\mathscr{S}\mathscr{B} \vdash p = q$.*

Proof. By induction on the sum of depths of p and q. If the sum is 0, then $p = q = \mathbf{0}$ and the thesis follows by rule 1 (reflexivity) in Table 4.1.

Otherwise, suppose $\mu.p'$ is a summand of p, and so $p \xrightarrow{\mu} p'$. As $p \sim q$, also $q \xrightarrow{\mu} q'$ with $p' \sim q'$. Since q is a normal form, $\mu.q'$ must be a summand of q, by Proposition 4.15. Observe that the sum of depths of p' and q' is strictly decreased, hence induction can be applied in order to get $\mathscr{S}\mathscr{B} \vdash p' = q'$. By rule 4 (substitutivity) of Table 4.1, then also $\mathscr{S}\mathscr{B} \vdash \mu.p' = \mu.q'$ is derivable. Hence for any summand $\mu.p'$ of p, we have found a summand $\mu.q'$ of q so that the two are equated by the axioms. Symmetrically, we can prove that for any summand $\mu.q'$ of q, there exists a summand $\mu.p'$ of p such that $\mathscr{S}\mathscr{B} \vdash \mu.p' = \mu.q'$ is derivable.

Hence, putting all the summands together (via substitutivity w.r.t. $+$), we have $\mathscr{S}\mathscr{B} \vdash p = q$, by using the axioms **A4** (for removing possible duplicates) and **A1–A2** (for rearranging the remaining summands). □

Now we want to prove that for any process p there exists a normal form q such that $\mathscr{S}\mathscr{B} \vdash p = q$. We need two auxiliary lemmata.

Lemma 4.3. *If p and q are normal forms, then there exists a normal form r such that $\mathscr{S}\mathscr{B} \vdash p \,|\, q = r$.*

Proof. By induction on the sum of depths of p and q. If the sum is 0, then $p = q = \mathbf{0}$ and $\mathbf{Exp} \vdash \mathbf{0} \,|\, \mathbf{0} = \mathbf{0}$ (when n and m are 0). Otherwise, by **Exp** we get

$$p \,|\, q \;=\; \Sigma_i \mu_i.(p_i \,|\, q) \;+\; \Sigma_j \mu'_j.(p \,|\, q_j) \;+\; \Sigma_{\overline{\mu_i} = \mu'_j} \tau.(p_i \,|\, q_j),$$

where each term of the form $p_i \,|\, q$, $p \,|\, q_j$ and $p_i \,|\, q_j$ (for $i = 1, \ldots n$ and $j = 1, \ldots m$) has strictly decreased sum of depths, hence induction applies: there exist normal forms r_i, t_j and s_{ij} for $i = 1, \ldots n$ and $j = 1, \ldots m$, such that $\mathscr{S}\mathscr{B} \vdash p_i \,|\, q = r_i$, $\mathscr{S}\mathscr{B} \vdash p \,|\, q_j = t_j$, and $\mathscr{S}\mathscr{B} \vdash p_i \,|\, q_j = s_{ij}$. By rule 4 (substitutivity) of Table 4.1,

$$p \,|\, q \;=\; \Sigma_i \mu_i.r_i \;+\; \Sigma_j \mu'_j.t_j \;+\; \Sigma_{\overline{\mu_i} = \mu'_j} \tau.s_{ij},$$

which is a normal form. □

Lemma 4.4. *If p is a normal form, then there exists a normal form r such that $\mathscr{S}\mathscr{B} \vdash (\nu a)p = r$.*

Proof. *By induction on the depth of p. If it is 0, then* $p = \mathbf{0}$, *and the thesis follows by axiom* **R1**. *Otherwise, if* $p = \Sigma_{i \in I} \mu_i.p_i$ *(with* $I \neq \emptyset$*), then*

$$\begin{aligned} \mathscr{S}\mathscr{B} \vdash (\nu a)p &= (\nu a)(\Sigma_{i \in I} \mu_i.p_i) \\ &= \Sigma_{i \in I} (\nu a)(\mu_i.p_i) && \textit{Axiom } \mathbf{R4} \\ &= \Sigma_{\mu_i \notin \{a, \bar{a}\}} \mu_i.(\nu a)(p_i) && \textit{Axioms } \mathbf{R2} \textit{ and } \mathbf{R3} \\ &= \Sigma_{\mu_i \notin \{a, \bar{a}\}} \mu_i.r_i && \textit{by induction, for } r_i \textit{ in normal form} \end{aligned}$$

Hence $\mathscr{S}\mathscr{B} \vdash (\nu a)p = \Sigma_{\mu_i \notin \{a, \bar{a}\}} \mu_i.r_i$, *which is a normal form.* □

Proposition 4.17. (Reduction to normal form) *For any process p there exists a normal form q such that* $\mathscr{S}\mathscr{B} \vdash p = q$.

Proof. *By structural induction on p. If* $p = \mathbf{0}$, *then p is already a normal form. If* $p = \mu.p'$, *then by induction we know that there exists a normal form* q' *such that* $\mathscr{S}\mathscr{B} \vdash p' = q'$. *Hence, by substitutivity,* $\mathscr{S}\mathscr{B} \vdash \mu.p' = \mu.q'$, *where* $\mu.q'$ *is in normal form. If* $p = p_1 + p_2$, *then by induction we know that* $\mathscr{S}\mathscr{B} \vdash p_i = q_i$ *with* q_i *in normal form for* $i = 1, 2$. *Hence, by substitutivity,* $\mathscr{S}\mathscr{B} \vdash p_1 + p_2 = q_1 + q_2$, *where* $q_1 + q_2$ *is in normal form. If* $p = p_1 \mid p_2$, *then by induction we know that* $\mathscr{S}\mathscr{B} \vdash p_i = q_i$ *with* q_i *in normal form for* $i = 1, 2$. *Hence, by substitutivity,* $\mathscr{S}\mathscr{B} \vdash p_1 \mid p_2 = q_1 \mid q_2$. *By Lemma 4.3, there exists a normal form r such that* $\mathscr{S}\mathscr{B} \vdash q_1 \mid q_2 = r$, *hence the thesis by transitivity. If* $p = (\nu a)p'$, *then by induction we know that there exists a normal form* q' *such that* $\mathscr{S}\mathscr{B} \vdash p' = q'$. *Hence, by substitutivity,* $\mathscr{S}\mathscr{B} \vdash (\nu a)p' = (\nu a)q'$. *By Lemma 4.4, there exists a normal form r such that* $\mathscr{S}\mathscr{B} \vdash (\nu a)q' = r$, *hence the thesis by transitivity.* □

Theorem 4.9. (Completeness) *If* $p \sim q$, *then* $\mathscr{S}\mathscr{B} \vdash p = q$.

Proof. *By Proposition 4.17, there exist two normal forms s and t such that* $\mathscr{S}\mathscr{B} \vdash p = s$ *and* $\mathscr{S}\mathscr{B} \vdash q = t$. *By soundness, we know that* $p \sim s$ *and* $q \sim t$. *Since* $p \sim q$, *by transitivity also* $s \sim t$. *By Proposition 4.16 (completeness on normal forms), we have that* $\mathscr{S}\mathscr{B} \vdash s = t$. *Hence, the thesis* $\mathscr{S}\mathscr{B} \vdash p = q$ *follows by transitivity.* □

Corollary 4.2. *For any processes p and q,* $p \sim q$ *if and only if* $\mathscr{S}\mathscr{B} \vdash p = q$.

Proof. *Soundness (i.e., the implication for right to left) is due to Theorem 4.8, and completeness (i.e., the implication from left to right) is due to Theorem 4.9.* □

Exercise 4.38. (Simulation equivalence) Prove that the equational theory $(\Sigma_{CCS}, \mathscr{S})$, with $\mathscr{S} = \mathscr{S}\mathscr{B} \cup \{\mathbf{S}\}$, where the additional axiom (schema) **S** is

$$\mathbf{S} \quad \mu.(x+y) = \mu.(x+y) + \mu.y$$

is sound and complete for simulation equivalence over finite CCS. □

Exercise 4.39. (Trace equivalence) Prove that the equational theory $(\Sigma_{CCS}, \mathscr{T})$, with $\mathscr{T} = \mathscr{S}\mathscr{B} \cup \{\mathbf{T}\}$, where the additional axiom (schema) **T** of distributivity of prefixing w.r.t. summation is in Table 4.2, is a sound and complete axiomatization for trace equivalence over finite CCS. □

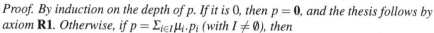

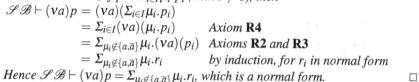

P1	$\mathbf{0} \mid x = x$	
P2	$x \mid y = y \mid x$	
P3	$(x+y) \mid z = x \mid z + y \mid z$	
P4	$\mu.x \mid \mu'.y = \mu.(x \mid \mu'.y) + \mu'.(\mu.x \mid y)$	if $\mu' \neq \overline{\mu}$
P5	$\alpha.x \mid \overline{\alpha}.y = \alpha.(x \mid \overline{\alpha}.y) + \overline{\alpha}.(\alpha.x \mid y) + \tau.(x \mid y)$	

Table 4.4 Axioms for trace equivalence in place of **Exp**

Remark 4.8. **(Finite axiomatization for trace equivalence)** It is possible to prove that the equational theory $(\Sigma_{CCS}, \mathscr{F}\mathscr{T})$, with $\mathscr{F}\mathscr{T}$ composed of the axioms **A1–A4** and **T** of Table 4.2, **R1–R4** of Table 4.3, together with the axioms **P1–P5** in Table 4.4 is a sound and complete *finite* axiomatization for trace equivalence over finite CCS. Following Remark 4.7, note that the infinitary axiom schematon **Exp** is removed in favor of the axioms (or finitary axiom schemata) in Table 4.4.[3]

Such a finite axiomatization cannot be adapted to the case of (bi)simulation equivalence, as axiom **P3** of distributivity of parallel composition w.r.t. summation is not sound for such an equivalence, as discussed in Exercise 4.7. □

4.3.3 Axiomatizing Weak Equivalences

Consider first weak trace equivalence (Definition 2.17). It is not difficult to prove that it is a congruence for the operators of finite CCS. Also the axiomatization $\mathscr{W}\mathscr{T}$ is rather intuitive: it is enough to add to set $\mathscr{T}$ of axioms for trace equivalence (Exercise 4.39) also axiom **WT** of Table 4.5, which essentially states that all occurrences of τ's can be canceled. If you have solved Exercise 4.39, then it will not be difficult to see that $\mathscr{W}\mathscr{T}$ is a sound and complete axiomatization for weak trace equivalence over finite CCS.[4]

Similarly, weak simulation equivalence is a congruence for finite CCS. Continuing Exercise 4.38, it is not difficult to see that a sound and complete axiomatization for weak simulation equivalence can be obtained by adding axiom **WT** of Table 4.5 to the set $\mathscr{S} = \mathscr{S}\mathscr{B} \cup \{\mathbf{S}\}$. The interested reader can find more details in [AF+14].

More complex is the case of rooted weak bisimilarity $\approx^c$. The axiomatization $\mathscr{W}\mathscr{B}$ is obtained by adding to the axiomatization $\mathscr{S}\mathscr{B}$ of strong bisimulation the three axioms **W1**, **W2**, **W3** of Table 4.5, the so-called τ-laws.[5]

[3] Note that an equivalent set of axioms can be obtained by removing the commutativity axiom **P2** and by duplicating axioms **P1** and **P3** in their symmetric form.

[4] It is also possible to define a *finite* sound and complete axiomatization $\mathscr{F}\mathscr{W}\mathscr{T}$ for weak trace equivalence by adding axiom **WT** to the axiomatization $\mathscr{F}\mathscr{T}$ of Remark 4.8.

[5] Note that axioms **W1** and **W3** are actually axiom schemata, i.e., there is an instance of these axioms for any possible action prefix $\mu \in Act$.

WT	$\tau.x = x$
W1	$\mu.\tau.x = \mu.x$
W2	$x+\tau.x = \tau.x$
W3	$\mu.(x+\tau.y) = \mu.(x+\tau.y)+\mu.y$
B	$\mu.(\tau.(x+y)+y) = \mu.(x+y)$

Table 4.5 Axioms for the silent action τ

For *soundness*, we have to prove that, for any pair of closed terms p and q, if $\mathscr{W}\mathscr{B} \vdash p = q$, then $p \approx^c q$. This derives from the soundness of all the axioms in $\mathscr{S}\mathscr{B}$ — proven in Theorem 4.8 w.r.t. strong bisimulation equivalence $\sim$, hence valid also w.r.t. rooted weak bisimilarity $\approx^c$ — and from the soundness of the three axioms **W1–W3**, proven in Proposition 4.13: indeed, any of these axioms, say $\mu.\tau.x = \mu.x$, is sound if for any process p we have that $\mu.\tau.p \approx^c \mu.p$.

Theorem 4.10. (Soundness) *For any pair of closed terms p and q, if* $\mathscr{W}\mathscr{B} \vdash p = q$, *then* $p \approx^c q$.

Proof. It follows by Theorem 4.8 and Proposition 4.13. □

For *completeness*, we have to resort to a new normal form, called *saturated normal form*, and to prove first the completeness result for saturated normal forms only. Then, we extend the result to all processes by proving that any process can be reduced to a saturated normal form. The details of the proofs are based on [Mil89].

Definition 4.7. A normal form $p = \Sigma_{i \in I} \mu_i . p_i$ is a *saturated* normal form if whenever $p \stackrel{\mu}{\Longrightarrow} p'$, then there exists an index i such that $\mu = \mu_i$ and $p' = p_i$ and, additionally, each p_i is a saturated normal form. □

A saturated normal form p is a normal form such that for any weak move $p \stackrel{\mu}{\Longrightarrow} p'$ there is also a corresponding strong move $p \stackrel{\mu}{\longrightarrow} p'$, i.e., $\mu.p'$ is a summand of p. Note that any process p can be reduced to a normal form p' due to Proposition 4.17. Therefore, if we prove that any normal form p' can be reduced to a saturated normal form p'', we can conclude that any process p can be reduced to a saturated normal form p''.

Lemma 4.5. (Saturation Lemma)
Given a normal form p, if $p \stackrel{\mu}{\Longrightarrow} p'$, *then* $\mathscr{W}\mathscr{B} \vdash p = p + \mu.p'$.

Proof. By induction on the length of $p \stackrel{\varepsilon}{\Longrightarrow} \stackrel{\mu}{\longrightarrow} \stackrel{\varepsilon}{\Longrightarrow} p'$. *The base case is when* $p \stackrel{\mu}{\longrightarrow} p'$; *this means that* $\mu.p'$ *is a summand of p. Then, by axiom* **A4** *the thesis follows.*

Inductively, we have two cases: either $p \xrightarrow{\mu} q \stackrel{\tau}{\Longrightarrow} p'$ *or* $p \xrightarrow{\tau} q \stackrel{\mu}{\Longrightarrow} p'$. *In the former case, we have that* $\mu.q$ *is a summand of* p. *Moreover, by induction (as the computation is shorter) we know that* $\mathscr{WB} \vdash q = q + \tau.p'$. *Hence, we conclude that*

$$\begin{aligned} \mathscr{WB} \vdash p &= p + \mu.q && \textit{axiom } \mathbf{A4} \\ &= p + \mu.(q + \tau.p') && \textit{by induction and substitutivity} \\ &= p + \mu.(q + \tau.p') + \mu.p' && \textit{axioms } \mathbf{W3} \\ &= p + \mu.p' && \textit{by previous steps reversed} \end{aligned}$$

In the latter case, we have that $\tau.q$ *is a summand of* p. *Moreover, by induction, we know that* $\mathscr{WB} \vdash q = q + \mu.p'$. *Hence, we conclude that*

$$\begin{aligned} \mathscr{WB} \vdash p &= p + \tau.q && \textit{axiom } \mathbf{A4} \\ &= p + \tau.q + q && \textit{axioms } \mathbf{W2} \\ &= p + \tau.q + q + \mu.p' && \textit{by induction and substitutivity} \\ &= p + \mu.p' && \textit{by previous steps reversed} \end{aligned}$$

and this concludes the proof. □

Proposition 4.18. (Saturation of normal forms) *For any normal form* p, *there exists a saturated normal form* q *of equal depth such that* $\mathscr{WB} \vdash p = q$.

Proof. By induction on the depth of p. *If it is* 0, *then* $p = \mathbf{0}$, *which is a saturated normal form. Otherwise, assume by induction that for any summand* $\mu_i.p_i$ *of* p, $\mathscr{WB} \vdash p_i = q_i$ *where* q_i *is a saturated normal form such that* $depth(p_i) = depth(q_i)$. *By substitutivity, we have* $\mathscr{WB} \vdash \mu_i.p_i = \mu_i.q_i$. *Let* $q' = \Sigma_i \mu_i.q_i$. *Then, by substitutivity,* $\mathscr{WB} \vdash p = q'$. *Process* q' *is a normal form of equal depth but not saturated yet because for some* i, μ_i *can be* τ. *Now we consider the set* $I = \{(\mu'_k, p'_k) \mid q' \stackrel{\mu'_k}{\Longrightarrow} p'_k$ *but not* $q' \xrightarrow{\mu'_k} p'_k\}$. *Then, by Lemma 4.5, if* $|I| = m$, $\mathscr{WB} \vdash q' = q' + \mu'_1.p'_1 + \ldots + \mu'_m.p'_m$, *which is a saturated normal form. Note that this saturated normal form has the same depth as* q' *because any summand* $\mu'_k.p'_k$ *has smaller depth (the maximal paths are shorter).* □

Theorem 4.11. (Reduction to saturated normal form) *For any process* p, *there exists a saturated normal form* q *such that* $\mathscr{WB} \vdash p = q$.

Proof. By Proposition 4.17, any process can be reduced to a normal form, and by Proposition 4.18, any normal form can be reduced to a saturated normal form. □

Proposition 4.19. (Completeness for saturated normal forms) *If* p *and* q *are saturated normal forms such that* $p \approx^c q$, *then* $\mathscr{WB} \vdash p = q$.

Proof. By induction on the sum of the depths of p *and* q. *If the sum is* 0, *then* $p = q = \mathbf{0}$ *and the thesis follows by rule 1 (reflexivity) in Table 4.1.*

Otherwise, suppose $\mu.p'$ *is a summand of* p, *hence* $p \xrightarrow{\mu} p'$. *As* $p \approx^c q$, *we have that* $q \stackrel{\mu}{\Longrightarrow} q'$ *with* $p' \approx q'$. *Since* q *is a saturated normal form, we have* $q \xrightarrow{\mu} q'$, *i.e.,* $\mu.q'$ *is a summand of* q. *Summing up, for each summand* $\mu.p'$ *of* p, *we have a summand* $\mu.q'$ *of* q *such that* $p' \approx q'$. *Now, by Lemma 4.1 (Hennessy Lemma), we know that*

$$p' \approx q' \textit{ iff } (p' \approx^c q' \textit{ or } p' \approx^c \tau.q' \textit{ or } \tau.p' \approx^c q').$$

We have three cases.

(1) If $p' \approx^c q'$ then, as p' and q' are saturated normal forms, by induction (the sum of the depths is strictly decreased), $\mathscr{W}\mathscr{B} \vdash p' = q'$, hence $\mathscr{W}\mathscr{B} \vdash \mu.p' = \mu.q'$ by substitutivity.

(2) If $p' \approx^c \tau.q'$, we have first of all to reduce $\tau.q'$ to a saturated normal form. By Proposition 4.18 we have that there exists a saturated normal form q'' of equal depth such that $\mathscr{W}\mathscr{B} \vdash \tau.q' = q''$. By Theorem 4.10, we have that $\tau.q' \approx^c q''$, hence $p' \approx^c q''$ by transitivity. Since $p' \approx^c q''$ and the sum of depth of p' and q'' is one less than that of p and q, we can apply induction and derive that $\mathscr{W}\mathscr{B} \vdash p' = q''$, hence $\mathscr{W}\mathscr{B} \vdash p' = \tau.q'$ by transitivity, and $\mathscr{W}\mathscr{B} \vdash \mu.p' = \mu.\tau.q'$ by substitutivity, and $\mathscr{W}\mathscr{B} \vdash \mu.p' = \mu.q'$ by axiom **W1** *and transitivity.*

(3) If $\tau.p' \approx^c q'$, then we can proceed as for case (2) above.

In all the three cases above, for each summand $\mu.p'$ of p, we have a summand $\mu.q'$ of q such that $\mathscr{W}\mathscr{B} \vdash \mu.p' = \mu.q'$. Symmetrically, it can be proved that for each summand $\mu.q'$ of q, we have a summand $\mu.p'$ of p such that $\mathscr{W}\mathscr{B} \vdash \mu.p' = \mu.q'$.

Therefore, $\mathscr{W}\mathscr{B} \vdash p = q$ by substitutivity and possible applications of axioms **A4** *(for removing possible duplicates) and* **A1–A2** *(for rearranging the remaining summands).* □

Exercise 4.40. Prove completeness for arbitrary processes, following analogous steps as in the proof of Theorem 4.9. □

Corollary 4.3. *For any processes p and q, $p \approx^c q$ if and only if $\mathscr{W}\mathscr{B} \vdash p = q$.*

Proof. *Soundness (i.e., the implication from right to left) is due to Theorem 4.10, and completeness (i.e., the implication from left to right) is due to the exercise above.* □

Let us now consider rooted branching bisimulation congruence $\approx^c_{br}$. The axiomatization $\mathscr{R}\mathscr{B}$ is obtained by adding to the axiomatization $\mathscr{S}\mathscr{B}$ of strong bisimulation the single axiom **B** of Table 4.5.

Soundness of $\mathscr{R}\mathscr{B}$ is obtained by soundness (Theorem 4.8) of $\mathscr{S}\mathscr{B}$ w.r.t. strong bisimulation $\sim$ (hence also w.r.t. $\approx^c_{br}$), and for axiom **B**, by Proposition 4.14.

Theorem 4.12. (Soundness) *For any pair of closed terms p and q, if $\mathscr{R}\mathscr{B} \vdash p = q$, then $p \approx^c_{br} q$.* □

As already discussed in Section 4.1.3, of the three τ-laws for rooted weak bisimulation congruence, only **W1** is sound for $\approx^c_{br}$.

Exercise 4.41. Prove that axiom **B** (together with **A3**) subsumes **W1**, and can be derived from **W1** and **W2** (together with **A4**). □

As a consequence of the exercise above, in the following proofs, we sometimes take the liberty to use also **W1** as an axiom for $\approx^c_{br}$.

For completeness, we resort to a technique in [Den07], based on the so-called *Promotion Lemma*, which plays the same role as the Hennessy Lemma (Lemma 4.1) in the completeness of $\mathscr{W}\mathscr{B}$. Indeed, we cannot resort to the Hennessy Lemma, as it is invalid for branching bisimilarity (Exercise 4.24). The proof is based on induction on a different measure of a normal form, called *size*. The size of a normal form is the total number of prefixes in its syntax; formally, it is defined as follows:

$$size(\Sigma_{i\in I}\mu_i.p_i) = \Sigma_{i\in I} size(\mu_i.p_i)$$
$$size(\mu.p) = 1 + size(p)$$

with the proviso that $size(\Sigma_{i\in I}\mu_i.p_i) = 0$ if and only if $I = \emptyset$.

Lemma 4.6. (Promotion Lemma) *If $p \approx_{br} q$, then $\mathscr{RB} \vdash \tau.p = \tau.q$.*

Proof. *By Proposition 4.17 we can assume that p and q are in normal form. The proof is by induction on the sum of sizes of p and q. If it is 0, then $p = q = \mathbf{0}$ and the thesis trivially follows $\mathscr{RB} \vdash \tau.p = \tau.q$ by reflexivity. Otherwise, as $p \approx_{br} q$, by the Deng Lemma (Lemma 4.2), one of the following three cases holds:*

(1) $p \xrightarrow{\tau} p'$ and $p' \approx_{br} q$. Hence, $\tau.p'$ is a summand of p, i.e., $p = \tau.p' + r$ for some r. Since $\tau.p' + r \approx_{br} q \approx_{br} p'$, it follows that $p' + r \approx_{br} p' \approx_{br} q$, by Exercise 4.26. As $size(p') + size(q) < size(p) + size(q)$, we can apply induction and conclude that $\mathscr{RB} \vdash \tau.p' = \tau.q$. Analogously, as $size(p'+r) + size(q) < size(p) + size(q)$, we can apply induction and conclude that $\mathscr{RB} \vdash \tau.(p'+r) = \tau.q$, hence also $\mathscr{RB} \vdash \tau.(p'+r) = \tau.p'$ by transitivity. So, we can derive the thesis as follows:

$$\begin{aligned}
\mathscr{RB} \vdash \tau.p &= \tau.(\tau.p' + r) \\
&= \tau.(\tau.(p'+r) + r) && \text{by induction and substitutivity} \\
&= \tau.(p'+r) && \text{axioms } \mathbf{B} \\
&= \tau.q && \text{by induction}
\end{aligned}$$

(2) $q \xrightarrow{\tau} q'$ and $p \approx_{br} q'$. This case is symmetric to the above, hence omitted.

(3) $p \approx^c_{br} q$. We aim at proving that each summand of p can be absorbed by q. Let $\mu.p'$ be one such summand; hence $p \xrightarrow{\mu} p'$. As for rooted branching bisimulation, the first move is to be matched as in strong bisimulation, we have $q \xrightarrow{\mu} q'$ with $p' \approx_{br} q'$; hence also $\mu.q'$ is a summand of q. Since the sum of sizes of p' and q' is decreased, we can apply induction and derive that $\mathscr{RB} \vdash \tau.p' = \tau.q'$. So, we can derive the following:

$$\begin{aligned}
\mathscr{RB} \vdash q + \mu.p' &= q + \mu.\tau.p' && \text{axiom } \mathbf{W1} \text{ and substitutivity} \\
&= q + \mu.\tau.q' && \text{by induction and substitutivity} \\
&= q + \mu.q' && \text{axioms } \mathbf{W1} \text{ and substitutivity} \\
&= q && \text{axiom } \mathbf{A4}
\end{aligned}$$

that is, any summand of p is absorbed by q. This means that $\mathscr{RB} \vdash q + p = q$. Symmetrically, we can prove in the same way that $\mathscr{RB} \vdash p + q = p$. Therefore, $\mathscr{RB} \vdash \tau.p = \tau.(p+q) = \tau.(q+p) = \tau.q$, as required. □

Exercise 4.42. Show that the Promotion Lemma holds also for weak bisimilarity, i.e., if $p \approx q$, then $\mathscr{WB} \vdash \tau.p = \tau.q$, by exploiting the fact that the Deng Lemma holds also for weak bisimilarity (Exercise 4.25). □

Proposition 4.20. (Completeness for normal forms) *If p and q are normal forms such that $p \approx^c_{br} q$, then $\mathscr{RB} \vdash p = q$.*

Proof. *By induction on the sum of the sizes of p and q. If the sum is 0, then* $p = q = \mathbf{0}$ *and the thesis follows by rule 1 (reflexivity) in Table 4.1.*

Otherwise, let $\mu.p'$ *be a summand of p; then,* $p \xrightarrow{\mu} p'$, *which is (strongly) matched by* $q \xrightarrow{\mu} q'$ *with* $p' \approx_{br} q'$, *because* $p \approx_{br}^{c} q$. *Hence, also* $\mu.q'$ *is a summand of q and, by the Promotion Lemma, we have* $\mathscr{RB} \vdash \tau.p' = \tau.q'$. *So, we can derive the following:*

$$\begin{aligned} \mathscr{RB} \vdash q + \mu.p' &= q + \mu.\tau.p' && \text{axiom } \mathbf{W1} \text{ and substitutivity} \\ &= q + \mu.\tau.q' && \text{by substitutivity} \\ &= q + \mu.q' && \text{axioms } \mathbf{W1} \text{ and substitutivity} \\ &= q && \text{axiom } \mathbf{A4} \end{aligned}$$

that is, any summand of p is absorbed by q. This means that $\mathscr{RB} \vdash q + p = q$. *Symmetrically, we can prove in the same way that* $\mathscr{RB} \vdash p + q = p$. *Therefore,* $\mathscr{RB} \vdash p = q$, *as required.* □

It is interesting to note that this proof technique for completeness can be also used as an alternative completeness proof for rooted weak bisimilarity given in Proposition 4.19. As a matter of fact, Exercise 4.42 ensures that the analogous of the Promotion Lemma holds for weak bisimilarity; then, it is enough to follow the proof of Proposition 4.20 considering *saturated* normal forms p and q (instead of generic normal forms).

Exercise 4.43. Prove completeness for arbitrary processes, following steps analogous to those in the proof of Theorem 4.9. □

Corollary 4.4. *For any processes p and q,* $p \approx_{br}^{c} q$ *if and only if* $\mathscr{RB} \vdash p = q$.

Proof. *Soundness (i.e., the implication from right to left) is due to Theorem 4.12 and completeness (i.e., the implication from left to right) is due to the exercise above.* □

4.3.4 Left Merge and Communication Merge

Axiom **Exp** in Table 4.3 is not an axiom, but an axiom schema, which stands for an infinite collection of axioms, each one for the fixed choice of n, m and for the choice of the initial prefixes μ_i and μ'_j. This means that $\mathscr{SB}$ is not a finite axiomatization of strong bisimilarity $\sim$ over finite CCS.[6] However, a finite axiomatization for $\sim$ over finite CCS can be obtained with the use of the auxiliary operators of *left merge* and *communication merge* of ACP [BK84a, BK85], hence extending the syntax of CCS. Moller in [Mol90a, Mol90b] proved that $\sim$ is not finitely axiomatizable without introducing such auxiliary operators. The technical presentation outlined here follows [Ace94], where the interested reader can find more details.

The left merge operator $\rfloor$ takes two processes p and q and generates a new process $p \rfloor q$ which is similar to the parallel composition $p \,|\, q$, but with the distinguishing

[6] Remember that a finite axiomatization for trace equivalence can be produced by removing axiom **Exp** in favor of the axioms **P1–P5** of Table 4.4, as discussed in Remark 4.8.

feature that the first transition is to be taken from p. Operationally, the only SOS rule describing the behavior of $p \rfloor q$ is

$$\text{(Left)}\ \frac{p \xrightarrow{\mu} p'}{p \rfloor q \xrightarrow{\mu} p' \mid q}$$

Observe that the reached state is $p' \mid q$, i.e., left merge is turned into parallel composition in one step.

The communication merge operator $\|$ takes two processes p and q and generates a new process $p \| q$ which is similar to the parallel composition $p \mid q$, but with the distinguishing feature that the first transition is to be a synchronization between p and q. Operationally, the only SOS rule describing the behavior of $p \| q$ is

$$\text{(Merge)}\ \frac{p \xrightarrow{\alpha} p' \qquad q \xrightarrow{\overline{\alpha}} q'}{p \| q \xrightarrow{\tau} p' \mid q'}$$

Observe that the reached state is $p' \mid q'$, i.e., communication merge is turned into parallel composition in one step.

Exercise 4.44. Prove that, if $p \sim q$, then the following hold for all $r \in \mathscr{P}$:

1. $p \rfloor r \sim q \rfloor r$
2. $r \rfloor p \sim r \rfloor q$
3. $p \| r \sim q \| r$
4. $r \| p \sim r \| q$. □

The exercise above states that $\sim$ is preserved by the left merge and the communication merge. So, it is a congruence not only for the operators of finite CCS, but also for these auxiliary operators. Hence, it can be axiomatized. The axiomatization of the parallel composition operator is given by axiom **Par** of Table 4.6, together with the axioms **L1–L3** for left merge and **C1–C5** for communication merge. Let $\mathscr{ASB}$ be the set of axioms **A1–A4** of Table 4.2, **R1–R4** of Table 4.3, together with the nine axioms of Table 4.6. It is possible to prove that the equational theory $(\Sigma_{CCS}, \mathscr{ASB})$ is a sound and complete *finite* axiomatization of $\sim$. This result derives from the fact that with this finite set of axioms it is possible to prove the analogue of Proposition 4.17, ensuring that any process can be reduced to a normal form (see Definition 4.6). The crucial property that holds for these auxiliary operators is distributivity w.r.t. sum (axioms **L3** and **C5**), which does not hold for parallel composition (see Exercise 4.7 and Remark 4.8).

Exercise 4.45. Prove that, if p and q are normal forms then there exist normal forms r, r', r'' such that $\mathscr{ASB} \vdash p \rfloor q = r$, $\mathscr{ASB} \vdash p \| q = r'$ and $\mathscr{ASB} \vdash p \mid q = r''$. (*Hint:* By induction on the sum of the depths of p and q; one has to prove all the three statements at the same time.) □

Par		$x \mid y = x \rfloor y + y \rfloor x + x \| y$
L1		$\mathbf{0} \rfloor y = \mathbf{0}$
L2		$(\mu.x) \rfloor y = \mu.(x \mid y)$
L3		$(x+y) \rfloor z = x \rfloor z + y \rfloor z$
C1		$x \| y = y \| x$
C2		$\mathbf{0} \| y = \mathbf{0}$
C3	if $\mu_1 = \overline{\mu_2}$	$(\mu_1.x) \| (\mu_2.y) = \tau.(x \mid y)$
C4	if $\mu_1 \neq \overline{\mu_2}$	$(\mu_1.x) \| (\mu_2.y) = \mathbf{0}$
C5		$(x+y) \| z = x \| z + y \| z$

Table 4.6 Axioms for the auxiliary operators

Exercise 4.46. Prove, by structural induction (following the same idea of the proof of Proposition 4.17), that for any process p there exists a normal form r such that $\mathscr{ASB} \vdash p = r$. □

Let us now consider rooted branching bisimulation $\approx_{br}^c$. The axiomatization $\mathscr{ARB}$ is obtained by adding to the axiomatization $\mathscr{ASB}$ of strong bisimulation the single axiom **B** of Table 4.5. It can be proved that $\mathscr{ARB}$ is a sound and complete finite axiomatization for $\approx_{br}^c$ over finite CCS.

Exercise 4.47. Prove that $\approx_{br}^c$ is a congruence for both left merge and communication merge. Prove also that all the axioms in $\mathscr{ARB}$ are sound for $\approx_{br}^c$. □

A similar result does not hold for rooted weak bisimilarity $\approx^c$. In fact, it happens that $\approx^c$ is not even preserved by $\|$. The equality $\tau.\alpha.\mathbf{0} + \alpha.\mathbf{0} \approx^c \tau.\alpha.\mathbf{0}$ is not preserved by the context $\mathscr{C}[-] = -\|\overline{\alpha}.\mathbf{0}$: we have that $\tau.\mathbf{0} \approx^c (\tau.\alpha.\mathbf{0} + \alpha.\mathbf{0}) \| \overline{\alpha}.\mathbf{0} \not\approx^c \tau.\alpha.\mathbf{0} \| \overline{\alpha}.\mathbf{0} \approx^c \mathbf{0}$. Hence, rooted weak bisimilarity $\approx^c$ is not a congruence for communication merge.

Exercise 4.48. Prove that weak bisimilarity $\approx$ is not a congruence for left merge as well. (*Hint:* Consider $\tau.\mathbf{0} \approx \mathbf{0}$ and the left merge context $a.\mathbf{0}$.) □

Exercise 4.49. Prove that rooted weak bisimilarity $\approx^c$ is a congruence for left merge. (This means that, by closing $\approx$ w.r.t. summation (obtaining $\approx^c$), we get a congruence also w.r.t. left merge!) □

Therefore, it is necessary to look for an alternative auxiliary communication merge operator that can be anyway useful for reducing any process p into a normal form p', but such that rooted weak bisimilarity $\approx^c$ is a congruence also for it. With abuse of notation, we call this new operator *communication merge* as well, and denote it by the same symbol $-\|-$; but now the new SOS rule is

$$\text{(WMerge)}\ \frac{p \stackrel{\varepsilon}{\Longrightarrow} p' \stackrel{\alpha}{\longrightarrow} p'' \qquad q \stackrel{\varepsilon}{\Longrightarrow} q' \stackrel{\overline{\alpha}}{\longrightarrow} q'}{p \| q \stackrel{\tau}{\longrightarrow} p'' \,|\, q''}$$

where, in the premise, we make use of the weak transition relation $\stackrel{\varepsilon}{\Longrightarrow}$, hence allowing for the execution of initial τ transitions. Hence, a single-step synchronization can be derived if p and q can perform weakly complementary actions. With this new rule, $(\tau.\alpha.\mathbf{0} + \alpha.\mathbf{0}) \| \overline{\alpha}.\mathbf{0} \approx^c \tau.\alpha.\mathbf{0} \| \overline{\alpha}.\mathbf{0} \approx^c \tau.\mathbf{0}$.

Exercise 4.50. Prove that $\approx^c$ is a congruence for the communication merge defined by rule (WMerge). □

A sound (and complete) finite axiomatization $\mathscr{AWB}$ of $\approx^c$ can be now obtained by adding to the axiomatization $\mathscr{ASB}$ of strong bisimulation the three axioms **W1, W2, W3** of Table 4.5, but with the proviso of replacing axiom **C4** with the following two:

C4.1 if $\tau \neq \mu_1 \neq \overline{\mu_2} \neq \tau$ $\quad (\mu_1.x) \| (\mu_2.y) = \mathbf{0}$
C4.2 $\quad (\tau.x) \| y = x \| y$

Chapter 5
Additional Operators

Abstract Some additional operators are investigated: the internal choice operator and the hiding one of CSP, the relabeling operator occurring in early versions of CCS, the sequential composition operator and the iteration operator of ACP, the replication operator of the π-calculus, and, finally, CSP parallel composition with multiway synchronization. The main aim of this chapter is to show that most of them are encodable into CCS, hence proving that CCS is reasonably expressive.

5.1 Internal Choice

The binary, infix operator $-\oplus-$ of internal choice, originally introduced in TCSP [Hoa85], can be added to the syntax of CCS, to get CCS$^{\oplus}$, as follows:[1]

$$p ::= \mathbf{0} \mid p+p \mid p|p \mid p\oplus p \mid (\nu a)p \mid C$$

with the intuition that $p\oplus q$ can perform the actions prescribed by the following SOS operational rules:

$$(\text{IntChoice}_1)\ \frac{}{p\oplus q \xrightarrow{\tau} p} \qquad (\text{IntChoice}_2)\ \frac{}{p\oplus q \xrightarrow{\tau} q}$$

where the choice between p and q is taken internally, without any intervention of the environment. $\mathscr{P}_{CCS^{\oplus}}$ denotes the set of CCS$^{\oplus}$ processes.

Example 5.1. Consider process $Coin \stackrel{def}{=} head.Coin \oplus tail.Coin$, i.e., a process offering the purely nondeterministic behavior of coin flipping. We can define:

$$GoodLuck \stackrel{def}{=} (\nu head, tail)(Coin \mid Lucky) \qquad Lucky \stackrel{def}{=} \overline{head}.win.Lucky$$

[1] Here, and in the rest of this section, we often consider the richer version of CCS which allows for unguarded sum, for simplicity's sake.

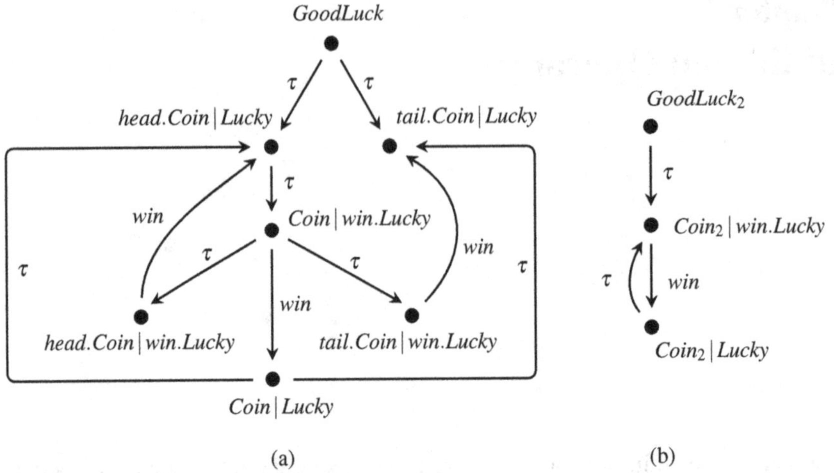

Fig. 5.1 The LTS for the *GoodLuck* process and for *GoodLuck*2 (restrictions omitted)

Note that the synchronization between *Coin* and *Lucky* may occur only after the flipping of the coin has been executed. Process *GoodLuck* may perform a sequence of lucky flippings and will possibly end in deadlock, just after the flipping result is *tail*. Figure 5.1(a) illustrates the resulting labeled transition system, where, for simplicity's sake, we have omitted the restriction operator $(\nu head, tail)\_$ in the labeling of the states.

To clarify the difference between internal choice and the CCS choice operator, consider processes

$$Coin2 \stackrel{def}{=} head.Coin2 + tail.Coin2 \qquad GoodLuck2 \stackrel{def}{=} (\nu head, tail)(Coin2 \,|\, Lucky)$$

whose LTS is depicted in Figure 5.1(b). Since *Lucky* never offers the output $\overline{tail}$, *GoodLuck*2 will always synchronize *Coin*2 and *Lucky* on action *head*, resulting in a cycling system where *win* is always executable. Of course, *Coin*2 is not a good model for coin flipping. □

Exercise 5.1. Show that $a.\mathbf{0} + b.\mathbf{0}$ and $a.\mathbf{0} \oplus b.\mathbf{0}$ are not weakly bisimilar, while they are weakly simulation equivalent. □

Exercise 5.2. (Algebraic properties) Prove that the following algebraic properties hold for internal choice: for all $p, q, r \in \mathscr{P}_{CCS^{\oplus}}$

- $p \oplus q \sim q \oplus p$, i.e., commutativity holds for strong bisimilarity;
- $p \oplus (q \oplus r) \simeq (p \oplus q) \oplus r$, i.e., associativity holds for weak similarity;
- $p \oplus p \approx p$, i.e., idempotence holds for weak bisimilarity;
- $p \oplus \mathbf{0} \simeq p$, i.e., identity ($\mathbf{0}$-absorption) holds for weak similarity.

Show also that $p \oplus p \approx^c \tau.p$ as well as $(\nu a)(p \oplus q) \sim (\nu a)p \oplus (\nu a)q$. Prove also that $p \oplus (q \oplus r) \not\approx (p \oplus q) \oplus r$ and that $p \oplus \mathbf{0} \not\approx p$. Finally, chech that $(a.p \oplus a.q) \oplus a.(p \oplus q) \approxeq a.(p \oplus q)$. □

Exercise 5.3. (Congruence) Show that if $p \sim q$, then $p \oplus r \sim q \oplus r$ for all $p, q, r \in \mathscr{P}_{CCS^{\oplus}}$, i.e., strong bisimilarity is a congruence for internal choice. Show also that weak bisimilarity $\approx$ is a congruence for internal choice. □

Internal choice is a derived operator in CCS: given a CCS$^{\oplus}$ process p, we can find a CCS process q such that $p \sim q$. This result is obtained by an encoding $[\![\text{-}]\!]$ from CCS$^{\oplus}$ into CCS, defined as

$$
\begin{aligned}
[\![\mathbf{0}]\!] &= \mathbf{0} \\
[\![\mu.p]\!] &= \mu.[\![p]\!] \\
[\![p_1 + p_2]\!] &= [\![p_1]\!] + [\![p_2]\!] & \qquad [\![p_1 \oplus p_2]\!] &= \tau.[\![p_1]\!] + \tau.[\![p_2]\!] \\
[\![p_1 \,|\, p_2]\!] &= [\![p_1]\!] \,|\, [\![p_2]\!] \\
[\![(\nu a)p]\!] &= (\nu a)[\![p]\!] \\
[\![A]\!] &= A' & \qquad \text{where } A' &\stackrel{def}{=} [\![q]\!] \text{ if } A \stackrel{def}{=} q
\end{aligned}
$$

where we assume that for each constant A used in CCS$^{\oplus}$ there exists a new constant A' for CCS. Observe that $[\![\text{-}]\!]$ is *homomorphic* w.r.t. all the CCS operators (left column), and it is not only for internal choice (right column).

Example 5.2. Consider process constant $B \stackrel{def}{=} a.A \oplus b.B$, with $A \stackrel{def}{=} c.A \oplus d.B$. The encoding of B, $[\![B]\!]$, gives rise to a new constant $B' \stackrel{def}{=} [\![a.A \oplus b.B]\!] = \tau.a.A' + \tau.b.B'$, with $A' \stackrel{def}{=} \tau.c.A' + \tau.d.B'$. □

It is an easy task to prove that $R = \{(p, [\![p]\!]) \mid p \text{ is a CCS}^{\oplus} \text{ process}\}$ is a strong bisimulation. One has to prove that $p \xrightarrow{\mu} p'$ implies $[\![p]\!] \xrightarrow{\mu} [\![p']\!]$ and, conversely, that $[\![p]\!] \xrightarrow{\mu} q$ implies there exists p' such that $q = [\![p']\!]$ and $p \xrightarrow{\mu} p'$. This can be proved by induction on the proof of the transitions — starting from the SOS axioms (Pref), (IntChoice$_1$) and (IntChoice$_2$) — and is left as an exercise for the reader.

5.2 Hiding

The hiding operator $(\iota\_)\_$, originally introduced in TCSP [Hoa85] and widely used in ACP [BK85, BBR10], takes an observable (input) action a and a process p to build a new process $(\iota a)p$, whose semantics is that of p, where actions a and $\bar{a}$ are internalized, i.e., transformed into τ. Hence, like the restriction operator, also the hiding operator is a *scoping* operator, but much weaker, as the actions the former prevents are simply internalized by the latter. The language CCS^{hide} is defined thus:

$$p ::= \mathbf{0} \mid \mu.p \mid p+p \mid p|p \mid (\iota a)p \mid (\nu a)p \mid C$$

with the intuition that $(\iota a)p$ can execute the actions prescribed by the following SOS operational rules:

$$(\mathrm{H}_1)\ \frac{p \xrightarrow{\mu} p'}{(\iota a)p \xrightarrow{\mu} (\iota a)p'}\ \mu, \overline{\mu} \neq a \qquad (\mathrm{H}_2)\ \frac{p \xrightarrow{\mu} p'}{(\iota a)p \xrightarrow{\tau} (\iota a)p'}\ \mu = a \text{ or } \overline{\mu} = a$$

$\mathscr{P}_{CCS^{hide}}$ denotes the set of CCS^{hide} processes.

Exercise 5.4. (Algebraic properties for $\sim$) Prove that the following algebraic properties of the hiding operator hold for strong bisimilarity,[2] where $fn((\iota a)p) = fn(p) \setminus \{a\}$. For all $p, q \in \mathscr{P}_{CCS^{hide}}$

$$\begin{array}{lrl}
(i) & (\iota a)p \sim p & \text{if } a \notin fn(p) \\
(ii) & (\iota a)((\iota b)p) \sim (\iota b)((\iota a)p) & \\
(iii) & (\iota a)((\iota a)p) \sim (\iota a)p & \\
(iv) & (\iota a)(\mu.p) \sim & \begin{cases} \tau.(\iota a)p \text{ if } \mu = a \text{ or } \mu = \overline{a} \\ \mu.(\iota a)p \text{ otherwise} \end{cases} \\
(v) & (\iota a)(p+q) \sim (\iota a)p + (\iota a)q & \\
(vi) & (\iota a)p \mid q \sim (\iota a)(p \mid q) & \text{if } a \notin fn(q) \\
(vii) & p \mid (\iota a)q \sim (\iota a)(p \mid q) & \text{if } a \notin fn(p) \\
(viii) & (\iota a)p \sim (\iota b)(p\{b/a\}) & \text{if } b \notin fn(p) \cup bn(p) \\
(ix) & (\iota a)p + q \sim (\iota a)(p+q) & \text{if } a \notin fn(q) \\
(x) & p + (\iota a)q \sim (\iota a)(p+q) & \text{if } a \notin fn(p) \\
(xi) & (\iota a)((\nu a)p) \sim (\nu a)p & \\
(xii) & (\nu a)((\iota a)p) \sim (\iota a)p & \\
(xiii) & (\iota a)((\nu b)p) \sim (\nu b)((\iota a)p) & \text{if } a \neq b \\
(xiv) & (\iota a)A \sim & \begin{cases} A \quad \text{if } a \notin fn(A) \\ A^a \quad \text{otherwise, where } A^a \stackrel{def}{=} (\iota a)q \text{ if } A \stackrel{def}{=} q \end{cases}
\end{array}$$

□

The second law above states that the actual order of hidden names is inessential. Therefore, we can generalize the hiding operator over a set of names; e.g., for the case of this law, we can use the notation $(\iota a, b)p$.

Exercise 5.5. (Algebraic property for $\approx^c$) Prove that the following algebraic property holds for (rooted) weak bisimilarity: for all $p, q \in \mathscr{P}_{CCS^{hide}}$

$$(\iota a)(p \mid q) \approx^c (\iota a)p \mid (\iota a)q$$

□

Exercise 5.6. (Congruence) Prove that, for all $p, q \in \mathscr{P}_{CCS^{hide}}$, if $p \sim q$, then $(\iota a)p \sim (\iota a)q$ for all $a \in \mathscr{L}$, i.e., strong bisimilarity is a congruence for the hiding operator.

Prove also that if $p \approx q$, then $(\iota a)p \approx (\iota a)q$ for all $a \in \mathscr{L}$, i.e., also weak bisimilarity is a congruence for the hiding operator. Finally, prove that rooted weak bisimilarity is a congruence for the hiding operator. □

[2] Some of these laws appeared originally, in the form of conditional axioms, in [BBK87].

The hiding operator is a *derived operator* in CCS. This means that we can encode any process p in CCS^{hide} into a CCS process q (with no occurrence of the hiding operator) such that $p \sim q$, hence showing that the extension is inessential as it adds no expressive power to CCS. The formal encoding $[\![-]\!]_1$ is defined homomorphically for all the CCS operators (as done in the previous section) and for the hiding operator as follows:

$$[\![(\iota a)p]\!]_1 = (\nu a)([\![p]\!]_1 \,|\, A_a) \qquad \text{where } A_a \stackrel{def}{=} a.A_a + \overline{a}.A_a$$

Example 5.3. For instance, if $B \stackrel{def}{=} a.(\iota a)(b.a.\mathbf{0} \,|\, \overline{a}.B)$, then the corresponding CCS process is $[\![B]\!]_1 = B'$ where $B' \stackrel{def}{=} [\![a.(\iota a)(b.a.\mathbf{0} \,|\, \overline{a}.B)]\!]_1 = a.[\![(\iota a)(b.a.\mathbf{0} \,|\, \overline{a}.B)]\!]_1 = a.(\nu a)([\![b.a.\mathbf{0} \,|\, \overline{a}.B]\!]_1 \,|\, A_a) = a.(\nu a)(b.a.\mathbf{0} \,|\, \overline{a}.B' \,|\, A_a)$. □

It is not difficult to prove that relation

$$R_1 = \{(p, [\![p]\!]_1) \mid p \text{ is a CCS}^{hide} \text{ process}\}$$

is a strong bisimulation by induction on the proof of transitions from p and $[\![p]\!]_1$. In particular, take the pair $((\iota a)p, [\![(\iota a)p]\!]_1) \in R_1$. If $(\iota a)p \xrightarrow{\mu} q$ with $\mu \neq a, \overline{a}$, then this can be due only to SOS rule (H$_1$), with $q = (\iota a)p'$ and $p \xrightarrow{\mu} p'$. By induction, we have that $[\![p]\!]_1 \xrightarrow{\mu} [\![p']\!]_1$ and so $[\![(\iota a)p]\!]_1 = (\nu a)([\![p]\!]_1 \,|\, A_a) \xrightarrow{\mu} (\nu a)([\![p']\!]_1 \,|\, A_a) = [\![(\iota a)p']\!]_1$, by SOS rules (Par$_1$) and (Res), and the reached states form a pair in R_1. If $(\iota a)p \xrightarrow{\tau} q$, then this is due either to SOS rule (H$_1$), and this case is as above, or to SOS rule (H$_2$), with $q = (\iota a)p'$ and $p \xrightarrow{\alpha} p'$, $\alpha = a$ or $\alpha = \overline{a}$. By induction, $[\![p]\!]_1 \xrightarrow{\alpha} [\![p']\!]_1$ and so $[\![(\iota a)p]\!]_1 = (\nu a)([\![p]\!]_1 \,|\, A_a) \xrightarrow{\tau} (\nu a)([\![p']\!]_1 \,|\, A_a) = [\![(\iota a)p']\!]_1$, by SOS rules (Com) and (Res), and the reached states form a pair in R_1. Symmetrically, if $[\![(\iota a)p]\!]_1$ moves first.

In case the hiding operator is applied to a set $L = \{a_1, \ldots, a_n\}$ of actions, the encoding can be generalized as follows: $[\![(\iota L)p]\!]_1 = (\nu L)([\![p]\!]_1 \,|\, A_L)$ where constant A_L (not in use by p) is the one-state process

$$A_L \stackrel{def}{=} \Sigma_{1 \leq i \leq n}(a_i.A_L + \overline{a}_i.A_L).$$

An alternative way of encoding CCS^{hide} into CCS, up to rooted weak bisimilarity $\approx^c$, is to exploit syntactic substitution (see Definition 4.4). The formal encoding $[\![-]\!]_2$ is defined homomorphically for all CCS operators, with the new rule for hiding defined as follows:

$$[\![(\iota a)p]\!]_2 = [\![p]\!]_2\{\tau/a\}$$

To be precise, we are a bit more liberal, as in Definition 4.4 we require that also the replacement be a *visible* action, but it is clear that such a definition of syntactic substitution can be trivially extended to the case of action τ (see below for details).

Example 5.4. Continuing Example 5.3, consider again $B \stackrel{def}{=} a.(\iota a)(b.a.\mathbf{0} \mid \overline{a}.B)$. According to this second encoding, the corresponding CCS process is $[\![B]\!]_2 = B'$ where $B' \stackrel{def}{=} [\![a.(\iota a)(b.a.\mathbf{0} \mid \overline{a}.B)]\!]_2 = a.[\![(\iota a)(b.a.\mathbf{0} \mid \overline{a}.B)]\!]_2 = a.([\![b.a.\mathbf{0} \mid \overline{a}.B]\!]_2)\{\tau/a\}$ $= a.(b.a.\mathbf{0} \mid \overline{a}.B')\{\tau/a\} = a.(b.\tau.\mathbf{0} \mid \tau.(B'\{\tau/a\})) = a.(b.\tau.\mathbf{0} \mid \tau.B'_{\{\tau/a\}})$, where

$B'_{\{\tau/a\}} \stackrel{def}{=} (a.(b.\tau.\mathbf{0} \mid \tau.B'_{\{\tau/a\}}))\{\tau/a\} = (\tau.(b.\tau.\mathbf{0} \mid \tau.B'_{\{\tau/a\}})$. □

It is not difficult to see that relation

$$R_2 = \{(p, [\![p]\!]_2) \mid p \text{ is a CCS}^{hide} \text{ process}\}$$

is a rooted weak bisimulation by induction on the proofs of transitions from p and $[\![p]\!]_2$. An alternative proof is by inspecting the definition of syntactic substitution and the algebraic properties of hiding. In fact, the definition of syntactic substitution (left column below) parallels these algebraic properties (right column), up to rooted weak bisimilarity:

$$\begin{array}{rclrcll}
\mathbf{0}\{\tau/a\} &=& \mathbf{0} & (\iota a)\mathbf{0} &\sim& \mathbf{0} & \\
(a.p)\{\tau/a\} &=& \tau.(p\{\tau/a\}) & (\iota a)((a.p)) &\sim& \tau.((\iota a)p) & \\
(\overline{a}.p)\{\tau/a\} &=& \tau.(p\{\tau/a\}) & (\iota a)((\overline{a}.p)) &\sim& \tau.((\iota a)p) & \\
(\mu.p)\{\tau/a\} &=& \mu.(p\{\tau/a\}) & (\iota a)((\mu.p)) &\sim& \mu.((\iota a)p) & \text{if } \mu \neq a, \overline{a} \\
(p+p')\{\tau/a\} &=& p\{\tau/a\} + p'\{\tau/a\} & (\iota a)(p+p') &\sim& (\iota a)p + (\iota a)p' & \\
(p \mid p')\{\tau/a\} &=& p\{\tau/a\} \mid p'\{\tau/a\} & (\iota a)(p \mid p') &\approx^c& (\iota a)p \mid (\iota a)p' & \\
((\nu b)p)\{\tau/a\} &=& (\nu b)(p\{\tau/a\}) & (\iota a)((\nu b)p) &\sim& (\nu b)((\iota a)p) & \text{if } b \neq a \\
((\nu a)p)\{\tau/a\} &=& (\nu a)p & (\iota a)((\nu a)p) &\sim& (\nu a)p &
\end{array}$$

and for constants,

$$A\{\tau/a\} = \begin{cases} A & \text{if } a \notin fn(A) \\ A_{\{\tau/a\}} & \text{otherwise, where } A_{\{\tau/a\}} \stackrel{def}{=} q\{\tau/a\} \text{ if } A \stackrel{def}{=} q \end{cases}$$

$$(\iota a)A \sim \begin{cases} A & \text{if } a \notin fn(A) \\ A^a & \text{otherwise, where } A^a \stackrel{def}{=} (\iota a)q \text{ if } A \stackrel{def}{=} q \end{cases}$$

The two different encodings differ not only for the equivalence they respect — $\sim$ for $[\![\text{-}]\!]_1$ and $\approx^c$ for $[\![\text{-}]\!]_2$ — but may differ also for the size of generated CCS process, as the following exercise shows.

Exercise 5.7. Consider process constant $C \stackrel{def}{=} (\iota a)(a.C)$. (*i*) Show that C generates an LTS with infinitely many states. (*ii*) Compute $[\![C]\!]_1$ and show that its LTS is infinite as well. (*iii*) Compute $[\![C]\!]_2$ and show that its LTS is finite-state. □

5.3 Relabeling

The relabeling operator $\_[b/a]$ takes a unary substitution b/a (hence, $a \neq b$), and a process p to construct a new process $p[b/a]$, whose semantics is that of p, where action a ($\overline{a}$) is turned into b ($\overline{b}$). The language CCS^{rel} is defined as

$$p ::= \mathbf{0} \mid \mu.p \mid p+p \mid p|p \mid p[b/a] \mid (\nu a)p \mid C$$

with the intuition that $p[b/a]$ can execute the actions prescribed by the following SOS operational rules:

$$(\text{Rel}_1)\ \frac{p \xrightarrow{\mu} p'}{p[b/a] \xrightarrow{\mu} p'[b/a]}\ \mu \neq a \wedge \mu \neq \overline{a}$$

$$(\text{Rel}_2)\ \frac{p \xrightarrow{\alpha} p'}{p[b/a] \xrightarrow{\beta} p'[b/a]}\ (\alpha = a \wedge \beta = b) \vee (\alpha = \overline{a} \wedge \beta = \overline{b})$$

$\mathscr{P}_{CCS^{rel}}$ denotes the set of CCS^{rel} processes.

The relabeling operator may be useful when simple variants of a given pattern are to be defined, hence economizing on the number of definitions. For instance, consider the counter C, studied in Example 3.17. In Section 3.5.2, we had to define three registers R_1, R_2 and R_3, as a slight variation on the counter. These can be more conveniently defined as follows, with the help of the relabeling operator:

$$R_j \stackrel{def}{=} ((C[zero_j/zero])[inc_j/inc])[dec_j/dec] \quad \text{for } j = 1,2,3$$

Exercise 5.8. (Congruence) Prove that, for all $p, q \in \mathscr{P}_{CCS^{rel}}$ and for all unary substitutions b/a, if $p \sim q$, then $p[b/a] \sim q[b/a]$, i.e., $\sim$ is a congruence for the relabeling operator. □

Exercise 5.9. (Algebraic properties for $\sim$) Prove that the following properties of the relabeling operator hold for strong bisimilarity, where $fn(p[b/a]) = fn(p)$ if $a \notin fn(p)$, $fn(p[b/a]) = (fn(p) \setminus \{a\}) \cup \{b\}$ otherwise. For all $p, q \in \mathscr{P}_{CCS^{rel}}$

(i) $\quad p[b/a] \sim p \;$ if $a \notin fn(p)$

(ii) $\quad (p[b/a])[c/a] \sim p[b/a]$

(iii) $\quad (p[b/a])[d/c] \sim (p[d/c])[b/a] \;$ if $a \neq d$, $c \neq b$ and $a \neq c$

(iv) $\quad (\mu.p)[b/a] \sim \begin{cases} b.(p[b/a]) \text{ if } \mu = a \\ \overline{b}.(p[b/a]) \text{ if } \mu = \overline{a} \\ \mu.(p[b/a]) \text{ otherwise} \end{cases}$

(v) $\quad (p+q)[b/a] \sim p[b/a] + q[b/a]$

(vi) $\quad p \mid (q[b/a]) \sim (p \mid q)[b/a] \;$ if $a \notin fn(p)$

(vii) $\quad (p[b/a]) \mid q \sim (p \mid q)[b/a] \;$ if $a \notin fn(q)$

$(viii)$ $\quad p[b/a] + q \sim (p+q)[b/a] \;$ if $a \notin fn(q)$

(ix) $\quad p + q[b/a] \sim (p+q)[b/a] \;$ if $a \notin fn(p)$

(x) $\quad ((\nu a)p)[b/a] \sim (\nu a)p$

(xi) $\quad ((\nu c)p)[b/a] \sim (\nu c)(p[b/a]) \;$ if $c \neq a, b$

(xii) $\quad ((\nu b)p)[b/a] \sim \begin{cases} (\nu b)p & \text{if } a \notin fn(p) \\ (\nu c)((p\{c/b\})[b/a]) & \text{o.w., with } c \notin fn(p) \cup bn(p) \end{cases}$

$$(xiii) \qquad A[b/a] \sim \begin{cases} A & \text{if } a \notin fn(A) \\ A_{[b/a]} & \text{otherwise, where } A_{[b/a]} \stackrel{def}{=} q[b/a] \text{ if } A \stackrel{def}{=} q \end{cases} \qquad \square$$

Note that one important law is missing: $(p\,|\,q)[b/a] \sim p[b/a]\,|\,q[b/a]$. Unfortunately, this law is invalid in general. For instance, consider $p = a.\mathbf{0}$ and $q = \overline{b}.\mathbf{0}$. Of course, the set of completed traces of $(p\,|\,q)[b/a]$ is $\{b\overline{b}, \overline{b}b\}$, while the set of completed traces for $p[b/a]\,|\,q[b/a]$ includes additionally trace τ.

Exercise 5.10. Take inspiration from the example above to study sufficient conditions on the sorts of p and q so that $(p\,|\,q)[b/a] \sim p[b/a]\,|\,q[b/a]$ is valid. □

As a consequence, we cannot use an encoding $[\![-]\!]_3$ based on the same idea of the encoding $[\![-]\!]_2$ for hiding, i.e., homomorphically defined for all CCS operators, with the new rule for relabeling defined as

$$[\![p[b/a]]\!]_3 = [\![p]\!]_3\{b/a\}$$

However, such an encoding is correct, up to $\sim$, for CCS subcalculi that are not using parallel composition, as the following example illustrates. Moreover, it may be even used for processes of CCS, including parallel composition, when the conditions to be studied in Exercise 5.10 are satisfied.

Example 5.5. The language FS-CCS^{rel} is defined as

$$p ::= \mathbf{0} \;\big|\; \mu.p \;\big|\; p+p \;\big|\; p[b/a] \;\big|\; C,$$

essentially extending finite-state CCS with the relabeling operator. We can use $[\![\_]\!]_3$ to map FS-CCS^{rel} onto finite-state CCS, up to $\sim$. Observe that a process in FS-CCS^{rel} may be infinite-state; for instance, $A \stackrel{def}{=} (a.A)[b/a]$ generates the LTS

$$A \xrightarrow{b} A[b/a] \xrightarrow{b} (A[b/a])[b/a] \xrightarrow{b} ((A[b/a])[b/a])[b/a] \xrightarrow{b} \ldots.$$

However, $[\![A]\!]_3$ gives rise to the following finite-state CCS constants $A' \stackrel{def}{=} b.A_{\{b/a\}}$ and $A_{\{b/a\}} \stackrel{def}{=} b.A_{\{b/a\}}$. It can be easily proved that

$$R_3 = \{(p, [\![p]\!]_3) \;\big|\; p \text{ is a FS-CCS}^{rel} \text{ process}\}$$

is a bisimulation by induction on the proofs of transitions from p and $[\![p]\!]_3$. □

Exercise 5.11. The language FS-CCS^{rel,res} is defined as

$$p ::= \mathbf{0} \;\big|\; \mu.p \;\big|\; p+p \;\big|\; p[b/a] \;\big|\; (\nu a)p \;\big|\; C,$$

essentially extending finite-state CCS (FS-CCS) with the relabeling operator and the restriction one. Define a suitable encoding $[\![\_]\!]_4$, up to strong bisimilarity $\sim$, to map FS-CCS^{rel,res} onto FS-CCS, (*Hint:* You may articulate the encoding as the composition of two encodings: the former, from FS-CCS^{rel,res} onto FS-CCS^{res}, is essentially

$[\![\text{-}]\!]_3$; the latter, from FS-CCS^{res} onto FS-CCS, is $[\![-]\!]$, defined homomorphically, except for restriction: $[\![(\nu a)p]\!] = [\![p]\!]_{\{a,\overline{a}\}}$, where $[\![p]\!]_A$ is homomorphic on all the operators except prefixing, as it prunes all the prefixes, and the terms they guard, starting with an action in A.) □

CCS was equipped in [Mil80, Mil89] with a more general form of relabeling $p[f]$, taking as a parameter a function $f : \mathscr{L} \to \mathscr{L}$ — extended to $Act = \mathscr{L} \cup \overline{\mathscr{L}} \cup \{\tau\}$ in such a way that $f(\overline{\alpha}) = \overline{f(\alpha)}$ and $f(\tau) = \tau$ — which has the effect of renaming the actions of p according to f. (However, note that the relabeling operator was removed altogether in subsequent versions of CCS [Mil99] and its mobile evolution, the π-calculus [MPW92].) Its operational rule is

$$(\text{Rel})\ \frac{p \xrightarrow{\mu} p'}{p[f] \xrightarrow{f(\mu)} p'[f]}$$

This new relabeling operator is more powerful than the basic version we have described above, as it may be used to generate a process that is not bisimulation equivalent to any finitary CCS process. Indeed, let us assume that $Act = \{a_i \mid i \in \mathbb{N}\}$ and consider f defined thus: $f(a_i) = a_{i+1}$ for all $i \in \mathbb{N}$. The constant $A \stackrel{def}{=} a_0.(A[f])$ gives rise to an LTS with infinitely many states:

$$A \xrightarrow{a_0} A[f] \xrightarrow{a_1} (A[f])[f] \xrightarrow{a_2} ((A[f])[f])[f] \xrightarrow{a_3} \dots$$

Note that A is not equivalent to any finitary CCS process p, as $sort(A) = Act$, while $sort(p)$ is always a finite set (see Corollary 4.1). Hence, finitary CCS with the (full) relabeling operator is more powerful than finitary CCS.

5.4 Sequential Composition

The binary, infix operator $\_\cdot\_$ of sequential composition, largely studied by the ACP community (see, e.g., [BK84a, BW90, Fok00, BBR10]), takes two processes, p and q, to build a new process $p \cdot q$, whose behavior, intuitively, is composed of the behavior of p first, followed by that of q, if and when p terminates successfully.

This intuition can be formalized only if we allow for a semantic model more generous than LTSs, namely LTSs with two types of states: the *final* states, which can immediately terminate successfully their execution, and the *nonfinal* states, which cannot immediately terminate successfully. The behavioral equivalences, studied in Chapter 2, are to be adapted accordingly in order to take into account also the state type. This extended model, called LTSF, is the subject of the next subsection.

We then discuss five calculi of increasing expressive power, based on sequential composition. The first is finite BPA, where BPA stands for Basic Process Algebra, which includes, besides sequential composition, also the CCS choice operator as well as two distinct constants: **0** for unsuccessful termination, and **1** for successful termination (on the contrary CCS has one single constant **0** that denotes successful termination). Finite BPA can represent any finite LTSF, up to strong bisimilarity. The second language is BPA* which extends finite BPA with the iteration operator p^*. BPA* is essentially the language of regular expressions (see Section 1.3.2). Not surprisingly, any regular language can be represented by a suitable BPA* process; surprisingly, there exist finite-state LTSFs that are not bisimilar to any BPA* process [Mil84]; hence, BPA* is unsatisfactory in this respect. The problem can be overcome by replacing iteration with the more powerful recursion construct, such as the one in CCS implemented by means of process constants. The resulting calculus is called BPA, which can represent any finite-state LTSF, as well as a large class of infinite state systems. In particular, we will show that any context-free language can be represented by a suitable BPA process, as context-free grammars (see Section 1.3.3) can be easily represented as BPA processes. Bisimulation equivalence is decidable for BPA [BBK93], even if a real counter that tests for 0 can be represented in BPA. Then, we will present PA, an extension of BPA with the BPP parallel operator (i.e., the CCS parallel operator without communication). To prove or disprove that bisimulation equivalence is decidable for PA is a long-lasting open problem. Then, we introduce a slight extension of PA, called PAER, for PA with External Restriction (and communication). PAER is Turing complete because the CCS modeling of counter machines we have presented in Section 3.5.2 can be easily adapted to PAER: the only critical component is the finitary CCS process for the counter, which uses restriction inside the body of three recursively defined constants; however, it can be safely replaced in PAER by the simple BPA process for the counter, not using any restriction at all. As a consequence, bisimulation equivalence is undecidable for PAER. This section ends with a possible implementation of sequential composition within CCS, so that it turns out to be a derived operator. Interestingly enough, this implementation is obtained by means of parallel composition and restriction, by using two bound names only.

5.4.1 Transition Systems with Final States

Definition 5.1. (Lts's with final states) A labeled transition system with final states (LTSF for short) is a tuple $TS = (Q, A, \rightarrow, F)$ where $(Q, A, \rightarrow)$ is an LTS and $F \subseteq Q$ is the set of *final* states. A *rooted* LTSF is a pair (TS, q_0) where $TS = (Q, A, \rightarrow, F)$ is an LTSF and $q_0 \in Q$ is the initial state. Sometimes we write $TS = (Q, A, \rightarrow, F, q_0)$ for a rooted LTSF. □

Graphically, a final state is represented by a circle around the state, as is customary in automata theory. Figure 5.2 shows some LTSFs. For instance, q_1 and q_2 are final states, while q_0 is not.

Behavioral semantics over this enriched model can be given by adapting the definitions of Chapter 2. To start, besides traces (Definition 2.9), we can define *f-traces*, i.e., traces that end in a final state. Similarly, for *weak f-traces*.

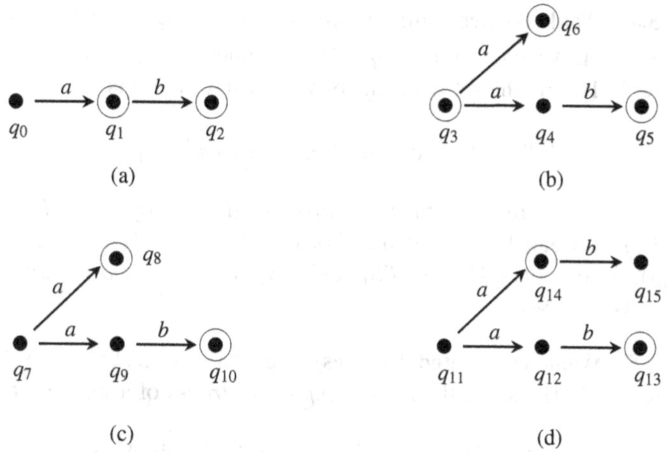

Fig. 5.2 Some LTSs with final states

Definition 5.2. (F-trace equivalence) Let $(Q,A,\to,F)$ be a LTSF and let $q \in Q$. A *f-trace* of q is a sequence $\sigma \in A^*$ such that $q \xrightarrow{\sigma}{}^* q'$ for some $q' \in F$. Hence, the set of *f-traces* of q is

$$Tr_f(q) = \{\sigma \in A^* \mid \exists q' \in F.\, q \xrightarrow{\sigma}{}^* q'\}.$$

Two states $q_1, q_2 \in Q$ are *f-trace equivalent* if $Tr_f(q_1) = Tr_f(q_2)$. This is sometimes denoted with $q_1 =_{trf} q_2$.

The set $Tr_f(TS)$ of f-traces of the rooted LTSF $TS = (Q,A,\to,F,q_0)$ is $Tr_f(q_0)$. Two rooted LTSFs, TS_1 and TS_2, are *f-trace equivalent* if $Tr_f(TS_1) = Tr_f(TS_2)$. □

Definition 5.3. (Completed f-trace equivalence) A *completed f-trace* for q is a sequence $\sigma \in A^*$ such that $q \xrightarrow{\sigma}{}^* q'$ for some deadlock state $q' \in F$. Hence, the set of *completed f-traces* of q is

$$CTr(q) = \{\sigma \in A^* \mid \exists q' \in F.\, q \xrightarrow{\sigma}{}^* q' \wedge q' \not\to\}.$$

Two states $q_1, q_2 \in Q$ are completed f-trace equivalent if $Tr_f(q_1) = Tr_f(q_2)$ and $CTr_f(q_1) = CTr_f(q_2)$. □

Example 5.6. Consider the LTSFs in Figure 5.2. Observe that $Tr(q_0) = \{\varepsilon, a, ab\}$, while $Tr_f(q_0) = \{a, ab\}$; hence, $Tr_f(q_0)$ is not prefix-closed. In general, while the set of traces of a TS (with final states) is prefix-closed, this may be not the case for the set of its f-traces. Note that $Tr_f(q_3) = \{\varepsilon, a, ab\}$, hence, in this case, the set of f-traces is prefix-closed. Of course, q_0 and q_3 are not f-trace equivalent. On the contrary, q_0 and q_7 are f-trace equivalent: $q_0 =_{trf} q_7$. Note that also $q_{11} =_{trf} q_0$.

Observe that $CTr_f(q_0) = \{ab\} = CTr_f(q_{11})$. Hence, q_0 and q_{11} are completed f-trace equivalent. On the contrary, $CTr_f(q_3) = \{a, ab\} = CTr_f(q_7)$, but q_3 and q_7 are not completed f-trace equivalent because they are not f-trace equivalent. □

Definition 5.4. (Weak f-trace equivalence) Let $TS = (Q, A \cup \{\tau\}, \rightarrow, F)$ be an LTSF, where $\tau \notin A$. A *weak f-trace* of $q \in Q$ is a sequence $\sigma \in A^*$ such that $q \stackrel{\sigma}{\Longrightarrow} q'$ for some $q' \in F$. Hence, the set $WTr_f(q)$ of weak f-traces of q is

$$WTr_f(q) = \{\sigma \in A^* \mid \exists q' \in F.\, q \stackrel{\sigma}{\Longrightarrow} q'\}.$$

Two states $q_1, q_2 \in Q$ are *weak f-trace equivalent* if $WTr_f(q_1) = WTr_f(q_2)$. The set $WTr_f(TS)$ of weak f-traces of the rooted LTSF $TS = (Q, A \cup \{\tau\}, \rightarrow, F, q_0)$ is $WTr_f(q_0)$. Two rooted LTSFs, TS_1 and TS_2, are *weak f-trace equivalent* if $WTr(TS_1) = WTr(TS_2)$. □

Definition 5.5. (Weak completed f-traces) Let $TS = (Q, A \cup \{\tau\}, \rightarrow, F)$ be an LTSF, where $\tau \notin A$. The set of the *weak completed f-traces* of a state $q \in Q$ is

$$WCTr_f(q) = \{\sigma \in A^* \mid \exists q' \in F.\, q \stackrel{\sigma}{\Longrightarrow} q' \wedge q' \stackrel{\alpha}{\not\Longrightarrow} \text{ for all observable } \alpha \in A\}.$$

Two states $q_1, q_2 \in Q$ are *weak completed f-trace equivalent* if $WTr_f(q_1) = WTr_f(q_2)$ and $WCTr_f(q_1) = WCTr_f(q_2)$. The set $WCTr_f(TS)$ of weak completed f-traces of the rooted LTSF $TS = (Q, A \cup \{\tau\}, \rightarrow, F, q_0)$ is $WCTr_f(q_0)$. TS_1 and TS_2 are weak completed f-trace equivalent if $WTr_f(TS_1) = WTr_f(TS_2)$ and $WCTr_f(TS_1) = WCTr_f(TS_2)$. □

Of course, if $Tr_f(q_1) = Tr_f(q_2)$ then also $WTr_f(q_1) = WTr_f(q_2)$. Similarly, if q_1 and q_2 are completed f-trace equivalent, then they are also weak completed f-trace equivalent.

Remark 5.1. **(NFAs and finite-state LTSFs)** Note that rooted, finite-state LTSFs are exactly finite automata (NFAs), as defined in Section 1.3.4, with the only proviso that ε-labeled transitions of an NFA are τ-labeled transitions in an LTSF. Once a rooted LTSF $TS = (Q, A, \rightarrow, F, q_0)$ is interpreted as an NFA (Q, A, δ, F, q_0), [3] we can easily observe that the definition of $WTr_f(TS) = \{\sigma \in A^* \mid \exists q' \in F.\, q \stackrel{\sigma}{\Longrightarrow} q'\}$ coincides with the definition of recognized language $L[TS] = \{w \in A^* \mid \exists q \in F.(q_0, w) \longrightarrow^* (q, \varepsilon)\}$, with relation $\longrightarrow^*$ as defined in Section 1.3.4.

As NFAs recognize the class of regular languages, hence also (rooted) finite-state LTSFs recognize the same class. Note that in Remark 2.7 we proved that also lts's (*without* final states) can represent the class of regular languages; however, that characterization is based on weak *completed* traces, while the one proposed here is based on the notion of weak f-trace. □

Also, bisimulation-based semantics are to be adapted properly to take into account the state type. As a matter of fact, if two states q and q' are related by an *f-bisimulation* relation, then they not only satisfy the bisimulation conditions but also are of the same type.

[3] A transition relation $\rightarrow \subseteq Q \times (A \cup \{\tau\}) \times Q$ can be equivalently represented as a transition function $\delta : Q \times (A \cup \{\tau\}) \rightarrow \wp(Q)$, and vice versa.

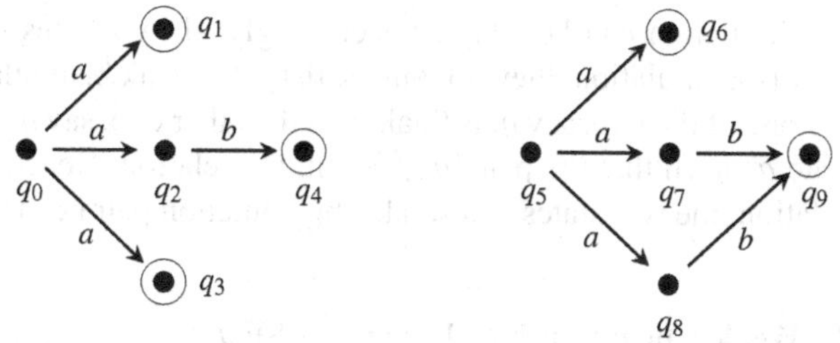

Fig. 5.3 Two f-bisimilar LTSs with final states

Definition 5.6. (F-bisimulation) Let $TS = (Q, A, \to, F)$ be an LTSF. An *f-bisimulation* is a relation $R \subseteq Q \times Q$ such that:

- R is a bisimulation on the underlying LTS $(Q, A, \to)$, and
- if $(q, q') \in R$ then $q \in F$ iff $q' \in F$.

Two states q and q' are f-bisimilar, denoted $q \sim_f q'$, if there exists an f-bisimulation R such that $(q, q') \in R$. □

Example 5.7. Consider the two LTSFs in Figure 5.3. It is easy to check that relation $R = \{(q_0, q_5), (q_1, q_6), (q_3, q_6), (q_2, q_7), (q_2, q_8), (q_4, q_9)\}$ is an f-bisimulation. Note that these two LTSFs are also f-bisimilar to the LTSF in Figure 5.2(c). However, it is never the case that two of the LTSFs in Figure 5.2 are f-bisimilar. In particular, the LTSF in (a) is not f-bisimilar to the LTSF in (b) because the two are not bisimilar (consider the pair (q_1, q_6)); differently, the LTSF in (a) is bisimilar to the LTSF in (d), but the two are not f-bisimilar: to transition $q_{11} \xrightarrow{a} q_{12}$, state q_0 can only reply with $q_0 \xrightarrow{a} q_1$, but q_{12} and q_1 are of different type. Similarly, the LTSF in (b) is not f-bisimilar to the one in (c) because, even if the two are bisimilar, the two roots, namely q_3 and q_7, are not of the same type. □

Exercise 5.12. Prove that for any LTSF $TS = (Q, A, \to, F)$, the following hold:

1. the identity relation $\mathscr{I} = \{(q,q) \mid q \in Q\}$ is an f-bisimulation;
2. the inverse relation $R^{-1} = \{(q', q) \mid (q, q') \in R\}$ of an f-bisimulation R is an f-bisimulation;
3. the relational composition $R_1 \circ R_2 = \{(q, q'') \mid \exists q'.(q, q') \in R_1 \wedge (q', q'') \in R_2\}$ of two f-bisimulations R_1 and R_2 is an f-bisimulation.
4. the union $\bigcup_{i \in I} R_i$ of f-bisimulations R_i is an f-bisimulation.

These properties justify the fact that $\sim_f$, defined as

$$\sim_f = \bigcup \{R \subseteq Q \times Q \mid R \text{ is a f-bisimulation,}\},$$

is the largest f-bisimulation and an equivalence relation. □

Exercise 5.13. Let $TS = (Q, A, \to, F)$ be an LTSF. Prove that, if $p \sim_f q$, then $Tr_f(p) = Tr_f(q)$, for all $p, q \in Q$. (*Hint:* Follow the idea in the proof of Proposition 2.1, or adapt Exercise 2.41 to this setting.) □

Also weak bisimulation can be adapted accordingly: if two states q and q' are related by a weak f-bisimulation, they not only satisfy the weak bisimulation conditions, but also, if one of the two, say q, is final, then the other one, say q', can silently reach a final state q'' such that the pair (q, q'') is in the relation. Hence, contrary to strong f-bisimulation, the two states in a weak f-bisimulation pair need not be of the same type.

Definition 5.7. (Weak f-bisimulation) For any LTSF $TS = (Q, A \cup \{\tau\}, \rightarrow, F)$, a *weak f-bisimulation* is a relation $R \subseteq (Q \times Q)$ such that:

- R is a weak bisimulation over the underlying LTS $(Q, A \cup \{\tau\}, \rightarrow)$;
- if $(q, q') \in R$ and $q \in F$, then $\exists q'' \in F$ such that $q' \stackrel{\varepsilon}{\Longrightarrow} q''$ and $(q, q'') \in R$;
- if $(q, q') \in R$ and $q' \in F$, then $\exists q'' \in F$ such that $q \stackrel{\varepsilon}{\Longrightarrow} q''$ and $(q'', q') \in R$.

State q is weakly f-bisimilar to q', denoted $q \approx_f q'$, if there exists a weak f-bisimulation R such that $(q, q') \in R$. □

Exercise 5.14. Show that

$$\approx_f = \bigcup \{R \subseteq Q \times Q \mid R \text{ is a weak f-bisimulation}\}$$

is the largest weak f-bisimulation and an equivalence relation. (*Hint:* Be careful in proving that the relational composition of two weak f-bisimulations is a weak f-bisimulation.) □

Of course, if $q_1 \sim_f q_2$, then also $q_1 \approx_f q_2$.

Exercise 5.15. Following the same idea behind the definition of weak f-bisimulation, define also branching f-bisimulation. □

A bit of care is needed for the definition of rooted weak f-bisimilarity. As its definition is driven by the need of finding the coarsest congruence for the choice operator, we should require a stricter requirement: $q_1 \approx_f^c q_2$ is true not only if to an initial τ-move of q_1, q_2 replies by performing at least one τ (as for rooted weak bisimilarity), but also q_1 and q_2 are of the same type. The need for this requirement will be clearer in the next subsection (Theorem 5.4).

Definition 5.8. (Rooted weak f-bisimilarity) Given an LTSF $(Q, A \cup \{\tau\}, \rightarrow, F)$, q_1 and q_2 are rooted weak f-bisimilar, denoted $q_1 \approx_f^c q_2$, if for all $\mu \in A \cup \{\tau\}$

- $\forall q_1'$ such that $q_1 \stackrel{\mu}{\longrightarrow} q_1'$, $\exists q_2'$ such that $q_2 \stackrel{\mu}{\Longrightarrow} q_2'$ and $q_1' \approx_f q_2'$
- $\forall q_2'$ such that $q_2 \stackrel{\mu}{\longrightarrow} q_2'$, $\exists q_1'$ such that $q_1 \stackrel{\mu}{\Longrightarrow} q_1'$ and $q_1' \approx_f q_2'$
- $q_1 \in F$ iff $q_2 \in F$. □

Exercise 5.16. Argue that if $q_1 \approx_f^c q_2$, then $q_1 \approx_f q_2$. □

Exercise 5.17. Following the same idea behind the definition of rooted weak f-bisimilarity, define also rooted branching f-bisimilarity. □

$\dfrac{}{\mathbf{1}\downarrow}$	$\dfrac{p\downarrow}{(p+q)\downarrow}$	$\dfrac{q\downarrow}{(p+q)\downarrow}$	$\dfrac{p\downarrow \quad q\downarrow}{(p\cdot q)\downarrow}$

Table 5.1 Final states: processes that can terminate properly

Definition 5.9. (F-isomorphism) Let $TS_1 = (Q_1, A_1, \to_1, F_1)$ and $TS_2 = (Q_2, A_2, \to_2, F_2)$ be two labeled transition systems with final states. An *f-isomorphism* is a bijection $f : Q_1 \to Q_2$ such that it is an isomorphism on the underlying lts's and that, additionally, it preserves the type states, i.e., $q \in F_1$ iff $f(q) \in F_2$. If there exists an f-isomorphism between TS_1 and TS_2 then we say that TS_1 and TS_2 are *f-isomorphic*, denoted $TS_1 \cong_f TS_2$.

This definition can be lifted to rooted LTSFs by requiring that the f-isomorphism f preserves also the initial states, i.e., $f(q_1) = q_2$ if q_1 and q_2 are the initial states of TS_1 and TS_2, respectively. □

5.4.2 Finite BPA

In this section we start the presentation of five calculi, based on sequential composition $\_\cdot\_$ and on the presence of special constants $\mathbf{1}$ and $\mathbf{0}$ denoting, respectively, a *properly* terminated process and an *improperly* terminated one. Some aspects of the technical development are inspired by [BBR10].

The language *finite BPA* is defined as

$$p ::= \mathbf{0} \mid \mathbf{1} \mid \mu \mid p+p \mid p\cdot p$$

where μ is any action in *Act*, $\_+\_$ is the CCS choice operator and $\_\cdot\_$ is the sequential composition operator. We assume that $\_\cdot\_$ binds tighter than $\_+\_$, so that $a\cdot b+c$ denotes $(a\cdot b)+c$. We will use brackets wherever necessary to disambiguate. $\mathscr{P}_{finBPA}$ is the set of finite BPA processes.

In order to define the operational semantics for finite BPA, we need first to identify which states are *final*, i.e., which states represent a process that may terminate properly. We use notation $p\downarrow$ to state that p is a final state. Formally, $\_\downarrow$ is the minimal predicate on processes satisfying the axiom and rules of Table 5.1. For example, the following are final states: $\mathbf{1}$, $\mathbf{0}+\mathbf{1}$, $\mathbf{1}\cdot\mathbf{1}$, $b\cdot c+\mathbf{1}$, $(a+\mathbf{1})\cdot(b\cdot c+\mathbf{1})$. While the following are not final: $\mathbf{0}$, a, $\mathbf{1}\cdot\mathbf{0}$, $a+b\cdot c$.

Finite BPA operational semantics is defined by the SOS rules in Table 5.2. Axiom (Act) states that process μ reaches the properly terminated state $\mathbf{1}$ by executing action μ. Rule (Seq$_1$) states that $p\cdot q$ can execute any initial action p can perform, and then the computation will continue with the residual of p followed by q. Rule

$$
\text{(Act)} \quad \frac{}{\mu \xrightarrow{\mu} \mathbf{1}}
$$

$$
\text{(Seq}_1\text{)} \quad \frac{p \xrightarrow{\mu} p'}{p \cdot q \xrightarrow{\mu} p' \cdot q} \qquad \text{(Seq}_2\text{)} \quad \frac{p \downarrow \quad q \xrightarrow{\mu} q'}{p \cdot q \xrightarrow{\mu} q'}
$$

$$
\text{(Sum}_1\text{)} \quad \frac{p \xrightarrow{\mu} p'}{p + q \xrightarrow{\mu} p'} \qquad \text{(Sum}_2\text{)} \quad \frac{q \xrightarrow{\mu} q'}{p + q \xrightarrow{\mu} q'}
$$

Table 5.2 SOS rules for finite BPA

(Seq$_2$) states that if p may properly terminate, then $p \cdot q$ can execute any initial action q can perform and p is discarded. As an example, consider process $(a+\mathbf{1}) \cdot b$. Its associated LTSF is composed of the three transitions below.

$$
\text{(Seq}_1\text{)} \; \cfrac{\text{(Sum}_1\text{)} \; \cfrac{\text{(Act)} \; \cfrac{}{a \xrightarrow{a} \mathbf{1}}}{a + \mathbf{1} \xrightarrow{a} \mathbf{1}}}{(a+\mathbf{1}) \cdot b \xrightarrow{a} \mathbf{1} \cdot b} \qquad \text{(Seq}_2\text{)} \; \cfrac{\cfrac{}{\mathbf{1}\downarrow} \qquad \text{(Act)} \; \cfrac{}{b \xrightarrow{b} \mathbf{1}}}{\mathbf{1} \cdot b \xrightarrow{b} \mathbf{1}}
$$

$$
\text{(Seq}_2\text{)} \; \cfrac{\cfrac{\cfrac{}{\mathbf{1}\downarrow}}{a + \mathbf{1}\downarrow} \qquad \text{(Act)} \; \cfrac{}{b \xrightarrow{b} \mathbf{1}}}{(a+\mathbf{1}) \cdot b \xrightarrow{b} \mathbf{1}}
$$

Note that $Tr_f((a+\mathbf{1}) \cdot b) = \{ab, b\}$, because only state $\mathbf{1}$ is final.

Exercise 5.18. Compute the LTSF associated to the following finite BPA processes: $a \cdot (\mathbf{1}+b)$, $a \cdot b + a + \mathbf{1}$, $a + a \cdot b$ and $a \cdot b + a \cdot (b \cdot \mathbf{0} + \mathbf{1})$. Compare the resulting LTSFs with those in Figure 5.2. □

Remark 5.2. **(Representability up to $\sim_f$)** Exercise 5.18 is an instance of a more general result: any finite LTSF can be represented, up to $\sim_f$, by a finite BPA process. As a matter of fact, given a rooted LTSF $(Q, A, \rightarrow, F, q_0)$, we can define a finite BPA process p_0 such that $q_0 \sim_f p_0$. The procedure is very simple: start from the deadlock states, each one represented either by an instance of $\mathbf{0}$ if it is not final, or by an instance of $\mathbf{1}$ if it is final; then, a state q such that $T(q) = \{(q, \mu, q_k) \mid \exists \mu \in A, \exists q_k \in Q.\, q \xrightarrow{\mu} q_k\}$, which reaches only states q_k that are already represented by suitable finite BPA terms p_k, originates the finite BPA process $\Sigma_{(q,\mu,q_k) \in T(q)} \mu \cdot p_k$, if q is not final, or $(\Sigma_{(q,\mu,q_k) \in T(q)} \mu \cdot p_k) + \mathbf{1}$ if q is final. It is then quite obvious to observe that if q is not final then $q \sim_f \Sigma_{(q,\mu,q_k) \in T(q)} \mu \cdot p_k$, or if q is final then $q \sim_f (\Sigma_{(q,\mu,q_k) \in T(q)} \mu \cdot p_k) + \mathbf{1}$. □

Exercise 5.19. (Algebraic properties of choice) Prove that, for any $p,q,r \in \mathscr{P}_{finBPA}$, the following hold:

$$\begin{aligned} p+(q+r) &\sim_f (p+q)+r \\ p+q &\sim_f q+p \\ p+\mathbf{0} &\sim_f p \\ p+p &\sim_f p \end{aligned}$$

Note that $\mathbf{0} \sim \mathbf{1}$, but $\mathbf{0} \not\sim_f \mathbf{1}$. Similarly, $p+\mathbf{1} \sim p$, but $p+\mathbf{1} \not\sim_f p$ in general, e.g., $a+\mathbf{1} \not\sim_f a$, because only $a+\mathbf{1}$ is final. □

Exercise 5.20. (Algebraic properties of sequential composition for $\sim_f$) Prove that, for any $p,q,r \in \mathscr{P}_{finBPA}$, the following hold:

$$\begin{array}{ll} (i) & p\cdot(q\cdot r) \sim_f (p\cdot q)\cdot r \\ (ii) & \mathbf{0}\cdot p \sim_f \mathbf{0} \\ (iii) & \mathbf{1}\cdot p \sim_f p \\ (iv) & p\cdot \mathbf{1} \sim_f p \\ (v) & (p+q)\cdot r \sim_f p\cdot r + q\cdot r \end{array}$$

(*Hint:* For (i), relation $R = \{(p\cdot(q\cdot r),(p\cdot q)\cdot r) \mid p,q,r \in \mathscr{P}_{finBPA}\} \cup \mathscr{I}$ is an f-bisimulation. As a matter of fact, if $p\cdot(q\cdot r) \xrightarrow{\mu} s$, then, by rules (Seq$_1$) and (Seq$_2$), this can be due to one of the following three cases: $p \xrightarrow{\mu} p'$ and $s = p'\cdot(q\cdot r)$, *or* to $p\downarrow$, $q \xrightarrow{\mu} q'$ and $s = q'\cdot r$, *or* to $p\downarrow$, $q\downarrow$, $r \xrightarrow{\mu} r'$ and $s = r'$. In any case, one can prove by means of (Seq$_1$) and (Seq$_2$) that $(p\cdot q)\cdot r \xrightarrow{\mu} s'$, with $(s,s') \in R$.) □

Note that $p\cdot\mathbf{0} \not\sim_f \mathbf{0}$, i.e., $\mathbf{0}$ does not act as a zero on the right (while (ii) above states that $\mathbf{0}$ acts as a zero on the left). E.g., $a\cdot\mathbf{0} \not\sim_f \mathbf{0}$ because only $a\cdot\mathbf{0}$ can execute a. Moreover, even if $p\cdot\mathbf{0} \sim p$, it may hold that $p\cdot\mathbf{0} \not\sim_f p$: E.g., $a\cdot\mathbf{0} \not\sim_f a$ because, after performing a, we have that $\mathbf{1}\cdot\mathbf{0} \not\sim_f \mathbf{1}$. On the contrary, for f-trace equivalence, we have that $Tr_f(p\cdot\mathbf{0}) = Tr_f(\mathbf{0})$, because each process $p'\cdot\mathbf{0}$ is not final, for any p' reachable from p. This is the subject of the following exercise.

Exercise 5.21. (Algebraic properties of sequential composition for $=_{trf}$) Prove that, for any $p,q,r \in \mathscr{P}_{finBPA}$, the following additional properties hold for f-trace equivalence:

$$\begin{array}{ll} (i) & p\cdot\mathbf{0} =_{trf} \mathbf{0} \\ (ii) & r\cdot(p+q) =_{trf} r\cdot p + r\cdot q \end{array}$$

Show an example proving that law (ii) is invalid for f-bisimulation equivalence. □

Theorem 5.1. (Congruence of $\sim_f$) *If $p \sim_f q$, then the following hold:*

1) $p+r \sim_f q+r$ *for all* $r \in \mathscr{P}_{finBPA}$,
2) $p\cdot r \sim_f q\cdot r$ *for all* $r \in \mathscr{P}_{finBPA}$,
3) $r\cdot p \sim_f r\cdot q$ *for all* $r \in \mathscr{P}_{finBPA}$,

Proof. *Assume R is an f-bisimulation such that $(p,q) \in R$.*

For case 1, it is easy to see that $R_1 = \{(p+r,q+r) \mid r \in \mathscr{P}_{finBPA}\} \cup R \cup \mathscr{I}$ *is an f-bisimulation, where* $\mathscr{I} = \{(r,r) \mid r \in \mathscr{P}_{finBPA}\}$. *We simply observe that, since p is final iff q is final because $(p,q) \in R$ and R is an f-bisimulation, it is also* *the case that* $p+r$ *is final iff* $q+r$ *is final, for any* $r \in \mathscr{P}_{finBPA}$. *Hence, relation* R_1 *satisfies the type condition on related states. Checking the bisimulation condition for the pair* $(p+r, q+r)$ *is as usual (see Theorem 4.1), hence omitted.*

For case 2, we show that $R_2 = \{(p' \cdot r, q' \cdot r) \mid (p', q') \in R \wedge r \in \mathscr{P}_{finBPA}\} \cup \mathscr{I}$ *is an f-bisimulation. First of all, for any* $(p', q') \in R$, *we have that* p' *is final iff* q' *is final; hence, it is also the case that* $p' \cdot r$ *is final iff* $q' \cdot r$ *is final, for any* $r \in \mathscr{P}_{finBPA}$. *Therefore, relation* R_2 *satisfies the type condition on related states. If* $p' \cdot r \xrightarrow{\mu} s$, *then this is due either to* $p' \xrightarrow{\mu} p''$ *and* $s = p'' \cdot r$ *(by rule (Seq*$_1$*)), or to* $p' \downarrow$, $r \xrightarrow{\mu} r'$ *and* $s = r'$ *(by rule (Seq*$_2$*)). In the former case, since* $(p', q') \in R$, *we have that* $q' \xrightarrow{\mu} q''$ *with* $(p'', q'') \in R$, *and by rule (Seq*$_1$*) also that* $q' \cdot r \xrightarrow{\mu} q'' \cdot r$ *with* $(p'' \cdot r, q'' \cdot r) \in R_2$. *In the latter case, since* $(p', q') \in R$, *also* q' *must be final, and so by rule (Seq*$_2$*) we have that* $q' \cdot r \xrightarrow{\mu} r'$ *with* $(r', r') \in R_2$. *The symmetric case, when* $q' \cdot r$ *moves first, is analogous, hence omitted.*

For case 3, the reader may check that $R_3 = \{(r' \cdot p, r' \cdot q) \mid r' \in \mathscr{P}_{finBPA}\} \cup R$ *is an f-bisimulation* □

Proposition 5.1. (τ-laws for rooted weak f-bisimilarity) *For any processes* p, q *and for any* $\mu \in Act$, *the following hold:*

$$
\begin{array}{ll}
(i) & \mu \cdot \tau \cdot p \approx_f^c \mu \cdot p \\
(ii) & p + \tau \cdot p \approx_f^c \begin{cases} \tau \cdot p + \mathbf{1} & \text{if } p\downarrow \\ \tau \cdot p & \text{otherwise} \end{cases} \\
(iii) & \mu \cdot (p + \tau \cdot q) \approx_f^c \mu \cdot (p + \tau \cdot q) + \mu \cdot q
\end{array}
$$

Proof. The proof is very similar to that for Proposition 4.13. Note that property (ii) is slightly different to take into account the state type. □

Theorem 5.2. *If* $p \approx_f q$, *then the following hold:*
1) $p \cdot r \approx_f q \cdot r$ *for all* $r \in \mathscr{P}_{finBPA}$,
2) $r \cdot p \approx_f r \cdot q$ *for all* $r \in \mathscr{P}_{finBPA}$.

Proof. Assume R is a weak f-bisimulation such that $(p, q) \in R$.

For case 1, relation $R_1 = \{(p' \cdot r, q' \cdot r) \mid (p', q') \in R \wedge r \in \mathscr{P}_{finBPA}\} \cup \mathscr{I}$ is a weak f-bisimulation. If $p' \cdot r \downarrow$, then both $p' \downarrow$ and $r \downarrow$. As $(p', q') \in R$, then $q' \stackrel{\varepsilon}{\Longrightarrow} q''$ with $q'' \downarrow$ and $(p', q'') \in R$. Hence, by (possibly repeated applications of) rule (Seq$_1$), also $q' \cdot r \stackrel{\varepsilon}{\Longrightarrow} q'' \cdot r$, with $q'' \cdot r \downarrow$ and $(p' \cdot r, q'' \cdot r) \in R_1$. Hence, the type condition is satisfied. If $p' \cdot r \xrightarrow{\mu} s$, then this can be due either to $p' \xrightarrow{\mu} p''$ and $s = p'' \cdot r$, or to $p' \downarrow$, $r \xrightarrow{\mu} r'$ and $s = r'$. In the former case, we have that either $q' \stackrel{\alpha}{\Longrightarrow} q''$ (if $\mu = \alpha$) or $q' \stackrel{\varepsilon}{\Longrightarrow} q''$ (if $\mu = \tau$) with $(p'', q'') \in R$. By (possibly repeated applications of) rule (Seq$_1$), also $q' \cdot r \stackrel{\alpha}{\Longrightarrow} q'' \cdot r$ or $q' \cdot r \stackrel{\varepsilon}{\Longrightarrow} q'' \cdot r$ with $(p'' \cdot r, q'' \cdot r) \in R_1$. In the latter case, as $(p', q') \in R$ and $p' \downarrow$, we have that $q' \stackrel{\varepsilon}{\Longrightarrow} q''$, with $q'' \downarrow$. Hence, to move $p' \cdot r \xrightarrow{\mu} r'$, process $q' \cdot r$ can respond with $q' \cdot r \stackrel{\varepsilon}{\Longrightarrow} q'' \cdot r \xrightarrow{\mu} r'$, with $(r', r') \in R_1$, *as required. The symmetric cases when $q' \cdot r$ is final and moves first are analogous, hence omitted.*

For case 2, the reader may check that $R_2 = \{(r' \cdot p, r' \cdot q) \mid r' \in \mathscr{P}_{finBPA}\} \cup R$ is a weak f-bisimulation. □

As expected, $\approx_f$ is not a congruence for $+$. As a matter of fact, $\tau \cdot a \approx_f a$, but $\tau \cdot a + b \not\approx_f a + b$, because $\tau \cdot a + b \xrightarrow{\tau} a$ and the only silent step that can be taken from $a+b$ is $a+b \stackrel{\varepsilon}{\Longrightarrow} a+b$, but a and $a+b$ are not weakly f-bisimilar. As a further example, $\tau \approx_f \mathbf{1}$; however, $\tau + a \not\approx_f \mathbf{1} + a$, because $\tau + a \xrightarrow{\tau} \mathbf{1}$ and the only silent reply can be $\mathbf{1} + a \stackrel{\varepsilon}{\Longrightarrow} \mathbf{1} + a$, but $\mathbf{1}$ and $\mathbf{1} + a$ are not weakly f-bisimilar.

Nonetheless, rooted weak f-bisimulation equivalence $\approx_f^c$ is a congruence for both finite BPA operators.

Theorem 5.3. *If $p \approx_f^c q$, then the following hold:*

1) $p \cdot r \approx_f^c q \cdot r$ *for all* $r \in \mathscr{P}_{finBPA}$,
2) $r \cdot p \approx_f^c r \cdot q$ *for all* $r \in \mathscr{P}_{finBPA}$,
3) $p + r \approx_f^c q + r$ *for all* $r \in \mathscr{P}_{finBPA}$.

Proof. *For case 1, if $p \cdot r \downarrow$, then both $p \downarrow$ and $r \downarrow$; since $p \approx_f^c q$, then also $q \downarrow$, and therefore $q \cdot r \downarrow$, as required. If $p \cdot r \xrightarrow{\mu} s$, this can be due either to $p \xrightarrow{\mu} p'$ and $s = p' \cdot r$, or to $p \downarrow$, $r \xrightarrow{\mu} r'$ and $s = r'$. In the former case, since $p \approx_f^c q$, we have that $q \stackrel{\mu}{\Longrightarrow} q'$, with $p' \approx_f q'$. Hence, by (possibly repeated applications of) rule (Seq$_1$), we can derive $q \cdot r \stackrel{\mu}{\Longrightarrow} q' \cdot r$ with $p' \cdot r \approx_f q' \cdot r$ by congruence of sequential composition w.r.t. $\approx_f$ (Theorem 5.2(1)). In the latter case, since $p \approx_f^c q$, then also $q \downarrow$. Hence, by rule (Seq$_2$) we can derive $q \cdot r \xrightarrow{\mu} r'$, with $r' \approx_f r'$. The symmetric cases when $q \cdot r$ is final or moves first are analogous, hence omitted.*

For case 2, if $r \cdot p \downarrow$, then both $r \downarrow$ and $p \downarrow$; since $p \approx_f^c q$, then also $q \downarrow$, and therefore $r \cdot q \downarrow$, as required. If $r \cdot p \xrightarrow{\mu} s$, this can be due either to $r \xrightarrow{\mu} r'$ and $s = r' \cdot p$, or to $r \downarrow$, $p \xrightarrow{\mu} p'$ and $s = p'$. In the former case, by rule (Seq$_1$), we can derive $r \cdot q \xrightarrow{\mu} r' \cdot q$ and $r' \cdot p \approx_f r' \cdot q$ by congruence of sequential composition w.r.t. $\approx_f$ (Theorem 5.2(2)), as by Exercise 5.16 we know that $p \approx_f q$. In the latter case, since $p \approx_f^c q$, we have $q \stackrel{\mu}{\Longrightarrow} q'$ with $p' \approx_f q'$. Hence, by induction on the length of the path associated to $q \stackrel{\mu}{\Longrightarrow} q'$ and by applying rule (Seq$_2$) on the first transition on that path, also $r \cdot q \stackrel{\mu}{\Longrightarrow} q'$ with $p' \approx_f q'$, as required. The symmetric cases when $r \cdot q$ is final or moves first are analogous, hence omitted.

For case 3, if $p + r \downarrow$, then either $p \downarrow$ or $r \downarrow$. In the former case, since $p \approx_f^c q$, then also $q \downarrow$, and therefore $q + r \downarrow$, as required. In the latter case, trivially also $q + r \downarrow$. If $p + r \xrightarrow{\mu} s$, then this is due to either $p \xrightarrow{\mu} p'$ and $s = p'$ (rule (Sum$_1$)), or $r \xrightarrow{\mu} r'$ and $s = r'$ (rule (Sum$_2$)). In the former case, since $p \approx_f^c q$, we have that $q \stackrel{\mu}{\Longrightarrow} q'$, with $p' \approx_f q'$. Hence, by induction on the length of the path associated to $q \stackrel{\mu}{\Longrightarrow} q'$ and by applying rule (Sum$_1$) on the first transition on that path, we can derive $q + r \stackrel{\mu}{\Longrightarrow} q'$

with $p' \approx_f q'$. *In the latter case, by rule (Sum*$_2$*) we can derive* $q+r \xrightarrow{\mu} r'$, *hence* $q+r \stackrel{\mu}{\Longrightarrow} r'$ *with* $r' \approx_f r'$. *The symmetric cases when* $q+r$ *is final or moves first are analogous, hence omitted.* □

Actually, one can prove that $\approx_f^c$ is the *coarsest congruence* (i.e., the congruence that identifies as much as possible) contained in weak f-bisimulation equivalence. This is made precise by the following theorem that justifies rooted weak f-bisimilarity as the compositional refinement of weak f-bisimilarity.

Theorem 5.4. *Assume that* $fn(p) \cup fn(q) \neq \mathscr{L}$. *Then* $p \approx_f^c q$ *if and only if, for all* $r \in \mathscr{P}_{finBPA}$, $p+r \approx_f q+r$.

Proof. *The implication from left to right follows by Theorem 5.3(3) and Exercise 5.16.*

For the implication from right to left, let $p+r \approx_f q+r$ *for all* $r \in \mathscr{P}_{finBPA}$. *Take any action* $a \in \mathscr{L}$ *such that* $a \notin fn(p) \cup fn(q)$[4] *and consider* $p+a$. *If* $p\downarrow$, *then also* $p+a\downarrow$. *As* $p+a \approx_f q+a$, *then* $q+a \stackrel{\varepsilon}{\Longrightarrow} s$ *such that* $s\downarrow$ *and* $p+a \approx_f s$. *Of course, if* $q+a \stackrel{\tau}{\Longrightarrow} s$ *truly originates from* q, *i.e.,* $q \stackrel{\tau}{\Longrightarrow} s$, *then* s *cannot be weak f-bisimilar to* $p+a$, *because* s *cannot execute* a. *Hence, the only possibility left is that* $q+a \stackrel{\varepsilon}{\Longrightarrow} q+a$ *with* $q+a\downarrow$; *this is possible only if* $q\downarrow$ *as* $a \not\downarrow$. *Summing up, under the condition that for all* $r \in \mathscr{P}_{finBPA}$, $p+r \approx_f q+r$, *if* $p\downarrow$ *then also* $q\downarrow$, *as required by the definition of rooted weak f-bisimilarity.*

Now assume $p \xrightarrow{\mu} p'$. *Then also* $p+a \xrightarrow{\mu} p'$ *(by rule (Sum*$_1$*)). As* $p+a \approx_f q+a$, *then also* $q+a$ *must respond to this transition. We have to examine two different cases:* either $\mu = \tau$ *and* $q+a \stackrel{\varepsilon}{\Longrightarrow} q+a$ *with* $p' \approx_f q+a$, or *the transition truly originates from* q, *i.e.,* $q+a \stackrel{\mu}{\Longrightarrow} q'$ *(*μ *can be* τ*), because* $q \stackrel{\mu}{\Longrightarrow} q'$, *with* $p' \approx_f q'$. *The former case is impossible:* p' *cannot be weakly f-bisimilar to* $q+a$, *as* p' *cannot perform* a. *Hence, the second case must be true; but this is indeed what is requested by rooted weak f-bisimulation: if* $p \xrightarrow{\mu} p'$, *then* $q \stackrel{\mu}{\Longrightarrow} q'$ *with* $p' \approx_f q'$.

The symmetric cases when $q\downarrow$ *or* q *moves first are analogous, hence omitted.* □

5.4.3 BPA*: Finite BPA with Iteration

Finite BPA can be extended with one additional, unary, postfix operator, $\_^*$, called the *iteration* operator, that takes a process p to build its iterate p^*, with the intuition that p^* can repeat the behavior of p any number of times. The resulting language is called BPA* and is exactly the language of *regular expressions* (see Section 1.3.2),

$$p ::= \mathbf{0} \mid \mathbf{1} \mid \mu \mid p+p \mid p \cdot p \mid p^*$$

[4] The assumption that $\mathscr{L}$ is not covered by the free names of p and q is obviously satisfied when p and q are finite BPA processes, and such a constraint holds for all the calculi discussed in this section, too.

whose operational semantics is defined by means of the rules in Tables 5.1 and 5.2, complemented by the following:

$$\frac{p \in \mathscr{P}_{BPA^*}}{p^* \downarrow} \qquad (\text{Star}) \frac{p \xrightarrow{\mu} p'}{p^* \xrightarrow{\mu} p' \cdot p^*}$$

where $\mathscr{P}_{BPA^*}$ denotes the set of BPA* processes. The iteration construct allows for the definition of infinite behavior, yet only cyclic, hence finite-state. As an example of SOS derivations, consider process $(a \cdot b)^*$. Its LTSF is composed of three states and of the following three transitions only:

$$(a \cdot b)^* \xrightarrow{a} (\mathbf{1} \cdot b) \cdot (a \cdot b)^* \xrightarrow{b} \mathbf{1} \cdot (a \cdot b)^* \xrightarrow{a} (\mathbf{1} \cdot b) \cdot (a \cdot b)^*$$

where only $(\mathbf{1} \cdot b) \cdot (a \cdot b)^*$ is not final. Similarly, the LTSF for $a \cdot (b \cdot a)^*$ is determined by the following three transitions:

$$a \cdot (b \cdot a)^* \xrightarrow{a} \mathbf{1} \cdot (b \cdot a)^* \xrightarrow{b} (\mathbf{1} \cdot a) \cdot (b \cdot a)^* \xrightarrow{a} \mathbf{1} \cdot (b \cdot a)^*$$

where only $\mathbf{1} \cdot (b \cdot a)^*$ is final.

Exercise 5.22. Build the LTSF for $(a \cdot b)^* + a \cdot (b \cdot a)^*$ and check that it is f-isomorphic to the LTSF in Figure 5.4(b). □

Exercise 5.23. (i) Prove that if q is reachable from p^*, then $q = p' \cdot p^*$ for some p' reachable from p. (ii) Given r reachable from p, prove that if $r \cdot p^* \longrightarrow^* q$, then $q = p' \cdot p^*$ for some p' reachable from p. (*Hint:* The proof is by induction on the length of the reachability path.) □

Proposition 5.2. (Congruence of $\sim_f$ for iteration) *If $p \sim_f q$, then $p^* \sim_f q^*$.*

Proof. *Assume R is an f-bisimulation such that $(p,q) \in R$. It is not difficult to check that $R_1 = \{(p^*, q^*)\} \cup \{(p' \cdot p^*, q' \cdot q^*) \mid (p', q') \in R\}$ is an f-bisimulation.*

First, consider pair (p^*, q^*); observe that both are final, hence the type condition is satisfied. Then, by rule (Star), $p^* \xrightarrow{\mu} p' \cdot p^*$ if $p \xrightarrow{\mu} p'$. Since $(p,q) \in R$, then also $q \xrightarrow{\mu} q'$ with $(p',q') \in R$. By rule (Star), also $q^* \xrightarrow{\mu} q' \cdot q^*$ is derivable, with $(p' \cdot p^*, q' \cdot q^*) \in R_1$. Symmetrically, if q^* moves first.

Now, consider a generic pair of the form $(p' \cdot p^*, q' \cdot q^*)$ such that $(p', q') \in R$. If $p' \cdot p^* \downarrow$, then $p' \downarrow$; since $(p', q') \in R$, also $q' \downarrow$ and so $q' \cdot q^* \downarrow$; hence the type condition is satisfied. If $p' \cdot p^* \xrightarrow{\mu} s$, this can be due either to $p' \xrightarrow{\mu} p''$ and $s = p'' \cdot p^*$, or to $p' \downarrow$, $p^* \xrightarrow{\mu} p_1 \cdot p^*$ and $s = p_1 \cdot p^*$. In the former case, since $(p', q') \in R$, also $q' \xrightarrow{\mu} q''$ with $(p'', q'') \in R$. By rule (Seq$_1$), also $q' \cdot q^* \xrightarrow{\mu} q'' \cdot q^*$ is derivable, with $(p'' \cdot p^*, q'' \cdot q^*) \in R_1$. In the latter case, since $(p', q') \in R$, also $q' \downarrow$; moreover, as $(p^*, q^*) \in R_1$, we can conclude that $q^* \xrightarrow{\mu} q_1 \cdot q^*$ is derivable, with $(p_1 \cdot p^*, q_1 \cdot q^*) \in R_1$. Hence, $q' \cdot q^* \xrightarrow{\mu} q_1 \cdot q^*$, with the reached states forming a pair in R_1. □

Exercise 5.24. (Algebraic properties of iteration) Prove that, for any $p,q,r \in \mathscr{P}_{BPA^*}$, the following hold:

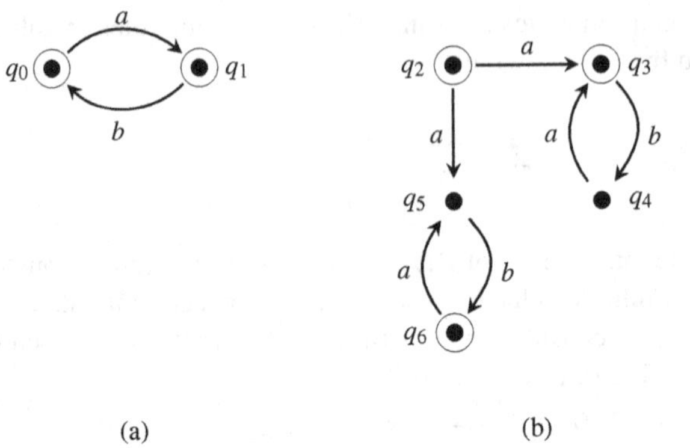

Fig. 5.4 An LTSF not representable in BPA*, up to $\sim_f$, in (a); and one of its f-trace equivalent LTSFs in (b)

(i) $\quad p^* \sim_f p \cdot p^* + \mathbf{1}$
(ii) $\quad p^* \sim_f (p + \mathbf{1})^*$
(iii) $(p+q)^* \sim_f p^* \cdot (q \cdot (p+q)^* + \mathbf{1})$

Note that by (i) we have that $\mathbf{0}^* \sim_f \mathbf{0} \cdot \mathbf{0}^* + \mathbf{1}$. By the law (ii) in Exercise 5.20, $\mathbf{0} \cdot \mathbf{0}^* \sim_f \mathbf{0}$, and by the laws in Exercise 5.19, $\mathbf{0} + \mathbf{1} \sim_f \mathbf{1}$. Hence, by congruence (Proposition 5.1), $\mathbf{0} \cdot \mathbf{0}^* + \mathbf{1} \sim_f \mathbf{0} + \mathbf{1} \sim_f \mathbf{1}$: summing up, $\mathbf{0}^* \sim_f \mathbf{1}$. Then, prove, by using (ii) with $p = \mathbf{0}$, that $\mathbf{1}^* \sim_f \mathbf{1}$. □

In order to prove that any BPA* process generates a finite-state LTSF, we first introduce function *s-size* on BPA* processes, which gives un upper bound on the number of reachable states. Formally:

$$\begin{array}{ll} \textit{s-size}(\mathbf{0}) = 1 & \textit{s-size}(\mathbf{1}) = 1 \\ \textit{s-size}(\mu) = 2 & \textit{s-size}(p_1 + p_2) = \textit{s-size}(p_1) + \textit{s-size}(p_2) + 1 \\ \textit{s-size}(p^*) = \textit{s-size}(p) + 1 & \textit{s-size}(p_1 \cdot p_2) = \textit{s-size}(p_1) + \textit{s-size}(p_2) \end{array}$$

Function *s-size*, applied to p, gives an intuitively correct upper bound on the number of states reachable from p. For the only nontrivial case of iteration, one can get convinced by Exercise 5.23.

Proposition 5.3. (BPA* processes are finite-state LTSFs) *For any BPA* process p, its associated LTSF is finite-state.*

Proof. It is readily observed, by structural induction on p, that s-size(p) *is a finite number for any p.* □

Now a natural question is whether any finite-state LTSF can be represented, up to $\sim_f$, by a suitable BPA* process. Unfortunately, the answer is negative. Milner in [Mil84] observed that iteration is not expressive enough to represent all finite-state LTSFs, up to $\sim_f$. As an instance, it can be proved, by using techniques in [Mil84, BFP01, BCG07], that no BPA* process is f-bisimilar to the LTSF in Figure

5.4(a). In the next section, we will see that by replacing iteration with recursion, as in CCS, we will be able to represent all finite-state LTSFs, up to f-isomorphism.

Nonetheless, BPA* can represent any regular language. Any BPA* process p can be seen both as a process, with an associated LTSF, and as a regular expression; we will prove that the set of f-traces of process p, $Tr_f(p)$, equals the language associated to the regular expression p, $\mathscr{L}[p]$, as defined in Section 1.3.2. For instance, a BPA* process f-trace equivalent to the one in Figure 5.4(a) is $(a \cdot b)^* + a \cdot (b \cdot a)^*$, whose associated LTSF is f-isomorphic to the LTSF in Figure 5.4(b).

Lemma 5.1. *For any $p, q, r \in \mathscr{P}_{BPA^*}$, the following hold:*

(i) $Tr_f(\mathbf{0}) = \emptyset$
(ii) $Tr_f(\mathbf{1}) = \{\varepsilon\}$
(iii) $Tr_f(\mu) = \{\mu\}$
(iv) $Tr_f(p+q) = Tr_f(p) \cup Tr_f(q)$
(v) $Tr_f(p \cdot q) = Tr_f(p) \cdot Tr_f(q)$
(vi) $Tr_f(p^*) = (Tr_f(p))^*$

where the operators on the right-hand-side of the equations are the semantic operations over languages defined in Section 1.3.2.

Proof. By Definition 5.2, $Tr_f(q) = \{\sigma \in A^* \mid \exists q' \in F.\ q \xrightarrow{\sigma}{}^* q'\}$, where relation $\longrightarrow^*$ is reported in Definition 2.4.

Case (i): since $\mathbf{0}$ is a deadlock and $\mathbf{0}$ is not final, $Tr_f(\mathbf{0}) = \emptyset$.

Case (ii): similarly, since $\mathbf{1}$ is a deadlock, but $\mathbf{1}$ is final, then $\mathbf{1} \xrightarrow{\varepsilon}{}^* \mathbf{1}$, hence $Tr_f(\mathbf{1}) = \{\varepsilon\}$.

Case (iii): since $\mu \xrightarrow{\mu} \mathbf{1}$, $\mathbf{1}\downarrow$ and $\mu \not\downarrow$, then $Tr_f(\mu) = \{\mu\}$.

Case (iv): note that $p+q\downarrow$ iff $p\downarrow$ or $q\downarrow$. Hence, $\varepsilon \in Tr_f(p+q)$ iff $\varepsilon \in Tr_f(p) \cup Tr_f(q)$. Derivation $p+q \xrightarrow{\mu} s \xrightarrow{\sigma}{}^* s'$ with $s'\downarrow$ is possible iff either $p \xrightarrow{\mu} s \xrightarrow{\sigma}{}^* s'$ or $q \xrightarrow{\mu} s \xrightarrow{\sigma}{}^* s'$; therefore, $\mu\sigma \in Tr_f(p+q)$ iff $\mu\sigma \in Tr_f(p) \cup Tr_f(q)$.

Case (v): note that $p \cdot q\downarrow$ iff $p\downarrow$ and $q\downarrow$. Hence, $\varepsilon \in Tr_f(p \cdot q)$ iff $\varepsilon \in Tr_f(p) \cdot Tr_f(q)$. Any derivation $p \cdot q \xrightarrow{\sigma}{}^* q'$ with $q'\downarrow$ is due to a derivation $p \cdot q \xrightarrow{\sigma_1}{}^* p' \cdot q$ with $p'\downarrow$, where all the transitions involved in its proof are proved by rule (Seq_1) at the top level, followed by a derivation $p' \cdot q \xrightarrow{\sigma_2}{}^* q'$ with $q'\downarrow$, where the first transition involved (if present) is derived by rule (Seq_2) at the top level, with $\sigma = \sigma_1\sigma_2$. Derivation $p \cdot q \xrightarrow{\sigma_1}{}^* p' \cdot q$ is possible iff $p \xrightarrow{\sigma_1}{}^* p'$; similarly, $p' \cdot q \xrightarrow{\sigma_2}{}^* q'$ is possible iff $q \xrightarrow{\sigma_2}{}^* q'$; therefore, $\sigma_1\sigma_2 \in Tr_f(p \cdot q)$ iff $\sigma_1 \in Tr_f(p)$ and $\sigma_2 \in Tr_f(q)$ iff $\sigma_1\sigma_2 \in Tr_f(p) \cdot Tr_f(q)$.

Case (vi): By law (i) of Exercise 5.24, $p^* \sim_f p \cdot p^* + \mathbf{1}$, hence also $Tr_f(p^*) = Tr_f(p \cdot p^* + \mathbf{1}) = Tr_f(p) \cdot Tr_f(p^*) \cup \{\varepsilon\}$, by the properties (ii), (iv) and (v) above. Note that $\{\varepsilon\} = (Tr_f(p))^0$. If we then replace again $Tr_f(p^*)$, we get $Tr_f(p^*) = Tr_f(p) \cdot [Tr_f(p) \cdot Tr_f(p^*) \cup \{\varepsilon\}] \cup \{\varepsilon\} = (Tr_f(p))^2 \cdot Tr_f(p^*) \cup Tr_f(p) \cup \{\varepsilon\}$. This is a clearly a recursive equation whose (minimal) solution is exactly $(Tr_f(p))^* = \bigcup_{n \in \mathbb{N}} (Tr_f(p))^n$. $\square$

Theorem 5.5. *For any any $p \in \mathscr{P}_{BPA^*}$, $Tr_f(p) = \mathscr{L}[p]$.*

$$\frac{p\downarrow}{C\downarrow}\; C \stackrel{def}{=} p \qquad (\text{Cons})\; \frac{p \xrightarrow{\mu} p'}{C \xrightarrow{\mu} p'}\; C \stackrel{def}{=} p$$

Table 5.3 SOS rules for recursion

Proof. *By structural induction on p, by using Lemma 5.1. The base cases are trivial:*

- $Tr_f(\mathbf{0}) = \emptyset = \mathscr{L}[\mathbf{0}]$
- $Tr_f(\mathbf{1}) = \{\varepsilon\} = \mathscr{L}[\mathbf{1}]$
- $Tr_f(\mu) = \{\mu\} = \mathscr{L}[\mu]$

For a compound process, say $p \cdot q$, the inductive hypothesis ensures that the thesis holds for the constituents p and q.

- $Tr_f(p+q) = Tr_f(p) \cup Tr_f(q) = \mathscr{L}[p] \cup \mathscr{L}[q] = \mathscr{L}[p+q]$
- $Tr_f(p \cdot q) = Tr_f(p) \cdot Tr_f(q) = \mathscr{L}[p] \cdot \mathscr{L}[q] = \mathscr{L}[p \cdot q]$
- $Tr_f(p^*) = (Tr_f(p))^* = (\mathscr{L}[p])^* = \mathscr{L}[p^*]$

Therefore, we can conclude that, for any $p \in \mathscr{P}_{BPA^*}$, $Tr_f(p) = \mathscr{L}[p]$. □

5.4.4 BPA: Finite BPA with Recursion

Finite BPA can be extended with recursively defined constants, as we did for CCS. The resulting language is called *BPA*:

$$p ::= \mathbf{0} \mid \mathbf{1} \mid \mu \mid p+p \mid p \cdot p \mid C$$

where C is any process constant in the set $\mathscr{C}ons$. A BPA term p is a BPA process if the set $Const(p)$[5] of the constants it uses is finite and, for each $C \in Const(p)$, C is *defined* (i.e., equipped with a defining equation $C \stackrel{def}{=} p$) and *guarded*. The definition of guarded constant is a bit more involved than in Section 3.1.2, because of sequential composition: e.g., constant C is guarded in $a \cdot C$, but not in $\mathbf{1} \cdot C$. More specifically, we say that a constant $C \stackrel{def}{=} p$ is guarded if p can be adapted, by applications of the algebraic laws in Exercises 5.19 and 5.20, to the form $q = \alpha_1 \cdot p_1 + \ldots + \alpha_n \cdot p_n\{+\mathbf{1}\}$ (optional presence of a summand $\mathbf{1}$), with the proviso that $q = \mathbf{0}$ if $n = 0$ and the optional summand $\mathbf{1}$ is missing. $\mathscr{P}_{BPA}$ denotes the set of BPA processes.

BPA operational semantics is defined by means of the rules in Tables 5.1 and 5.2, complemented by those listed in Table 5.3.

It can be proved that f-bisimilarity $\sim_f$ is a congruence for recursion, by adapting the idea in Section 4.2.2.

[5] The extension of function $Const(-)$ as in Definition 3.1 to terms of the form $p \cdot q$ is as follows: $Const(p \cdot q) = Const(p) \cup Const(q)$.

The recursive construct allows for the definition of never-ending behavior, not only cyclic (hence finite-state), but even infinite-state.

Example 5.8. **(Unboundedly-branching processes)** Of course, by guardedness of process constants, any BPA process generates a finitely-branching LTSF: this can be proved by following an argument similar to the one in Section 3.3.1. However, because of the presence of **1**, it may be the case that a BPA process p generates an unboundedly-branching LTSF; by Exercise 2.7(ii), it follows that such a p cannot be finite-state. For instance, let us consider the following BPA process [BLT12]:

$$C \stackrel{def}{=} a \cdot C \cdot (\mathbf{1} + c) + b$$

It is interesting to observe that the repeated execution of a's generates a number of sequentialized subterms $(\mathbf{1}+c)$ guarded by C:

$$C \xrightarrow{a} C \cdot (\mathbf{1}+c) \xrightarrow{a} C \cdot (\mathbf{1}+c) \cdot (\mathbf{1}+c) \xrightarrow{a} C \cdot (\mathbf{1}+c) \cdot (\mathbf{1}+c) \cdot (\mathbf{1}+c) \ldots$$

When C then executes b, the reached state is of the form $\mathbf{1} \cdot (\mathbf{1}+c) \cdot \ldots \cdot (\mathbf{1}+c)$. From such a state, we have a number of distinct c-labeled transitions that equals the number of executed a's. For instance, from state $\mathbf{1} \cdot (\mathbf{1}+c) \cdot (\mathbf{1}+c) \cdot (\mathbf{1}+c)$, reachable by executing trace $aaab$ starting from C, we have three distinct c-labeled transitions to states $\mathbf{1} \cdot (\mathbf{1}+c) \cdot (\mathbf{1}+c)$, $\mathbf{1} \cdot (\mathbf{1}+c)$ and $\mathbf{1}$, respectively. Hence, C is unboundedly-branching because unbounded is the number of occurrences of action a that C can perform initially. □

Exercise 5.25. Let BPA$^-$ denote BPA without **1**. Prove that for any BPA$^-$ process p, its associated LTS is boundedly-branching. □

Example 5.9. **(The language** ww^R**)** In Exercise 3.64, we described how to model the language $L = \{ww^R \mid w \in \{a,b\}^*\}$ in finitary CCS. L is a context-free language because it may be generated by the context-free grammar $G = (\{S\}, \{a,b\}, S, \{S \to aSa, S \to bSb, S \to \varepsilon\})$. It is an easy exercise to check that $L = L(G)$. By mimicking grammar G, it is not difficult to produce a BPA process constant S such that its f-traces are exactly the strings in L:

$$S \stackrel{def}{=} a \cdot S \cdot a + b \cdot S \cdot b + \mathbf{1}$$

where each right-hand side of a production for S, say aSa, gives rise to a process term $a \cdot S \cdot a$, and the various terms are then composed with the choice operator. The following is the derivation of the f-trace $abba$:

$$S \xrightarrow{a} \mathbf{1} \cdot S \cdot a \xrightarrow{b} \mathbf{1} \cdot S \cdot b \cdot a \xrightarrow{b} \mathbf{1} \cdot a \xrightarrow{a} \mathbf{1}$$ □

Exercise 5.26. (The language $a^n b^n$**)** In Example 3.15, we presented a system whose weak completed traces are of the form $a^n b^n$, where $a = inc$ and $b = dec$. Define a BPA process constant S such that $Tr_f(S) = \{a^n b^n \mid n \in \mathbb{N}\}$. (*Hint:* Consider the context-free grammar G_2 outlined in Section 1.3.3.) □

Remark 5.3. **(Context-free languages are in BPA languages)** Example 5.9 and Exercise 5.26 are instances of a more general result [BCT08]: for any context-free grammar G, one can build a BPA process p such that $L(G) = Tr_f(p)$. Since BPA processes are guarded, we may restrict our attention to context-free grammars in Greibach normal form (see Section 1.3.3), as the right-hand side of their productions starts with an action. Since for any context-free grammar G, there exists a context-free grammar G' in Greibach normal form such that $L(G) = L(G')$, we lose nothing by restricting our attention to this subclass of context-free grammars.

Given a Greibach normal form grammar $G = (N, T, S, P)$, it is enough to define a BPA constant V in correspondence with each nonterminal $V \in N$; if nonterminal V has productions $V \to a_1\gamma_1 \ldots V \to a_k\gamma_k$, then constant V has k summands, each one corresponding to the obvious translation of a sequence $a_i\gamma_i = a_iX_1 \ldots X_{n_i}$ into the BPA process $a_i \cdot X_1 \cdot \ldots \cdot X_{n_i}$ (if $V \to \varepsilon$, then the corresponding BPA process constant V has a summand $\mathbf{1}$ in its body). The process p such that $L(G) = Tr_f(p)$ is simply constant S, as S is the initial nonterminal symbol of G.

It is interesting to observe that the resulting BPA constant S is mimicking only the *left-most derivations* (i.e., those derivations where only the left-most nonterminal symbol is rewritten), starting from the initial nonterminal symbol S, as described by the following left-most derivation rule:

$$\frac{S \longrightarrow^* \alpha V \beta \qquad V \to \delta \in P}{S \longrightarrow^* \alpha \delta \beta} \quad \text{where } \beta \in (T \cup N)^* \text{ and } \alpha \in T^*$$

However, this does not change the equality $L(G) = Tr_f(p)$ because for any generic derivation $S \longrightarrow^* w$ (as defined in Section 1.3.3), there is another derivation for w from S which is left-most, according to the left-most rule above [HMU01].

Since for any context-free grammar $G = (N, T, S, P)$ in Greibach normal form, the BPA process constant S is such that $L(G) = Tr_f(S)$, we can conclude that BPA can express all context-free languages. □

Example 5.10. **(Extracting a context-free grammar from a BPA process)** We can do also the reverse: a context-free grammar G can be extracted from a BPA process p in such a way that $L(G) = Tr_f(p)$. For instance, consider the BPA process constants C and D defined as follows:

$C \stackrel{def}{=} (a+b) \cdot D \cdot c + \mathbf{0}$

$D \stackrel{def}{=} a \cdot (\mathbf{0} + \mathbf{1}) \cdot C + \mathbf{1}$

By resorting to the algebraic laws discussed in Exercises 5.19, 5.20 and 5.21, we are able to define f-trace equivalent process constants C' and D', respectively. Of particular interest are the two distributivity laws: $(p+q) \cdot r \sim_f p \cdot r + q \cdot r$ that holds for f-bisimilarity, and $r \cdot (p+q) =_{trf} r \cdot p + r \cdot q$ that holds for f-trace equivalence. By using these laws, we can equivalently write:

$C' \stackrel{def}{=} a \cdot D' \cdot c + b \cdot D' \cdot c + \mathbf{0}$

$D' \stackrel{def}{=} a \cdot \mathbf{0} \cdot C' + a \cdot \mathbf{1} \cdot C' + \mathbf{1}$

By using the law $p \cdot \mathbf{0} =_{trf} \mathbf{0}$, $\mathbf{0} \cdot p \sim_f \mathbf{0}$, $p \cdot \mathbf{1} \sim_f p$ and $p + \mathbf{0} \sim_f p$, we can then obtain the f-trace equivalent forms:

$C'' \stackrel{def}{=} a \cdot D'' \cdot c + b \cdot D'' \cdot c$

$D'' \stackrel{def}{=} a \cdot C'' + \mathbf{1}$

Now, any summand in the body of C'' as well as of D'' is without any occurrence of the choice operator, or $\mathbf{0}$ or $\mathbf{1}$ (except for the summand $\mathbf{1}$ itself). Hence, we can define a grammar $G = (N, T, S, P)$, where $N = \{S, A\}$, $T = \{a, b, c\}$ and P is given by the following productions:

$S \to aAc \quad S \to bAc$

$A \to aS \quad A \to \varepsilon$

It is easy to observe that the context-free grammar G is such that $L(G) = Tr_f(C'') = Tr_f(C)$. □

Exercise 5.27. Consider the constant C of Example 5.8. As done in the example above, define a constant D, such that $Tr_f(C) = Tr_f(D)$ and each summand in the body of D does not contain occurrences of the choice operator or $\mathbf{1}$. Then extract from the definition of D a context-free grammar G such that $L(G) = Tr_f(D)$. □

Remark 5.4. **(BPA languages are in context-free languages)** Example 5.10 is an instance of a more general result: for any BPA process p, its set of f-traces $Tr_f(p)$ is a context-free language. The proof of this fact is actually based on the algebraic laws in Exercises 5.19, 5.20 and 5.21. In conjunction with Remark 5.3, we can conclude then that BPA languages are exactly context-free languages. □

Example 5.11. **(Counter)** In Example 3.17, we defined a real counter that tests for zero in finitary CCS by a clever use of restrictions inside three recursively defined constants. Here, we show that a counter can be defined in BPA very simply by means of two constants only, as follows:

$$BC \stackrel{def}{=} zero \cdot BC + inc \cdot (S \cdot BC)$$

$$S \stackrel{def}{=} dec + inc \cdot (S \cdot S)$$

It is not too difficult to prove that $BC \sim_f Counter_0$, where $Counter_0$ is defined in Example 3.17. (Note that it is enough to prove bisimilarity $\sim$, as no reachable state from BC or from $Counter_0$ is properly terminated.) As a consequence, we have that $BC \approx_f C$, where C is defined in Example 3.17, too. □

Trace equivalence is undecidable for BPA, as the problem of language equivalence for context-free grammars is so [BPS61]. On the contrary, bisimulation equivalence (as well as f-bisimulation equivalence) is decidable for BPA processes. This was first proved in [BBK93], for *normed* BPA processes, i.e., for those BPA processes that can always terminate successfully (e.g., constant C in Example 5.8), and then extended for the whole BPA in [CHS95]. Unfortunately, the best known algorithm is doubly exponential [Jan13] for the general case, even if efficient, polynomial algorithms are available for the normed case [HJM96]. The problem of deciding weak bisimilarity for BPA is open.

Remark 5.5. **(Representability of all finite-state LTSFs, up to f-isomorphism)** The following subcalculus of BPA, called *finite-state BPA*, is expressive enough to represent all finite-state LTSFs,

$$p ::= \mathbf{0} \mid \mathbf{1} \mid C \mid \Sigma_{j \in J} \mu_j \cdot p_j$$

where we assume that the set of constants $Const(p)$ used for any finite-state BPA process p is finite and that each constant is defined (guardedness is ensured by syntactic construction). The main syntactic restriction is that the first argument of the sequential composition operator must be an action, as for action prefixing; indeed, finite-state BPA is very similar to finite-state CCS, except for the presence of the additional termination state $\mathbf{1}$.

Analogously to Theorem 3.2, it is not difficult to prove the following *Representability Theorem*: for any reduced finite-state rooted LTSF $TS = (Q, A, \to_1, q_0, F)$, there exists a finite-state BPA process p such that the reachable LTSF $\mathscr{C}_p = (\mathscr{P}_p, sort(p), \to_p, p, \mathscr{F}_p)$ — where $\mathscr{F}_p$ contains the final states in $\mathscr{P}_p$ — is f-isomorphic to TS. The proof is as follows. Let $Q = \{q_0, q_1, \ldots, q_n\}$. We take a constant C_i in correspondence of each state q_i, for $i = 0, 1, , \ldots, n$, defined as follows: if q_i is a deadlock, then $C_i \stackrel{def}{=} \mathbf{1}$ in case $q_i \in F$, and $C_i \stackrel{def}{=} \mathbf{0}$ otherwise; if $T(q_i) = \{(q_i, \mu, q_k) \mid \exists \mu \in A, \exists q_k \in Q.\ q_i \stackrel{\mu}{\longrightarrow}_1 q_k\}$, then $C_i = \Sigma_{(q_i, \mu, q_k) \in T(q_i)} \mu \cdot C_k \{+\mathbf{1}\}$, where the optional summand $\mathbf{1}$ is present only in case $q_i \in F$. Let us consider $\mathscr{C}_{C_0} = (\mathscr{P}_{C_0}, sort(C_0), \to_2, \mathbf{1} \cdot C_0, \mathscr{F}_{C_0})$, where $\mathscr{P}_{C_0} = \{\mathbf{1} \cdot C_0, \mathbf{1} \cdot C_1 \ldots, \mathbf{1} \cdot C_n\}$ (because TS is reduced) and $\mathbf{1} \cdot C_i \in \mathscr{F}_{C_0}$ iff the body of C_i contains the summand $\mathbf{1}$. Hence, the bijection we are looking for is $f : Q \to \mathscr{P}_{C_0}$, defined as $f(q_i) = \mathbf{1} \cdot C_i$. It is also readily observed that the three conditions of f-isomorphism are satisfied:

- $\mathbf{1} \cdot C_0 = f(q_0)$,
- $q_i \stackrel{\mu}{\longrightarrow}_1 q_k$ iff $f(q_i) \stackrel{\mu}{\longrightarrow}_2 f(q_k)$, and
- $q_i \in F$ iff $f(q_i) = \mathbf{1} \cdot C_i \in \mathscr{F}_{C_0}$.

Hence, f is indeed an f-isomorphism.

As an instance, consider the LTSF depicted in Figure 5.4(a). We argued that no BPA* process p is f-bisimilar to q_0. But, following the construction above, we can build a finite-state BPA process, with associated LTSF f-isomorphic to it, as follows:

$C_0 \stackrel{def}{=} a \cdot C_1 + \mathbf{1}$
$C_1 \stackrel{def}{=} b \cdot C_0 + \mathbf{1}$

It is indeed readily seen that, by using the SOS operational rules, $\mathbf{1} \cdot C_0$ generates a LTSF f-isomorphic to Figure 5.4(a). □

Exercise 5.28. Build the finite-state BPA process whose associated LTSF is f-isomorphic to Figure 5.4(b). □

Exercise 5.29. (Encoding the iteration operator) Let BPA^{+*} denote the language BPA enriched with the iteration construct. Show that BPA^{+*} can be implemented into BPA by showing a simple encoding $[\![-]\!]$ of p^* by means of a recursively defined constant. (*Hint:* Look at the algebraic property $p^* \sim_f p \cdot p^* + \mathbf{1}$.) □

$$\frac{p\downarrow \quad q\downarrow}{(p\,|\,q)\downarrow} \qquad (\mathrm{Par}_1)\,\frac{p \xrightarrow{\mu} p'}{p\,|\,q \xrightarrow{\mu} p'\,|\,q} \qquad (\mathrm{Par}_2)\,\frac{q \xrightarrow{\mu} q'}{p\,|\,q \xrightarrow{\mu} p\,|\,q'}$$

Table 5.4 SOS rules for asynchronous parallel composition

5.4.5 PA and PAER

PA is the language obtained by extending BPA with the operator of asynchronous parallel composition (i.e., without synchronization) of BPP. Its syntax is

$$p ::= \mathbf{0} \mid \mathbf{1} \mid \mu \mid p+p \mid p \cdot p \mid p\,|\,p \mid C$$

A PA term p is a PA process if the set $Const(p)$ of the constants it uses is finite and, for each $C \in Const(p)$, C is defined and guarded. We say that a constant $C \stackrel{def}{=} p$ is guarded if p can be adapted, by applications of the algebraic laws in Exercises 5.19 and 5.20, to a process term q such that each occurrence of C occurs within a subprocess $\mu \cdot q'$ of q and any other constant D occurring in p is guarded. $\mathscr{P}_{PA}$ denotes the set of PA processes.

PA operational semantics is defined by means of the rules in Tables 5.1, 5.2, and 5.3, complemented by those described in Table 5.4.

Exercise 5.30. (Algebraic properties of parallel composition) Prove that, for any $p, q, r \in \mathscr{P}_{PA}$, the following hold:

$p\,|\,(q\,|\,r) \sim_f (p\,|\,q)\,|\,r$
$p\,|\,q \sim_f q\,|\,p$
$p\,|\,\mathbf{1} \sim_f p$
$p\,|\,\mathbf{0} \sim_f p \cdot \mathbf{0}$

Note that $p\,|\,\mathbf{0} \sim p$, but $p\,|\,\mathbf{0} \not\sim_f p$ in general (why?). □

Of course, also the analogous of the Expansion Law (Proposition 4.3) holds: given $p = \Sigma_{i=1}^{n} \mu_i \cdot p_i\{+\mathbf{1}\}$ and $q = \Sigma_{j=1}^{m} \mu'_j \cdot q_j\{+\mathbf{1}\}$,

$$p\,|\,q \sim_f \Sigma_{i=1}^{n} \mu_i \cdot (p_i\,|\,q) + \Sigma_{j=1}^{m} \mu'_j \cdot (p\,|\,q_j)\ \{+\mathbf{1}\},$$

where the optional summand $\{+\mathbf{1}\}$ on the right-hand side is present iff it is present in both p and q. Similarly, distributivity of choice w.r.t. parallel composition holds for f-trace equivalence: $(p+q)\,|\,r =_{trf} p\,|\,r + q\,|\,r$.

PA is a syntactic extension of BPA, because asynchronous parallel composition is added to PA, and also of BPP, as sequential composition is a generalization of action prefixing. Not surprisingly, PA is strictly more expressive than both BPA and BPP. On the one hand, there are context-free languages, such as $L = \{a^n c b^n \mid n \geq 0\}$ — discussed in Exercise 3.60 — that are not BPP languages, even if they are

$$P \stackrel{def}{=} (\nu\alpha,\beta)(X\,|\,A\,|\,B)$$
$$X \stackrel{def}{=} g.(X\,|\,A\,|\,B) + c.Y + \overline{\alpha}.d.Z + \overline{\beta}.d.Z$$
$$Y \stackrel{def}{=} \overline{\alpha}.a.Y + \overline{\alpha}.d.Z + \overline{\beta}.b.Y + \overline{\beta}.d.Z \qquad Z \stackrel{def}{=} \mathbf{0}$$
$$A \stackrel{def}{=} \alpha.\mathbf{0} \qquad B \stackrel{def}{=} \beta.\mathbf{0}$$

Table 5.5 A finite-net CCS process whose set of weak completed traces is not a PA language

BPA languages (by Remark 5.3), and hence also PA languages. On the other hand, PA languages include also context-dependent languages, such as the BPP language $L' = Tr_f(B)$ for $B \stackrel{def}{=} a \cdot (B\,|\,b) + c \cdot (B\,|\,d) + e$ — discussed in Example 3.13 — which are not BPA languages (by Remark 5.4).

PA is incomparable with respect to finite-net CCS, not only syntactically (as finite-net CCS allows for a limited use of restriction as well as of communication, while PA allows for the use of sequential composition in place of action prefixing), but also semantically. On the one hand, language $L'' = \{ww^R \mid w \in \{a,b\}^*\}$ is easily representable by a PA process (see Example 5.9), but, as we discussed after Exercise 3.61, such a language is not representable in finite-net CCS. On the other hand, the following example, adapted from [May00], shows a finite-net CCS process P whose set of weak completed traces is a language not representable by any PA process.

Example 5.12. **(A finite-net CCS language, which is not a PA language)** In [May00], Mayr presented a marked Petri net [Rei85] N with the following features:

(i) its set $L[N]$ of completed traces is $\{g^m c\sigma \mid m \geq 0 \wedge \sigma \in (a+b)^* \wedge \sharp(a,\sigma) = m+1 \wedge \sharp(b,\sigma) = m+1\} \cup$ $\{g^m d \mid m \geq 0\} \cup$ $\{g^m c\sigma d \mid m \geq 0 \wedge \sigma \in (a+b)^* \wedge \sharp(a,\sigma) \leq m+1 \wedge \sharp(b,\sigma) \leq m+1 \wedge \sharp(a,\sigma) + \sharp(b,\sigma) \leq 2m+1\}$.

(ii) $L[N] \cap g^*ca^*b^* = \{g^m ca^{m+1}b^{m+1} \mid m \geq 0\}$ is a context-dependent language, hence also $L[N]$ is context-dependent. (Remember that the intersection of a context-free language with a regular language produces a context-free language.)

(iii) a proof that if N were bisimilar to a process in a proper extension of PA, called PAD, then N would also be bisimilar (thus also completed trace equivalent) to a process in the language PDA, a syntactic extension of BPA which does not increase the class of representable languages (i.e., the PDA languages are the context-free languages).

As a conclusion, Mayr states that no PAD process (hence no PA process) can represent language $L[N]$ because, if this were the case, then $L[N]$ would be context-free, which is a contradiction with (ii) above. To conclude the example, Table 5.5 shows a finite-net CCS representation of net N. It is not too difficult to get convinced that the set of weak completed traces for P is exactly $L[N]$ in (i). □

(Com) $\dfrac{p \xrightarrow{\alpha} p' \quad q \xrightarrow{\overline{\alpha}} q'}{p\,|\,q \xrightarrow{\tau} p'\,|'q}$	$\dfrac{p\downarrow}{(\nu a)p\downarrow}$	(Res) $\dfrac{p \xrightarrow{\mu} p'}{(\nu a)p \xrightarrow{\mu} (\nu a)p'}\ \mu \neq a,\overline{a}$

Table 5.6 Rules for communication and restriction

PA is not a Turing-complete language: in [May00] it is proved that the reachability problem is still decidable for PA.

The problem of deciding bisimilarity for PA is open, though for normed PA processes a positive decidability result already exists [HJ99]. Checking weak bisimilarity for PA is undecidable [Sr02].

A natural extension of PA is PAER, obtained by adding the communication capability to the parallel operator and an external operator of restriction, as we did when extending BPP to finite-net CCS. PAER is defined by the abstract syntax

$$\begin{aligned} p &::= \mathbf{0} \mid \mathbf{1} \mid \mu \mid p+p \mid p\cdot p \mid p|p \mid C \\ q &::= p \mid (\nu a)q \end{aligned}$$

where, as usual, we assume that process constants are always defined and guarded and that $Const(p)$ is finite.[6] $\mathscr{P}_{PAER}$ denotes the set of PAER processes.

PAER operational semantics is defined by means of the rules in Tables 5.1, 5.2, 5.3 and 5.4, complemented by those listed in Table 5.6.

PAER is a Turing-complete formalism, as any counter machine M can be encoded in PAER. It is enough to adapt the construction described in Section 3.5.2 for finitary CCS: each instruction $(i : I_i)$ gives rise to a recursive process constant P_i, whose body is a (finite-state) BPA process. For instance, $(i : Inc(r_j))$ is modeled as:

$$P_i \stackrel{def}{=} p_i \cdot P_i' \qquad P_i' \stackrel{def}{=} \overline{inc_j} \cdot \overline{p_{i+1}} \cdot P_i$$

The bootstrapping process $B_{(v_1,v_2,v_3)}$ is described as a (finite-state) BPA process, too:

$$B_{(v_1,v_2,v_3)} \stackrel{def}{=} \underbrace{\overline{inc_1} \cdot \ldots \cdot \overline{inc_1}}_{v_1 \text{ times}} \cdot \underbrace{\overline{inc_2} \cdot \ldots \cdot \overline{inc_2}}_{v_2 \text{ times}} \cdot \underbrace{\overline{inc_3} \cdot \ldots \cdot \overline{inc_3}}_{v_3 \text{ times}} \cdot \overline{p_1} \cdot \mathbf{1}$$

Each counter R_j, for $j = 1,2,3$, can be described as an instantiation of the BPA counter of Example 5.11:

$$\begin{aligned} R_j &\stackrel{def}{=} zero_j \cdot R_j + inc_j \cdot (S_j \cdot R_j) \\ S_j &\stackrel{def}{=} dec_j + inc_j \cdot (S_j \cdot S_j) \end{aligned}$$

[6] Note that the body of a constant C cannot contain occurrences of the restriction operator. However, for convenience, we will sometimes make use of constants that are non-recursively defined over the general class of PAER processes, as we did for finite-net CCS, see Remark 3.13.

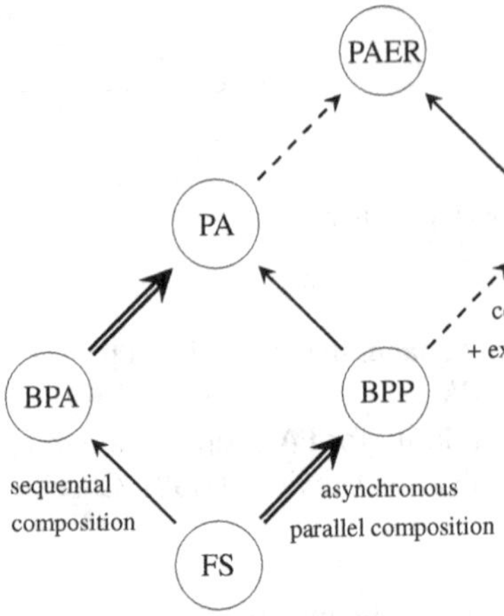

Fig. 5.5 Syntactic and semantic classification of some calculi (FS stands for finite-state CCS, FN for finite-net CCS)

Hence, process $CM_{M(v_1,v_2,v_3)}$ modeling the counter machine M is:

$$CM_{M(v_1,v_2,v_3)} \stackrel{def}{=} (\nu L)(P_1 \,|\, \ldots \,|\, P_m \,|\, R_1 \,|\, R_2 \,|\, R_3 \,|\, B_{(v_1,v_2,v_3)})$$

where the set L of restricted action is $\{p_i \mid \text{index } i \text{ occurs in some instruction}\} \cup \{inc_j, zero_j, dec_j \mid 1 \leq j \leq 3\}$. As a matter of fact, such a process is a PAER process, and so counter machines can be modeled faithfully in PAER. As a consequence, the reachability problem and all the behavioral equivalences are undecidable for PAER, following similar arguments reported in Section 3.5.3 for the case of finitary CCS.

Figure 5.5 summarizes the relationships among most of the calculi we have discussed in this section. Finite-state CCS (FS in the picture), which is essentially finite-state BPA of Remark 5.5, can be extended by replacing action prefixing with the more general sequential composition operator; the resulting calculus is BPA. Orthogonally, finite-state CCS can be extended by adding asynchronous parallel composition, hence obtaining BPP. The join of these two calculi is PA. BPP can be extended with communication and external restriction, obtaining finite-net CCS (FN in the picture). A similar extension to PA yields PAER, which is the only Turing-complete formalism in this picture. All the six calculi have different expressive power, so that the classification by syntactic containment is also a semantic classification. Bisimulation equivalence is decidable for FS, BPA and BPP, but it is not for FN and PAER, while its decidability is an open problem for PA. To complete the picture, recall that BPA* is less expressive than finite-state BPA, as only the latter can represent any finite-state LTSF up to $\sim_f$ (actually, up to f-isomorphism).

$$\frac{}{\mathbf{0}\downarrow} \qquad \frac{p\downarrow \quad q\downarrow}{(p\cdot q)\downarrow} \qquad \frac{p\downarrow \quad q\downarrow}{(p\,|\,q)\downarrow} \qquad \frac{p\downarrow}{(\nu a)p\downarrow} \qquad \frac{p\downarrow}{C\downarrow}\ C\overset{def}{=}p$$

Table 5.7 CCS^{seq} final states: processes that can terminate properly

5.4.6 Derived Operator

We have already seen some derived operators in CCS: the internal choice operator $p \oplus q$ (Section 5.1), the hiding operator $\iota a(p)$ (Section 5.2) and, to some extent, also the relabeling operator $p[a/b]$ (Section 5.3). Here we present a more substantial example of derived operator: sequential composition $p \cdot q$.

CCS^{seq} is the calculus obtained by enriching CCS — as specified by the syntax in Remark 3.5 — with sequential composition. Its abstract syntax is

$$p ::= \Sigma_{j\in J}\mu_j.p_j \;\big|\; p\,|\,p \;\big|\; p\cdot p \;\big|\; (\nu a)p \;\big|\; C$$

with the assumption that J is finite and $\Sigma_{j\in J}\mu_j.p_j = \mathbf{0}$ when $J=\emptyset$. Note that it is not permissible to write terms of the form $a.\mathbf{0}+\mathbf{0}$, i.e., $\mathbf{0}$ is not allowed as a summand. As usual, for any p, set *Const(p)* is finite, and each constant is defined and guarded. $\mathscr{P}_{CCS^{seq}}$ is the set of CCS^{seq} processes.

Contrary to BPA, $\mathbf{0}$ denotes *proper* termination, i.e., $\mathbf{0}$ is a final state. There is no basic process denoting unsuccessful termination. A typical example of a deadlock, non-final state is $(\nu a)(a.\mathbf{0})$. As usual, notation $p\downarrow$ is used to state that p is a final state. Formally, $\_\downarrow$ is the minimal predicate on processes satisfying the axiom and rules of Table 5.7. Note that a term of the form $p+q$ cannot be final: by syntactic construction, both p and q must start with a prefix, hence they are not final.

The operational semantics for CCS^{seq} is given by the set of rules outlined in Table 3.1, complemented by the following rules for sequential composition:

$$(\mathrm{Seq}_1)\ \frac{p \xrightarrow{\mu} p'}{p\cdot q \xrightarrow{\mu} p'\cdot q} \qquad (\mathrm{Seq}_2)\ \frac{p\downarrow \quad q \xrightarrow{\mu} q'}{p\cdot q \xrightarrow{\mu} q'}$$

Exercise 5.31. Prove that, for any $p \in \mathscr{P}_{CCS^{seq}}$, if $p\downarrow$, then $p \not\rightarrow$. (*Hint:* By induction on the proof of $p\downarrow$.) Note that this fact is not true for BPA and related languages; e.g., the BPA process $p = a+\mathbf{1}$ is such that $p\downarrow$ and $p \xrightarrow{a}$. □

Observe that ordinary bisimilarity $\sim$ is not a congruence for sequential composition: even if $a.\mathbf{0} \sim (\nu c)(a.c.\mathbf{0})$, it is not the case that $a.\mathbf{0}\cdot b.\mathbf{0}$ is bisimilar to $(\nu c)(a.c.\mathbf{0})\cdot b.\mathbf{0}$. On the one hand, $a.\mathbf{0}\cdot b.\mathbf{0} \xrightarrow{a} \mathbf{0}\cdot b.\mathbf{0} \xrightarrow{b} \mathbf{0}$, hence trace ab is executable. On the other hand, the latter can only execute a reaching the state $(\nu c)(c.\mathbf{0})\cdot b.\mathbf{0}$, which is a deadlock as $(\nu c)(c.\mathbf{0})$ is not final, and so rule (Seq$_2$)

is not applicable. However, f-bisimilarity $\sim_f$ (Definition 5.6), as well as weak f-bisimilarity $\approx_f$ (Definition 5.7), is a congruence for sequential composition.

Exercise 5.32. (Congruence of sequential composition) By looking at the proofs of Theorems 5.1 and 5.2, prove the following:

1) If $p \sim_f q$, then $p \cdot r \sim_f q \cdot r$ and $r \cdot p \sim_f r \cdot q$, for all $r \in \mathscr{P}_{CCS^{seq}}$,
2) If $p \approx_f q$, then $p \cdot r \approx_f q \cdot r$ and $r \cdot p \approx_f r \cdot q$, for all $r \in \mathscr{P}_{CCS^{seq}}$. □

Exercise 5.33. (Congruence of parallel composition and restriction) Prove that the following hold:

1) If $p \sim_f q$, then $p \mid r \sim_f q \mid r$ for all $r \in \mathscr{P}_{CCS^{seq}}$,
2) If $p \approx_f q$, then $p \mid r \approx_f q \mid r$ for all $r \in \mathscr{P}_{CCS^{seq}}$,
3) If $p \sim_f q$, then $(\nu a)p \sim_f (\nu a)q$ for all $a \in \mathscr{L}$,
4) If $p \approx_f q$, then $(\nu a)p \approx_f (\nu a)q$ for all $a \in \mathscr{L}$. □

Exercise 5.34. (Algebraic properties for $\sim_f$) Prove that, for any $p, q, r \in \mathscr{P}_{CCS^{seq}}$, the following hold:

$$
\begin{array}{lll}
(i) & p \cdot (q \cdot r) \sim_f (p \cdot q) \cdot r & \\
(ii) & p \cdot q \sim_f q & \text{if } p\downarrow \\
(iii) & q \cdot p \sim_f q & \text{if } p\downarrow \\
(iv) & (p+q) \cdot r \sim_f p \cdot r + q \cdot r & \\
(v) & p \mid q \sim_f q & \text{if } p\downarrow \\
(vi) & (\nu a)p \sim_f p & \text{if } a \notin fn(p)
\end{array}
$$

(*Hint:* For (i), look at the solution to Exercise 5.20.) □

We are now ready to show that sequential composition is a derived operator for CCS, by showing an encoding $[\![-]\!]$ of CCS^{seq} into CCS, up to weak f-bisimilarity $\approx_f$. More explicitly, the encoding is a function $[\![-]\!] : \mathscr{P}_{CCS^{seq}} \to \mathscr{P}_{CCS}$ such that $p \approx_f [\![p]\!]$. The formal encoding, following ideas in [Mil89, Tau89, San12], is given in Table 5.8; it is essentially homomorphic on all the CCS operators, while for sequential composition, it is defined thus:

$$[\![p \cdot q]\!] = (\nu d)([\![p]\!]^e_d \mid \overline{d}.[\![q]\!]) \quad d, e \notin fn(p \cdot q) \cup bn(p \cdot q)^7$$

This encoding uses an auxiliary encoding $[\![p]\!]^e_d$, parametrized on two new, distinct names d and e, such that d is a free name and e a bound name. The intuition is that whenever p reaches a final state p', then $[\![p']\!]^e_d$ can perform d as its last action and then deadlocks; this is made clear by the rule $[\![\mathbf{0}]\!]^e_d = d.\mathbf{0}$. Such a d-labeled transition from $[\![p']\!]^e_d$ is to be synchronized with $\overline{d}.[\![q]\!]$, because of the presence of restriction on d, hence activating the continuation $[\![q]\!]$.

The auxiliary encoding of parallel composition reveals the need for the additional new bound name e:

[7] Free names and bound names for sequential composition are defined as follows: $fn(p \cdot q) = fn(p) \cup fn(q)$ and $bn(p \cdot q) = bn(p) \cup bn(q)$.

$[\![\mathbf{0}]\!] = \mathbf{0}$	$[\![\mu.p]\!] = \mu.[\![p]\!]$	$[\![p_1 + p_2]\!] = [\![p_1]\!] + [\![p_2]\!]$	$[\![p_1 \mid p_2]\!] = [\![p_1]\!] \mid [\![p_2]\!]$
$[\![(\nu a)p]\!] = (\nu a)[\![p]\!]$	$[\![A]\!] = A'$ where $A' \stackrel{def}{=} [\![p]\!]$ if $A \stackrel{def}{=} p$		
$[\![p \cdot q]\!] = (\nu d)([\![p]\!]^e_d \mid \overline{d}.[\![q]\!])$ $d, e \notin fn(p \cdot q) \cup bn(p \cdot q)$			
$[\![\mathbf{0}]\!]^e_d = d.\mathbf{0}$	$[\![\mu.p]\!]^e_d = \mu.[\![p]\!]^e_d$	$[\![p_1 + p_2]\!]^e_d = [\![p_1]\!]^e_d + [\![p_2]\!]^e_d$	
$[\![p_1 \mid p_2]\!]^e_d = (\nu e)(([\![p_1]\!]^d_e \mid [\![p_2]\!]^d_e) \mid \overline{e}.\overline{e}.d.\mathbf{0})$			
$[\![(\nu a)p]\!]^e_d = (\nu a)[\![p]\!]^e_d$	$[\![A]\!]^e_d = A_{ed}$ where $A_{ed} \stackrel{def}{=} [\![p]\!]^e_d$ if $A \stackrel{def}{=} p$		
$[\![p \cdot q]\!]^e_d = (\nu e)([\![p]\!]^d_e \mid \overline{e}.[\![q]\!]^e_d)$			

Table 5.8 Encoding CCS^{seq} into CCS

$$[\![p_1 \mid p_2]\!]^e_d = (\nu e)(([\![p_1]\!]^d_e \mid [\![p_2]\!]^d_e) \mid \overline{e}.\overline{e}.d.\mathbf{0})$$

On the one hand, $[\![p_1]\!]^d_e$ will possibly terminate its execution by performing e (using d as an auxiliary bound name) if p_1 terminates successfully; similarly, $[\![p_2]\!]^d_e$. On the other hand, by restricting on action e, two synchronizations with the two occurrences of action $\overline{e}$ in the third component are to be executed, with the effect of activating subprocess $d.\mathbf{0}$: hence, the last action performed by $[\![p_1 \mid p_2]\!]^e_d$ is indeed d, provided that both components p_1 and p_2 terminate successfully.

A similar inversion of the roles between d and e is present also in the auxiliary encoding of sequential composition:

$$[\![p \cdot q]\!]^e_d = (\nu e)([\![p]\!]^d_e \mid \overline{e}.[\![q]\!]^e_d)$$

where in $[\![p \cdot q]\!]^e_d$ action d is free and e is bound, while in $[\![p]\!]^d_e$ action e is free and d is bound. Process $[\![p]\!]^d_e$ will possibly end its execution by performing e, to be synchronized with $\overline{e}.[\![q]\!]^e_d$ because of the restriction on the auxiliary e. Then $[\![q]\!]^e_d$ will possibly end by performing d, as required by $[\![p \cdot q]\!]^e_d$.

Note that two new names, d and e, are enough: the two names are used in alternated restrictions in such a way that there is no unwanted capture of free names.

Remark 5.6. **(Alpha-conversion and substitution)** With the help of the algebraic law of alpha-conversion (see Section 4.1.2), one can be convinced that the following laws hold for any choice of distinct actions d, e, f new in p:

(1) $[\![p]\!]^e_d \sim_f [\![p]\!]^f_d$
(2) $[\![p]\!]^e_d\{f/d\} \sim_f [\![p]\!]^e_f$
(3) $[\![p]\!]^d_e \sim_f [\![p]\!]^e_d\{e/d\}$

As a matter of fact, law (1) holds because $[\![p]\!]^e_d$ and $[\![p]\!]^f_d$ differ only for the actual choice of bound names, hence it is possible to rename by alpha-conversion the bound names, up to $\sim_f$, to rewrite both of them to the same syntactic term (up to the actual names of constants). Also (2) holds because $[\![p]\!]^e_d$ and $[\![p]\!]^e_f$ differ only in the possible termination action (d in the former, f in the latter), which is renamed in the former accordingly by means of the substitution $\{f/d\}$. Law (3) holds because the substitution $\{e/d\}$ applied to $[\![p]\!]^e_d$ has the effect of renaming the free name d to e, by possibly alpha-converting bound names (the superscript name e for sure), hence the result is a term f-bisimilar to $[\![p]\!]^d_e$ by (1).

The formal proof of these obvious facts should be based on the explicit construction of three f-bisimulations R_i (for $i = 1,\dots,3$), each R_i containing the corresponding pair in law (i) (e.g., $R_1^{p(e,d,f,d)}$ contains the pair $([\![p]\!]^e_d, [\![p]\!]^f_d)$). This construction is not easy because of the treatment of constants. However, if we restrict our attention to finite CCS^{seq} processes, then a slightly simpler construction of the three bisimulations is possible, by structural induction.

For instance, $R_1^{\mathbf{0}(e,d,f,d)} = \{(d.\mathbf{0}, d.\mathbf{0})\}$, $R_3^{\mu.p(d,e,e,d)} = \{(\mu.[\![p]\!]^d_e, \mu.[\![p]\!]^e_d\{e/d\})\} \cup R_3^{p(d,e,e,d)}$, where $R_3^{p(d,e,e,d)}$ is the (inductive) f-bisimulation for the subterm p. The cases are not always that easy. As an instance of a more elaborate case, consider $R_1^{p_1\cdot p_2(e,d,f,d)}$. On the one hand, $[\![p_1 \cdot p_2]\!]^e_d = (\nu e)([\![p_1]\!]^d_e \,|\, \overline{e}.[\![p_2]\!]^e_d)$; on the other hand $[\![p_1 \cdot p_2]\!]^f_d = (\nu f)([\![p_1]\!]^d_f \,|\, \overline{f}.[\![p_2]\!]^f_d)$. Inductively, we can assume the existence of a f-bisimulation $R_2^{p_1(d,e,d,f)}$ proving that $[\![p_1]\!]^d_e\{f/e\} \sim_f [\![p_1]\!]^d_f$, as well as a f-bisimulation $R_1^{p_2(e,d,f,d)}$ proving that $[\![p_2]\!]^e_d \sim_f [\![p_2]\!]^f_d$. Hence, $R_1^{p_1\cdot p_2(e,d,f,d)} = S_1 \cup S_2$ where:

$$S_1 = \{((\nu e)(q_1 \,|\, \overline{e}.[\![p_2]\!]^e_d), (\nu f)(q_1' \,|\, \overline{f}.[\![p_2]\!]^f_d)) \mid (q_1\{f/e\}, q_1') \in R_2^{p_1(d,e,d,f)}\}$$
$$S_2 = \{((\nu e)(q_1 \,|\, q_2), (\nu f)(q_1' \,|\, q_2')) \mid q_1 \downarrow, q_1' \downarrow, (q_2, q_2') \in R_1^{p_2(e,d,f,d)}\}.$$

And similarly for the other cases. □

Exercise 5.35. (Counters, again) Consider the BPA counter BC of Example 5.11. It can be rephrased in CCS^{seq} as follows:

$$CC \stackrel{def}{=} zero.CC + inc.(S \cdot CC)$$
$$S \stackrel{def}{=} dec.\mathbf{0} + inc.(S \cdot S)$$

Compute $[\![CC]\!]$ and compare the resulting process with the finitary CCS counter C, defined in Example 3.17. □

The rest of this section is devoted to proving the correctness of the encoding. The proof is rather technical and so, on the one hand, it can be safely skipped if one is not interested in the technical detail. On the other hand, it is instructive on how to prove the correctness of a *compiler* from a language (in this case CCS^{seq}) into its kernel (in this case CCS).

We want to prove that $p \approx_f [\![p]\!]$ for all $p \in \mathscr{P}_{CCS^{seq}}$. This fact has the immediate consequence that the encoding preserves the (weak f-bisimilarity) equivalence classes: for all $p_1, p_2 \in \mathscr{P}_{CCS^{seq}}$ we have that $p_1 \approx_f p_2$ if and only if $[\![p_1]\!] \approx_f [\![p_2]\!]$, by transitivity.

To prove that $p \approx_f [\![p]\!]$ for all $p \in \mathscr{P}_{CCS^{seq}}$, one has to provide a weak f-bisimulation $R \subseteq \mathscr{P}_{CCS^{seq}} \times \mathscr{P}_{CCS}$ containing all the pairs of the form $(p, [\![p]\!])$. This is a hard work that may be alleviated if we could use the proof technique provided by weak bisimilarity up to $\approx$ (Definition 2.22), suitably adapted to take into account the state type. The candidate weak f-bisimilarity up to $\approx_f$ relation could be

$$R = \{(p, [\![p]\!]) \mid p \in \mathscr{P}_{CCS^{seq}}\}$$

Unfortunately, the up-to proof technique is not flexible enough to be used profitably in this setting. Indeed, we would need the unsound version of this proof technique that uses $\approx_f$ on both sides (see Exercise 2.75): if $[\![p]\!] \stackrel{\tau}{\longrightarrow} q$, by executing some internal synchronization on the auxiliary names e (or d), then p can idle, $p \stackrel{\varepsilon}{\Longrightarrow} p$, and the reached states are related: $p \approx_f pR[\![p]\!] \approx_f q$. As this version of the proof technique is unsound, we have to follow a different approach.

[AH92] introduces an efficiency preorder, called *expansion* and denoted $\precsim$, that lies in between strong bisimilarity and weak bisimilarity: $p \sim q$ implies $p \precsim q$ implies $p \approx q$. Similarly to strong bisimilarity $\sim$, expansion $\precsim$ enjoys a nice mathematical theory and in [SM92] it is shown that it also enjoys a nice, flexible up-to technique. Here we adapt the definitions outlined there to our setting with the state type distinction. We use notation $q \stackrel{\hat{\mu}}{\longrightarrow} q'$ to state that $q \stackrel{\mu}{\longrightarrow} q'$ if $\mu \neq \tau$, while when $\mu = \tau$, it states that $q = q'$ (no move) or $q \stackrel{\tau}{\longrightarrow} q'$. We also use notation $q \downarrow$ to mean q is a final state in F.

Definition 5.10. (Expansion and f-expansion) For any LTSF $(Q, A \cup \{\tau\}, \rightarrow, F)$, where $\tau \notin A$, an *expansion* is a relation R such that if $(q_1, q_2) \in R$ then for all $\mu \in A \cup \{\tau\}$

- $\forall q_1'$ such that $q_1 \stackrel{\mu}{\longrightarrow} q_1'$, $\exists q_2'$ such that $q_2 \stackrel{\mu}{\Longrightarrow} q_2'$ and $(q_1', q_2') \in R$,
- $\forall q_2'$ such that $q_2 \stackrel{\mu}{\longrightarrow} q_2'$, $\exists q_1'$ such that $q_1 \stackrel{\hat{\mu}}{\longrightarrow} q_1'$ and $(q_1', q_2') \in R$,

State q' is an expansion of q, denoted $q \precsim q'$, if there exists an expansion R such that $(q, q') \in R$. Hence, $\precsim = \bigcup\{R \subseteq Q \times Q \mid R \text{ is an expansion}\}$.

An *f-expansion* R is an expansion such that, additionally, whenever $(q_1, q_2) \in R$:

- if $q_1 \downarrow$, then $\exists q_2'$ such that $q_2 \stackrel{\varepsilon}{\Longrightarrow} q_2'$, $q_2' \downarrow$ and $(q_1, q_2') \in R$,
- if $q_2 \downarrow$, then $q_1 \downarrow$.

State q' is an f-expansion of q, denoted $q \precsim_f q'$, if there exists an f-expansion R such that $(q, q') \in R$. Hence, $\precsim_f = \bigcup\{R \subseteq Q \times Q \mid R \text{ is an f-expansion}\}$. □

Exercise 5.36. (*i*) Prove that the identity relation $\mathscr{I}$ is an f-expansion. (*ii*) Prove that the composition $R_1 \circ R_2$ of two f-expansions R_1 and R_2 is an f-expansion. (*iii*) Prove that the union $\bigcup_{j \in J} R_j$ of f-expansions R_j is an f-expansion. □

By means of the results in Exercise 5.36, it is easy to see that $\precsim_f$ is a preorder (reflexive and transitive) and the largest f-expansion. To see that $\precsim_f$ is not symmetric, consider $a.\mathbf{0}$ and $\tau.a.\mathbf{0}$; it is easy to see that $a.\mathbf{0} \precsim_f \tau.a.\mathbf{0}$, but $\tau.a.\mathbf{0} \not\precsim_f a.\mathbf{0}$.

Another easy observation is that any strong f-bisimulation is also an f-expansion, hence $\sim_f \subseteq \precsim_f$. Similarly, any f-expansion is also a weak f-bisimulation, hence $\precsim_f \subseteq \approx_f$. So if we prove that $p \precsim_f [\![p]\!]$, then we also get that $p \approx_f [\![p]\!]$.

The f-expansion preorder $\precsim_f$ enjoys also many algebraic properties. For instance, all those valid for $\sim_f$ (e.g., $(\nu a)p \sim_f p$ if $a \notin fn(p)$) are also valid for $\precsim_f$. Also $\precsim_f$ is a congruence for all CCS^{seq} operators, except $+$.

Exercise 5.37. (Congruence) Prove that $\precsim_f$ is a congruence for all the operators of CCS^{seq} but $+$. In particular, prove that if $p_1 \precsim_f p_2$, then

- $p_1 | q \precsim_f p_2 | q$, for all q, and
- $(\nu a)p_1 \precsim_f (\nu a)p_2$. □

We are now ready to introduce the up-to technique for f-expansions, which will be very useful in the following.

Definition 5.11. (F-expansion up to $\precsim_f$) For any LTSF $(Q, A \cup \{\tau\}, \rightarrow, F)$, an f-expansion up to $\precsim_f$ is a relation R such that if $(q_1, q_2) \in R$ then for all $\mu \in A \cup \{\tau\}$

- $\forall q_1'$ such that $q_1 \xrightarrow{\mu} q_1'$, $\exists q_2'$ such that $q_2 \stackrel{\mu}{\Longrightarrow} q_2'$ and $q_1' \sim_f R \precsim_f q_2'$,
- $\forall q_2'$ such that $q_2 \xrightarrow{\mu} q_2'$, $\exists q_1'$ such that $q_1 \xrightarrow{\hat{\mu}} q_1'$ and $q_1' \precsim_f R \precsim_f q_2'$,

and additionally

- if $q_1 \downarrow$, then $\exists q_2'$ such that $q_2 \stackrel{\varepsilon}{\Longrightarrow} q_2'$, $q_2' \downarrow$ and $q_1 \sim_f R \precsim_f q_2'$,
- if $q_2 \downarrow$, then $q_1 \downarrow$. □

Exercise 5.38. Prove that if R is an f-expansion up to $\precsim_f$, then $R \subseteq \precsim_f$. (*Hint:* Prove first that $\sim_f R \precsim_f$ and $\precsim_f R \precsim_f$ are both f-expansions, following the similar proof of Lemma 2.1. Hence, $\sim_f R \precsim_f \subseteq \precsim_f$, as well as $\precsim_f R \precsim_f \subseteq \precsim_f$ by definition of $\precsim_f$. As the identity relation $\mathscr{I}$ is a subset of both $\sim_f$ and $\precsim_f$, we have that relation $R = \mathscr{I} \circ R \circ \mathscr{I} \subseteq \sim_f R \precsim_f$, hence $R \subseteq \precsim_f$ by transitivity.) □

The exercise above states the correctness of the proof principle based on expansion up to: if $(q_1, q_2) \in R$ and R is an f-expansion up to $\precsim_f$, then $q_1 \precsim_f q_2$. So, if we prove that relation

$$R = \{(p, [\![p]\!]) \mid p \in \mathscr{P}_{CCS^{seq}}\}$$

is an f-expansion up to $\precsim_f$, we get the expected result that $p \precsim_f [\![p]\!]$, hence also that $p \approx_f [\![p]\!]$. The correctness proof is based on some auxiliary lemmata and propositions we are going to list. To simplify the approach, we use a notational convention: τ' is an additional unobservable action that we use to label transitions $q_1 \xrightarrow{\tau'} q_2$ due to a synchronization on the auxiliary new names e and d; hence, when we write $q_1 \xrightarrow{\tau} q_2$, we are assuming that the executed τ is either a prefix in q_1 or the result of a synchronization not on e or d. We also use μ to range over observable actions and τ, but often not τ'. Finally, $\stackrel{\varepsilon'}{\Longrightarrow}$ is used to denote a sequence of zero or more τ'-labeled transitions.

Lemma 5.2. *For any $p \in \mathscr{P}_{CCS^{seq}}$, $[\![p]\!]^e_d \precsim_f (\nu e)([\![p]\!]^e_d\{e/d\} \,|\, \overline{e}.d.\mathbf{0})$, for any choice of d, e new in p.*

Proof. Consider relation $R_p = S_1 \cup S_2 \cup S_3$, where

$S_1 = \{(q, (\nu e)(q\{e/d\} \,|\, \overline{e}.d.\mathbf{0})) \mid [\![p]\!]^e_d \longrightarrow^* q \text{ and } d \text{ has not been performed}\}$,

$S_2 = \{(q, (\nu e)(\overline{q}\{e/d\} \,|\, d.\mathbf{0})) \mid [\![p]\!]^e_d \longrightarrow^* q \xrightarrow{d} \overline{q}\}$, and

$S_3 = \{(\overline{q}, (\nu e)(\overline{q}\{e/d\} \,|\, \mathbf{0})) \mid [\![p]\!]^e_d \longrightarrow^* q \xrightarrow{d} \overline{q}\}$.

Note that $([\![p]\!]^e_d, (\nu e)([\![p]\!]^e_d\{e/d\} \,|\, \overline{e}.d.\mathbf{0})) \in S_1$ and so, if we prove that R_p is an expansion, we are done.

Let us consider a generic pair $(q, (\nu e)(q\{e/d\} \,|\, \overline{e}.d.\mathbf{0}))$ in S_1.

If $q \xrightarrow{\mu} q'$ (where μ is not d, but can also be τ'), then $q\{e/d\} \xrightarrow{\mu} q'\{e/d\}$ because $\mu \neq d$, and so also $(\nu e)(q\{e/d\} \,|\, \overline{e}.d.\mathbf{0}) \xrightarrow{\mu} (\nu e)(q'\{e/d\} \,|\, \overline{e}.d.\mathbf{0})$ with $(q', (\nu e)(q'\{e/d\} \,|\, \overline{e}.d.\mathbf{0})) \in S_1$. If $q \xrightarrow{d} q'$, then $q\{e/d\} \xrightarrow{e} q'\{e/d\}$ and so also $(\nu e)(q\{e/d\} \,|\, \overline{e}.d.\mathbf{0}) \stackrel{d}{\Longrightarrow} (\nu e)(q'\{e/d\} \,|\, \mathbf{0})$ with $(q', (\nu e)(q'\{e/d\} \,|\, \mathbf{0})) \in S_3$.

If $(\nu e)(q\{e/d\} \,|\, \overline{e}.d.\mathbf{0}) \xrightarrow{\mu} r$, then either $q\{e/d\} \xrightarrow{\mu} q'\{e/d\}$ with the reached state $r = (\nu e)(q'\{e/d\} \,|\, \overline{e}.d.\mathbf{0})$, or $q\{e/d\} \xrightarrow{e} q'\{e/d\}$ with $\mu = \tau'$ and the reached state $r = (\nu e)(q'\{e/d\} \,|\, d.\mathbf{0})$. In the former case, $q\{e/d\} \xrightarrow{\mu} q'\{e/d\}$ is possible only if $q \xrightarrow{\mu} q'$ and the pair $(q', (\nu e)(q'\{e/d\} \,|\, \overline{e}.d.\mathbf{0}))$ belongs to S_1. In the latter case, $q\{e/d\} \xrightarrow{e} q'\{e/d\}$ is possible only if $q \xrightarrow{d} q'$; in this case, q can idle, and the pair $(q, (\nu e)(q'\{e/d\} \,|\, d.\mathbf{0}))$ belongs to S_2.

The check of the pairs in S_2 or S_3 is analogous and so left as an exercise. □

Proposition 5.4. *Prove that for any $p \in \mathscr{P}_{CCS^{seq}}$, $[\![p]\!]^e_d \precsim_f (\nu e)([\![p]\!]^d_e \,|\, \overline{e}.d.\mathbf{0})$, for any choice of d, e new in p.*

Proof. By Lemma 5.2, we have that $[\![p]\!]^e_d \precsim_f (\nu e)([\![p]\!]^e_d\{e/d\} \,|\, \overline{e}.d.\mathbf{0})$. By Remark 5.6, we have that $[\![p]\!]^e_d\{e/d\} \sim_f [\![p]\!]^d_e$, and so also $[\![p]\!]^e_d\{e/d\} \precsim_f [\![p]\!]^d_e$. By congruence w.r.t. parallel composition and restriciton (Exercise 5.37), the thesis then follows by transitivity. □

Lemma 5.3. *For any $p \in \mathscr{P}_{CCS^{seq}}$, and for any choice of d, e new in p, if $[\![p]\!]^e_d \xrightarrow{d} q$, then $p\downarrow$ and $q\downarrow$.*

Proof. By induction on the proof of $[\![p]\!]^e_d \xrightarrow{d} q$. We proceed by case analysis.

If $p = \mathbf{0}$, then $[\![p]\!]^e_d = d.\mathbf{0} \xrightarrow{d} \mathbf{0}$, and the thesis follows trivially.

If $p = \mu.p'$, then $[\![p]\!]^e_d = \mu.[\![p']\!]^e_d \stackrel{d}{\not\longrightarrow}$, and so this case is empty. Similarly, the case $p = \Sigma_{j\in J}\mu_j.p_j$.

If $p = p_1 \,|\, p_2$, then $[\![p]\!]^e_d = (\nu e)([\![p_1]\!]^d_e \,|\, [\![p_2]\!]^d_e \,|\, \overline{e}.\overline{e}.d.\mathbf{0}) \stackrel{d}{\not\longrightarrow}$ and so this case is empty, as well as the case for $p = p_1 \cdot p_2$.

If $p = (\nu a)p_1$, then $[\![p]\!]^e_d = (\nu a)[\![p_1]\!]^e_d$, which can execute d, reaching q, only if $[\![p_1]\!]^e_d \xrightarrow{d} q_1$ with $q = (\nu a)q_1$. By induction, we have $p_1\downarrow$ and $q_1\downarrow$. Hence, $p\downarrow$ and $q\downarrow$ by definition of relation $\downarrow$.

If $p = C$*, where* $C \stackrel{def}{=} q$*, then* $[\![p]\!]^e_d = C_{ed}$ *with* $C_{ed} \stackrel{def}{=} [\![q]\!]^e_d$*. Transition* $C_{ed} \xrightarrow{d} r$ *is possible (by rule (Cons)) only if* $[\![q]\!]^e_d \xrightarrow{d} r$*; hence, by induction, we have* $q \downarrow$ *and* $r \downarrow$*. By definition of relation* $\downarrow$*, also* $C \downarrow$*.* □

Lemma 5.4. *For any* $p \in \mathscr{P}_{CCS^{seq}}$*, for any choice of* d, e *new in* p*, if* $[\![p]\!]^e_d \xrightarrow{\tau'} q$*, then there exists* p' *such that (i)* $p \sim_f p'$ *and (ii)* $[\![p']\!]^e_d \precapprox_f q$*.*

Proof. *By induction on the proof of* $[\![p]\!]^e_d \xrightarrow{\tau'} q$*. We proceed by cases.*

If $p = \mathbf{0}$*, then* $[\![p]\!]^e_d \not\xrightarrow{\tau'}$*, and so this case is empty, as well as the cases for* $p = \mu.p'$ *and* $p = \Sigma_{j \in J} \mu_j.p_j$*.*

If $p = p_1 \cdot p_2$*, then* $[\![p_1 \cdot p_2]\!]^e_d = (\nu e)([\![p_1]\!]^d_e \,|\, \overline{e}.[\![p_2]\!]^e_d) \xrightarrow{\tau'} q$ *is possible only if either (a)* $[\![p_1]\!]^d_e \xrightarrow{e} q_1$ *and* $q = (\nu e)(q_1 \,|\, [\![p_2]\!]^e_d)$*, or (b)* $[\![p_1]\!]^d_e \xrightarrow{\tau'} q_1$ *and* $q = (\nu e)(q_1 \,|\, \overline{e}.[\![p_2]\!]^e_d)$*.*

Case (a): by Lemma 5.3, $p_1 \downarrow$ *and* $q_1 \downarrow$*. In this case, we choose* $p' = p_2$*. So (i)* $p_1 \cdot p_2 \sim_f p_2$*, and (ii)* $[\![p_2]\!]^e_d \precapprox_f (\nu e)(q_1 \,|\, [\![p_2]\!]^e_d) = q$*, because this holds for* $\sim_f$*.*

Case (b): by induction, there exists p'_1 *such that (i)* $p_1 \sim_f p'_1$*, and (ii)* $[\![p'_1]\!]^e_d \precapprox_f q_1$*. Hence, in this case we choose* $p' = p'_1 \cdot p_2$*. So (i) by congruence* $p_1 \cdot p_2 \sim_f p'_1 \cdot p_2$*, and (ii)* $[\![p'_1 \cdot p_2]\!]^e_d = (\nu e)([\![p'_1]\!]^d_e \,|\, \overline{e}.[\![p_2]\!]^e_d) \precapprox_f (\nu e)(q_1 \,|\, \overline{e}.[\![p_2]\!]^e_d) = q$ *by congruence, as well.*

If $p = p_1 \,|\, p_2$*, then* $[\![p_1 \,|\, p_2]\!]^e_d = (\nu e)([\![p_1]\!]^d_e \,|\, [\![p_2]\!]^d_e \,|\, \overline{e}.\overline{e}.d.\mathbf{0}) \xrightarrow{\tau'} q$ *is possible only if either (a)* $[\![p_1]\!]^d_e \xrightarrow{e} q_1$ *and* $q = (\nu e)(q_1 \,|\, [\![p_2]\!]^d_e \,|\, \overline{e}.d.\mathbf{0})$*, or (b)* $[\![p_1]\!]^d_e \xrightarrow{\tau'} q_1$ *and* $q = (\nu e)(q_1 \,|\, [\![p_2]\!]^d_e \,|\, \overline{e}.\overline{e}.d.\mathbf{0})$ *(plus the two cases for the corresponding moves from* $[\![p_2]\!]^d_e$*, which are symmetric and so omitted).*

Case (a): by Lemma 5.3, $p_1 \downarrow$ *and* $q_1 \downarrow$*. In this case, we choose* $p' = p_2$*. So (i)* $p_1 \,|\, p_2 \sim_f p_2$*, and (ii) by Proposition 5.4,* $[\![p_2]\!]^e_d \precapprox_f (\nu e)([\![p_2]\!]^d_e \,|\, \overline{e}.d.\mathbf{0})$*, and* $(\nu e)([\![p_2]\!]^d_e \,|\, \overline{e}.d.\mathbf{0}) \sim_f (\nu e)(q_1 \,|\, [\![p_2]\!]^d_e \,|\, \overline{e}.d.\mathbf{0}) = q$*, by algebraic properties.*

Case (b): by induction, there exists p'_1 *such that (i)* $p_1 \sim_f p'_1$*, and (ii)* $[\![p'_1]\!]^e_d \precapprox_f q_1$*. Hence, here we choose* $p' = p'_1 \,|\, p_2$*. So (i)* $p_1 \,|\, p_2 \sim_f p'_1 \,|\, p_2$ *by congruence, and (ii)* $[\![p'_1 \,|\, p_2]\!]^e_d = (\nu e)([\![p'_1]\!]^d_e \,|\, [\![p_2]\!]^d_e \,|\, \overline{e}.\overline{e}.d.\mathbf{0}) \precapprox_f (\nu e)(q_1 \,|\, [\![p_2]\!]^d_e \,|\, \overline{e}.\overline{e}.d.\mathbf{0}) = q$*, by congruence as well.*

If $p = (\nu a)p_1$*, then* $[\![p]\!]^e_d = (\nu a)[\![p_1]\!]^e_d \xrightarrow{\tau'} q$ *is possible only if* $[\![p_1]\!]^e_d \xrightarrow{\tau'} q_1$ *with* $q = (\nu a)q_1$*. By induction, there exists a process* p'_1 *such that (i)* $p_1 \sim_f p'_1$*, and (ii)* $[\![p'_1]\!]^e_d \precapprox_f q_1$*. Hence, here we choose* $p' = (\nu a)p'_1$*. So, (i)* $(\nu a)p_1 \sim_f (\nu a)p'_1$ *by congruence, and (ii)* $[\![p']\!]^e_d = (\nu a)[\![p'_1]\!]^e_d \precapprox_f (\nu a)q_1 = q$*, by congruence as well.*

If $p = C$ *with* $C \stackrel{def}{=} r$*, then* $[\![p]\!]^e_d = C_{ed}$ *with* $C_{ed} \stackrel{def}{=} [\![r]\!]^e_d$*. If* $C_{ed} \xrightarrow{\tau'} q$*, then this is possible only if* $[\![r]\!]^e_d \xrightarrow{\tau'} q$*. By induction, there exists* p' *such that (i)* $r \sim_f p'$*, and (ii)* $[\![p']\!]^e_d \precapprox_f q$*. Since* $C \sim_f r$*, we get* $p' \sim_f C$*, and so the two theses hold.* □

Proposition 5.5. *For any* $p \in \mathscr{P}_{CCS^{seq}}$*, if* $[\![p]\!] \xrightarrow{\tau'} q$*, then there exists* p' *such that (i)* $p \sim_f p'$*, and (ii)* $[\![p']\!] \precapprox_f q$*.*

Proof. By induction on the proof of $[\![p]\!] \xrightarrow{\tau'} q$. We proceed by cases. The proof is very similar to that for Lemma 5.4 and so we discuss a few cases only.

If $p = p_1 \cdot p_2$, then $[\![p_1 \cdot p_2]\!] = (\nu d)([\![p_1]\!]^e_d \,|\, \overline{d}.[\![p_2]\!]) \xrightarrow{\tau'} q$ is possible only if either (a) $[\![p_1]\!]^e_d \xrightarrow{d} q_1$ and $q = (\nu d)(q_1 \,|\, [\![p_2]\!])$, or (b) $[\![p_1]\!]^e_d \xrightarrow{\tau'} q_1$ and $q = (\nu e)(q_1 \,|\, \overline{d}.[\![p_2]\!])$.

Case (a): by Lemma 5.3, $p_1 \downarrow$ and $q_1 \downarrow$. In this case, we choose $p' = p_2$. So (i) $p_1 \cdot p_2 \sim_f p_2$, and (ii) $[\![p_2]\!] \precapprox_f (\nu e)(q_1 \,|\, [\![p_2]\!]) = q$, because this holds for $\sim_f$.

Case (b): by Lemma 5.4, there exists p'_1 such that (i) $p_1 \sim_f p'_1$ and (ii) $[\![p'_1]\!]^e_d \precapprox_f q_1$. Hence, in this case we choose $p' = p'_1 \cdot p_2$. So (i) $p_1 \cdot p_2 \sim_f p'_1 \cdot p_2$ by congruence, and (ii) $[\![p'_1 \cdot p_2]\!] = (\nu e)([\![p'_1]\!]^d_e \,|\, \overline{e}.[\![p_2]\!]) \precapprox_f (\nu e)(q_1 \,|\, \overline{e}.[\![p_2]\!]) = q$ by congruence as well.

If $p = p_1 \,|\, p_2$, then $[\![p_1 \,|\, p_2]\!] = [\![p_1]\!] \,|\, [\![p_2]\!] \xrightarrow{\tau'} q$ is possible only if either (a) $[\![p_1]\!] \xrightarrow{\tau'} q_1$ and $q = q_1 \,|\, [\![p_2]\!]$, or (b) $[\![p_2]\!] \xrightarrow{\tau'} q_2$ and $q = [\![p_1]\!] \,|\, q_2$.

Case (a): by induction, there exists p'_1 such that (i) $p_1 \sim_f p'_1$ and (ii) $[\![p'_1]\!] \precapprox_f q_1$. Hence, in this case we choose $p' = p'_1 \,|\, p_2$. So, (i) $p_1 \,|\, p_2 \sim_f p'_1 \,|\, p_2$ by congruence, and (ii) $[\![p'_1 \,|\, p_2]\!] = [\![p'_1]\!] \,|\, [\![p_2]\!] \precapprox_f q_1 \,|\, [\![p_2]\!] = q$ by congruence as well.

Case (b) is symmetric to case (a) above, hence omitted. □

Proposition 5.6. *For any $p \in \mathscr{P}_{CCS^{seq}}$, if $[\![p]\!] \downarrow$, then $p \downarrow$.*

Proof. By induction on the proof of $[\![p]\!] \downarrow$. We proceed by cases.

If $p = \mathbf{0}$, then $[\![p]\!] = \mathbf{0}$ and the thesis trivially holds.

If $p = \mu.p'$, then $[\![p]\!] = \mu.[\![p']\!]$, which is not a final state, hence this case is empty. Similarly for $p = \Sigma_{j \in J} \mu_j.p_j$.

If $p = p_1 \cdot p_2$, then $[\![p_1 \cdot p_2]\!] = (\nu d)([\![p_1]\!]^e_d \,|\, \overline{d}.[\![p_2]\!])$, which is not final as well, hence also this case is empty.

If $p = p_1 \,|\, p_2$, then $[\![p_1 \,|\, p_2]\!] = [\![p_1]\!] \,|\, [\![p_2]\!]$ is final only if both $[\![p_1]\!]$ and $[\![p_2]\!]$ are final. By induction, we have that $p_1 \downarrow$ and $p_2 \downarrow$, so that $p_1 \,|\, p_2 \downarrow$.

If $p = (\nu a)p_1$, then $[\![p]\!] = (\nu a)[\![p_1]\!]$, which may be final only if $[\![p_1]\!] \downarrow$. By induction, we have that $p_1 \downarrow$, and so the thesis $p = (\nu a)p_1 \downarrow$ follows trivially.

If $p = C$ with $C \stackrel{def}{=} q$, then $[\![p]\!] = C'$ with $C' \stackrel{def}{=} [\![q]\!]$. C' may be final only if $[\![q]\!] \downarrow$. By induction, we have then that $q \downarrow$ and so the thesis $C \downarrow$. □

Lemma 5.5. *For any $p \in \mathscr{P}_{CCS^{seq}}$, if $p \downarrow$ then $[\![p]\!]^e_d \stackrel{d}{\Longrightarrow} p'$ with $p' \downarrow$, for any choice of d, e new in p.*

Proof. By induction on the proof of $p \downarrow$, according to the rules in Table 5.7. The base case is when $p = \mathbf{0}$. In such a case, for any choice of d and e, $[\![\mathbf{0}]\!]^e_d = d.\mathbf{0}$ and the thesis trivially holds: $d.\mathbf{0} \stackrel{d}{\Longrightarrow} \mathbf{0}$ with $\mathbf{0} \downarrow$. The inductive cases are as follows.

If $p = p_1 \,|\, p_2$ and $p \downarrow$, then necessarily $p_1 \downarrow$ and $p_2 \downarrow$. By induction, we have that $[\![p_1]\!]^d_e \stackrel{e}{\Longrightarrow} p'_1$[8] with $p'_1 \downarrow$, as well as $[\![p_2]\!]^d_e \stackrel{e}{\Longrightarrow} p'_2$ with $p'_2 \downarrow$. Hence by using the operational rules it is possible to prove that

[8] Note the inversion of d and e; induction is applicable because the thesis is is true for any choice of e and d, hence also when they are exchanged.

$[\![p_1 \mid p_2]\!]^e_d = (\nu e)(([\![p_1]\!]^d_e \mid [\![p_2]\!]^d_e) \mid \bar{e}.\bar{e}.d.\mathbf{0}) \stackrel{\varepsilon}{\Longrightarrow} (\nu e)((p'_1 \mid [\![p_2]\!]^d_e) \mid \bar{e}.d.\mathbf{0})$
$\stackrel{\varepsilon}{\Longrightarrow} (\nu e)((p'_1 \mid p'_2) \mid d.\mathbf{0}) \stackrel{d}{\longrightarrow} (\nu e)((p'_1 \mid p'_2) \mid \mathbf{0})$, *with* $p' = (\nu e)((p'_1 \mid p'_2) \mid \mathbf{0}) \downarrow$.

If $p = p_1 \cdot p_2$ *and* $p \downarrow$, *then* $p_1 \downarrow$ *and* $p_2 \downarrow$. *By induction, we have that* $[\![p_1]\!]^d_e \stackrel{e}{\Longrightarrow} p'_1$ *with* $p'_1 \downarrow$, *as well as* $[\![p_2]\!]^e_d \stackrel{d}{\Longrightarrow} p'_2$ *with* $p'_2 \downarrow$. *Hence by using the operational rules it is possible to prove that*

$[\![p_1 \cdot p_2]\!]^e_d = (\nu e)([\![p_1]\!]^d_e \mid \bar{e}.[\![p_2]\!]^e_d) \stackrel{\varepsilon}{\Longrightarrow} (\nu e)(p'_1 \mid [\![p_2]\!]^e_d) \stackrel{d}{\Longrightarrow} (\nu e)(p'_1 \mid p'_2)$
with $(\nu e)(p'_1 \mid p'_2) \downarrow$, *as required.*

If $p = (\nu a)p_1$ *and* $p \downarrow$, *then* $p_1 \downarrow$. *By induction we have* $[\![p_1]\!]^e_d \stackrel{d}{\Longrightarrow} p'_1$ *with* $p'_1 \downarrow$. *Hence by using the operational rule (Res) we can prove that* $[\![(\nu a)p_1]\!]^e_d = (\nu a)[\![p_1]\!]^e_d \stackrel{d}{\Longrightarrow} (\nu a)p'_1$ *with* $(\nu a)p_1 \downarrow$, *as required.*

If $p = C$ *(with* $C \stackrel{def}{=} q$*) and* $C \downarrow$, *then also* $q \downarrow$. *By induction we have* $[\![q]\!]^e_d \stackrel{d}{\Longrightarrow} q'$ *with* $q' \downarrow$. *As* $[\![C]\!]^e_d = C_{ed}$ *with* $C_{ed} \stackrel{def}{=} [\![q]\!]^e_d$, *by operation rule (Cons) we also have* $C_{ed} \stackrel{d}{\Longrightarrow} q'$, *with* $q' \downarrow$, *as required.* □

Proposition 5.7. *For any* $p \in \mathscr{P}_{CCS^{seq}}$, *if* $p \downarrow$ *then* $[\![p]\!] \stackrel{\varepsilon}{\Longrightarrow} p'$ *with* $p' \downarrow$.

Proof. *By induction on the proof of* $p \downarrow$, *according to the rules in Table 5.7. The proof follows the line of Lemma 5.5 above. Hence, we focus on a few cases only.*

The base case is when $p = \mathbf{0}$ *and the thesis trivially holds, as* $[\![\mathbf{0}]\!] = \mathbf{0}$, *by choosing* $p' = \mathbf{0}$.

Of the inductive cases we consider only parallel and sequential compositions.

If $p = p_1 \mid p_2$ *and* $p \downarrow$, *then necessarily* $p_1 \downarrow$ *and* $p_2 \downarrow$. *By induction, we have that* $[\![p_1]\!] \stackrel{\varepsilon}{\Longrightarrow} p'_1$ *with* $p'_1 \downarrow$, *as well as* $[\![p_2]\!] \stackrel{\varepsilon}{\Longrightarrow} p'_2$ *with* $p'_2 \downarrow$. *Hence by using the operational rules it is possible to prove that*

$[\![p_1 \mid p_2]\!] = [\![p_1]\!] \mid [\![p_2]\!] \stackrel{\varepsilon}{\Longrightarrow} p'_1 \mid [\![p_2]\!] \stackrel{\varepsilon}{\Longrightarrow} p'_1 \mid p'_2$ *with* $p'_1 \mid p'_2 \downarrow$.

If $p = p_1 \cdot p_2$ *and* $p \downarrow$, *then* $p_1 \downarrow$ *and* $p_2 \downarrow$. *By induction we have that* $[\![p_2]\!] \stackrel{\varepsilon}{\Longrightarrow} p'_2$ *with* $p'_2 \downarrow$. *By Lemma 5.5, we also have that* $[\![p_1]\!]^e_d \stackrel{d}{\Longrightarrow} p'_1$ *with* $p'_1 \downarrow$. *Hence,*

$[\![p_1 \cdot p_2]\!] = (\nu d)([\![p_1]\!]^e_d \mid \bar{d}.[\![p_2]\!]) \stackrel{\varepsilon}{\Longrightarrow} (\nu d)(p'_1 \mid [\![p_2]\!]) \stackrel{\varepsilon}{\Longrightarrow} (\nu d)(p'_1 \mid p'_2)$
with $(\nu d)(p'_1 \mid p'_2) \downarrow$. □

Lemma 5.6. *For any* $p \in \mathscr{P}_{CCS^{seq}}$, *if* $p \stackrel{\mu}{\longrightarrow} p'$, *then there exists* q *such that* $[\![p]\!]^e_d \stackrel{\mu}{\Longrightarrow} q$ *with* $q \sim_f [\![p']\!]^e_d$.

Proof. *By induction on the proof of* $p \stackrel{\mu}{\longrightarrow} p'$.

If $p = \mu.p'$, *then* $\mu.p' \stackrel{\mu}{\longrightarrow} p'$. *Then* $[\![\mu.p']\!]^e_d = \mu.[\![p']\!]^e_d \stackrel{\mu}{\longrightarrow} [\![p']\!]^e_d$, *and the thesis follows trivially, as* $q = [\![p']\!]^e_d$.

If $p = \Sigma_{j \in J} \mu_j.p_j$, *then* $\Sigma_{j \in J} \mu_j.p_j \stackrel{\mu_k}{\longrightarrow} p_k$ *for any* $k \in J$. *Also this case is trivial:* $[\![\Sigma_{j \in J} \mu_j.p_j]\!]^e_d = \Sigma_{j \in J} \mu_j.[\![p_j]\!]^e_d \stackrel{\mu_k}{\longrightarrow} [\![p_k]\!]^e_d$ *for any* $k \in J$.

If $p = p_1 \mid p_2$, *and* $p_1 \mid p_2 \stackrel{\mu}{\longrightarrow} p'$, *then we have three subcases:* $p_1 \stackrel{\mu}{\longrightarrow} p'_1$ *and* $p' = p'_1 \mid p_2$; or $p_2 \stackrel{\mu}{\longrightarrow} p'_2$ *and* $p' = p_1 \mid p'_2$; or $\mu = \tau$, $p_1 \stackrel{\alpha}{\longrightarrow} p'_1$, $p_2 \stackrel{\bar{\alpha}}{\longrightarrow} p'_2$ *and* $p' = p'_1 \mid p'_2$. *We consider only the last subcase, as the other two are simpler. By induction, we have that* $[\![p_1]\!]^e_d \stackrel{\alpha}{\Longrightarrow} q_1$ *with* $q_1 \sim_f [\![p'_1]\!]^e_d$*, as well as* $[\![p_2]\!]^e_d \stackrel{\overline{\alpha}}{\Longrightarrow} q_2$ *with* $q_2 \sim_f [\![p'_2]\!]^e_d$*. It is possible to prove by operational rules that*

$[\![p_1 \mid p_2]\!]^e_d = (\nu e)(([\![p_1]\!]^d_e \mid [\![p_2]\!]^d_e) \mid \overline{e}.\overline{e}.d.\mathbf{0}) \stackrel{\tau}{\Longrightarrow} (\nu e)((q_1 \mid q_2) \mid \overline{e}.\overline{e}.d.\mathbf{0})$
with $(\nu e)((q_1 \mid q_2) \mid \overline{e}.\overline{e}.d.\mathbf{0}) \sim_f (\nu e)(([\![p'_1]\!]^d_e \mid [\![p'_2]\!]^d_e) \mid \overline{e}.\overline{e}.d.\mathbf{0}) = [\![p'_1 \mid p'_2]\!]^e_d$ *by congruence, as required.*

If $p = p_1 \cdot p_2$ *and* $p_1 \cdot p_2 \stackrel{\mu}{\longrightarrow} p'$*, then we have two subcases:* either $p_1 \stackrel{\mu}{\longrightarrow} p'_1$ *and* $p' = p'_1 \cdot p_2$; or $p_1 \downarrow$, $p_2 \stackrel{\mu}{\longrightarrow} p'_2$ *and* $p' = p'_2$*. In the former case, by induction, we have that* $[\![p_1]\!]^e_d \stackrel{\mu}{\Longrightarrow} q_1$ *with* $q_1 \sim_f [\![p'_1]\!]^e_d$*. We can prove by operational rules that*

$[\![p_1 \cdot p_2]\!]^e_d = (\nu e)([\![p_1]\!]^d_e \mid \overline{e}.[\![p_2]\!]^e_d) \stackrel{\mu}{\Longrightarrow} (\nu e)(q_1 \mid \overline{e}.[\![p_2]\!]^e_d)$
with $(\nu e)(q_1 \mid \overline{e}.[\![p_2]\!]^e_d) \sim_f (\nu e)([\![p'_1]\!]^d_e \mid \overline{e}.[\![p_2]\!]^e_d) = [\![p'_1 \cdot p_2]\!]^e_d$ *by congruence, as required. In the latter case, as* $p_1 \downarrow$*, we have that* $[\![p_1]\!]^d_e \stackrel{e}{\Longrightarrow} q_1$ *with* $q_1 \downarrow$ *by Lemma 5.5. Moreover, as* $p_2 \stackrel{\mu}{\longrightarrow} p'_2$*, by induction we have that* $[\![p_2]\!]^e_d \stackrel{\mu}{\Longrightarrow} q_2$ *with* $q_2 \sim_f [\![p'_2]\!]^e_d$*. It is possible to prove by operational rules that*

$[\![p_1 \cdot p_2]\!]^e_d = (\nu e)([\![p_1]\!]^d_e \mid \overline{e}.[\![p_2]\!]^e_d) \stackrel{\tau}{\Longrightarrow} (\nu e)(q_1 \mid [\![p_2]\!]^e_d) \stackrel{\mu}{\Longrightarrow} (\nu e)(q_1 \mid q_2)$
with $(\nu e)(q_1 \mid q_2) \sim_f (\nu e)(q_1 \mid [\![p'_2]\!]^e_d)$ *by congruence, and* $(\nu e)(q_1 \mid [\![p'_2]\!]^e_d) \sim_f [\![p'_2]\!]^e_d$ *by algebraic properties (see Exercise 5.34), as required.*

If $p = (\nu a)p_1$ *and* $(\nu a)p_1 \stackrel{\mu}{\longrightarrow} p'$*, then this is possible only if* $p_1 \stackrel{\mu}{\longrightarrow} p'_1$ *and* $p' = (\nu a)p'_1$*. By induction, we have that* $[\![p_1]\!]^e_d \stackrel{\mu}{\Longrightarrow} q_1$ *with* $q_1 \sim_f [\![p'_1]\!]^e_d$*. Hence, by operational rules,* $[\![(\nu a)p_1]\!]^e_d = (\nu a)[\![p_1]\!]^e_d \stackrel{\mu}{\Longrightarrow} (\nu a)q_1$ *with* $(\nu a)q_1 \sim_f (\nu a)[\![p'_1]\!]^e_d = [\![(\nu a)p'_1]\!]^e_d$*, by congruence, as required.*

If $p = C$ *with* $C \stackrel{def}{=} p_1$ *and* $C \stackrel{\mu}{\longrightarrow} p'$*, then this is possible only if* $p_1 \stackrel{\mu}{\longrightarrow} p'$*. By induction,* $[\![p_1]\!]^e_d \stackrel{\mu}{\Longrightarrow} q_1$ *with* $q_1 \sim_f [\![p']\!]^e_d$*. Observe that* $[\![C]\!]^e_d = C_{ed}$*, with* $C_{ed} \stackrel{def}{=} [\![p_1]\!]^e_d$*. Hence, by operational rules, we have* $C_{ed} \stackrel{\mu}{\Longrightarrow} q_1$ *with* $q_1 \sim_f [\![p']\!]^e_d$*, as required.* □

Proposition 5.8. *For any* $p \in \mathscr{P}_{CCS^{seq}}$*, if* $p \stackrel{\mu}{\longrightarrow} p'$*, then there exists* q *such that* $[\![p]\!] \stackrel{\mu}{\Longrightarrow} q$ *with* $q \sim_f [\![p']\!]$*.*

Proof. *By induction on the proof of* $p \stackrel{\mu}{\longrightarrow} p'$*. The proof is very similar to the previous lemma. We focus on a few cases, only.*

If $p = \Sigma_{j\in J}\mu_j.p_j$*, then* $\Sigma_{j\in J}\mu_j.p_j \stackrel{\mu_k}{\longrightarrow} p_k$ *for any* $k \in J$*. This case is trivial:* $[\![\Sigma_{j\in J}\mu_j.p_j]\!] = \Sigma_{j\in J}\mu_j.[\![p_j]\!] \stackrel{\mu_k}{\longrightarrow} [\![p_k]\!]$ *for any* $k \in J$*.*

If $p = p_1 \mid p_2$ *and* $p_1 \mid p_2 \stackrel{\mu}{\longrightarrow} p'$*, then we have three subcases:* $p_1 \stackrel{\mu}{\longrightarrow} p'_1$ *and* $p' = p'_1 \mid p_2$; or $p_2 \stackrel{\mu}{\longrightarrow} p'_2$ *and* $p' = p_1 \mid p'_2$; or $\mu = \tau$, $p_1 \stackrel{\alpha}{\longrightarrow} p'_1$, $p_2 \stackrel{\overline{\alpha}}{\longrightarrow} p'_2$ *and* $p' = p'_1 \mid p'_2$*. We consider only the last one, as the other two are simpler. By induction, we have that* $[\![p_1]\!] \stackrel{\alpha}{\Longrightarrow} q_1$ *with* $q_1 \sim_f [\![p'_1]\!]$*, as well as* $[\![p_2]\!] \stackrel{\overline{\alpha}}{\Longrightarrow} q_2$ *with* $q_2 \sim_f [\![p'_2]\!]$*. Hence,* $[\![p_1 \mid p_2]\!] = [\![p_1]\!] \mid [\![p_2]\!] \stackrel{\varepsilon}{\Longrightarrow} q_1 \mid q_2$*, with* $q_1 \mid q_2 \sim_f [\![p'_1]\!] \mid [\![p'_2]\!] = [\![p'_1 \mid p'_2]\!]$*, as required.*

If $p = p_1 \cdot p_2$ *and* $p_1 \cdot p_2 \stackrel{\mu}{\longrightarrow} p'$*, then we have two subcases: either* $p_1 \stackrel{\mu}{\longrightarrow} p'_1$ *and* $p' = p'_1 \cdot p_2$*, or* $p_1 \downarrow$*,* $p_2 \stackrel{\mu}{\longrightarrow} p'_2$ *and* $p' = p'_2$*. In the former case, by Lemma 5.6, we have that* $[\![p_1]\!]^e_d \stackrel{\mu}{\Longrightarrow} q_1$ *with* $q_1 \sim_f [\![p'_1]\!]^e_d$*. We can prove by operational rules that*

$[\![p_1 \cdot p_2]\!] = (\nu d)([\![p_1]\!]_d^e \,|\, \overline{d}.[\![p_2]\!]) \stackrel{\mu}{\Longrightarrow} (\nu d)(q_1 \,|\, \overline{d}.[\![p_2]\!])$
with $(\nu d)(q_1 \,|\, \overline{d}.[\![p_2]\!]) \sim_f (\nu d)([\![p_1']\!]_d^e \,|\, \overline{d}.[\![p_2]\!]) = [\![p_1' \cdot p_2]\!]$ *by congruence, as required. In the latter case, as* $p_1 \downarrow$*, we have that* $[\![p_1]\!]_d^e \stackrel{d}{\Longrightarrow} q_1$ *with* $q_1 \downarrow$ *by Lemma 5.5. Moreover, as* $p_2 \stackrel{\mu}{\longrightarrow} p_2'$*, by induction we have that* $[\![p_2]\!] \stackrel{\mu}{\Longrightarrow} q_2$ *with* $q_2 \sim_f [\![p_2']\!]$*. It is possible to prove by operational rules that*
$[\![p_1 \cdot p_2]\!] = (\nu d)([\![p_1]\!]_d^e \,|\, \overline{d}.[\![p_2]\!]) \stackrel{\tau}{\Longrightarrow} (\nu d)(q_1 \,|\, [\![p_2]\!]) \stackrel{\mu}{\Longrightarrow} (\nu d)(q_1 \,|\, q_2)$
with $(\nu d)(q_1 \,|\, q_2) \sim_f (\nu d)(q_1 \,|\, [\![p_2']\!])$ *by congruence, and* $(\nu d)(q_1 \,|\, [\![p_2']\!]) \sim_f [\![p_2']\!]$ *by algebraic properties (see Exercise 5.34), as required.* □

Lemma 5.7. *For any* $p \in \mathscr{P}_{CCS^{seq}}$*, for any choice of* d, e *new in* p*, if* $[\![p]\!]_d^e \stackrel{\mu}{\longrightarrow} q$ *(where* $\mu \neq \tau'$*), then there exists* p' *such that* $p \stackrel{\mu}{\longrightarrow} p'$*, with* $q \sim_f [\![p']\!]_d^e$*.*

Proof. *By induction on the proof of* $[\![p]\!]_d^e \stackrel{\mu}{\longrightarrow} q$*, where* μ *is not* τ'*. We proceed by case analysis.*

$p = \mu.p_1$*. In such a case,* $[\![\mu.p_1]\!]_d^e = \mu.[\![p_1]\!]_d^e \stackrel{\mu}{\longrightarrow} [\![p_1]\!]_d^e$*, as well as* $\mu.p_1 \stackrel{\mu}{\longrightarrow} p_1$*, and* $[\![p_1]\!]_d^e \sim_f [\![p_1]\!]_d^e$*, as required.*

$p = \Sigma_{j \in J} \mu_j.p_j$*. Also this case is trivial. On the one hand,* $[\![\Sigma_{j \in J} \mu_j.p_j]\!]_d^e = \Sigma_{j \in J} \mu_j.[\![p_j]\!]_d^e \stackrel{\mu_k}{\longrightarrow} [\![p_k]\!]_d^e$ *for any* $k \in J$*. On the other hand,* $\Sigma_{j \in J} \mu_j.p_j \stackrel{\mu_k}{\longrightarrow} p_k$ *for any* $k \in J$*. And* $[\![p_k]\!]_d^e \sim_f [\![p_k]\!]_d^e$*, as required.*

$p = p_1 \,|\, p_2$*. In this case,* $[\![p_1 \,|\, p_2]\!]_d^e = (\nu e)(([\![p_1]\!]_e^d \,|\, [\![p_2]\!]_e^d) \,|\, \overline{e}.\overline{e}.d.\mathbf{0}) \stackrel{\mu}{\longrightarrow} q$*. There are three subcases:* $[\![p_1]\!]_e^d \stackrel{\mu}{\longrightarrow} q_1$ *and* $q = (\nu e)((q_1 \,|\, [\![p_2]\!]_e^d) \,|\, \overline{e}.\overline{e}.d.\mathbf{0})$; or $[\![p_2]\!]_e^d \stackrel{\mu}{\longrightarrow} q_2$ *and* $q = (\nu e)(([\![p_1]\!]_e^d \,|\, q_2) \,|\, \overline{e}.\overline{e}.d.\mathbf{0})$; or $\mu = \tau$, $[\![p_1]\!]_e^d \stackrel{\alpha}{\longrightarrow} q_1$, $[\![p_2]\!]_e^d \stackrel{\overline{\alpha}}{\longrightarrow} q_2$ *and* $q = (\nu e)((q_1 \,|\, q_2) \,|\, \overline{e}.\overline{e}.d.\mathbf{0})$*. In the first case, by induction, we have that* $p_1 \stackrel{\mu}{\longrightarrow} p_1'$*, with* $q_1 \sim_f [\![p_1']\!]_e^d$*. Hence, we also have* $p_1 \,|\, p_2 \stackrel{\mu}{\longrightarrow} p_1' \,|\, p_2$ *by rule (Par$_1$); moreover, by congruence,* $q = (\nu e)((q_1 \,|\, [\![p_2]\!]_e^d) \,|\, \overline{e}.\overline{e}.d.\mathbf{0}) \sim_f (\nu e)(([\![p_1']\!]_e^d \,|\, [\![p_2]\!]_e^d) \,|\, \overline{e}.\overline{e}.d.\mathbf{0}) = [\![p_1' \,|\, p_2]\!]_d^e$*, as required. The second case is symmetric, hence omitted. The third case is left as an exercise.*

$p = p_1 \cdot p_2$*. In this case, we have* $[\![p_1 \cdot p_2]\!]_d^e = (\nu e)([\![p_1]\!]_e^d \,|\, \overline{e}.[\![p_2]\!]_d^e) \stackrel{\mu}{\longrightarrow} q$*. We have only one possibility:* $[\![p_1]\!]_e^d \stackrel{\mu}{\longrightarrow} q_1$ *and* $q = (\nu e)(q_1 \,|\, \overline{e}.[\![p_2]\!]_d^e)$*. By induction, we have that* $p_1 \stackrel{\mu}{\longrightarrow} p_1'$*, with* $q_1 \sim_f [\![p_1']\!]_e^d$*. Hence, we also have* $p_1 \cdot p_2 \stackrel{\mu}{\longrightarrow} p_1' \cdot p_2$ *by rule (Seq$_1$); moreover, by congruence,* $q = (\nu e)(q_1 \,|\, \overline{e}.[\![p_2]\!]_d^e) \sim_f (\nu e)([\![p_1']\!]_e^d \,|\, \overline{e}.[\![p_2]\!]_d^e) = [\![p_1' \cdot p_2]\!]_d^e$*, as required.*

$p = (\nu a)p_1$*. In this case, we have* $[\![(\nu a)p_1]\!]_d^e = (\nu a)[\![p_1]\!]_d^e \stackrel{\mu}{\longrightarrow} q$*, which is possible only if* $[\![p_1]\!]_d^e \stackrel{\mu}{\longrightarrow} q_1$ *and* $q = (\nu a)q_1$*. By induction, we have that* $p_1 \stackrel{\mu}{\longrightarrow} p_1'$*, with* $q_1 \sim_f [\![p_1']\!]_d^e$*. Hence, we also have* $(\nu a)p_1 \stackrel{\mu}{\longrightarrow} (\nu a)p_1'$ *by rule (Res); moreover, by congruence,* $q = (\nu a)q_1 \sim_f (\nu a)[\![p_1']\!]_d^e = [\![(\nu a)p_1']\!]_d^e$*, as required.*

$p = C$ *with* $C \stackrel{def}{=} p_1$*. In this case, we have* $[\![C]\!]_d^e = C_{ed} \stackrel{\alpha}{\longrightarrow} q$*, which is possible only if* $[\![p_1]\!]_d^e \stackrel{\alpha}{\longrightarrow} q$*, as* $C_{ed} \stackrel{def}{=} [\![p_1]\!]_d^e$*. By induction, we have* $p_1 \stackrel{\alpha}{\longrightarrow} p_1'$*, with* $q \sim_f [\![p_1']\!]_d^e$*. Hence, we have* $C \stackrel{\alpha}{\longrightarrow} p_1'$ *by rule (Cons); moreover, we already have* $q \sim_f [\![p_1']\!]_d^e$*, as required.* □

Proposition 5.9. *For any $p \in \mathscr{P}_{CCS^{seq}}$, if $[\![p]\!] \xrightarrow{\mu} q$ (with $\mu \neq \tau'$), then there exists p' such that $p \xrightarrow{\mu} p'$, with $q \sim_f [\![p']\!]$.*

Proof. By induction on the proof of $[\![p]\!] \xrightarrow{\alpha} q$. The proof is very similar to that of Lemma 5.7. So we focus on a few cases only.

$p = p_1 \,|\, p_2$. In this case, $[\![p_1 \,|\, p_2]\!] = [\![p_1]\!] \,|\, [\![p_2]\!] \xrightarrow{\mu} q$. Three subcases are possible: $[\![p_1]\!] \xrightarrow{\mu} q_1$ and $q = q_1 \,|\, [\![p_2]\!]$; or $[\![p_2]\!] \xrightarrow{\mu} q_2$ and $q = [\![p_1]\!] \,|\, q_2$; or $[\![p_1]\!] \xrightarrow{\alpha} q_1$, $[\![p_2]\!] \xrightarrow{\overline{\alpha}} q_2$, $\mu = \tau$ and $q = q_1 \,|\, q_2$. We consider only the third case, as the other two are simpler. By induction, we have that $p_1 \xrightarrow{\alpha} p_1'$, with $q_1 \sim_f [\![p_1']\!]$, as well as $p_2 \xrightarrow{\overline{\alpha}} p_2'$, with $q_2 \sim_f [\![p_2']\!]$. Hence, we also have $p_1 \,|\, p_2 \xrightarrow{\tau} p_1' \,|\, p_2'$ by rule (Com); moreover, by congruence, we have $q = q_1 \,|\, q_2 \sim_f [\![p_1']\!] \,|\, [\![p_2']\!] = [\![p_1' \,|\, p_2']\!]$, as required.

$p = p_1 \cdot p_2$. In this case, we have $[\![p_1 \cdot p_2]\!] = (\nu e)([\![p_1]\!]_e^d \,|\, \overline{e}.[\![p_2]\!]) \xrightarrow{\mu} q$. We have only one possibility: $[\![p_1]\!]_e^d \xrightarrow{\mu} q_1$ and $q = (\nu e)(q_1 \,|\, \overline{e}.[\![p_2]\!])$. By Lemma 5.7, we have that $p_1 \xrightarrow{\alpha} p_1'$, with $q_1 \sim_f [\![p_1']\!]_e^d$. Hence, we also have $p_1 \cdot p_2 \xrightarrow{\alpha} p_1' \cdot p_2$ by rule (Seq$_1$); moreover, by congruence, $q = (\nu e)(q_1 \,|\, \overline{e}.[\![p_2]\!]) \sim_f (\nu e)([\![p_1']\!]_e^d \,|\, \overline{e}.[\![p_2]\!]) = [\![p_1' \cdot p_2]\!]$, as required. □

Theorem 5.6. (Correctness of the implementation) *For any $p \in \mathscr{P}_{CCS^{seq}}$, we have that $p \precsim_f [\![p]\!]$.*

Proof. Let us consider relation $R \subseteq \mathscr{P}_{CCS^{seq}} \times \mathscr{P}_{CCS}$ defined as follows:

$$R = \{(p, [\![p]\!]) \;\big|\; p \in \mathscr{P}_{CCS^{seq}}\}$$

If we prove that R is an f-expansion up to $\precsim_f$ (Definition 5.11), we get the thesis $p \precsim_f [\![p]\!]$ because $(p, [\![p]\!]) \in R$ and $R \subseteq \precsim_f$ (Exercise 5.38). We proceed by cases, according to Definition 5.11.

1) If $p \xrightarrow{\mu} p'$, then by Proposition 5.8, we have that there exists q such that $[\![p]\!] \xRightarrow{\mu} q$ with $q \sim_f [\![p']\!]$; hence, also $[\![p']\!] \precsim_f q$. Summing up, to move $p \xrightarrow{\mu} p'$, $[\![p]\!]$ replies with $[\![p]\!] \xRightarrow{\mu} q$ so that $p' \sim_f p' R [\![p']\!] \precsim_f q$, as required.

2.a) If $[\![p]\!] \xrightarrow{\mu} q$, with $\mu \neq \tau'$, then by Proposition 5.9, there exists p' such that $p \xrightarrow{\mu} p'$ with $q \sim_f [\![p']\!]$; hence, also $[\![p']\!] \precsim_f q$. Summing up, to move $[\![p]\!] \xrightarrow{\mu} q$, p replies with $p \xrightarrow{\mu} p'$ so that $p' \precsim_f p' R [\![p']\!] \precsim_f q$, as required.

2.b) If $[\![p]\!] \xrightarrow{\tau'} q$, then by Proposition 5.5 there exists p' such that (i) $p \sim_f p'$ (and so also $p \precsim_f p'$), and (ii) $[\![p']\!] \precsim_f q$. Summing up, to move $[\![p]\!] \xrightarrow{\tau'} q$, p replies by idling (no move) so that $p \precsim_f p' R [\![p']\!] \precsim_f q$, as required.

3) If $p\downarrow$, then by Proposition 5.7, there exists $q\downarrow$ such that $[\![p]\!] \xRightarrow{\varepsilon} q$. The thesis then follows as $p \sim_f \mathbf{0} R \mathbf{0} \precsim_f q$.

4) If $[\![p]\!]\downarrow$, then by Proposition 5.6, also $p\downarrow$, as required. □

Corollary 5.1. *For any $p \in \mathscr{P}_{CCS^{seq}}$, we have that $p \approx_f [\![p]\!]$.*

Proof. Since $\precsim_f \subseteq \approx_f$, the thesis follows directly from Theorem 5.6. □

Corollary 5.2. *For any* $p_1, p_2 \in \mathscr{P}_{CCS^{seq}}$, *we have that* $p_1 \approx_f p_2$ *if and only if* $[\![p_1]\!] \approx_f [\![p_2]\!]$. □

Exercise 5.39. Let CCS^{seq^*} be language extending CCS^{seq} with the iteration construct, discussed in Section 5.4.3. Prove that CCS^{seq^*} can be compiled into CCS, up to $\approx_f$, by showing an encoding $[\![-]\!]$ from the former language into the latter. (*Hint:* Define the encoding $[\![-]\!]$ as the composition of two encodings, the former from CCS^{seq^*} into CCS^{seq}, based on Exercise 5.29 (where the summand **1** should be $\tau.\mathbf{0}$), the latter from CCS^{seq} into CCS.) □

5.5 Replication

Replication is an elegant operator that allows for the definition of infinite-state behavior: with $!p$ we denote a process that, intuitively, can spawn an unbounded number of copies of process p at each step. Contrary to recursion, it allows for a structural induction-based proof technique: in proving a property for $!p$, one can assume that, by structural induction, that property holds for p. However, there are also cons for the use of replication, e.g., the fact that $!p$ generates a not-image-finite lts, in general, as illustrated in Example 5.13. Moreover, replication is less expressive than recursion, as we will see in the following.

Let $CCS_!$ be finite CCS enriched with the additional replication operator

$$p ::= \mathbf{0} \mid \mu.p \mid p+p \mid p|p \mid (\nu a)p \mid !p$$

where $!p$ can perform the actions prescribed by the SOS operational rule

$$(\text{Rep})\ \frac{p\,|\,!p \xrightarrow{\mu} p'}{!p \xrightarrow{\mu} p'}$$

which is not defined by structural induction, as the premise considers a source state, $p\,|\,!p$, which is more complex than the source state, $!p$, of the conclusion. Let $\mathscr{P}_{CCS_!}$ denote the set of $CCS_!$ processes.

Example 5.13. **(Replication determines not image-finite lts's)** Consider the simple $CCS_!$ process $!p$, where $p = a.\mathbf{0}$. According to the SOS rules, $!p \xrightarrow{a} \mathbf{0}\,|\,!p$ as well as $!p \xrightarrow{a} \prod_1^n p\,|\,\mathbf{0}\,|\,!p$ for any $n \geq 1$. □

Note the similarities between process $!(a.\mathbf{0})$ above and the unguarded process constant $C \overset{def}{=} a.\mathbf{0}\,|\,C$ discussed in Section 3.3.1. Indeed, replication can be implemented in CCS by means of unguarded recursion. We can define a simple encoding $[\![-]\!]_1$ from $CCS_!$ to CCS *with* unguarded recursion that is homomorphic on all the finite CCS operators and defined for replication as follows:

$$[\![!p]\!]_1 = A_p \qquad \text{where} \quad A_p \stackrel{def}{=} [\![p]\!]_1 | A_p$$

It is a trivial exercise to show that relation $R = \{(p, [\![p]\!]_1) \mid p \in \mathscr{P}_{CCS_!}\}$ is a strong bisimulation, hence proving the correctness of the implementation up to $\sim$. However, this solution is not very satisfactory, as we would like to implement CCS$_!$ into *finitary* CCS, hence by means of guarded recursion only, up to $\sim$. But before approaching this problem, let us describe some properties of replication.

Exercise 5.40. Prove that, for any $p \in \mathscr{P}_{CCS_!}$, the following property holds:

$$!p \sim p \,|\, !p$$

(*Hint:* Check that relation $R = \{(!p, p\,|\,!p)\} \cup \mathscr{I}$ is a bisimulation.) □

Proposition 5.10. *If* $!p \xrightarrow{\mu} p'$*, then* $p' = p'' \,|\, !p$*, where either* $p' \sim q \,|\, !p$ *with* $p \xrightarrow{\mu} q$*, or* $\mu = \tau$ *and* $p' \sim q \,|\, q' \,|\, !p$ *with* $p \xrightarrow{\alpha} q$ *and* $p \xrightarrow{\overline{\alpha}} q'$*.*

Proof. By induction on the proof of $!p \xrightarrow{\mu} p'$ and using, whenever necessary, the algebraic law $!p \sim p\,|\,!p$ and the congruence of $\sim$ w.r.t. parallel composition. By rule (Rep), such a transition is derivable only if $p\,|\,!p \xrightarrow{\mu} p'$. Then, we have three cases, corresponding to the three operational rules that can be used for parallel composition.

First, by rule (Par$_1$), if $p \xrightarrow{\mu} q$*, then* $p\,|\,!p \xrightarrow{\mu} q\,|\,!p$*, with* $p' = q\,|\,!p$*; the thesis holds:* $p' \sim q\,|\,!p$*. This is the base case.*

Second, by rule (Par$_2$), if $!p \xrightarrow{\mu} r$*, then* $p\,|\,!p \xrightarrow{\mu} p\,|\,r$*, with* $p' = p\,|\,r$*. By induction, we can assume that* $r = r'\,|\,!p$*, where* r *is bisimilar either to* $q\,|\,!p$ *with* $p \xrightarrow{\mu} q$*, or to* $q\,|\,q'\,|\,!p$ *with* $p \xrightarrow{\alpha} q$ *and* $p \xrightarrow{\overline{\alpha}} q'$ *and* $\mu = \tau$*. Summing up, transition* $p\,|\,!p \xrightarrow{\mu} p' = p\,|\,r$ *is such that the reached state is* $p' = p\,|\,r = p\,|\,r'\,|\,!p \sim r'\,|\,p\,|\,!p \sim r'\,|\,!p = r$*, and so* $p' \sim r$*. Then, the thesis follows by transitivity: either* $p' \sim q\,|\,!p$ *with* $p \xrightarrow{\mu} q$*, or* $p' \sim q\,|\,q'\,|\,!p$ *with* $p \xrightarrow{\alpha} t$ *and* $p \xrightarrow{\overline{\alpha}} t'$*, as required.*

Finally, the third case: by rule (Com), if $p \xrightarrow{\alpha} q$ *and* $!p \xrightarrow{\overline{\alpha}} r$*, then* $p\,|\,!p \xrightarrow{\tau} p'$*, with* $p' = q\,|\,r$*. By induction, we can assume that* $r = r'\,|\,!p$*, where* r *is bisimilar to* $q'\,|\,!p$ *with* $p \xrightarrow{\overline{\alpha}} q'$*. Summing up, transition* $p\,|\,!p \xrightarrow{\tau} p' = q\,|\,r$ *is such that the reached state is* $p' = q\,|\,r = q\,|\,r'\,|\,!p$ *and* $p' = q\,|\,r \sim q\,|\,q'\,|\,!p$*, with* $p \xrightarrow{\alpha} q$ *and* $p \xrightarrow{\overline{\alpha}} q'$*, as required.* □

Exercise 5.41. Prove that if $!p \longrightarrow^* p'$, then $\exists n \in \mathbb{N}$ such that $p' \sim \prod_{i=1}^{n} p_i \,|\, !p$, where for each $i = 1, \ldots, n$, $p \longrightarrow^* p_i$.

(*Hint:* By induction on the definition of the reachability relation $\longrightarrow^*$: (i) for each p, $p \longrightarrow^* p$, and (ii) if $p \longrightarrow^* p'$ and $p' \xrightarrow{\mu} p''$, then $p \longrightarrow^* p''$.) □

Exercise 5.42. (Congruence) Prove that for any $p_1, p_2 \in \mathscr{P}_{CCS_!}$, if $p_1 \sim p_2$ then $!p_1 \sim !p_2$. □

Let us now introduce a different, more restrictive, replication operator $?p$, that we call *duplication*. $CCS_?$ denotes the language obtained by enriching finite CCS with the additional duplication operator

$$p ::= \mathbf{0} \mid \mu.p \mid p+p \mid p|p \mid (\nu a)p \mid ?p$$

where $?p$ can perform the actions prescribed by the following SOS operational rules, defined by structural induction:

$$(\mathrm{Dup}_1)\ \frac{p \xrightarrow{\mu} p'}{?p \xrightarrow{\mu} p' \,|\, ?p} \qquad (\mathrm{Dup}_2)\ \frac{p \xrightarrow{\alpha} p' \quad p \xrightarrow{\overline{\alpha}} p''}{?p \xrightarrow{\tau} p' \,|\, p'' \,|\, ?p}$$

Let $\mathscr{P}_{CCS_?}$ denote the set of $CCS_?$ processes. The main distinguishing feature of $CCS_?$ is that its associated lts is finitely-branching. Indeed, the operational rules for duplication ensure that the number of transitions leaving state $?p$ is given by the number of transitions leaving p (by (Dup_1)), plus all the possible self-synchronizations of p with itself (by (Dup_2)).

Exercise 5.43. ($CCS_?$ is finitely-branching) Prove that, for any $p \in \mathscr{P}_{CCS_?}$, the set $T_p = \{p' \mid \exists \mu \in Act. p \xrightarrow{\mu} p'\}$ is finite. (*Hint:* By structural induction on p.) □

Now we want to show that $CCS_!$ and $CCS_?$ are equally expressive. This is achieved by providing two encodings in the two directions; the former $[\![-]\!]^! : \mathscr{P}_{CCS_!} \to \mathscr{P}_{CCS_?}$ such that $p \sim [\![p]\!]^!$ for all $p \in \mathscr{P}_{CCS_!}$; the latter $[\![-]\!]^? : \mathscr{P}_{CCS_?} \to \mathscr{P}_{CCS_!}$ such that $q \sim [\![q]\!]^?$ for all $q \in \mathscr{P}_{CCS_?}$.

The first encoding is homomorphic on all finite CCS operators and for replication is defined as follows: $[\![!p]\!]^! = ?[\![p]\!]^!$. It is not difficult to check that $R_1 = \{(p, [\![p]\!]^!) \mid p \in \mathscr{P}_{CCS_!}\}$ is a strong bisimulation up to $\sim$. In particular, the pair $(!p, ?[\![p]\!]^!)$ is a bisimulation pair, thanks to Proposition 5.10. If $!p$ performs μ by reaching a state p', then p' is bisimilar either to $q \,|\, !p$ with $p \xrightarrow{\mu} q$, or to $q \,|\, q' \,|\, !p$ with $p \xrightarrow{\alpha} q$ and $p \xrightarrow{\overline{\alpha}} q'$ and $\mu = \tau$.

In the former case, since $p \xrightarrow{\mu} q$, then by induction we can assume that $[\![p]\!]^! \xrightarrow{\mu} r$ with $q \sim \overline{q} \, R_1 \, [\![\overline{q}]\!]^! \sim r$. Therefore, by rule ($\mathrm{Dup}_1$), $?[\![p]\!]^! \xrightarrow{\mu} r \,|\, ?[\![p]\!]^!$ with $r \,|\, ?[\![p]\!]^! \sim [\![\overline{q}]\!]^! \,|\, ?[\![p]\!]^!$ by congruence, and $[\![\overline{q}]\!]^! \,|\, ?[\![p]\!]^! = [\![\overline{q} \,|\, !p]\!]^!$; summing up, to move $!p \xrightarrow{\mu} p'$ with $p' \sim q \,|\, !p$, process $?[\![p]\!]^!$ replies with $?[\![p]\!]^! \xrightarrow{\mu} r \,|\, ?[\![p]\!]^!$ so that $p' \sim q \,|\, !p \sim \overline{q} \,|\, !p \, R_1 \, [\![\overline{q} \,|\, !p]\!]^! = [\![\overline{q}]\!]^! \,|\, [\![!p]\!]^! \sim r \,|\, ?[\![p]\!]^!$, as required.

In the latter case, since $p \xrightarrow{\alpha} q$ and $p \xrightarrow{\overline{\alpha}} q'$, by induction, we can assume that $[\![p]\!]^! \xrightarrow{\alpha} r$ with $q \sim \overline{q} \, R_1 \, [\![\overline{q}]\!]^! \sim r$ as well as $[\![p]\!]^! \xrightarrow{\overline{\alpha}} r'$ with $q' \sim \overline{q}' \, R_1 \, [\![\overline{q}']\!]^! \sim r'$. Therefore, by rule ($\mathrm{Dup}_2$), we derive that $?[\![p]\!]^! \xrightarrow{\tau} r \,|\, r' \,|\, ?[\![p]\!]^!$, with $r \,|\, r' \,|\, ?[\![p]\!]^! \sim [\![\overline{q}]\!]^! \,|\, [\![\overline{q}']\!]^! \,|\, ?[\![p]\!]^!$, by congruence, and $[\![\overline{q}]\!]^! \,|\, [\![\overline{q}']\!]^! \,|\, ?[\![p]\!]^! = [\![\overline{q} \,|\, \overline{q}' \,|\, !p]\!]^!$; summing up, to move $!p \xrightarrow{\tau} p'$ with $p' \sim q \,|\, q' \,|\, !p$, process $?[\![p]\!]^!$ replies with $?[\![p]\!]^! \xrightarrow{\tau} r \,|\, r' \,|\, ?[\![p]\!]^!$ so that

$$p' \sim q \,|\, q' \,|\, !p \sim \overline{q} \,|\, \overline{q}' \,|\, !p \, R_1 \, [\![\overline{q} \,|\, \overline{q}' \,|\, !p]\!]^! = [\![\overline{q}]\!]^! \,|\, [\![\overline{q}']\!]^! \,|\, ?[\![p]\!]^! \sim r \,|\, r' \,|\, ?[\![p]\!]^!$$

as required. The case when $?[\![p]\!]^!$ moves first is not more difficult, and so left as an exercise for the reader.

Similarly, the encoding in the opposite direction, $[\![-]\!]^? : \mathscr{P}_{CCS_?} \to \mathscr{P}_{CCS_!}$, is defined homomorphically on finite CCS operators, while for duplication is $[\![?p]\!]^? = ![\![p]\!]^?$. Checking that $R_2 = \{(p, [\![p]\!]^?) \mid p \in \mathscr{P}_{CCS_?}\}$ is a strong bisimulation up to $\sim$ can be done similarly to the proof that R_1 is a strong bisimulation up to $\sim$. Therefore, we can conclude that $CCS_!$ and $CCS_?$ are equally expressive. Note that $[\![[\![p]\!]^!]\!]^? = p$ as well as $[\![[\![q]\!]^?]\!]^! = q$, i.e., the two encodings are one the inverse of the other.

Now we want to show that $CCS_?$ can be implemented into finitary CCS, by providing an encoding $[\![-]\!]_2 : \mathscr{P}_{CCS_?} \to \mathscr{P}_{CCS}$ defined homomorphically for all finite CCS operators, while for $?p$ it is defined as follows:

$$[\![?p]\!]_2 = B_p \text{ with } B_p \stackrel{def}{=} \Sigma_{(\mu_i,p_i)\in T([\![p]\!]_2)}\, \mu_i.(p_i \mid B_p) + \Sigma_{(p_i,p_i')\in S([\![p]\!]_2)}\, \tau.(p_i \mid p_i' \mid B_p),$$

where $T([\![p]\!]_2) = \{(\mu_i, p_i) \mid [\![p]\!]_2 \xrightarrow{\mu_i} p_i\}$ and $S([\![p]\!]_2) = \{(p_i, p_i') \mid \exists \alpha. [\![p]\!]_2 \xrightarrow{\alpha} p_i \wedge [\![p]\!]_2 \xrightarrow{\overline{\alpha}} p_i'\}$. It is not difficult to prove that $R_3 = \{(p, [\![p]\!]_2) \mid p \in \mathscr{P}_{CCS_?}\}$ is a strong bisimulation. In particular, pair $(?p, B_p)$ is a bisimulation pair. If $?p \xrightarrow{\mu} p'$, then p' can be either $q \mid ?p$ in case $p \xrightarrow{\mu} q$ (rule (Dup$_1$)), or $q \mid q' \mid ?p$ in case $\mu = \tau$, $p \xrightarrow{\alpha} q$ and $p \xrightarrow{\overline{\alpha}} q'$ (rule (Dup$_2$)). In the former case, by induction we can assume that $[\![p]\!]_2 \xrightarrow{\mu} [\![q]\!]_2$, hence $(\mu, [\![q]\!]_2) \in T([\![p]\!]_2)$. Therefore, by rule (Cons), $B_p \xrightarrow{\mu} [\![q]\!]_2 \mid B_p = [\![q \mid ?p]\!]_2$, as required. Similarly, for the latter case, we can assume by induction that $[\![p]\!]_2 \xrightarrow{\alpha} [\![q]\!]_2$ as well as $[\![p]\!]_2 \xrightarrow{\overline{\alpha}} [\![q']\!]_2$, hence $([\![q]\!]_2, [\![q']\!]_2) \in S([\![p]\!]_2)$. Therefore, by rule (Cons), $B_p \xrightarrow{\mu} [\![q]\!]_2 \mid [\![q']\!]_2 \mid B_p = [\![q \mid q' \mid ?p]\!]_2$, as required. The case when B_p moves first is analogous, hence omitted.

Summing up, we can define an encoding $[\![-]\!]$ from $CCS_!$ to finitary CCS by composing the encodings $[\![-]\!]^!$ and $[\![-]\!]_2$, up to $\sim$, hence proving that $CCS_!$ is not more expressive than finitary CCS.

As a matter of fact, it can be proven that finitary CCS is strictly more expressive than $CCS_!$, as the latter is not Turing-complete, so that no reasonable encoding of finitary CCS into $CCS_!$ is possible. A process p satisfies the termination property if all of its computations terminate. Formally, we can define the termination predicate $-\ddagger$ as the least subset of processes generated by the rule

$$\frac{\forall q' \forall \mu. (q \xrightarrow{\mu} q' \text{ implies } q'\ddagger)}{q\ddagger},$$

stating that if all the states q' reachable in one step are terminated, then also q is. Note that this rule subsumes the fact that if q is a deadlock, then the implication in the premise is vacuously true, and so $q\ddagger$.

In finitary CCS the termination predicate $-\ddagger$ is undecidable: if we could decide it, then we would solve the halting problem for counter machines, which is not possible for a Turing-complete formalism. As a matter of fact, following the presentation in Section 3.5.3, $TCM_{M_x(v_1,v_2,v_3)}$ represents a deterministic finitary CCS process modeling faithfully the three-counter machine M_x with inputs v_1, v_2 and v_3 for the three counters, respectively. Of course, $TCM_{M_x(v_1,v_2,v_3)}\ddagger$, i.e., it will terminate its unique, deterministic computation, if and only if $M_x(v_1,v_2,v_3)$ will terminate. Therefore, we could redefine the halting problem for counter machines as follows (where y stands for an encoding of the triple (v_1,v_2,v_3)):

$$halt(x,y) = \begin{cases} 1 & \text{if } TCM_{M_x(v_1,v_2,v_3)}\ddagger \\ 0 & \text{otherwise} \end{cases}$$

On the contrary, [BGZ09] proves that the termination property $-\ddagger$ is decidable for $CCS_?$, and so also for the equally expressive $CCS_!$. This means that $CCS_!$ cannot be Turing-complete: if such a language were expressive enough to model counter machines faithfully,[9] then we could solve the halting problem for them. In conclusion, no termination-preserving encoding of finitary CCS into $CCS_!$ is possible.

By the argument above, we may be induced to think that replication, albeit elegant and amenable for structural inductive reasoning, is a useless operator. This is not the case. Replication was introduced in the context of the π-calculus [MPW92, Par01, SW01], a calculus that originated from value-passing CCS, with the distinguishing capability that the transmittable values can be used as channel names. With this extra feature, it can be shown that recursion can be encoded into replication. For an explanation of this fact, we refer you to [Par01, SW01].

5.5.1 Guarded Replication

Replication enjoys some interesting properties that are at the base of a possible simplification of its definition which we will discuss in the following. The first proposition states that replication is an idempotent operator, i.e., if applied twice to p, it acts as if applied once: $!(!p) \sim !p$. Moreover, $!p$ is invariant by composition with copies of itself, i.e., $!p \,|\, !p \sim !p$.

Proposition 5.11. (Idempotence of replication) *Prove that for any $p \in \mathscr{P}_{CCS_!}$, the following hold:*

(i) $!p \sim !p \,|\, !p$
(ii) $!p \sim !(!p)$

Proof. *For case (i), check that* $R_1 = \{(!p, !p \,|\, !p)\} \cup \{(\prod_{i=1}^{n} p_i \,|\, !p, \prod_{i=1}^{n} p_i \,|\, !p \,|\, !p) \mid n \geq 1$ *and for each* $i = 1,\ldots,n, p \longrightarrow^* p_i\}$ *is a strong bisimulation up to* $\sim$.

For case (ii), check that $R_2 = \{(!p, !(!p))\} \cup \{(\prod_{i=1}^{n} p_i \,|\, !p, \prod_{i=1}^{n} p_i \,|\, !p \,|\, !(!p)) \mid n \geq 1$ *and for each* $i = 1,\ldots,n, p \longrightarrow^* p_i\}$ *is a strong bisimulation up to* $\sim$ *(by using also the law in (i)).* □

---

[9] By *faithfully*, we mean that the implementation of CMs is to be done in a *deterministic* manner. However, [BGZ09] shows that a nondeterministic, weak encoding of counter machines in $CCS_?$ is possible, with the feature that it does not preserve the termination property.

The following propositions state that the replication operator can be distributed over all the operators of finite CCS, except prefixing.

Proposition 5.12. (Distributivity of replication w.r.t. parallel composition) *Prove that for any $p,q \in \mathscr{P}_{CCS_!}$, the following hold:*

$(i)\ \ !(p\,|\,q) \sim p\,|\,!(p\,|\,q)$
$(ii)\ \ !(p\,|\,q) \sim q\,|\,!(p\,|\,q)$
$(iii)\ !(p\,|\,q) \sim\, !p\,|\,!q$

Proof. *For case (i), check that relation*

$$R_1 = \{(\Pi_{i=1}^{n} p_i \,|\, \Pi_{j=1}^{m} q_j \,|\, !(p\,|\,q), p\,|\,\Pi_{i=1}^{n} p_i \,|\, \Pi_{j=1}^{m} q_j \,|\, !(p\,|\,q)) \mid n,m \geq 0, p \longrightarrow^{*} p_i \forall i, q \longrightarrow^{*} q_j \forall j\} \ \cup$$
$$\{(\Pi_{i=1}^{n} p_i \,|\, \Pi_{j=1}^{m} q_j \,|\, q\,|\,!(p\,|\,q), \Pi_{i=1}^{n} p_i \,|\, \Pi_{j=1}^{m} q_j \,|\, !(p\,|\,q)) \mid n,m \geq 0, p \longrightarrow^{*} p_i \forall i, q \longrightarrow^{*} q_j \forall j\}$$

is a strong bisimulation up to $\sim$. Note that when $n,m = 0$, we get the thesis as follows: $!(p\,|\,q) \sim \mathbf{0}\,|\,\mathbf{0}\,|\,!(p\,|\,q)\ R_1\ p\,|\,\mathbf{0}\,|\,\mathbf{0}\,|\,!(p\,|\,q) \sim p\,|\,!(p\,|\,q)$.

For case (ii), a simple adaptation of R_1 gives the required relation R_2 (left as an exercise).

For case (iii), by using the laws (i) and (ii) above, one can prove that relation

$$R_3 = \{(\Pi_{i=1}^{n} p_i \,|\, \Pi_{j=1}^{m} q_j \,|\, !(p\,|\,q), \Pi_{i=1}^{n} p_i \,|\, \Pi_{j=1}^{m} q_j \,|\, !p\,|\,!q) \mid n,m \geq 0, p \longrightarrow^{*} p_i \forall i, q \longrightarrow^{*} q_j \forall j\}$$

is a strong bisimulation up to $\sim$. Note that when $n,m = 0$, we get the thesis as follows: $!(p\,|\,q) \sim \mathbf{0}\,|\,\mathbf{0}\,|\,!(p\,|\,q)\ R_3\ \mathbf{0}\,|\,\mathbf{0}\,|\,!p\,|\,!q \sim !p\,|\,!q$. □

Exercise 5.44. (Sum and parallel composition are interchangeable under replication) Prove that, for any $p,q \in \mathscr{P}_{CCS_!}$, $!(p+q) \sim !(p\,|\,q)$. (*Hint:* Adapt relation R_3 in the proof of Proposition 5.12, and make use of the laws (i) and (ii) in that proposition.) □

Proposition 5.13. (Distributivity of replication w.r.t. choice) *Prove that for any $p,q \in \mathscr{P}_{CCS_!}$, $!(p+q) \sim !p\,|\,!q$.*

Proof. *By Exercise 5.44, $!(p+q) \sim !(p\,|\,q)$, and by Proposition 5.12, $!(p\,|\,q) \sim !p\,|\,!q$; hence the thesis by transitivity.* □

Proposition 5.14. (Distributivity of replication w.r.t. restriction) *Prove that for any $p \in \mathscr{P}_{CCS_!}$, $!((\nu a)p) \sim (\nu a)(!p)$ if $\Pi_{i=1}^{n} (\nu a)p \sim (\nu a)(\Pi_{i=1}^{n} p)$ for any $n \geq 2$.*

Proof. *It is enough to check that relation $R = \{(\Pi_{i=1}^{n} (\nu a)p_i \,|\, !((\nu a)p), (\nu a)(\Pi_{i=1}^{n} p_i \,|\, !p)) \mid n \geq 0, p \longrightarrow^{*} p_i \forall i\}$ is a bisimulation up to $\sim$ under the assumption that $\Pi_{i=1}^{n} (\nu a)p \sim (\nu a)(\Pi_{i=}^{n} p)$ for any $n \geq 2$. Note that when $n = 0$, we get the thesis as follows:*

$!((\nu a)p) \sim (\nu a)\mathbf{0}\,|\,!((\nu a)p)\ R\ (\nu a)(\mathbf{0}\,|\,!p) \sim (\nu a)(!p)$. □

Note that the premise $\Pi_{i=1}^{n} (\nu a)p \sim (\nu a)(\Pi_{i=1}^{n} p)$ for any $n \geq 2$ is necessary: if we consider $p = b.a.c.\mathbf{0} + \bar{a}.\mathbf{0}$ (which does not satisfy the premise for $n = 2$), then $(\nu a)(!p)$ can perform the weak trace bc, while $!((\nu a)p)$ cannot. Nonetheless, this premise is not too restrictive, i.e., there are many nontrivial processes that satisfy it, e.g., $p = b.a.c.\mathbf{0} \,|\, b.\overline{a}.\mathbf{0}$.

By Proposition 5.11, 5.12, 5.13 and 5.14, it is clear that replication can be distributed over all the operators (with some limitations for restriction) till reaching prefixed subprocesses only. For instance,

$$\begin{array}{l} !((\nu b)(c.\mathbf{0}+b.\mathbf{0}) \,|\, !(a.\mathbf{0}+c.\mathbf{0})) \sim !(\nu b)(c.\mathbf{0}+b.\mathbf{0}) \,|\, !(!(a.\mathbf{0}+c.\mathbf{0})) \\ \sim (\nu b)!(c.\mathbf{0}+b.\mathbf{0}) \,|\, !(a.\mathbf{0}+c.\mathbf{0}) \sim (\nu b)(!c.\mathbf{0} \,|\, !b.\mathbf{0}) \,|\, (!a.\mathbf{0} \,|\, !c.\mathbf{0}) \end{array}$$

Therefore, we can safely consider a sublanguage of $\text{CCS}_!$, called $\text{CCS}_{g!}$, for finite CCS with *guarded replication*, defined as

$$p ::= \mathbf{0} \;\big|\; \mu.p \;\big|\; !\mu.p \;\big|\; p+p \;\big|\; p|p \;\big|\; (\nu a)p$$

which has essentially the same expressive power of $\text{CCS}_!$. Following the discussion about the merits of duplication over replication, the operational rule for guarded replication can be defined as a variation of rule (Dupl_1):

$$\text{(G-rep)} \quad \frac{}{!\mu.p \xrightarrow{\mu} p \,|\, !\mu.p}$$

This version of replication is sometimes used in the π-calculus [SW01].

5.6 Multi-party Synchronization

Another, quite interesting, operator is parallel composition for multi-party synchronization, originally introduced in CSP [Hoa85, Ros98]. While in CCS synchronization is point-to-point (strictly binary), in CSP it is multi-party, i.e., one single synchronization step may involve many different sequential processes. In CSP the set *Act* of actions is not partitioned into the subsets of input actions and output actions (co-actions), as in CCS; so actions have no type. The parallel composition of two processes, say p and q, is parametrized by a set of actions $A \subseteq Act \setminus \{\tau\}$, called *synchronization set*, and take the following syntactic form: $p \parallel_A q$. A synchronization between p and q can occur only if both are able to perform the very same action $a \in A$; the effect is that $p \parallel_A q$ also performs action a; thus, contrary to CCS, the result of a synchronization is observable and it can be used for further synchronization with other parallel subprocesses. Within $p \parallel_A q$, processes p and q cannot perform asynchronously any action belonging to the synchronization set A; on the contrary, p and q cannot synchronize on actions not belonging to A. Here are the operational rules prescribing the behavior of the parametrized CSP parallel operator:

$$(\text{Csp}_1)\ \frac{p \xrightarrow{\mu} p'}{p \parallel_A q \xrightarrow{\mu} p' \parallel_A q}\ \mu \notin A \qquad (\text{Csp}_2)\ \frac{q \xrightarrow{\mu} q'}{p \parallel_A q \xrightarrow{\mu} p \parallel_A q'}\ \mu \notin A$$

$$(\text{Csp}_3)\ \frac{p \xrightarrow{a} p' \quad q \xrightarrow{a} q'}{p \parallel_A q \xrightarrow{a} p' \parallel_A q'}\ a \in A$$

A simple example may help clarify how multi-party synchronization can take place in CSP. Let us consider process $(a.\mathbf{0} \parallel_{\{a\}} a.\mathbf{0}) \parallel_{\{a\}} a.\mathbf{0}$. The following proof tree shows how to derive transition $(a.\mathbf{0} \parallel_{\{a\}} a.\mathbf{0}) \parallel_{\{a\}} a.\mathbf{0} \xrightarrow{a} (\mathbf{0} \parallel_{\{a\}} \mathbf{0}) \parallel_{\{a\}} \mathbf{0}$:

$$(\text{Csp}_3)\ \dfrac{(\text{Csp}_3)\ \dfrac{(\text{Pref})\ \dfrac{}{a.\mathbf{0} \xrightarrow{a} \mathbf{0}} \quad (\text{Pref})\ \dfrac{}{a.\mathbf{0} \xrightarrow{a} \mathbf{0}}}{a.\mathbf{0} \parallel_{\{a\}} a.\mathbf{0} \xrightarrow{a} \mathbf{0} \parallel_{\{a\}} \mathbf{0}} \quad (\text{Pref})\ \dfrac{}{a.\mathbf{0} \xrightarrow{a} \mathbf{0}}}{(a.\mathbf{0} \parallel_{\{a\}} a.\mathbf{0}) \parallel_{\{a\}} a.\mathbf{0} \xrightarrow{a} (\mathbf{0} \parallel_{\{a\}} \mathbf{0}) \parallel_{\{a\}} \mathbf{0}}$$

Of course, as the result of a synchronization is observable, it is necessary to confine the visibility of names by means of the hiding operator $(\iota a)p$, discussed in Section 5.2. A process $(\iota a)(p_1 \parallel_{\{a\}} p_2) \parallel_{\{a\}} q$ is such that p_1 and p_2 can synchronize over action a, which is then turned into τ by hiding, so that q cannot interact with them over action a. For instance, if $p_1 = p_2 = q = a.\mathbf{0}$, the only possible transition is $(\iota a)(p_1 \parallel_{\{a\}} p_2) \parallel_{\{a\}} q \xrightarrow{\tau} ((\iota a)(\mathbf{0} \parallel_{\{a\}} \mathbf{0})) \parallel_{\{a\}} q$, while q is stuck: it cannot execute a asynchronously (the side condition of rule (Csp$_2$) is not satisfied), nor synchronously by rule (Csp$_3$) because $(\iota a)(p_1 \parallel_{\{a\}} p_2)$ cannot offer any a-labeled transition. Note that a similar effect of scoping is provided in CCS by restriction.

CSP parallel composition enjoys a few algebraic properties. It is commutative: $p \parallel_A q \sim q \parallel_A p$. It is associative when the indexing synchronization set is the same for both occurrences of the parallel operator — $p \parallel_A (q \parallel_A r) \sim (p \parallel_A q) \parallel_A r$ — but it is not associative in general; for instance, $(a.b.\mathbf{0} \parallel_{\{b\}} b.\mathbf{0}) \parallel_{\{a\}} a.b.\mathbf{0}$ can perform trace abb, while this is not possible for $a.b.\mathbf{0} \parallel_{\{b\}} (b.\mathbf{0} \parallel_{\{a\}} a.b.\mathbf{0})$. Moreover, the neutral element for parallel composition is not $\mathbf{0}$, but rather a recursive process constant parametrized on A: $p \parallel_A C_A \sim p$, where $C_A \stackrel{def}{=} \Sigma_{a\in A}\, a.C_A$. It is also easy to see that $p \parallel_A \mathbf{0} \sim (\nu A)p$.

A natural question arises about whether it is possible to encode CSP parallel composition into finitary CCS in a compositional way, up to any sensible notion of equivalence. The answer to this question is negative, as we will see in the next chapter. If we want to encode CSP parallel composition into CCS, we have to extend CCS's capabilities somehow. As a matter of fact, the next chapter introduces an extension to CCS, called Multi-CCS. Its main feature is that it includes one additional operator, called *strong prefixing*, that allows for the implementation of atomic sequences. In this setting, a multi-party synchronization can be implemented as an atomic sequence of binary CCS synchronizations. See Section 6.5.3 for more details about (a variant of) CSP and for an encoding of CSP parallel composition operator into Multi-CCS.

Chapter 6
Multi-CCS

Abstract We present Multi-CCS, an extension to CCS obtained by introducing one additional operator of prefixing, $\underline{\alpha}.p$, called *strong prefixing* (as opposed to normal prefixing, $\mu.p$), with the capability of expressing atomic sequences of actions and, together with parallel composition, also multi-party synchronization.

6.1 Lack of Expressiveness of CCS

As we have seen in Section 3.5, CCS is a Turing-complete formalism, i.e., it has the ability to compute all the computable functions. Therefore, one may think that it is able to solve any kind of problem. Unfortunately this is not the case: Turing-completeness is not enough to ensure the solvability of all the problems in concurrency theory. For instance, it is well known that a classic solution to the famous dining philosophers problem [Dij71] (see below for details) that assumes atomicity in the acquisition of forks (or, equivalently, that requires a three-way synchronization among one philosopher and the two forks), cannot be given in CCS. An extension to CCS able to solve this problem, among others, is the subject of this chapter.

6.1.1 Dining Philosophers Problem

This famous problem, proposed by Dijkstra in [Dij71] and then elaborated on by Hoare [Hoa85] in its current formulation, is defined as follows. Five philosophers sit at a round table, with a bowl of spaghetti in the middle; each philosopher is equipped with a private plate; there are only five forks, each one placed between two adjacent neighbors. Philosophers can think and eat; in order to eat spaghetti, a philosopher has to acquire both forks that he shares with his neighbors; of course, one fork (or even both) may be unavailable and so the philosopher can only think in the meanwhile. As a matter of fact, if a philosopher eats, then his two neighbors cannot eat: eating can happen in *mutual exclusion* w.r.t. the neighbors, so that at most two philosophers can eat at the same time. After a philosopher finishes eating, he has to put down both forks, so that they become available to his neighbors. The problem is to conceive a suitable algorithm that satisfies at least one of the following properties:

- *deadlock-freeness*: no reachable state is a deadlock; this ensures that all the computations can be extended *ad infinitum*;
- *weak non-starvation*: not only can all the computations be extended *ad infinitum*, but also for any never-ending computation there is at least one philosopher who eats infinitely often; this means that not all hungry philosophers starve;
- *strong non-starvation*: as above, with the additional constraint that for any computation each time a philosopher wants to eat, he will eat eventually; this means that no hungry philosopher will starve.

Of course, if an algorithm ensures strong non-starvation, then it also ensures weak non-starvation; analogously, weak non-starvation implies deadlock-freeness.

We are interested in a solution to this problem that satisfies the following constraints:

- *fully distributed*: there is no central memory to which all the philosophers can have access, nor a global scheduler that coordinates the activities of the philosophers;
- *symmetric*: all philosophers are identical.

A first tentative solution in CCS to this problem can be given as follows, where for simplicity's sake we consider the subproblem with two philosophers only. The two forks can be defined by the constants F_i,

$$F_i \stackrel{def}{=} \overline{up_i}.\overline{dn_i}.F_i \quad \text{for } i = 0,1,$$

where the complementary action up_i (pick fork$_i$ up) and dn_i (put fork$_i$ down) are to be performed by the philosopher willing to use that fork. The two philosophers can be described as

$$P_i \stackrel{def}{=} think.P_i + up_i.up_{i+1}.eat.dn_i.dn_{i+1}.P_i \quad \text{for } i = 0,1,$$

where $i+1$ is computed modulo 2 (i.e., $up_{2\,mod\,2}$ is up_0). A philosopher can think or can begin the procedure for the acquisition of both forks, starting from the one with his index (we may assume it is the one on his right); when he has got both forks, then he can eat, and when he has finished eating, he has to put down both forks in the same order in which they have been grabbed. The whole system is

$$DP \stackrel{def}{=} (\nu L)(((P_0 \,|\, P_1) \,|\, F_0) \,|\, F_1),$$

where $L = \{up_0, up_1, dn_0, dn_1\}$.

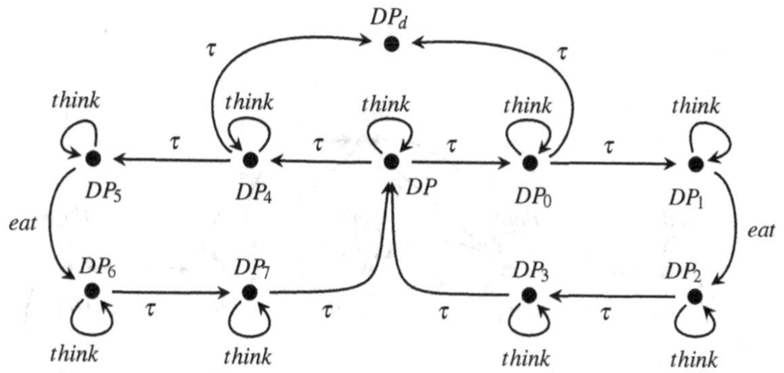

Fig. 6.1 The two dining philosophers in CCS, with deadlock

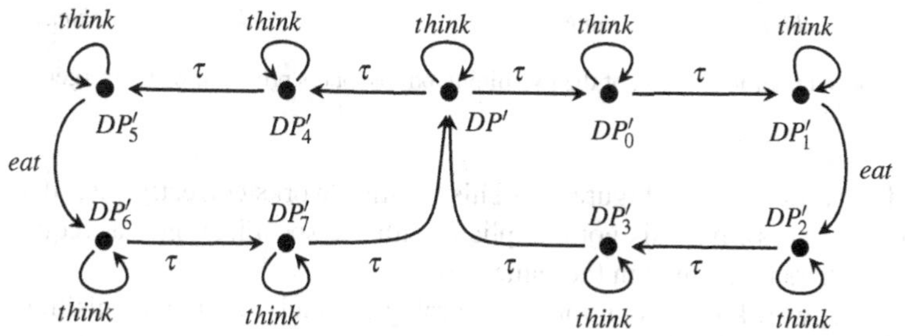

Fig. 6.2 The deadlock-free asymmetric solution of the two dining philosophers problem

Clearly this naïve solution would cause a deadlock when the two philosophers take the fork at their right at the same time and are waiting for the fork at their left. This is illustrated in Figure 6.1, where the state DP_d is a deadlock:

$$DP_d \stackrel{def}{=} (\nu L)(((P'_0 \mid P'_1) \mid F'_0) \mid F'_1)$$

$$P'_i \stackrel{def}{=} up_{i+1}.eat.dn_i.dn_{i+1}.P_i \quad \text{for } i = 0, 1$$

$$F'_i \stackrel{def}{=} \overline{dn_i}.F_i \quad \text{for } i = 0, 1$$

A well-known solution to this problem is breaking the symmetry by inverting the order of acquisition of the forks for the last philosopher. In our restricted case with two philosophers only, we have that

$$P''_0 \stackrel{def}{=} think.P''_0 + up_0.up_1.eat.dn_0.dn_1.P''_0$$

$$P''_1 \stackrel{def}{=} think.P''_1 + up_0.up_1.eat.dn_1.dn_0.P''_1$$

and the whole system is now

$$DP' \stackrel{def}{=} (\nu L)(((P''_0 \mid P''_1) \mid F_0) \mid F_1),$$

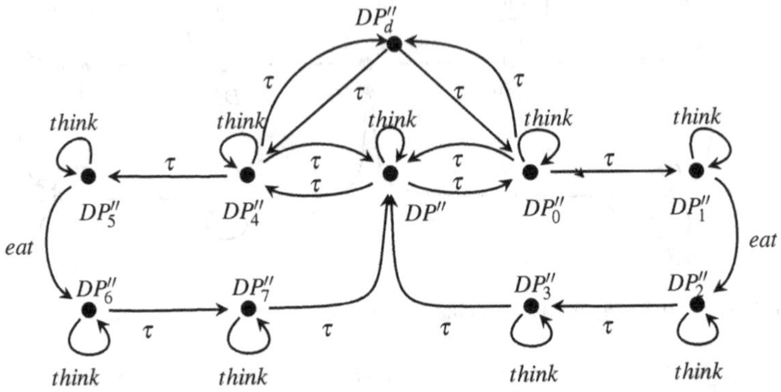

Fig. 6.3 The symmetric solution of the two dining philosophers problem, with divergence

whose LTS is depicted in Figure 6.2. This solution works correctly (i.e., it satisfies deadlock-freeness), but it is not compliant with the specification that requires that all philosophers be defined in the same way.

A simple, well-known solution is to forcing atomicity on the acquisition of the two forks so that either both are taken or none. This requirement can be satisified approximately in CCS by defining a philosopher as

$$P_i''' \stackrel{def}{=} think.P_i''' + up_i.(dn_i.P_i''' + up_{i+1}.eat.dn_i.dn_{i+1}.P_i''') \quad \text{for } i = 0, 1,$$

where, in case the second fork is unavailable, the philosopher may put down the first fork and return to his initial state. However, the new system

$$DP'' \stackrel{def}{=} (\nu L)(((P_0''' \,|\, P_1''') \,|\, F_0) \,|\, F_1),$$

whose LTS is depicted in Figure 6.3, even if deadlock-free, may now diverge: the two philosophers may be engaged in a never-ending divergent computation because the long operation of acquisition of the two forks may always fail.

Remark 6.1. **(Is DP'' a sensible solution?)** It is not difficult to prove that the deadlock-free, asymmetric solution DP' and the divergent, symmetric solution DP'' are weakly bisimilar, so that, to some extent, we could consider DP'' a reasonable solution as well. According to the discussion after Exercise 2.71, this equality makes sense under an assumption of fairness: if a philosopher has the possibility of acquiring his second fork infinitely often (i.e., he reaches state DP_0'' or DP_4'' infinitely often), then he will eventually grab it. This is indeed what the model in Figure 6.3 shows: any never-ending internal computation must pass through state DP_0'' or DP_4'' infinitely often so that, by fairness, we can assume that the other fork will be eventually grabbed. Nonetheless, divergence is a possible behavior and it is unrealistic in this setting to ignore it, as weak bisimilarity does. Moreover, in practice a deeper form of divergence is indeed possible when the two philosophers work somehow synchronously: both acquire the fork on their right at the same time and then, since the other fork is unavailable, both put down their fork at the same time, reaching the initial state and so possibly repeating this cycle forever: a sort of livelock due to synchronous timing of the philosophers' activities. This kind of behavior is not explicitly represented in the interleaving model of Figure 6.3: it could be represented by an additional internal transition from DP'' to DP''_d and by its reverse internal transition. Such a richer model may be obtained by the so-called *step semantics* we will discuss in Section 6.3.2. In this enriched model, the never-ending computation that alternates between states DP'' and DP''_d is such that both philosophers never have the possibility of grabbing the other fork: it is a sort of resource starvation. In conclusion, even though DP'' and DP' are weakly bisimilar, we argue that solution DP'' should not be considered feasible as divergence is an important aspect of its behavior. □

Unfortunately, a deadlock-free, divergence-free solution that implements correctly the atomic acquisition of the two forks cannot be programmed in CCS because it lacks any construct for atomicity that would also enable a multi-party synchronization between one philosopher and the two forks. Indeed, on the one hand, Francez and Rodeh proposed in [FR80] a symmetric, fully distributed, deterministic, deadlock-free solution to the dining philosophers problem in (the original version of) CSP [Hoa78] by exploiting its ability to test (and set) atomically many conditions in a guarded command, which has the same effect as allowing for multi-party synchronization (see also Section 6.5.3 for a simple CSP solution, which is also divergence-free, to this problem). On the other hand, Lehmann and Rabin demonstrated that a symmetric, fully distributed, deterministic, deadlock-free (and divergence-free) solution does not exist in a language with only binary synchronization such as CCS [LR81, RL94].[1] Hence, if we want to solve this problem, we have to extend the capabilities of CCS.

6.1.2 Strong Prefixing: An Operator for Atomicity

We enrich CCS with an additional operator, $\underline{\alpha}.p$, called *strong prefixing* (originally introduced in [GMM90]), where α is the first visible action of a *transaction* (i.e., an atomic sequence of visible actions) that continues with the *sequential* process p, provided that p can complete the transaction. Its operational SOS rule is

$$(\text{S-Pref})\ \frac{p \xrightarrow{\sigma} p'}{\underline{\alpha}.p \xrightarrow{\alpha\diamond\sigma} p'} \quad \text{where} \quad \alpha \diamond \sigma = \begin{cases} \alpha & \text{if } \sigma = \tau, \\ \alpha\sigma & \text{otherwise,} \end{cases}$$

[1] The divergence-free condition is not explicitly stated in their proof, but their argument can be adapted easily to take also this constraint into account.

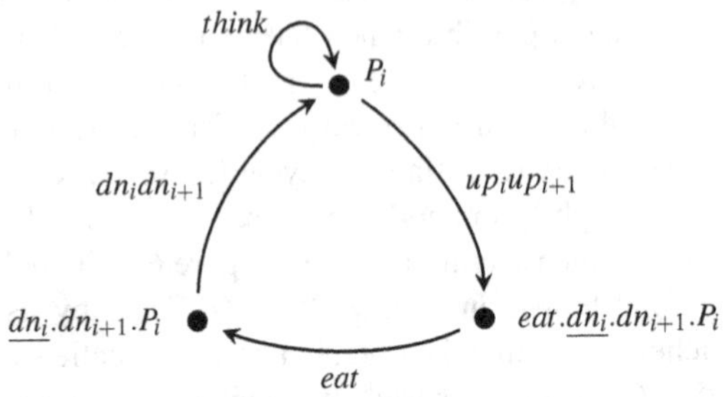

Fig. 6.4 The labeled transition system for philosopher P_i

where σ is either a single action, or a nonempty sequence of *visible* actions; more precisely, σ ranges over $\mathscr{A} = \{\tau\} \cup (\mathscr{L} \cup \overline{\mathscr{L}})^+$, and so $\alpha \diamond \sigma$ ranges over $(\mathscr{L} \cup \overline{\mathscr{L}})^+$.

Rule (S-Pref) allows for the creation of transitions labeled with a nonempty sequence of observable actions. For instance, $\underline{a}.(b.\mathbf{0}+c.\mathbf{0})$ can perform two transitions reaching state $\mathbf{0}$: one labeled with sequence ab, the other one with ac, as illustrated by the following proof trees:

$$\dfrac{(\text{S-Pref})\;\dfrac{(\text{Sum}_1)\;\dfrac{(\text{Pref})\;\overline{}}{b.\mathbf{0} \xrightarrow{b} \mathbf{0}}}{b.\mathbf{0}+c.\mathbf{0} \xrightarrow{b} \mathbf{0}}}{\underline{a}.(b.\mathbf{0}+c.\mathbf{0}) \xrightarrow{ab} \mathbf{0}} \qquad \dfrac{(\text{S-Pref})\;\dfrac{(\text{Sum}_2)\;\dfrac{(\text{Pref})\;\overline{}}{c.\mathbf{0} \xrightarrow{c} \mathbf{0}}}{b.\mathbf{0}+c.\mathbf{0} \xrightarrow{c} \mathbf{0}}}{\underline{a}.(b.\mathbf{0}+c.\mathbf{0}) \xrightarrow{ac} \mathbf{0}}$$

In order for $\underline{\alpha}.p$ to make a move, it is necessary that p be able to perform a transition, i.e., the rest of the transaction. Hence, if $p \xrightarrow{\sigma} p'$ then $\underline{\alpha}.p \xrightarrow{\alpha\diamond\sigma} p'$. Note that $\underline{\alpha}.\mathbf{0}$ cannot execute any action, as $\mathbf{0}$ is deadlocked. If a transition is labeled with $\sigma = \alpha_1 \dots \alpha_{n-1}\alpha_n$, then all the actions $\alpha_1 \dots \alpha_{n-1}$ are due to strong prefixes, while α_n is due to a normal prefix (or α_n is a strong prefix followed by a normal prefix τ).

Exercise 6.1. Show that $\underline{\alpha}.\mathbf{0} \sim \mathbf{0}$ and also that $\underline{a}.\tau.p \sim a.p$. Draw the LTS for $\underline{a}.b.\mathbf{0} + \underline{a}.c.\mathbf{0}$ and show that it is bisimilar to $\underline{a}.(b.\mathbf{0}+c.\mathbf{0})$. □

Example 6.1. **(Philosopher with atomic acquisition of forks)** With the help of strong prefixing, we can now describe the two philosophers as

$$P_i \stackrel{def}{=} think.P_i + \underline{up_i}.up_{i+1}.eat.\underline{dn_i}.dn_{i+1}.P_i \quad \text{for } i = 0, 1,$$

where $i+1$ is computed modulo 2 and the atomic sequence $up_i up_{i+1}$ models the atomic acquisition of the two forks. For simplicity, we assume also that the release of the two forks is atomic, but this is not necessary for correctness. The LTS for P_i is depicted in Figure 6.4. □

Is strong prefixing a good operator for atomicity? To answer this question we should first answer the following: what do we really mean by atomicity? The execution of a sequence of actions is atomic if such execution is

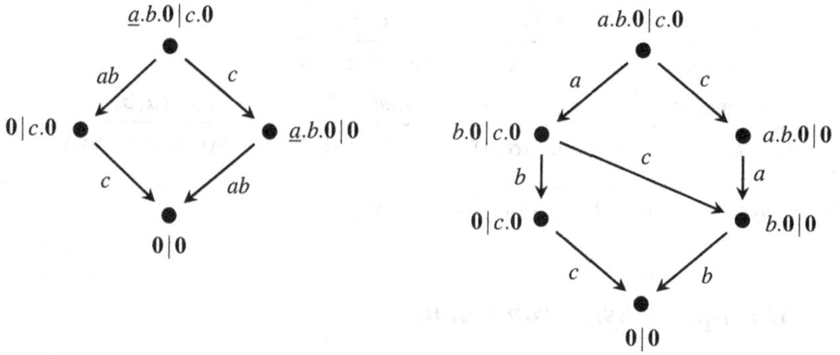

Fig. 6.5 Two labeled transition systems

- *all-or-nothing*: the sequence is either executed completely, or not at all; this implies that the process is not observable during the execution of such a sequence, but only before and after that; and
- *non-interruptible*: no other process can interrupt its execution; this implies that the atomic sequence is never interleaved with others.

Of course, strong prefixing ensures the all-or-nothing property: by rule (S-Pref), $\underline{a}.b.\mathbf{0}$ can only perform a transition to $\mathbf{0}$ labeled ab and its execution is either completed or not done at all; the semantic model does not represent the intermediate state of the execution where action a has been performed but b not yet. Moreover, the situation when the atomic execution may fail in the middle and needs recovery is not modeled at all.

And what about non-interruptibility? What happens when we put a process $\underline{\alpha}.p$ in parallel with another process? For instance, if we take $q = \underline{a}.b.\mathbf{0}\,|\,c.\mathbf{0}$, then the obvious generalization of the CCS operational rules (Par$_1$) and (Par$_2$) of Table 3.1 ensure that the LTS for q is the one shown on the left of Figure 6.5. If we compare this LTS with the one on the right of Figure 6.5 for the CCS process $q' = a.b.\mathbf{0}\,|\,c.\mathbf{0}$, we note that the sequence acb is a trace for q' but not for q: indeed, the atomic sequence ab cannot be interleaved with the action c of the other parallel component, hence non-interruptibility is ensured.

Summing up, strong prefixing is a good operator for atomicity, even if it allows for the execution of atomic sequences only. More general operators for atomicity are described in the literature: for instance, $\langle p \rangle$ means that process p as a whole is executed atomically (see, e.g., [DG91, GR01] and the references therein).

$$\frac{}{Sync(\alpha,\overline{\alpha},\tau)} \qquad \frac{\sigma \neq \varepsilon}{Sync(\alpha\sigma,\overline{\alpha},\sigma)} \qquad \frac{\sigma \neq \varepsilon}{Sync(\overline{\alpha},\alpha\sigma,\sigma)}$$

$$\frac{Sync(\sigma,\overline{\alpha},\tau)}{Sync(\beta\sigma,\overline{\alpha},\beta)} \qquad \frac{Sync(\overline{\alpha},\sigma,\tau)}{Sync(\overline{\alpha},\beta\sigma,\beta)} \qquad \frac{Sync(\sigma,\overline{\alpha},\sigma_1)}{Sync(\beta\sigma,\overline{\alpha},\beta\sigma_1)} \qquad \frac{Sync(\overline{\alpha},\sigma,\sigma_1)}{Sync(\overline{\alpha},\beta\sigma,\beta\sigma_1)}$$

Table 6.1 Synchronization relation *Sync*, where $\beta \neq \alpha$

6.1.3 Multi-party Synchronization

Rule (Com) of Table 3.1 must be extended, as now transitions are labeled on sequences of actions. The new rule is

$$\text{(S-Com)} \ \frac{p \xrightarrow{\sigma_1} p' \qquad q \xrightarrow{\sigma_2} q'}{p\,|\,q \xrightarrow{\sigma} p'\,|\,q'} \ Sync(\sigma_1,\sigma_2,\sigma),$$

which has a side condition on the possible synchronizability of sequences σ_1 and σ_2, whose result is σ. When should $Sync(\sigma_1,\sigma_2,\sigma)$ hold? As (S-Com) is a generalization of the CCS rule (Com), we should require that the basic condition expressed in rule (Com) be feasible: $Sync(\alpha,\overline{\alpha},\tau)$. We can generalize this idea by permitting a synchronization between an atomic sequence $\alpha\sigma$ (with $\sigma \neq \varepsilon$) and a single action $\overline{\alpha}$, as expressed by $Sync(\alpha\sigma,\overline{\alpha},\sigma)$ and $Sync(\overline{\alpha},\alpha\sigma,\sigma)$. Note that since the resulting σ is not τ, it can be used for a possible synchronization with an additional process r: since $p\,|\,q \xrightarrow{\sigma} p'\,|\,q'$, if $r \xrightarrow{\sigma'} r'$, then $(p\,|\,q)\,|\,r \xrightarrow{\sigma''} (p'\,|\,q')\,|\,r'$, provided that $Sync(\sigma,\sigma',\sigma'')$, hence representing a form of ternary synchronization, realized by means of an atomic sequence of binary synchronizations. Moreover, it is convenient to permit a synchronization of a sequence σ with an action $\overline{\alpha}$ even if action α occurs in σ, but not at the beginning; so, other rules are needed.

Relation *Sync* is formally defined by the axioms and the rules of Table 6.1. $Sync(\sigma_1,\sigma_2,\sigma)$ holds if at least one of the two sequences is a single action, say $\sigma_1 = \alpha$, to be synchronized with the first occurrence of its complementary co-action $\overline{\alpha}$ within the sequence σ_2. Note that it is not possible to synchronize two sequences. This means that, usually, a multi-party synchronization takes place among one *leader*, i.e., the process performing an atomic sequence σ, and a number of components (the *servants*) less than or equal to the length of σ, where each servant contributes one visible action only. This is strictly the case for so-called *well-formed processes*, i.e., processes that do not allow for the synchronization of two sequences, not even indirectly (see Section 6.2.2). However, more elaborate forms of synchronization are possible, as illustrated in Example 6.6, for non-well-formed processes. While the three axioms of Table 6.1 enable a synchronization at the beginning of a sequence, the four rules enable a synchronization in the middle of a sequence, as clarified by the following example.

Example 6.2. By using the rules in Table 6.1, we prove that $Sync(abac, \overline{a}, bac)$ and $Sync(abac, \overline{c}, aba)$ as follows:

$$\frac{bac \neq \varepsilon}{Sync(abac,\overline{a},bac)} \qquad \cfrac{\cfrac{\cfrac{}{Sync(c,\overline{c},\tau)}}{Sync(ac,\overline{c},a)}}{\cfrac{Sync(bac,\overline{c},ba)}{Sync(abac,\overline{c},aba)}}$$

Note that $Sync(abac, \overline{a}, abc)$ is not derivable because the synchronization may take place with the first occurrence of a only, as shown above; indeed, the four rules of the second row of Table 6.1 are applicable only if $\beta \neq \alpha$.

This example also shows that *Sync* is a (partial) function of its first two arguments, as the result of the synchronization of a sequence with an action, if defined, is unique. □

We list some useful properties of the synchronization relation *Sync*.

Proposition 6.1. **(*Sync* is deterministic)** *For any $\sigma' \in \mathscr{A}$ which contains at least one occurrence of action α, there exists exactly one sequence $\sigma'' \in \mathscr{A}$ such that $Sync(\sigma', \overline{\alpha}, \sigma'')$ and $Sync(\overline{\alpha}, \sigma', \sigma'')$.*

Proof. *By induction on the length of σ'.* □

Exercise 6.2. **(Commutativity of *Sync*)** Prove that for any $\sigma_1, \sigma_2, \sigma \in \mathscr{A}$ such that $Sync(\sigma_1, \sigma_2, \sigma)$, also $Sync(\sigma_2, \sigma_1, \sigma)$ holds. □

Proposition 6.2. **(Swap of synchronizations)** *For any $\sigma_1, \sigma_2, \sigma_3 \in \mathscr{A}$, if we have $Sync(\sigma_1, \sigma_2, \sigma')$ and $Sync(\sigma', \sigma_3, \sigma)$, then there exists a sequence σ'' such that either $Sync(\sigma_1, \sigma_3, \sigma'')$ and $Sync(\sigma_2, \sigma'', \sigma)$, or $Sync(\sigma_2, \sigma_3, \sigma'')$ and $Sync(\sigma_1, \sigma'', \sigma)$.*

Proof. *By definition of Sync, σ' must be a sequence (possibly of length one) of visible actions; so, either σ_1 or σ_2 (but not both) is an atomic sequence (of length two or more), while the other is a single action. W.l.o.g., assume $\sigma_1 = \overline{\sigma}$ and $\sigma_2 = \overline{\alpha_2}$; since the two can synchronize, $\overline{\sigma}$ must contain an occurrence of α_2; by Proposition 6.1, there is only one σ' such that $Sync(\sigma_1, \sigma_2, \sigma')$, the one obtained from σ_1 by canceling the first occurrence of α_2. Since $Sync(\sigma', \sigma_3, \sigma)$, σ_3 must be a single action, say $\overline{\alpha_3}$, and σ' must contain an occurrence of α_3. By Proposition 6.1, the resulting σ is obtained from σ' by canceling the first occurrence of α_3. Of course, also σ_1 must contain an occurrence of α_3. As a consequence, $Sync(\sigma_1, \sigma_3, \sigma'')$, where σ'' is obtained from σ_1 by canceling the first occurrence of α_3; hence, σ'' must contain an occurrence of α_2, so that $Sync(\sigma_2, \sigma'', \sigma)$.* □

Example 6.3. **(Dining Philosophers with multi-party synchronization)** Continuing Example 6.1, we define the complete two dining philosophers system *DP* as

$$DP \stackrel{def}{=} (\nu L)(((P_0 \,|\, P_1) \,|\, F_0) \,|\, F_1),$$

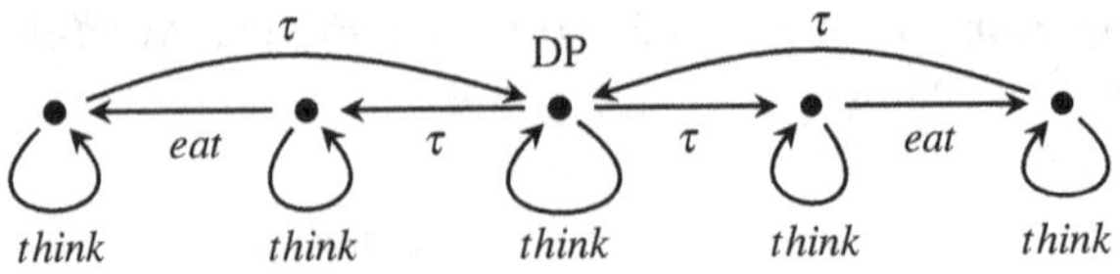

Fig. 6.6 The solution with atomic acquisition (and atomic release) of the two forks

$$\frac{\frac{\frac{\frac{\frac{\overline{up_1.P'_0 \xrightarrow{up_1} P'_0}}{\underline{up_0}.up_1.P'_0 \xrightarrow{up_0 up_1} P'_0}}{think.P_0 + \underline{up_0}.up_1.P'_0 \xrightarrow{up_0 up_1} P'_0}}{P_0 \xrightarrow{up_0 up_1} P'_0}}{P_0 \,|\, P_1 \xrightarrow{up_0 up_1} P'_0 \,|\, P_1} \qquad \frac{\overline{\overline{up_0}.F'_0 \xrightarrow{\overline{up_0}} F'_0}}{F_0 \xrightarrow{\overline{up_0}} F'_0}}{(P_0 \,|\, P_1) \,|\, F_0 \xrightarrow{up_1} (P'_0 \,|\, P_1) \,|\, F'_0}$$

$$\frac{(P_0 \,|\, P_1) \,|\, F_0 \xrightarrow{up_1} (P'_0 \,|\, P_1) \,|\, F'_0 \qquad \frac{\overline{\overline{up_1}.F'_1 \xrightarrow{\overline{up_1}} F'_1}}{F_1 \xrightarrow{\overline{up_1}} F'_1}}{\frac{((P_0 \,|\, P_1) \,|\, F_0) \,|\, F_1 \xrightarrow{\tau} ((P'_0 \,|\, P_1) \,|\, F'_0) \,|\, F'_1}{\frac{(\nu L)(((P_0 \,|\, P_1) \,|\, F_0) \,|\, F_1) \xrightarrow{\tau} (\nu L)(((P'_0 \,|\, P_1) \,|\, F'_0) \,|\, F'_1)}{DP \xrightarrow{\tau} (\nu L)(((P'_0 \,|\, P_1) \,|\, F'_0) \,|\, F'_1)}}}$$

Table 6.2 Multi-party synchronization among dining philosopher P_0 and the two forks

where $L = \{up_0, up_1, dn_0, dn_1\}$ and F_0, F_1 are described as in Section 6.1.1. The operational semantics generates a finite-state LTS for *DP*, depicted in Figure 6.6. Here we want to show how the multi-party synchronization of a philosopher with the two forks takes place. The transition

$$DP \xrightarrow{\tau} (\nu L)(((P'_0 \,|\, P_1) \,|\, F'_0) \,|\, F'_1),$$

where $P'_i = eat.\underline{dn_i}.dn_{i+1}.P_i$ and $F'_i = \overline{dn_i}.F_i$, is proved in Table 6.2. □

Remark 6.2. **(Is *DP* a fully distributed solution?)** It may debatable if this solution, based on multi-party synchronization, is to be considered fully distributed. On the one hand, the Multi-CCS specification *DP*, presented in Example 6.3, is not using any shared memory or global coordinator, so we can say that the *specification DP* is distributed at this abstract level of description. On the other hand, a truly distributed, deterministic *implementation* of the multi-party synchronization mechanism seems not to exist [LR81, RL94, FR80]. □

Remark 6.3. **(Is *DP* a correct solution?)** Looking at Figure 6.6, one may immediately note that *DP* is deadlock-free and divergence-free. However, this solution does not ensure *weak non-starvation* because there are never-ending computations in which no philosopher eats infinitely often: e.g., one philosopher, after having

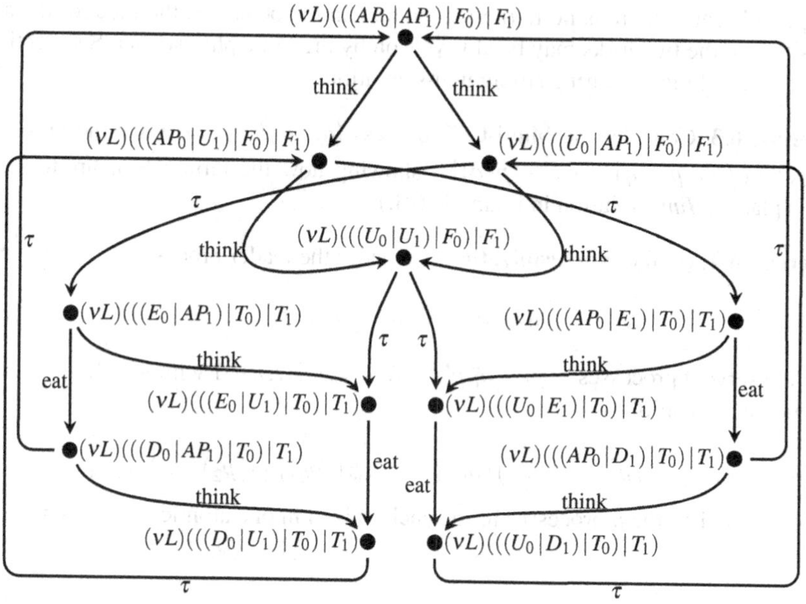

Fig. 6.7 The LTS for *ADP*

eaten, never releases the forks and the other thinks forever. In order to overcome this problem, one can define a more realistic version of the philosopher which alternates between the two phases of thinking and eating. The *alternating* philosopher AP_i is defined as

$$AP_i \stackrel{def}{=} think.U_i \qquad U_i \stackrel{def}{=} \underline{up_i}.up_{i+1}.E_i$$
$$E_i \stackrel{def}{=} eat.D_i \qquad D_i \stackrel{def}{=} \underline{dn_i}.dn_{i+1}.AP_i$$

where he first thinks, then acquires the forks, then eats, then releases the forks in a cyclic, never-ending behavior. Contrary to what happens for P_i, progress is not always ensured: if the forks are unavailable, the alternating philosopher is stuck. The whole system *ADP* can be defined as

$$ADP \stackrel{def}{=} (\nu L)(((AP_0 \mid AP_1) \mid F_0) \mid F_1),$$

where $L = \{up_0, up_1, dn_0, dn_1\}$ and the fork F_i is defined as usual

$$F_i \stackrel{def}{=} \overline{up_i}.T_i \qquad T_i \stackrel{def}{=} \overline{dn_i}.F_i$$

The operational semantics generates the finite-state, deadlock-free LTS, without self-loop transitions, of Figure 6.7. It is not difficult to see that *ADP* satisfies the weak non-starvation property: in any never-ending computation at least one philosopher eats infinitely often. However, it does not satisfy the *strong non-starvation* property because there is no fairness guarantee in the behavior: the race for the acquisition of the two forks may be always won by the same philosopher. See Section 6.4.2 for a solution ensuring strong non-starvation. □

Exercise 6.3. Consider the Multi-CCS process $(\underline{a}.b.p \,|\, \overline{b}.q) \,|\, \overline{a}.r$. Give the proof of transition $(\underline{a}.b.p \,|\, \overline{b}.q) \,|\, \overline{a}.r \xrightarrow{\tau} (p \,|\, q) \,|\, r$, showing how the ternary synchronization takes place. (*Hint:* Solution in Example 6.4.) □

Exercise 6.4. (***n*-ary synchronization**) Consider the leader process

$$p_0 = \underline{a_1}.\underline{a_2}.\ldots.\underline{a_{n-2}}.a_{n-1}.p_0'$$

and the servant processes $\quad p_i = \overline{a_i}.p_i', \quad$ for $i = 1, \ldots, n-1$ with $n \geq 3$.
Argue that transition

$$(\ldots((p_0 \,|\, p_1) \,|\, p_2) \,|\, \ldots) \,|\, p_{n-1} \xrightarrow{\tau} (\ldots((p_0' \,|\, p_1') \,|\, p_2') \,|\, \ldots) \,|\, p_{n-1}'$$

is derivable, i.e., the n processes have synchronized in one atomic transaction. □

6.2 Syntax and Operational Semantics

As for CCS, assume we have a denumerable set $\mathscr{L}$ of channel names (inputs), its complementary set $\overline{\mathscr{L}}$ of co-names (outputs), the set $\mathscr{L} \cup \overline{\mathscr{L}}$ (ranged over by $\alpha, \beta, \ldots$) of visible actions and the set of all actions $Act = \mathscr{L} \cup \overline{\mathscr{L}} \cup \{\tau\}$, such that $\tau \notin \mathscr{L} \cup \overline{\mathscr{L}}$, ranged over by μ. The Multi-CCS process terms are generated by the following abstract syntax, where we are using two syntactic categories: p, to range over sequential processes (i.e., processes that start sequentially), and q, to range over any kind of processes,

$$p ::= \mathbf{0} \mid \mu.q \mid \underline{\alpha}.p \mid p+p \quad \textit{sequential processes}$$
$$q ::= p \mid q \,|\, q \mid (\nu a)q \mid C \quad \textit{processes},$$

where the only new operator is strong prefixing, which takes a visible action α and a sequential process p to produce process $\underline{\alpha}.p$. We denote with $\mathscr{P}_M$ the set of Multi-CCS processes, containing any Multi-CCS term p such that each process constant in the set $Const(p)$ is defined and guarded. To be precise, the definition of guardedness for Multi-CCS constants is the same as for CCS, given in Definition 3.3, in that it considers only *normal* prefixes, and not strong prefixes; see also Remark 6.5 in Section 6.2.3. Note that Multi-CCS is a proper syntactic extension to CCS, i.e., any CCS process is also a Multi-CCS process: $\mathscr{P} \subseteq \mathscr{P}_M$. In the following, we will restrict our attention to *finitary* Multi-CCS, i.e., to the set $\{p \in \mathscr{P}_M \mid Const(p) \text{ is finite}\}$, where function $Const(-)$ is extended to strong prefixing in the obvious way, $Const(\underline{\alpha}.p) = Const(p)$.

(Pref)	$\dfrac{}{\mu.p \xrightarrow{\mu} p}$	(Cong)	$\dfrac{p \equiv p' \xrightarrow{\sigma} q' \equiv q}{p \xrightarrow{\sigma} q}$
(Sum$_1$)	$\dfrac{p \xrightarrow{\sigma} p'}{p+q \xrightarrow{\sigma} p'}$	(Sum$_2$)	$\dfrac{q \xrightarrow{\sigma} q'}{p+q \xrightarrow{\sigma} q'}$
(Par$_1$)	$\dfrac{p \xrightarrow{\sigma} p'}{p \mid q \xrightarrow{\sigma} p' \mid q}$	(Par$_2$)	$\dfrac{q \xrightarrow{\sigma} q'}{p \mid q \xrightarrow{\sigma} p \mid q'}$
(S-Pref)	$\dfrac{p \xrightarrow{\sigma} p'}{\underline{\alpha}.p \xrightarrow{\alpha \diamond \sigma} p'}$		$\alpha \diamond \sigma = \begin{cases} \alpha & \text{if } \sigma = \tau, \\ \alpha\sigma & \text{otherwise} \end{cases}$
(S-Res)	$\dfrac{p \xrightarrow{\sigma} p'}{(\nu a)p \xrightarrow{\sigma} (\nu a)p'}$		$a, \bar{a} \notin n(\sigma)$
(S-Com)	$\dfrac{p \xrightarrow{\sigma_1} p' \quad q \xrightarrow{\sigma_2} q'}{p \mid q \xrightarrow{\sigma} p' \mid q'}$		$Sync(\sigma_1, \sigma_2, \sigma)$

Table 6.3 Operational rules for Multi-CCS

We extend the definition of free names $F(p,I)$ (Definition 4.1) and bound names $B(p,I)$ (Definition 4.2) to strong prefixing as follows:

$$F(\underline{a}.p,I) = F(\underline{\bar{a}}.p,I) = F(p,I) \cup \{a\}$$
$$B(\underline{\alpha}.p,I) = B(p,I)$$

The operational semantics for Multi-CCS is given by the labelled transition system $TS_M = (\mathscr{P}_M, \mathscr{A}, \longrightarrow)$, where the states are the processes in $\mathscr{P}_M$, $\mathscr{A} = \{\tau\} \cup (\mathscr{L} \cup \overline{\mathscr{L}})^+$ is the set of labels (ranged over by σ), and $\longrightarrow \subseteq \mathscr{P}_M \times \mathscr{A} \times \mathscr{P}_M$ is the minimal transition relation generated by the rules listed in Table 6.3.[2]

The new rules (S-Pref) and (S-Com) have been already discussed. Rule (S-Res) is slightly different, as it requires that no action in σ can be a or $\bar{a}$. With $n(\sigma)$ we denote the set of all actions occurring in σ. Formally: $n(\mu) = \{\mu\}$, $n(\alpha\sigma) = \{\alpha\} \cup n(\sigma)$. For instance, $n(a\bar{b}ac) = \{a, \bar{b}, c\}$ and $n(\tau) = \{\tau\}$.

There is one further new rule, called (Cong), which makes use of a structural congruence $\equiv$, which is needed to overcome a shortcoming of parallel composition: without rule (Cong), parallel composition is not associative.

Example 6.4. **(Parallel composition is not associative, without (Cong))** Consider process $(\underline{a}.b.p \mid \bar{b}.q) \mid \bar{a}.r$ of Exercise 6.3. You should have already observed that the ternary synchronization among them, $(\underline{a}.b.p \mid \bar{b}.q) \mid \bar{a}.r \xrightarrow{\tau} (p \mid q) \mid r$, can take place, as proved in Table 6.4. However, if we consider the very similar process $\underline{a}.b.p \mid (\bar{b}.q \mid \bar{a}.r)$, then we can see that $\underline{a}.b.p$ is not able to synchronize with both $\bar{b}.q$

[2] Strictly speaking, according to Definition 2.2, TS_M is not a labeled transition system because its set of labels is not a subset of Act. In this chapter we use a generalization of LTS where the set of its labels can be much richer, e.g., $\mathscr{A}$ for TS_M; see also Section 6.3.2.

$$\dfrac{\dfrac{\dfrac{\dfrac{}{b.p \xrightarrow{b} p}}{\underline{a}.b.p \xrightarrow{ab} p} \quad \dfrac{}{\overline{b}.q \xrightarrow{\overline{b}} q}}{\underline{a}.b.p \,|\, \overline{b}.q \xrightarrow{a} p \,|\, q} \quad \dfrac{}{\overline{a}.r \xrightarrow{\overline{a}} r}}{(\underline{a}.b.p \,|\, \overline{b}.q) \,|\, \overline{a}.r \xrightarrow{\tau} (p \,|\, q) \,|\, r}$$

Table 6.4 Multi-party synchronization among three processes

and $\overline{a}.r$ at the same time! Indeed, on the one hand $\underline{a}.b.p \xrightarrow{ab} p$; on the other hand, $\overline{b}.q \,|\, \overline{a}.r$ can only offer either $\overline{b}$ (if the left component moves) or $\overline{a}$ (if the right component moves), but there is no way to offer both the needed actions in one transition, and so no ternary synchronization can take place. This means that parallel composition is not associative, unless a suitable structural congruence $\equiv$ is introduced, together with the operational rule (Cong) (see Example 6.5 for the continuation of this example).

Similarly, in Example 6.3, we have shown that the ternary synchronization between the philosopher P_0 and the two forks can really take place:

$$DP \xrightarrow{\tau} (\nu L)(((P_0' \,|\, P_1) \,|\, F_0') \,|\, F_1')$$

However, if we consider the slightly different system

$$DP_{as} \stackrel{def}{=} (\nu L)(((P_0 \,|\, P_1) \,|\, (F_0 \,|\, F_1))$$

then we can see that there is no way for a philosopher to synchronize with both forks! Indeed, $(F_0 \,|\, F_1)$ can only offer either $\overline{up}_0$ or $\overline{up}_1$. Hence, also this example shows that parallel composition is not associative without rule (Cong). □

Since associativity is an important property that any natural parallel composition operator should enjoy, we have to overcome this shortcoming by introducing a suitable structural congruence $\equiv$ and an associated operational rule (Cong).

Given a set of axioms E, the structural congruence $\equiv_E \subseteq \mathscr{P} \times \mathscr{P}$ is the congruence induced by the axioms in E. In other words, $p \equiv_E q$ if and only if $E \vdash p = q$, i.e., p can be proved equal to q by means of the equational deductive system $D(E)$, composed of the rules in Table 4.1. of Section 4.3.1.

Rule (Cong) makes use of the structural congruence $\equiv$ on process terms induced by the five equations in Table 6.5.

Axioms **E1** and **E2** are for associativity and commutativity, respectively, of the parallel operator. Axiom **E3** is for unfolding and explains why we have no explicit operational rule for handling constants in Table 6.3: the transitions derivable from C are those transitions derivable from the structurally congruent term p if $C \stackrel{def}{=} p$. As a matter of fact, the operational rule (Cons) for constants is subsumed by the following instance (Cong-c) of rule (Cong):

E1	$(p \mid q) \mid r = p \mid (q \mid r)$	
E2	$p \mid q = q \mid p$	
E3	$A = q$	if $A \stackrel{def}{=} q$
E4	$(\nu a)(p \mid q) = p \mid (\nu a)q$	if $a \notin fn(p)$
E5	$(\nu a)p = (\nu b)(p\{b/a\})$	if $b \notin fn(p) \cup bn(p)$

Table 6.5 Axioms generating the structural congruence $\equiv$

$$\text{(Cons)}\ \frac{p \xrightarrow{\sigma} p'}{C \xrightarrow{\sigma} p'}\ C \stackrel{def}{=} p \qquad \text{(Cong-c)}\ \frac{C \equiv p \xrightarrow{\sigma} p' \equiv p'}{C \xrightarrow{\sigma} p'}$$

Axiom **E4** allows for enlargement of the scope of restriction; its symmetric version,

E4′ $\quad (\nu a)(p \mid q) = (\nu a)p \mid q \quad$ if $a \notin fn(q)$,

is subsumed by axioms **E2** and **E4**. The last axiom **E5** is the so-called law of *alpha-conversion*, using syntactic substitution, defined in Section 4.1.2 for the CCS operators, while for strong prefixing it is defined as follows:

$$\begin{aligned} (\underline{a}.p)\{b/a\} &= \underline{b}.(p\{b/a\}) \\ (\underline{\overline{a}}.p)\{b/a\} &= \underline{\overline{b}}.(p\{b/a\}) \\ (\underline{\alpha}.p)\{b/a\} &= \underline{\alpha}.(p\{b/a\}) \qquad \text{if } \alpha \neq a, \overline{a} \end{aligned}$$

Rule (Cong) enlarges the set of transitions derivable from a given process p, as the following examples and exercises show. The intuition is that, given a process p, a transition is derivable from p if it is derivable from any p' obtained as a rearrangement in any order (or association) of all of its sequential subprocesses.

Example 6.5. **(Associativity)** Continuing Exercise 6.3 and Example 6.4, let us consider the Multi-CCS process $(\underline{a}.b.p \mid \overline{b}.q) \mid \overline{a}.r$. Example 6.4 shows the proof of transition $(\underline{a}.b.p \mid \overline{b}.q) \mid \overline{a}.r \xrightarrow{\tau} (p \mid q) \mid r$.

Now, consider the process $\underline{a}.b.p \mid (\overline{b}.q \mid \overline{a}.r)$, that differs from the above for the different association of its three sequential subprocesses. We have already noticed that $\underline{a}.b.p \mid (\overline{b}.q \mid \overline{a}.r) \stackrel{\tau}{\not\longrightarrow} p \mid (q \mid r)$ if rule (Cong) is not available. However, with the help of rule (Cong), $\underline{a}.b.p$ is now able to synchronize with both $\overline{b}.q$ and $\overline{a}.r$ at the same time as follows:

$$\frac{\underline{a}.b.p \mid (\overline{b}.q \mid \overline{a}.r) \equiv (\underline{a}.b.p \mid \overline{b}.q) \mid \overline{a}.r \xrightarrow{\tau} (p \mid q) \mid r \equiv p \mid (q \mid r)}{\underline{a}.b.p \mid (\overline{b}.q \mid \overline{a}.r) \xrightarrow{\tau} p \mid (q \mid r)}$$

Note that the needed structural congruence uses only the axiom for associativity. □

Example 6.6. **(Commutativity)** For an example where the commutativity axiom **E2** may be useful, consider process $p = (\underline{a}.c.\mathbf{0} \mid b.\mathbf{0}) \mid (\overline{a}.\mathbf{0} \mid \underline{\overline{b}}.\overline{c}.\mathbf{0})$. Such a process can do a four-way synchronization, labeled τ, to $q = (\mathbf{0} \mid \mathbf{0}) \mid (\mathbf{0} \mid \mathbf{0})$, because

$$\dfrac{\dfrac{\dfrac{\dfrac{}{c.\mathbf{0}\xrightarrow{c}\mathbf{0}}}{\underline{a}.c.\mathbf{0}\xrightarrow{ac}\mathbf{0}} \qquad \dfrac{}{\overline{a}.\mathbf{0}\xrightarrow{\overline{a}}\mathbf{0}}}{\underline{a}.c.\mathbf{0}\,|\,\overline{a}.\mathbf{0}\xrightarrow{c}\mathbf{0}\,|\,\mathbf{0}} \qquad \dfrac{\dfrac{}{b.\mathbf{0}\xrightarrow{b}\mathbf{0}} \qquad \dfrac{\dfrac{}{\overline{c}.\mathbf{0}\xrightarrow{\overline{c}}\mathbf{0}}}{\underline{b}.\overline{c}.\mathbf{0}\xrightarrow{b\overline{c}}\mathbf{0}}}{b.\mathbf{0}\,|\,\underline{b}.\overline{c}.\mathbf{0}\xrightarrow{\overline{c}}\mathbf{0}\,|\,\mathbf{0}}}{(\underline{a}.c.\mathbf{0}\,|\,\overline{a}.\mathbf{0})\,|\,(b.\mathbf{0}\,|\,\underline{b}.\overline{c}.\mathbf{0})\xrightarrow{\tau}(\mathbf{0}\,|\,\mathbf{0})\,|\,(\mathbf{0}\,|\,\mathbf{0})}$$

Table 6.6 Hierarchical multi-party synchronization among four processes

$p' = (\underline{a}.c.\mathbf{0}\,|\,\overline{a}.\mathbf{0})\,|\,(b.\mathbf{0}\,|\,\underline{b}.\overline{c}.\mathbf{0})$, which is structurally congruent to p, can perform τ reaching q. The proof of transition $(\underline{a}.c.\mathbf{0}\,|\,\overline{a}.\mathbf{0})\,|\,(b.\mathbf{0}\,|\,\underline{b}.\overline{c}.\mathbf{0}) \xrightarrow{\tau} (\mathbf{0}\,|\,\mathbf{0})\,|\,(\mathbf{0}\,|\,\mathbf{0})$ is reported in Table 6.6. This proof shows a different form of synchronization, where there is no leader, but rather there are two sub-transactions (each one with a sub-leader) synchronizing at the end. Without rule (Cong), process p could not perform such a hierarchical multi-party synchronization. □

Example 6.7. **(Unfolding)** In order to see that also the unfolding axiom **E3** of Table 6.5 is useful, consider $\underline{a}.c.\mathbf{0}\,|\,A$, where $A \stackrel{def}{=} \overline{a}.\mathbf{0}\,|\,\overline{c}.\mathbf{0}$. Clearly $\underline{a}.c.\mathbf{0}\,|\,A$ cannot perform a τ-labeled transition without rule (Cong). However, $\underline{a}.c.\mathbf{0}\,|\,A \equiv \underline{a}.c.\mathbf{0}\,|\,(\overline{a}.\mathbf{0}\,|\,\overline{c}.\mathbf{0})$ by axiom **E3**, and $\underline{a}.c.\mathbf{0}\,|\,(\overline{a}.\mathbf{0}\,|\,\overline{c}.\mathbf{0}) \equiv (\underline{a}.c.\mathbf{0}\,|\,\overline{a}.\mathbf{0})\,|\,\overline{c}.\mathbf{0}$ by axiom **E1**. Note that $(\underline{a}.c.\mathbf{0}\,|\,\overline{a}.\mathbf{0})\,|\,\overline{c}.\mathbf{0} \xrightarrow{\tau} (\mathbf{0}\,|\,\mathbf{0})\,|\,\mathbf{0}$, so that, with rule (Cong), it is now possible to derive $\underline{a}.c.\mathbf{0}\,|\,A \xrightarrow{\tau} \mathbf{0}\,|\,(\mathbf{0}\,|\,\mathbf{0})$ as follows:

$$\text{(Cong)}\ \dfrac{\text{(S-Com)}\ \dfrac{\text{(S-Com)}\ \dfrac{\text{(S-Pref)}\ \dfrac{\text{(Pref)}\ \dfrac{}{c.\mathbf{0}\xrightarrow{c}\mathbf{0}}}{\underline{a}.c.\mathbf{0}\xrightarrow{ac}\mathbf{0}} \qquad \text{(Pref)}\ \dfrac{}{\overline{a}.\mathbf{0}\xrightarrow{\overline{a}}\mathbf{0}}}{\underline{a}.c.0|\overline{a}.0\xrightarrow{c}\mathbf{0}\,|\,\mathbf{0}} \qquad \text{(Pref)}\ \dfrac{}{\overline{c}.\mathbf{0}\xrightarrow{\overline{c}}\mathbf{0}}}{\underline{a}.c.0|A \equiv (\underline{a}.c.0|\overline{a}.0)\,|\,\overline{c}.0 \xrightarrow{\tau} (\mathbf{0}\,|\,\mathbf{0})\,|\,\mathbf{0} \equiv \mathbf{0}\,|\,(\mathbf{0}\,|\,\mathbf{0})}}{\underline{a}.c.0|A \xrightarrow{\tau} \mathbf{0}\,|\,(\mathbf{0}\,|\,\mathbf{0})}$$

Exercise 6.5. (Scope enlargement) Consider process $Q = \underline{b}.c.p_1\,|\,(\nu a)(\overline{b}.p_2\,|\,\overline{c}.p_3)$ and assume $a \notin fn(p_1)$. Show that $Q \equiv Q'$, where $Q' = (\nu a)((\underline{b}.c.p_1\,|\,\overline{b}.p_2)\,|\,\overline{c}.p_3)$. (*Hint:* You need axioms **E4** for scope enlargement and also **E1**.) Show also that $Q \xrightarrow{\tau} Q''$, where $Q'' = p_1\,|\,(\nu a)(p_2\,|\,p_3)$. □

Exercise 6.6. (Alpha-conversion) Consider $Q = \underline{b}.c.p_1\,|\,(\nu a)(\overline{b}.p_2\,|\,\overline{c}.p_3)$ of Exercise 6.5, but now let $a \in fn(p_1)$. Show that, by taking a new d not occurring free or bound in Q, $Q \equiv (\nu d)((\underline{b}.c.p_1\,|\,(\overline{b}.p_2)\{d/a\})\,|\,(\overline{c}.p_3)\{d/a\})$. (*Hint:* You need axioms **E5**, **E4** and **E1**.) Show also that $Q \xrightarrow{\tau} Q''$, where $Q'' = p_1\,|\,(\nu a)(p_2\,|\,p_3)$. □

The structural congruence $\equiv$ allows for a transformation of any process p into a sort of canonical form q_p, its *pool form*, such that q_p can be naturally seen as a sort of pool, whose boundaries are defined by restrictions, containing a soup of sequential processes, that are free to float around thanks to the axioms of associativity **E1** and commutativity **E2** of the structural congruence $\equiv$.

Definition 6.1. (Pool form) A Multi-CCS process p is with *guarded restriction* if each occurrence of the restriction operator in p occurs within a subterm of the form $\mu.q'$ of p. Of course, any restriction-free process is also with guarded restriction, as well as $\underline{a}.(b.(\nu b)(b.\mathbf{0} \mid \underline{c}.\mathbf{0}) + c.\mathbf{0})$.

A Multi-CCS process p is with *prefixed constants* if each occurrence of any constant A in p is strongly guarded in p, i.e., it occurs within a subterm of the form $\mu.q'$ of p. For instance, $a.A$ is with prefixed constants, while $b.B \mid B$ is not.

A Multi-CCS process p is in *pool form* if p is either with *guarded restriction* and with *prefixed constants*, or of the form $(\nu a)q$ with q in pool form. □

In other words, a process p is in pool form if all of its *active* (i.e., not guarded) occurrences of the restriction operator are at the top level. All the sequential subprocesses within the scope of the external restrictions can be associated in any order inside the pool. For instance, $(\nu d)((a.(\nu b)(\underline{c}.b.\mathbf{0} \mid c.\mathbf{0}) \mid c.\mathbf{0}) \mid a.\mathbf{0})$ is in pool form, as well as $(\nu d)((c.\mathbf{0} \mid a.(\nu b)(\underline{c}.b.\mathbf{0} \mid c.\mathbf{0})) \mid a.\mathbf{0})$. Both can be seen as a pool form for $(\nu d)(a.(\nu b)(\underline{c}.b.\mathbf{0} \mid c.\mathbf{0})) \mid (a.\mathbf{0} \mid c.\mathbf{0})$, which is not in pool form. Now we see that any process can be transformed into a pool form by means of structural congruence $\equiv$.

Lemma 6.1. *Any sequential $p \in \mathscr{P}_M$ is in pool form.*

Proof. *By induction on the definition of sequential process.* $\mathbf{0}$ *is in pool form, as well as $\mu.q$. By induction, if p' is a sequential pool form, then $\underline{\alpha}.p'$ is in pool form; similarly, if p_1 and p_2 are sequential pool forms, then also $p_1 + p_2$ is.* □

Proposition 6.3. *For any $p \in \mathscr{P}_M$, there exists a Multi-CCS process q in pool form such that $p \equiv q$.*

Proof. *By induction on the structure of p, and then by induction on the number of unguarded occurrences of the restriction operator in p, denoted* r-size(p) *and defined as follows:*

$$\begin{aligned}
\text{r-size}(\mathbf{0}) &= 0 & \text{r-size}(\mu.p) &= 0\\
\text{r-size}(\underline{\alpha}.p) &= 0 & \text{r-size}(p_1 + p_2) &= 0\\
\text{r-size}(p_1 \mid p_2) &= \text{r-size}(p_1) + \text{r-size}(p_2) & \text{r-size}((\nu a)p) &= 1 + \text{r-size}(p)\\
\text{r-size}(A) &= \text{r-size}(p) \ \ \textit{if}\ A \overset{def}{=} p
\end{aligned}$$

Note that, since we are assuming that constants are guarded and finitely many, then function r-size(p) *is well defined.*

Any sequential process p is already in pool form, by Lemma 6.1. As a matter of fact, the definition of r-size(p) *reflects this fact.*

If $p = p_1 \mid p_2$, then by induction there exist pool forms q_1 and q_2 such that $p_1 \equiv q_1$ and $p_2 \equiv q_2$. The proof now proceeds by induction on r-size$(q_1 \mid q_2)$*. The base case is when* r-size$(q_1 \mid q_2) = 0$*, i.e., both q_1 and q_2 are with guarded restriction. In* *such a case,* $p = p_1 \,|\, p_2 \equiv q_1 \,|\, q_2$ *by structural congruence, where* $q_1 \,|\, q_2$ *is with guarded restriction, too, hence already in pool form. Otherwise, w.l.o.g., assume that* $q_1 = (\nu a)q_1'$ *is in pool form and* q_2 *is in pool form. Then either* $(\nu a)q_1' \,|\, q_2 \equiv (\nu a)(q_1' \,|\, q_2)$ *if* $a \not\in fn(q_2)$ *(axioms* **E2** *and* **E4**)*, or* $(\nu a)q_1' \,|\, q_2 \equiv (\nu a')(q_1'\{a'/a\} \,|\, q_2)$ *if* $a \in fn(q_2)$ *(axioms* **E5**, **E4** *and* **E2**)*. In the former case,* r-size$(q_1' \,|\, q_2) <$ r-size$(q_1 \,|\, q_2)$*, hence the inductive hypothesis can be applied to conclude that there exists a pool form* $\overline{q}$ *such that* $\overline{q} \equiv q_1' \,|\, q_2$*. By structural congruence, we can derive that* $p = p_1 \,|\, p_2 \equiv q_1 \,|\, q_2 = (\nu a)q_1' \,|\, q_2 \equiv (\nu a)(q_1' \,|\, q_2) \equiv (\nu a)\overline{q}$*, the last term being a pool form. Similarly, for the latter case. The symmetric case when* $q_2 = (\nu a)q_2'$ *is in pool form and* q_1 *is in pool form is analogous, hence omitted.*

If $p = (\nu a)p'$*, then by induction we can assume there exists a pool form* $q' \equiv p'$*, hence* $p = (\nu a)p' \equiv (\nu a)q'$*, the latter being a pool form.*

If $p = A$*, with* $A \stackrel{def}{=} r$*, then we have that* $A \equiv r$ *by axiom* **E3***. Note that* A *is not a process with prefixed constants, hence to get its pool form we have to resort to its body* r*. Here, the argument is not by structural induction: take the pool form* q *associated to* r *by the procedure above, so that* $A \equiv r \equiv q$*. Such a procedure will end eventually because we are assuming that constants are guarded and are finitely many.* □

Note that, given a pool form q, if $q \xrightarrow{\sigma} q'$, then q' may be not in pool form. For instance, $\underline{a}.(c.(\nu b)(b.\mathbf{0} \,|\, d.\mathbf{0})) \,|\, \overline{a}.b.\mathbf{0}$ is in pool form, but

$\underline{a}.(c.(\nu b)(b.\mathbf{0} \,|\, d.\mathbf{0})) \,|\, \overline{a}.b.\mathbf{0} \xrightarrow{c} (\nu b)(b.\mathbf{0} \,|\, d.\mathbf{0}) \,|\, b.\mathbf{0}$

and the reached state is not in pool form, but it is congruent to the pool form $(\nu b')((b'.\mathbf{0} \,|\, d.\mathbf{0}) \,|\, b.\mathbf{0})$.

6.2.1 Conservative Extension

From a syntactical point of view, any CCS process is also a Multi-CCS process, i.e., $\mathscr{P} \subseteq \mathscr{P}_M$. Hence, we may wonder also if the operational semantics rules of Table 6.3, when applied to CCS processes, generate an LTS bisimilar to the one the rules of Table 3.1 would generate. If this is the case, we may conclude that Multi-CCS is a conservative extension to CCS, up to $\sim$.

Let us denote by $TS_1 = (\mathscr{P}, Act, \rightarrow_1)$ the LTS for CCS generated by the axiom and rules in Table 3.1; and by $TS_2 = (\mathscr{P}, Act, \rightarrow_2)$ the LTS for CCS generated by the axiom and rules in Table 6.3.[3] Then we can prove the following:

Proposition 6.4. *For any* $p \in \mathscr{P}$*, if* $p \xrightarrow{\mu}_1 p'$*, then* $p \xrightarrow{\mu}_2 p'$*.*

Proof. The proof is by easy induction on the proof of $p \xrightarrow{\mu}_1 p'$*. From the proof tree of transition* $p \xrightarrow{\mu}_1 p'$*, we can reconstruct the proof tree of transition* $p \xrightarrow{\mu}_2 p'$*. It*

[3] Note that the set of labels for TS_2 is Act because strong prefixing is not available and so no transition can be labeled with a (tagged) sequence of actions.

is enough to observe that each operational rule of Table 3.1 has an obvious corresponding one in Table 6.3; the only nonobvious case is for the absence of rule (Cons). However, we have already discussed that (Cong) with axiom **E3** *is able to simulate rule (Cons) — rule (Cong-c), discussed before Example 6.5.* □

In the other direction, even if rule (Cong) does enlarge the set of possible transitions of a CCS process, such an extension is inessential for bisimilarity: any such an additional transition would reach a process structurally congruent to one of those already reached by means of the rules in Table 3.1. We want to prove that, given a transition $p \xrightarrow{\mu}_2 q$, we can find a process q' such that $p \xrightarrow{\mu}_1 q'$ with $q' \equiv q$. The proof of this fact needs an auxiliary lemma.

Lemma 6.2. *For any $p, q \in \mathscr{P}$ such that $p \equiv q$, if $p \xrightarrow{\mu}_1 p'$, then there exists q' such that $q \xrightarrow{\mu}_1 q'$ with $p' \equiv q'$.*

Proof. The proof is by induction on the proof of $p \equiv q$. The base cases are for the five axioms **E1–E5**. *These are rather simple. We work out only two cases.*

Case **E2**: $p = p_1 \mid p_2$ *and* $q = p_2 \mid p_1$. *According to the SOS rules in Table 3.1, transition* $p = p_1 \mid p_2 \xrightarrow{\mu}_1 p'$ *can be due to one of the following three cases:* $p_1 \xrightarrow{\mu}_1 p_1'$ *and* $p' = p_1' \mid p_2$ *— rule (Par₁); or* $p_2 \xrightarrow{\mu}_1 p_2'$ *and* $p' = p_1 \mid p_2'$ *— rule (Par₂); or* $\mu = \tau$, $p_1 \xrightarrow{\alpha}_1 p_1'$, $p_2 \xrightarrow{\overline{\alpha}}_1 p_2'$ *and* $p' = p_1' \mid p_2'$ *— rule (Com). In the first case, by (Par₂),* $q = p_2 \mid p_1 \xrightarrow{\mu}_1 p_2 \mid p_1' \equiv p'$; *in the second case, by (Par₁),* $q = p_2 \mid p_1 \xrightarrow{\mu}_1 p_2' \mid p_1 \equiv p'$; *in the third case, by rule (Com),* $q = p_2 \mid p_1 \xrightarrow{\tau}_1 p_2' \mid p_1' \equiv p'$. *Symmetrically if q moves first.*

Case **E4**: $s = p \mid (\nu a)q$ *and* $t = (\nu a)(p \mid q)$, *with* $a \notin \mathrm{fn}(p)$. *According to the rules in Table 3.1, transition* $s \xrightarrow{\mu} s'$ *can be due to one of the following three cases:* $p \xrightarrow{\mu} p'$ *and* $s' = p' \mid (\nu a)q$ *– rule (Par₁); or* $(\nu a)q \xrightarrow{\mu} (\nu a)q'$ *and* $s' = p \mid (\nu a)q'$ *– rule (Par₂); or* $\mu = \tau$, $p \xrightarrow{\alpha} p'$, $(\nu a)q \xrightarrow{\overline{\alpha}} (\nu a)q'$ *and* $s' = p' \mid (\nu a)q'$ *– rule (Com). In the first case, by (Par₁) and (Res) (since* $\mu \neq a, \overline{a}$*),* $t \xrightarrow{\mu} (\nu a)(p' \mid q) \equiv s'$. *In the second case,* $(\nu a)q \xrightarrow{\mu} (\nu a)q'$ *is derivable only from* $q \xrightarrow{\mu} q'$ *with* $\mu \neq a, \overline{a}$; *hence, by rule (Par₂) and (Res),* $t \xrightarrow{\mu} (\nu a)(p \mid q') \equiv s'$. *The third case is similar, and so omitted. Symmetrically, any transition* $t \xrightarrow{\mu} t'$ *can be due only to transition* $p \mid q \xrightarrow{\mu} t''$, *with* $t' = (\nu a)t''$ *and* $\mu \neq a, \overline{a}$. *Again, three cases are possible:* $p \xrightarrow{\mu} p'$ *and* $t'' = p' \mid q$ *– rule (Par₁); or* $q \xrightarrow{\mu} q'$ *and* $t'' = p \mid q'$ *– rule (Par₂); or* $\mu = \tau$, $p \xrightarrow{\alpha} p'$, $q \xrightarrow{\overline{\alpha}} q'$ *and* $t'' = p' \mid q'$ *– rule (Com). In the first case, by rule (Par₁),* $s \xrightarrow{\mu} p' \mid (\nu a)q \equiv t'$. *In the second case, by (Res),* $(\nu a)q \xrightarrow{\mu} (\nu a)q'$, *and so, by (Par₂),* $s \xrightarrow{\mu} p \mid (\nu a)q' \equiv t'$. *The third case is similar, hence omitted; you should simply note that* $\overline{\alpha} \neq a, \overline{a}$, *because p cannot perform an action of that (complementary) type, as* $a \notin fn(p)$.

The inductive cases are the rules of Table 4.1. Even if long, the proofs of these cases are obvious. In particular, the substitutivity rule

$$\frac{t_i = t_i'}{f(t_1,\ldots,t_i,\ldots,t_k) = f(t_1,\ldots,t_i',\ldots,t_k)} \quad \textit{for any } f \textit{ and } 1 \leq i \leq n,$$

where f is a static operator, is such that if the transition from $f(t_1,\ldots t_j,\ldots,t_i,\ldots,t_k)$ is due to a premise $t_j \xrightarrow{\mu}_1 t_j'$ $(j \neq i)$, then the states reached in the conclusions are $f(t_1,\ldots t_j',\ldots,t_i,\ldots,t_k)$ and $f(t_1,\ldots t_j',\ldots,t_i',\ldots,t_k)$ and these two states are congruent. Instead, if the transition from $f(t_1,\ldots,t_i,\ldots,t_k)$ is due to a premise $t_i \xrightarrow{\mu}_1 s$, then by induction $t_i' \xrightarrow{\mu}_1 s'$ with $s \equiv s'$ and so the states reached in the conclusions are $f(t_1,\ldots,s,\ldots,t_k)$ and $f(t_1,\ldots,s',\ldots,t_k)$ and these two states are congruent. □

Proposition 6.5. *For any $p \in \mathscr{P}$, if $p \xrightarrow{\mu}_2 q$, then there exists $\overline{q}$ such that $p \xrightarrow{\mu}_1 \overline{q}$ with $\overline{q} \equiv q$.*

Proof. *By induction on the proof of $p \xrightarrow{\mu}_2 q$. All the cases are obvious, except for rule (Cong):*

$$\frac{p \equiv p' \xrightarrow{\mu}_2 q' \equiv q}{p \xrightarrow{\mu}_2 q}$$

By induction, we can assume that, since $p' \xrightarrow{\mu}_2 q'$, there exists a state $\overline{q}'$ such that $p' \xrightarrow{\mu}_1 \overline{q}'$ with $q' \equiv \overline{q}'$. By Lemma 6.2, since $p \equiv p'$ and $p' \xrightarrow{\mu}_1 \overline{q}'$, there exists $\overline{q}''$ such that $p \xrightarrow{\mu}_1 \overline{q}''$ with $\overline{q}' \equiv \overline{q}''$. Summing up, if $p \xrightarrow{\mu}_2 q$, then there exists $\overline{q}''$ such that $p \xrightarrow{\mu}_1 \overline{q}''$ and $\overline{q}'' \equiv \overline{q}' \equiv q' \equiv q$, as required. □

Proposition 6.6. *For any $p, p' \in \mathscr{P}$ such that $p \equiv p'$, if $p' \xrightarrow{\mu}_2 q$, then $p \xrightarrow{\mu}_2 q$.*

Proof. *Trivial application of rule (Cong), with $p' \xrightarrow{\mu}_2 q$ as its premise.* □

Therefore, we can easily prove that relation $R = \{(p,q) \mid p,q \in \mathscr{P} \wedge p \equiv q\}$ is a strong bisimulation between the two LTSs TS_1 and TS_2. Let us consider a generic pair $(p,q) \in R$.

If $p \xrightarrow{\mu}_1 p'$, then $p \xrightarrow{\mu}_2 p'$ by Proposition 6.4; by Proposition 6.6 also $q \xrightarrow{\mu}_2 p'$ with $(p',p') \in R$, as required.

Instead, if $q \xrightarrow{\mu}_2 q'$, then by Proposition 6.5 there exists $\overline{q}'$ such that $q \xrightarrow{\mu}_1 \overline{q}'$ with $q' \equiv \overline{q}'$; by Lemma 6.2, as $p \equiv q$, we have that there exists p' such that $p \xrightarrow{\mu}_1 p'$ with $\overline{q}' \equiv p'$. Summing up, to move $q \xrightarrow{\mu}_2 q'$, p replies with transition $p \xrightarrow{\mu}_1 p'$ so that $(p',q') \in R$, as required.

Hence, R is a strong bisimulation indeed. As R contains the identity relation $\mathscr{I} = \{(p,p) \mid p \in \mathscr{P}\}$, we can conclude that Multi-CCS is a conservative extension of CCS, up to $\sim$.

6.2.2 Well-Formed Processes

We propose a syntactic condition on a process p, ensuring that, during its execution, p is unable to synchronize two atomic sequences, not even indirectly; a process satisfying such a syntactic condition will be called *well-formed*. The restriction to well-formed processes will be crucial in some technical proofs of the following sections.

The definition of relation $Sync(\sigma_1, \sigma_2, \sigma)$ requires that at least one of σ_1 or σ_2 be a single action; this is not enough to prevent that two sequences may synchronize, even if indirectly. For instance, assume we have three processes $p_1 = \underline{a}.b.\mathbf{0}$, $p_2 = \overline{a}.\mathbf{0}$ and $p_3 = \underline{\overline{b}}.c.\mathbf{0}$, which may perform the sequences $ab, \overline{a}, \overline{b}c$, respectively; then a ternary synchronization is possible, because first we synchronize p_1 and p_2, by $Sync(ab, \overline{a}, b)$, getting a single action b, which can be then used for a synchronization with p_3, by $Sync(b, \overline{b}c, c)$; in such a way, the two atomic sequences ab and $\overline{b}c$ have been synchronized, by means of the single action $\overline{a}$. So, we would like to mark $(p_1 \,|\, p_2) \,|\, p_3$ as not well-formed.

In order to define well-formed Multi-CCS processes, some auxiliary definitions are needed.

Definition 6.2. (Initials for sequential processes) For any sequential Multi-CCS process p, $In(p) \subseteq \mathscr{A}$ is the set of *initials* of p, defined inductively as

$$\begin{aligned} In(\mathbf{0}) &= \emptyset & In(\mu.p) &= \{\mu\} \\ In(\underline{\alpha}.p) &= \alpha \diamond In(p) & In(p_1 + p_2) &= In(p_1) \cup In(p_2) \end{aligned}$$

where $\alpha \diamond In(p) = \{\alpha \diamond \sigma \mid \sigma \in In(p)\}$. □

Exercise 6.7. Prove that, for any sequential Multi-CCS process p, if $p \equiv q$, then $In(p) = In(q)$.

(*Hint:* By induction on the proof of $p \equiv q$. Note that, since p is sequential, none of the axioms **E1–E5** is applicable, as they are all about non-sequential processes. For the rules of equational deduction, the only nontrivial case is the rule of substitutivity, that has to be proved only for the operators of prefixing, strong prefixing and choice, as p is sequential.) □

Exercise 6.8. Let p be a sequential Multi-CCS process. If $p \xrightarrow{\sigma} p'$, then $\sigma \in In(p)$. (*Hint:* By induction on the proof of $p \xrightarrow{\sigma} p'$.) □

Definition 6.3. (Names in sequences of a process) Let $ns(p) \subseteq \mathscr{L} \cup \overline{\mathscr{L}}$ be the set of (free) names occurring in sequences *of length two or more* of p. Set $ns(p)$ is defined as the least set of visible actions such that the following equations are satisfied:

$$\begin{aligned} ns(\mathbf{0}) &= \emptyset & ns(\mu.p) &= ns(p) \\ ns((\nu a)p) &= ns(p) \setminus \{a, \overline{a}\} & ns(p_1 + p_2) &= ns(p_1) \cup ns(p_2) \\ ns(A) &= ns(q) \;\text{ if } A \stackrel{def}{=} q & ns(p_1 \,|\, p_2) &= ns(p_1) \cup ns(p_2) \\ ns(\underline{\alpha}.p) &= ns(p) \cup \{\alpha\} \cup \textstyle\bigcup_{\sigma \in In(p) \wedge \sigma \neq \tau} n(\sigma) \end{aligned}$$

For finitary processes, $ns(p)$ can be computed by means of the auxiliary function $\gamma(p,\emptyset)$, whose second parameter is a set of already known constants, initially empty.

$$\gamma(\mathbf{0},I) = \emptyset \qquad \gamma(\mu.p,I) = \gamma(p,I)$$
$$\gamma((\nu a)p,I) = \gamma(p,I) \setminus \{a,\bar{a}\} \qquad \gamma(p_1+p_2,I) = \gamma(p_1,I) \cup \gamma(p_2,I)$$
$$\gamma(A,I) = \begin{cases} \emptyset & A \in I, \\ \gamma(p, I \cup \{A\}) & A \notin I \wedge A \overset{def}{=} p \end{cases} \qquad \gamma(p_1 \,|\, p_2,I) = \gamma(p_1,I) \cup \gamma(p_2,I)$$
$$\gamma(\underline{\alpha}.p,I) = \gamma(p,I) \cup \{\alpha\} \cup \bigcup_{\sigma \in In(p) \wedge \sigma \neq \tau} n(\sigma)$$ □

An interesting property of structurally congruent processes is that they have the same set of names in sequences. Some auxiliary results are needed to prove this. Given a set of labels $A \subseteq \mathscr{A}$ and a substitution $\{b/a\}$, we define

$$A\{b/a\} = \begin{cases} A & \text{if } a,\bar{a} \notin A, \\ A \setminus \{a\} \cup \{b\} & \text{if } a \in A \wedge \bar{a} \notin A, \\ A \setminus \{\bar{a}\} \cup \{\bar{b}\} & \text{if } \bar{a} \in A \wedge a \notin A, \\ A \setminus \{a,\bar{a}\} \cup \{b,\bar{b}\} & \text{if } a,\bar{a} \in A. \end{cases}$$

Exercise 6.9. For any sequential Multi-CCS process p and for any $b \notin fn(p) \cup bn(p)$, $In(p\{b/a\}) = In(p)\{b/a\}$. (*Hint:* By induction on the structure of the sequential process p.) □

Lemma 6.3. *For any finitary Multi-CCS process p and for any $b \notin fn(p) \cup bn(p)$, we have $ns(p\{b/a\}) = ns(p)\{b/a\}$.*

Proof. For simplicity's sake, the proof is given for finite Multi-CCS processes only, i.e., for processes not using any constant. The proof is by induction on the structure of p. We proceed by case analysis.

If $p = \mathbf{0}$, then $\mathbf{0}\{b/a\} = \mathbf{0}$, and the thesis trivially holds.

If $p = \mu.q$, then $(\mu.q)\{b/a\} = \begin{cases} \mu.(q\{b/a\}) & \text{if } \mu \neq a,\bar{a}, \\ b.(q\{b/a\}) & \text{if } \mu = a, \\ \bar{b}.(q\{b/a\}) & \text{if } \mu = \bar{a}. \end{cases}$

Let us consider only the first option (as the other two are similar): $(\mu.q)\{b/a\} = \mu.(q\{b/a\})$. Then, $ns((\mu.q)\{b/a\}) = ns(\mu.(q\{b/a\})) = ns(q\{b/a\})$, as $ns(\mu.p) = ns(p)$ by Definition 6.3. By induction, $ns(q\{b/a\}) = ns(q)\{b/a\}$, and so we have $ns(q)\{b/a\} = ns(\mu.q)\{b/a\}$ by Definition 6.3. The thesis follows by transitivity.

If $p = \underline{\alpha}.p_1$, then $(\underline{\alpha}.p_1)\{b/a\} = \begin{cases} \underline{\alpha}.(p_1\{b/a\}) & \text{if } \alpha \neq a,\bar{a}, \\ \underline{b}.(p_1\{b/a\}) & \text{if } \alpha = a, \\ \underline{\bar{b}}.(p_1\{b/a\}) & \text{if } \alpha = \bar{a}. \end{cases}$

Let us consider the first option only (as the other two are similar): $(\underline{\alpha}.p_1)\{b/a\} = \underline{\alpha}.(p_1\{b/a\})$. Then, $ns((\underline{\alpha}.p_1)\{b/a\}) = ns(\underline{\alpha}.(p_1\{b/a\})) = ns(p_1\{b/a\}) \cup \{\alpha\} \cup \bigcup_{\sigma \in In(p_1\{b/a\}) \wedge \sigma \neq \tau} n(\sigma)$, by Definition 6.3. By induction, we have $ns(p_1\{b/a\}) = ns(p_1)\{b/a\}$. By Exercise 6.9, we have $In(p_1\{b/a\}) = In(p_1)\{b/a\}$. Moreover, $\bigcup_{\sigma \in In(p_1)\{b/a\} \wedge \sigma \neq \tau} n(\sigma) = (\bigcup_{\sigma \in In(p_1) \wedge \sigma \neq \tau} n(\sigma))\{b/a\}$. Therefore, $ns(p_1)\{b/a\} \cup$

$\{\alpha\} \cup (\bigcup_{\sigma \in In(p_1) \wedge \sigma \neq \tau} n(\sigma))\{b/a\} = (ns(p_1) \cup \{\alpha\} \cup \bigcup_{\sigma \in In(p_1) \wedge \sigma \neq \tau} n(\sigma))\{b/a\} = ns(\underline{\alpha}.p_1)\{b/a\}$. *The thesis follows by transitivity.*

If $p = p_1 + p_2$, *then* $(p_1 + p_2)\{b/a\} = p_1\{b/a\} + p_2\{b/a\}$. *Therefore,* $ns((p_1 + p_2)\{b/a\}) = ns(p_1\{b/a\} + p_2\{b/a\}) = ns(p_1\{b/a\}) \cup ns(p_2\{b/a\})$. *By induction, we have* $ns(p_1\{b/a\}) = ns(p_1)\{b/a\}$ *and* $ns(p_2\{b/a\}) = ns(p_2)\{b/a\}$. *Therefore, we can conclude that* $ns(p_1\{b/a\}) \cup ns(p_2\{b/a\}) = ns(p_1)\{b/a\} \cup ns(p_2)\{b/a\} = (ns(p_1) \cup ns(p_2))\{b/a\} = ns(p_1 + p_2)\{b/a\}$. *The thesis follows by transitivity.*

The case when $p = q_1 \mid q_2$ *is as above, hence omitted.*

If $p = (\nu c)q$, *then either* $((\nu c)q)\{b/a\} = (\nu c)q$ *in case* $a = c$; *or* $((\nu c)q)\{b/a\} = (\nu c)(q\{b/a\})$ *in case* $a \neq c$; *note that no other case is possible, since* b *is a new name and so* $b \neq c$. *In the former case,* $ns(((\nu a)q)\{b/a\}) = ns((\nu a)q)$. *As* $a, \bar{a} \notin ns((\nu a)q)$, *it follows that* $ns((\nu a)q) = ns((\nu a)q)\{b/a\}$; *hence, the thesis follows by transitivity. In the latter case,* $ns(((\nu c)q)\{b/a\}) = ns((\nu c)(q\{b/a\}))$. *By Definition 6.3,* $ns((\nu c)(q\{b/a\})) = ns(q\{b/a\}) \setminus \{c, \bar{c}\}$. *By induction,* $ns(q\{b/a\}) = ns(q)\{b/a\}$. *Moreover, since* $a \neq c$ *and* $b \neq c$, *we have that* $ns(q)\{b/a\} \setminus \{c, \bar{c}\} = (ns(q) \setminus \{c, \bar{c}\})\{b/a\}$. *Therefore, by Definition 6.3, we have* $(ns(q) \setminus \{c, \bar{c}\})\{b/a\} = ns((\nu c)q)\{b/a\}$. *Then, the thesis follows by transitivity.* □

Proposition 6.7. *Let* p, q *be two Multi-CCS processes. If* $p \equiv q$, *then* $ns(p) = ns(q)$.

Proof. *By induction on the proof of* $p \equiv q$. *First, we have to prove the thesis for the five axioms of Table 6.5. For axiom* **E1**, *note that* $ns((p \mid q) \mid r) = ns(p \mid (q \mid r)) = ns(p) \cup ns(q) \cup ns(r)$; *similarly, for axiom* **E2**. *For axiom* **E3**, $A \equiv q$ *if* $A \stackrel{def}{=} q$; *then, the thesis follows because* $ns(A) = ns(q)$ *by Definition 6.3. For axiom* **E4**, *we have*

$$\begin{aligned} ns(p_1 \mid (\nu a)p_2) &= ns(p_1) \cup ns((\nu a)p_2) \\ &= ns(p_1) \setminus \{a, \bar{a}\} \cup ns(p_2) \setminus \{a, \bar{a}\} \qquad \text{because } a \notin fn(p_1) \\ &= (ns(p_1) \cup ns(p_2)) \setminus \{a, \bar{a}\} \\ &= ns(p_1 \mid p_2) \setminus \{a, \bar{a}\} \\ &= ns((\nu a)(p_1 \mid p_2)) \end{aligned}$$

For axiom **E5**, *we have*

$$\begin{aligned} ns((\nu a)p) &= ns(p) \setminus \{a, \bar{a}\} \\ &= ns(p)\{b/a\} \setminus \{b, \bar{b}\} \qquad \text{because } b \notin fn(p) \cup bn(p) \\ &= ns(p\{b/a\}) \setminus \{b, \bar{b}\} \qquad \text{by Lemma 6.3} \\ &= ns((\nu b)(p\{b/a\})) \end{aligned}$$

Then, one has to prove the thesis for all the rules of equational deduction. The only nontrivial case is about substitutivity for strong prefixing: $p = \underline{\alpha}.p_1$ *and* $q = \underline{\alpha}.q_1$, *with* $p_1 \equiv q_1$. *By induction, we have that* $ns(p_1) = ns(q_1)$, *and by Exercise 6.7,* $In(p_1) = In(q_1)$ *Therefore,*

$$\begin{aligned} ns(p) &= ns(p_1) \cup \{\alpha\} \cup \textstyle\bigcup_{\sigma \in In(p_1) \wedge \sigma \neq \tau} n(\sigma) \\ &= ns(q_1) \cup \{\alpha\} \cup \textstyle\bigcup_{\sigma \in In(q_1) \wedge \sigma \neq \tau} n(\sigma) \; = \; ns(q). \end{aligned}$$ □

The following proposition clarifies that the label of a transition from p, which involves some sequence of length two or more, is composed of actions in $ns(p)$. This is somehow the generalization of Proposition 4.6, stating that any action performed by a CCS process p is included in the set of its free names, $fn(p)$.

Proposition 6.8. *For any Multi-CCS process p, if $p \xrightarrow{\sigma} p'$ and either $|\sigma| \geq 2$, or $\sigma \neq \tau$ and there exists a transition label σ' in its proof tree with $|\sigma'| \geq 2$, then $n(\sigma) \subseteq ns(p)$.*

Proof. By induction on the proof of $p \xrightarrow{\sigma} p'$. We proceed by case analysis.

If $p = \mu.q$, then $p \xrightarrow{\sigma} p'$ is possible only if $\sigma = \mu$ and $p' = q$, by axiom (Pref). This case is vacuous as the only transition label in the proof tree is μ.

If $p = \underline{\alpha}.p_1$, then $p \xrightarrow{\sigma} p'$ is possible only if $p_1 \xrightarrow{\sigma'} p'$ with $\sigma = \alpha \diamond \sigma'$. If $|\sigma'| \geq 2$, then induction can be applied to conclude that $n(\sigma') \subseteq ns(p_1)$; as $n(\sigma) = \{\alpha\} \cup n(\sigma')$ and $ns(p_1) \cup \{\alpha\} \subseteq ns(p)$, the thesis follows trivially. If $|\sigma'| = 1$, then two further subcases are possible: either $\sigma' = \tau$ or $\sigma' = \beta$; in the former subcase, this is possible only if p_1, being sequential, has performed a prefix τ via (Pref), so that no transition label in the proof tree is longer than one, hence this subcase is vacuous; in the latter subcase, $\sigma = \alpha\beta$ and $n(\sigma) \subseteq ns(p)$, because, by Exercise 6.8, $\beta \in In(p_1)$.

If $p = p_1 + p_2$, then $p \xrightarrow{\sigma} p'$ is possible only if $p_1 \xrightarrow{\sigma} p'$ or $p_2 \xrightarrow{\sigma} p'$. W.l.o.g., assume that $p_1 \xrightarrow{\sigma} p'$; then, if the hypothesis holds for this premise, by induction, we have $n(\sigma) \subseteq ns(p_1)$. Since $ns(p_1) \subseteq ns(p)$, the thesis follows by transitivity.

If $p = q_1 \,|\, q_2$, then $p \xrightarrow{\sigma} p'$ is possible only if $q_1 \xrightarrow{\sigma} q_1'$ and $p' = q_1' \,|\, q_2$; or $q_2 \xrightarrow{\sigma} q_2'$ and $p' = q_1 \,|\, q_2'$; or $q_1 \xrightarrow{\sigma_1} q_1'$, $q_2 \xrightarrow{\sigma_2} q_2'$, $Sync(\sigma_1, \sigma_2, \sigma)$ and $p' = q_1' \,|\, q_2'$. In the first subcase, if the hypothesis holds for this premise, by induction, we have $n(\sigma) \subseteq ns(q_1)$; as $ns(q_1) \subseteq ns(p)$, the thesis follows by transitivity. The second subcase is symmetric, hence omitted. In the third subcase, as by hypothesis $\sigma \neq \tau$, then σ_1 or σ_2 must be of length greater than one. W.l.o.g., assume $|\sigma_1| \geq 2$. Since $Sync(\sigma_1, \sigma_2, \sigma)$, necessarily $n(\sigma) \subseteq n(\sigma_1)$. By induction, $n(\sigma_1) \subseteq ns(q_1)$; as $ns(q_1) \subseteq ns(p)$, the thesis follows by transitivity.

If $p = (\nu a)q$, then $p \xrightarrow{\sigma} p'$ is possible only if $q \xrightarrow{\sigma} q'$, with $a, \bar{a} \notin n(\sigma)$. If the hypothesis holds for the premise, then, by induction, $n(\sigma) \subseteq ns(q)$. As $ns((\nu a)q) = ns(q) \setminus \{a, \bar{a}\}$ and $a, \bar{a} \notin n(\sigma)$, it follows that also $n(\sigma) \subseteq ns((\nu a)q)$, as required.

If the last rule applied in deriving $p \xrightarrow{\sigma} p'$ is (Cong), then the premise is $p \equiv q \xrightarrow{\sigma} q' \equiv p'$. If the hypothesis holds for the premise $q \xrightarrow{\sigma} q'$, then $n(\sigma) \subseteq ns(q)$. By Proposition 6.7, $ns(q) = ns(p)$, and so the thesis follows trivially. □

Exercise 6.10. Prove that for any Multi-CCS process p, if $p \xrightarrow{\sigma} p'$, then $ns(p') \subseteq ns(p)$. (*Hint:* By induction on the proof of the transition.) □

Now we can define the well-formedness predicate for a process p; if a process p is well-formed, we say that $wf(p)$ holds, and $wf(p)$ holds if the auxiliary relation (with the same name, with abuse of notation) $wf(p, \emptyset)$ holds; the auxiliary relation $wf(p, I)$, where the second parameter is a set of constants, is defined as the least relation induced by the axioms and rules of Table 6.7. The assumption that any process uses finitely many constants ensures that the well-formedness predicate is well-defined.

$$\frac{}{wf(\mathbf{0},I)} \qquad \frac{wf(q,I)}{wf(\mu.q,I)} \qquad \frac{wf(p,I) \quad \nexists\beta.\beta\in ns(\underline{\alpha}.p)\wedge\overline{\beta}\in ns(\underline{\alpha}.p)}{wf(\underline{\alpha}.p,I)}$$

$$\frac{wf(p,I)}{wf((\nu a)p,I)} \qquad \frac{A\in I}{wf(A,I)} \qquad \frac{wf(p,I\cup\{A\}) \quad A\stackrel{def}{=}p \quad A\notin I}{wf(A,I)}$$

$$\frac{wf(p_1,I) \quad wf(p_2,I) \quad \nexists\beta.\beta\in ns(p_1)\wedge\overline{\beta}\in ns(p_2)}{wf(p_1\,|\,p_2,I) \quad wf(p_1+p_2,I)}$$

Table 6.7 Well-formedness predicate

Example 6.8. Let us consider three processes $p_1 = \underline{a}.b.\mathbf{0}$, $p_2 = \overline{a}.\mathbf{0}$ and $p_3 = \underline{\overline{b}}.c.\mathbf{0}$. Note that $wf(p_2)$, because $wf(\mathbf{0})$ holds; similarly, $wf(b.\mathbf{0})$; as a consequence, $wf(p_1)$ holds, because $ns(p_1) = \{a,b\}$ does not contain a pair of complementary actions. In the same way, we can prove that $wf(p_3)$ holds, with $ns(p_3) = \{\overline{b},c\}$. We also have that $wf(p_1\,|\,p_2)$, as no action of $ns(p_1)$ occurs complemented in $ns(p_2) = \emptyset$. However, it is not the case that $wf((p_1\,|\,p_2)\,|\,p_3)$, because there exists an action, namely b, such that $b \in ns(p_1\,|\,p_2)$ and $\overline{b} \in ns(p_3)$. □

Exercise 6.11. Consider process $p = (\underline{a}.c.\mathbf{0}\,|\,b.\mathbf{0})\,|\,(\overline{a}.\mathbf{0}\,|\,\underline{\overline{b}}.\overline{c}.\mathbf{0})$ of Example 6.6. Show that p is not well-formed. □

Exercise 6.12. If $wf(p)$, then

(*i*) $\nexists\beta.\beta\in ns(p)\wedge\overline{\beta}\in ns(p)$; and
(*ii*) $wf(p^n)$ for any $n \geq 1$, where $p^1 = p$ and $p^{n+1} = p\,|\,p^n$.

(*Hint:* Item (*i*) can be proved by induction on the proof of $wf(p)$; item (*ii*) follows directly from (*i*).) □

We will prove (see Theorem 6.2) that the well-formedness condition ensures that there is no potential synchronizability between two (not necessarily different) sequences of p.

Remark 6.4. **(Alternative, more restrictive, well-formedness condition)** A simpler, but more restrictive, definition of well-formedness requires that syntactically the strong prefixing operator takes the form $\underline{a}.p$, where a is any input, and, additionally, that atomic sequences can only be composed of inputs, by forcing that the normal prefix after a strong prefix be an input too. Formally, *strong well-formedness* of p, denoted by $swf(p)$, can be easily defined by means of an auxiliary relation $swf(p,I)$, in analogy to Table 6.7, where the two interesting rules are

$$\frac{swf(p,I) \quad In(p) \subseteq \mathcal{L}^+ \cup \{\tau\}}{swf(\underline{a}.p,I)} \qquad \frac{swf(p_1,I) \quad swf(p_2,I)}{swf(p_1 \,|\, p_2,I) \quad swf(p_1 + p_2,I)}$$

In this way, atomic sequences are composed of inputs only, and cannot synchronize. However, strong well-formedness is more restrictive than the one in Table 6.7 and would discard some useful processes, such as those in Section 6.4.2 and Section 6.5.3, which prove the great expressive power of well-formed Multi-CCS. □

Exercise 6.13. Prove that if $swf(p)$, then $wf(p)$. (*Hint:* First prove the auxiliary lemma that if $swf(p)$, then $ns(p) \subseteq \mathcal{L}$. Then, the proof is by induction on the proof of $swf(p)$.) □

We now want to prove that if $wf(p)$ holds and $p \equiv q$, then also $wf(q)$ holds. In order to prove this, we need an auxiliary lemma.

Lemma 6.4. *If $wf(p)$ holds and $b \notin fn(p) \cup bn(p)$, then $wf(p\{b/a\})$ holds.*

Proof. *For simplicity's sake, we restrict ourselves to finite Multi-CCS processes. The proof is by structural induction on p. We proceed by case analysis. As the proof is very similar to that of Lemma 6.3, we provide only a few cases.*

$$\text{If } p = \underline{\alpha}.p_1, \text{ then } (\underline{\alpha}.p_1)\{b/a\} = \begin{cases} \underline{\alpha}.(p_1\{b/a\}) & \text{if } \alpha \neq a, \bar{a}, \\ \underline{b}.(p_1\{b/a\}) & \text{if } \alpha = a, \\ \underline{\bar{b}}.(p_1\{b/a\}) & \text{if } \alpha = \bar{a}. \end{cases}$$

Let us consider the first option only (as the other two are similar): $(\underline{\alpha}.p_1)\{b/a\} = \underline{\alpha}.(p_1\{b/a\})$. Since $wf(\underline{\alpha}.p_1)$, we have $wf(p_1)$ and $\nexists \beta.\beta \in ns(\underline{\alpha}.p_1) \wedge \overline{\beta} \in ns(\underline{\alpha}.p_1)$. By induction, $wf(p_1\{b/a\})$ holds. Moreover, as b is a new name and $\alpha \neq a, \bar{a}$, it follows that $\nexists \beta.\beta \in ns(\underline{\alpha}.p_1)\{b/a\} \wedge \overline{\beta} \in ns(\underline{\alpha}.p_1)\{b/a\}$. By Lemma 6.3, we have that $ns(\underline{\alpha}.p_1)\{b/a\} = ns((\underline{\alpha}.p_1)\{b/a\}) = ns(\underline{\alpha}.(p_1\{b/a\}))$; therefore, we have that $\nexists \beta.\beta \in ns(\underline{\alpha}.(p_1\{b/a\})) \wedge \overline{\beta} \in ns(\underline{\alpha}.(p_1\{b/a\}))$. Hence, $wf(\underline{\alpha}.(p_1\{b/a\}))$ holds, as required.

If $p = p_1 \,|\, p_2$, then $(p_1 \,|\, p_2)\{b/a\} = p_1\{b/a\} \,|\, p_2\{b/a\}$. Since $wf(p_1 \,|\, p_2)$, we have $wf(p_1)$, $wf(p_2)$ and $\nexists \beta.\beta \in ns(p_1) \wedge \overline{\beta} \in ns(p_2)$. As b is a new name, it follows that $\nexists \beta.\beta \in ns(p_1)\{b/a\} \wedge \overline{\beta} \in ns(p_2)\{b/a\}$. By induction, $wf(p_1\{b/a\})$ and $wf(p_2\{b/a\})$ hold. By Lemma 6.3, it follows that $ns(p_1)\{b/a\} = ns(p_1\{b/a\})$ as well as $ns(p_2)\{b/a\} = ns(p_2\{b/a\})$. Hence, we also have that $\nexists \beta.\beta \in ns(p_1\{b/a\}) \wedge \overline{\beta} \in ns(p_2\{b/a\})$, and so $wf(p_1\{b/a\} \,|\, p_2\{b/a\})$ holds, as required. □

Theorem 6.1. *If $wf(p)$ and $p \equiv q$, then $wf(q)$.*

Proof. *By induction on the proof of $p \equiv q$. First, we have to prove the thesis for the five axioms of Table 6.5.*

For axiom **E1**, *$wf(p_1 \,|\, (p_2 \,|\, p_3))$ holds if $wf(p_1)$, $wf(p_2 \,|\, p_3)$ and $\nexists \beta.\beta \in ns(p_1) \wedge \overline{\beta} \in ns(p_2 \,|\, p_3)$. Moreover, $wf(p_2 \,|\, p_3)$ holds if $wf(p_2)$, $wf(p_3)$ and $\nexists \beta.\beta \in ns(p_2) \wedge \overline{\beta} \in ns(p_3)$. Hence, $wf(p_1 \,|\, p_2)$ holds as $wf(p_1)$, $wf(p_2)$ and $\nexists \beta.\beta \in ns(p_1) \wedge \overline{\beta} \in ns(p_2 \,|\, p_3)$, as $ns(p_2) \subseteq ns(p_2 \,|\, p_3)$. Finally, as required, also $wf((p_1 \,|\, p_2) \,|\, p_3)$ holds, because $wf(p_1 \,|\, p_2)$, $wf(p_3)$ and $\nexists \beta.\beta \in ns(p_1 \,|\, p_2) \wedge \overline{\beta} \in ns(p_3)$; the last condition,* *which is equivalent to* $\nexists\beta.\beta \in ns(p_1) \cup ns(p_2) \wedge \overline{\beta} \in ns(p_3)$, *holds because, on the one hand, we have that* $\nexists\beta.\beta \in ns(p_2) \wedge \overline{\beta} \in ns(p_3)$ *and, on the other hand, the condition* $\nexists\beta.\beta \in ns(p_1) \wedge \overline{\beta} \in ns(p_2 \,|\, p_3)$ *ensures that* $\nexists\beta.\beta \in ns(p_1) \wedge \overline{\beta} \in ns(p_3)$. *The cases of axioms* **E2** *and* **E3** *are trivial. For axiom* **E4**, $wf(p \,|\, (\nu a)q)$ *holds only if* $wf(p)$, $wf((\nu a)q)$ *and* $\nexists\beta.\beta \in ns(p) \wedge \overline{\beta} \in ns((\nu a)q)$. *As* $wf((\nu a)q)$ *holds, it follows that* $wf(q)$ *holds, too; moreover,* $ns((\nu a)q) = ns(q) \setminus \{a, \overline{a}\}$. *Therefore, as* $a \notin fn(p)$, *we have that* $\nexists\beta.\beta \in ns(p) \wedge \overline{\beta} \in ns(q)$. *It follows that* $wf(p \,|\, q)$ *holds, and so* $wf((\nu a)(p \,|\, q))$, *as required. For axiom* **E5**, $wf((\nu a)p)$ *holds only if* $wf(p)$ *holds, and so, by Lemma 6.4, also* $wf(p\{b/a\})$ *holds, for any* $b \notin fn(p) \cup bn(p)$. *Hence, also* $wf((\nu b)(p\{b/a\}))$ *holds, as required.*

Then, one has to prove the thesis for all the rules of equational deduction. The only nontrivial case is about substitutivity; we discuss two cases only. For strong prefixing, we assume we have $p = \underline{\alpha}.p_1$ *and* $q = \underline{\alpha}.q_1$, *with* $p_1 \equiv q_1$ *and* $wf(p)$; *hence,* $wf(p_1)$ *and* $\nexists\beta.\beta \in ns(\underline{\alpha}.p_1) \wedge \overline{\beta} \in ns(\underline{\alpha}.p_1)$ *By induction, we have that* $wf(q_1)$; *by Proposition 6.7, we have that* $ns(p_1) = ns(q_1)$, *and, finally, by Exercixe 6.7, also that* $In(p_1) = In(q_1)$. *Therefore, it follows that* $ns(\underline{\alpha}.p_1) = ns(\underline{\alpha}.q_1)$, *and so we have that* $\nexists\beta.\beta \in ns(\underline{\alpha}.q_1) \wedge \overline{\beta} \in ns(\underline{\alpha}.q_1)$. *Hence,* $wf(\underline{\alpha}.q_1)$ *holds, as required. For parallel composition, we assume we have* $p = p_1 \,|\, r$ *and* $q = p_2 \,|\, r$, *with* $p_1 \equiv p_2$ *and* $wf(p)$; *therefore,* $wf(p_1)$ *and* $wf(r)$ *hold and* $\nexists\beta.\beta \in ns(p_1) \wedge \overline{\beta} \in ns(r)$. *By induction, we have that* $wf(p_2)$. *By Proposition 6.7, we have that* $ns(p_1) = ns(p_2)$; *hence, we have that* $\nexists\beta.\beta \in ns(p_2) \wedge \overline{\beta} \in ns(r)$. *Therefore,* $wf(p_2 \,|\, r)$ *holds, as required.* □

Exercise 6.14. Prove that, if $swf(p)$ and $p \equiv q$, then $swf(q)$. □

Proposition 6.9. *Given a well-formed Multi-CCS process* p, *if* $p \xrightarrow{\sigma} p'$, *then* $wf(p')$.

Proof. *By induction on the proof of* $p \xrightarrow{\sigma} p'$. *We proceed by case analysis.*

If $p = \mu.p'$, *then* $p \xrightarrow{\mu} p'$; *the thesis follows because* $wf(\mu.p')$ *holds only if* $wf(p')$.

If $p = \underline{\alpha}.p_1$, *then* $p \xrightarrow{\sigma} p'$ *is possible only if* $p_1 \xrightarrow{\sigma'} p'$ *and* $\sigma = \alpha \diamond \sigma'$. *Since* $wf(p)$, *we also have that* $wf(p_1)$, *so that induction can be applied to conclude that* $wf(p')$.

If $p = p_1 + p_2$, *then* $p \xrightarrow{\sigma} p'$ *is possible only if* $p_1 \xrightarrow{\sigma} p'$ *or* $p_2 \xrightarrow{\sigma} p'$. *W.l.o.g., assume that* $p_1 \xrightarrow{\sigma} p'$. *Since* $wf(p)$, *we also have that* $wf(p_1)$, *so that induction can be applied to conclude that* $wf(p')$.

If $p = q_1 \,|\, q_2$, *then* $p \xrightarrow{\sigma} p'$ *is possible only if* $q_1 \xrightarrow{\sigma} q_1'$ *and* $p' = q_1' \,|\, q_2$; *or* $q_2 \xrightarrow{\sigma} q_2'$ *and* $p' = q_1 \,|\, q_2'$; *or* $q_1 \xrightarrow{\sigma_1} q_1'$, $q_2 \xrightarrow{\sigma_2} q_2'$, $Sync(\sigma_1, \sigma_2, \sigma)$ *and* $p' = q_1' \,|\, q_2'$. *Note that, by definition,* $wf(q_1 \,|\, q_2)$ *holds if* $wf(q_1)$, $wf(q_2)$ *and* $\nexists\beta.\beta \in ns(q_1) \wedge \overline{\beta} \in ns(q_2)$. *In the first case, by induction, we have that* $wf(q_1')$; *moreover, by Exercise 6.10,* $ns(q_1') \subseteq ns(q_1)$, *so that* $\nexists\beta.\beta \in ns(q_1') \wedge \overline{\beta} \in ns(q_2)$. *Hence,* $wf(q_1' \,|\, q_2)$. *The other two cases are similar, hence omitted.*

If $p = (\nu a)q$, *then* $p \xrightarrow{\sigma} p'$ *is possible only if* $q \xrightarrow{\sigma} q'$, *with* $a, \overline{a} \notin n(\sigma)$ *and* $p' = (\nu a)q'$. *Note that, by definition,* $wf((\nu a)q)$ *holds only if* $wf(q)$. *By induction, we have* $wf(q')$; *hence, also* $wf((\nu a)q')$.

If the last rule applied in deriving $p \xrightarrow{\sigma} p'$ *is (Cong), then the premise is* $p \equiv q \xrightarrow{\sigma} q' \equiv p'$. *By Theorem 6.1, we have* $wf(q)$, *so that induction can be applied to conclude that* $wf(q')$; *hence,* $wf(p')$ *by Theorem 6.1.* □

By iterating the result of the proposition above, we get that all the processes reachable from a well-formed process are well-formed.

Exercise 6.15. Prove that if $swf(p)$ and $p \xrightarrow{\sigma} p'$, then $swf(p')$. □

To conclude this section, we state the theorem justifying the well-formedness predicate, as defined in Table 6.7: if a process is well-formed, it is not possible to synchronize two sequences, not even indirectly. As a matter of fact, two sequences cannot be synchronized directly, as relation $Sync(\sigma_1, \sigma_2, \sigma)$ requires that at least one of σ_1 and σ_2 be a single action. However, indirectly, two sequences can be synchronized, as discussed at the beginning of this section and also in Example 6.6.

Theorem 6.2. (Well-formedness implies no synchronization of sequences) *If* $wf(p)$ *and* $p \xrightarrow{\sigma} p'$, *then in the proof of such a transition it is not possible that two sequences are synchronized, not even indirectly.*

Proof. *By induction on the proof of* $p \xrightarrow{\sigma} p'$. *All the cases are trivial — if the thesis holds for the premise transition, then it holds also for the conclusion — except when rule (S-Com) is used. In such a case, assume we have* $p = p_1 \,|\, p_2$, $p_1 \xrightarrow{\sigma_1} p_1'$, $p_2 \xrightarrow{\sigma_2} p_2'$, $Sync(\sigma_1, \sigma_2, \sigma)$ *and* $p' = p_1' \,|\, p_2'$. *Since* $wf(p_1 \,|\, p_2)$, *then also* $wf(p_1)$ *and* $wf(p_2)$. *Hence, by induction, we can assume that the thesis holds for the two premise transitions. So, it remains to prove that, since* $wf(p_1 \,|\, p_2)$ *holds, no indirect synchronization of two sequences is performed in the last proof step of transition* $p_1 \,|\, p_2 \xrightarrow{\sigma} p_1' \,|\, p_2'$. *By definition of Sync, at least one of* σ_1 *and* σ_2 *is a single action. W.l.o.g., assume* $\sigma_2 = \overline{\alpha}$. *We have two cases.*

(a) If no transition label σ_2', *with* $|\sigma_2'| \geq 2$, *occurs in the proof tree of* $p_2 \xrightarrow{\sigma_2} p_2'$, *then such a transition is obtained by no synchronization of a sequence; therefore, also the conclusion* $p \xrightarrow{\sigma} p'$ *does not synchronize two sequences.*

(b) If a transition label σ_2', *with* $|\sigma_2'| \geq 2$ *occurs in the proof tree of* $p_2 \xrightarrow{\sigma_2} p_2'$, *then by Proposition 6.8, we are sure that* $\sigma_2 = \overline{\alpha} \in ns(p_2)$. *Since* $wf(p_1 \,|\, p_2)$ *holds, we know that* $\nexists \beta . \beta \in ns(p_1) \wedge \overline{\beta} \in ns(p_2)$. *Since* $Sync(\sigma_1, \sigma_2, \sigma)$ *and* $\sigma_2 = \overline{\alpha}$, *it is necessary that* $\alpha \in n(\sigma_1)$. *Two subcases are now in order, depending on the possible presence of a label* σ_1', *with* $|\sigma_1'| \geq 2$, *in the proof tree of* $p_1 \xrightarrow{\sigma_1} p_1'$.

(i) *If such a sequence is not present, then* $p_1 \xrightarrow{\alpha} p_1'$ *is obtained by no synchronization of a sequence; therefore, also the conclusion* $p \xrightarrow{\sigma} p'$ *does not synchronize two sequences.*

(ii) *On the contrary, if such a sequence is present, then by Proposition 6.8,* $n(\sigma_1) \subseteq ns(p_1)$, *hence contradicting the well-formedness condition of* $p_1 \,|\, p_2$: *in fact, on the one hand, as action* $\alpha \in n(\sigma_1)$, *also* $\alpha \in ns(p_1)$; *on the other hand, we argued that* $\overline{\alpha} \in ns(p_2)$. *Therefore, as we assumed* $wf(p_1 \,|\, p_2)$, *this second subcase is not possible.*

Summing up, $wf(p_1 \mid p_2)$ ensures that no indirect synchronization of two sequences may take place in the proof of $p_1 \mid p_2 \xrightarrow{\sigma} p_1' \mid p_2'$. □

6.2.3 Some Subclasses of Multi-CCS Processes

As done for CCS in Section 3.4, we may introduce a hierarchy of six subclasses of Multi-CCS. The first subclass, called *finite* Multi-CCS, is obtained by forbidding the use of process constants. Of course, from a practical point of view, its interest is rather small. However, finite Multi-CCS is expressive enough to describe some simple, useful examples for discriminating between behavioral equivalences.

Finite-state Multi-CCS is obtained by restricting the syntax to allow only for action prefixing, strong prefixing, choice and process constants. Formally, it is generated by the abstract syntax

$$p ::= \mathbf{0} \mid \mu.p \mid \underline{\alpha}.p \mid p+p \mid C,$$

where it is assumed that the number of process constants is finite and that each process constant is (normally) guarded. This subclass is interesting as it allows for the modeling of any finite-state transition system labeled with sequences of actions, as discussed in the following remark.

Remark 6.5. **(Guardedness prevents infinitely branching sequential processes)** We have assumed that each process constant in the body of a defining equation occurs inside a normally prefixed subprocess $\mu.q$. This prevents even sequential processes from being infinitely branching. For instance, the constant $A \stackrel{def}{=} \underline{a}.A + b.\mathbf{0}$ is unguarded because the occurrence of A in the body is not normally prefixed. According to the operational rules, A has infinitely many transitions leading to $\mathbf{0}$, each of the form $a^n b$, for $n \geq 0$. So A is not finitely branching.

In fact, under the guardedness assumption, it is possible to prove, by following the proof of Theorem 3.2, that finite-state Multi-CCS defines, up to isomorphism, the set of transition systems whose transitions are labeled on $\mathscr{A}$ with finitely many states and transitions. □

Regular Multi-CCS extends finite-state Multi-CCS in the same way regular CCS extends finite-state CCS. It allows for a limited use of parallel composition and restriction, but its processes still generate finite-state LTSs. This class is the most useful from an applicative point of view, as it allows for the modeling of finite-state systems in a modular way.

In the same way, one can define *Multi-BPP* as the Multi-CCS variant of BPP, as well as *finite-net* Multi-CCS as the Multi-CCS variant of finite-net CCS. Formally, finite-net Multi-CCS processes are generated by the following abstract syntax:

$$
\begin{aligned}
s &::= \mathbf{0} \mid \mu.t \mid \underline{\alpha}.s \mid s+s \\
t &::= s \mid t|t \mid C \\
p &::= t \mid (\nu a)p
\end{aligned}
$$

This class is interesting because in [Gor15] (based on ideas in [GV10]) it is shown that its processes are in close connection with finite P/T Petri nets [Pet81, Rei85, DesRei98]. Indeed, in one direction, a *well-formed*, finite-net Multi-CCS process is given semantics in terms of a finite P/T Petri net. Conversely, for any finite P/T net N, we can find a *well-formed*, finite-net Multi-CCS process p such that the net of p is isomorphic to N. Hence, (well-formed) finite-net Multi-CCS processes are for Petri nets as fundamental as finite-state CCS processes are for LTSs.

Finally, *finitary* Multi-CCS is the calculus where the only syntactical constraint is that, for any $p \in \mathscr{P}_M$, the set *Const(p)* is finite. This is the actual calculus we are using in this chapter.

6.3 Behavioral Semantics

Ordinary bisimulation equivalence, called *interleaving* bisimulation equivalence in this context, enjoys some expected algebraic properties, but unfortunately it is not a congruence for parallel composition. In order to find a suitable compositional semantics for Multi-CCS, we define an alternative operational semantics, where transitions are labeled with a multiset of concurrently executable sequences. Ordinary bisimulation equivalence over this enriched transition system is called *step bisimulation* equivalence, which is proved to be a congruence for the Multi-CCS operators.

6.3.1 Interleaving Semantics

Strong bisimilarity was defined in Definition 2.14 over transition systems whose set of labels is any set A, where $A \subseteq Act = \mathscr{L} \cup \overline{\mathscr{L}} \cup \{\tau\}$. However, it is obviously possible to generalize its definition to take into account any arbitrary set of labels. In our setting, the set of labels is $\mathscr{A} = \{\tau\} \cup (\mathscr{L} \cup \overline{\mathscr{L}})^+$ and the labeled transition system is $TS_M = (\mathscr{P}_M, \mathscr{A}, \longrightarrow)$, where $\longrightarrow \subseteq \mathscr{P}_M \times \mathscr{A} \times \mathscr{P}_M$ is the minimal transition relation generated by the rules listed in Table 6.3.

So, a strong bisimulation over TS_M is a relation $R \subseteq \mathscr{P}_M \times \mathscr{P}_M$ such that if $(q_1, q_2) \in R$ then for all $\sigma \in \mathscr{A}$

- $\forall q_1'$ such that $q_1 \xrightarrow{\sigma} q_1'$, $\exists q_2'$ such that $q_2 \xrightarrow{\sigma} q_2'$ and $(q_1', q_2') \in R$
- $\forall q_2'$ such that $q_2 \xrightarrow{\sigma} q_2'$, $\exists q_1'$ such that $q_1 \xrightarrow{\sigma} q_1'$ and $(q_1', q_2') \in R$.

Two Multi-CCS processes p and q are *interleaving* bisimilar, written $p \sim q$, if there exists a strong bisimulation R over $\mathscr{P}_M$ such that $(p, q) \in R$.

Interleaving bisimulation equivalence $\sim$ enjoys some expected algebraic properties. The first is that two structurally congruent Multi-CCS processes are bisimilar.

Proposition 6.10. *Let $p,q \in \mathscr{P}_M$ be Multi-CCS processes. If $p \equiv q$, then $p \sim q$.*

Proof. It is enough to check that relation $R = \{(p,q) \mid p \equiv q\}$ is a bisimulation. If $(p,q) \in R$ and $p \xrightarrow{\sigma} p'$, then by rule (Cong) also $q \xrightarrow{\sigma} p'$ and $(p',p') \in R$. Symmetrically, if q moves first. □

Note that an obvious consequence of the above proposition is that the following algebraic laws hold for strong bisimilarity $\sim$, for all $p,q,r \in \mathscr{P}_M$:

$$
\begin{array}{lll}
(1) & p\,|\,(q\,|\,r) \sim (p\,|\,q)\,|\,r & \\
(2) & p\,|\,q \sim q\,|\,p & \\
(3) & C \sim p & \text{if } C \stackrel{def}{=} p \\
(4) & (\nu a)(p\,|\,q) \sim p\,|\,(\nu a)q & \text{if } a \notin fn(p) \\
(5) & (\nu a)p \sim (\nu b)(p\{b/a\}) & \text{if } b \notin fn(p) \cup bn(p)
\end{array}
$$

Other expected properties hold for interleaving bisimilarity, as the following propositions show.

Proposition 6.11. *Let $p,q,r \in \mathscr{P}_M$ be processes. Then the following hold:*

$$
\begin{array}{llll}
(6) & (p+q)+r \sim p+(q+r) & (7) & p+q \sim q+p \\
(8) & p+\mathbf{0} \sim p & (9) & p+p \sim p
\end{array}
$$

Proof. The proof is standard and is similar to that of Propositions 4.1. For instance, for property (7) it is enough to prove that $R = \{((p+q),(q+p)) \mid p,q \in \mathscr{P}_M\} \cup \{(p,p) \mid p \in \mathscr{P}_M\}$ is a strong bisimulation. □

Proposition 6.12. *Let $p \in \mathscr{P}_M$ be a Multi-CCS process. Then the following hold:*

$$
\begin{array}{llll}
(10) & p\,|\,\mathbf{0} \sim p & (11) & (\nu a)(\nu b)p \sim (\nu b)(\nu a)p \\
(12) & (\nu a)\mathbf{0} \sim \mathbf{0} & (13) & (\nu a)(\mu.p) \sim \begin{cases} \mathbf{0} & \text{if } \mu = a \text{ or } \mu = \overline{a} \\ \mu.(\nu a)p & \text{otherwise} \end{cases}
\end{array}
$$

Proof. The proof is standard and is similar to those of Propositions 4.2 and 4.4. □

Exercise 6.16. Prove the laws (6)–(13) above, by providing a suitable bisimulation relation for each law. □

Exercise 6.17. Prove that also the scope-enlargement laws for dynamic operators, listed in Proposition 4.12, hold for Multi-CCS. □

Strong prefixing enjoys some algebraic properties listed below.

Proposition 6.13. *Let $p,q \in \mathscr{P}_M$ be sequential processes. The following laws hold:*

$$
\begin{array}{ll}
(i) & \underline{\alpha}.\mathbf{0} \sim \mathbf{0} \\
(ii) & \underline{\alpha}.(p+q) \sim \underline{\alpha}.p + \underline{\alpha}.q \\
(iii) & \underline{\alpha}.\tau.p \sim \alpha.p
\end{array}
$$

Proof. *For (i), it is trivial to see that relation* $R_1 = \{(\underline{\alpha}.\mathbf{0},\mathbf{0}) \mid \alpha \in \mathscr{L} \cup \overline{\mathscr{L}}\}$ *is a bisimulation.*

For (ii), relation $R_2 = \{(\underline{\alpha}.(p+q), \underline{\alpha}.p + \underline{\alpha}.q) \mid \alpha \in \mathscr{L} \cup \overline{\mathscr{L}} \wedge p,q \in \mathscr{P}_M\} \cup \{(p,p) \mid p \in \mathscr{P}_M\}$ *is a bisimulation. Transition* $\underline{\alpha}.(p+q) \xrightarrow{\alpha\diamond\sigma} r$ *is possible, by rule (S-Pref), only if* $p+q \xrightarrow{\sigma} r$*, and so only if either* $p \xrightarrow{\sigma} r$ *(rule (Sum*$_1$*)) or* $q \xrightarrow{\sigma} r$ *(rule (Sum*$_2$*)). Hence, by rule (S-Pref) also transitions* $\underline{\alpha}.p \xrightarrow{\alpha\diamond\sigma} r$ *or* $\underline{\alpha}.q \xrightarrow{\alpha\diamond\sigma} r$ *are derivable and, finally,* $\underline{\alpha}.p + \underline{\alpha}.q \xrightarrow{\alpha\diamond\sigma} r$ *with* $(r,r) \in R_2$*. Symmetrically if* $\underline{\alpha}.p + \underline{\alpha}.q$ *moves first.*

For (iii), $R_3 = \{(\underline{\alpha}.\tau.p, \alpha.p) \mid \alpha \in \mathscr{L} \cup \overline{\mathscr{L}} \wedge p \in \mathscr{P}_M\} \cup \{(p,p) \mid p \in \mathscr{P}_M\}$ *is a bisimulation. The only transition from* $\underline{\alpha}.\tau.p$ *is* $\underline{\alpha}.\tau.p \xrightarrow{\alpha\diamond\sigma} p'$*, which is possible only if* $\tau.p \xrightarrow{\sigma} p'$*; by (Pref), we have that* $p' = p$ *and* $\sigma = \tau$*, so* $\alpha \diamond \tau = \alpha$*. This is matched by the only transition from* $\alpha.p$*,* $\alpha.p \xrightarrow{\alpha} p$*, with* $(p,p) \in R_3$*.* □

Remark 6.6. **(Interleaving law)** By the strong prefixing law above, in particular (ii) (distributivity of stron prefixing w.r.t. the choice operator), any sequential term p is bisimulation equivalent to a sumform $\Sigma_{i=1}^{n} \sigma_i.p_i$, where with the notation $\sigma_i.p_i$ we mean $\underline{\alpha_1}.\underline{\alpha_2}.\ldots.\underline{\alpha_{n-1}}.\alpha_n.p_i$, if $\sigma_i = \alpha_1\alpha_2\ldots\alpha_n$.

Let $p = \Sigma_{i=1}^{n} \sigma_i.p_i$ and $q = \Sigma_{j=1}^{m} \sigma'_j.q_j$. Then, it is easy to observe that

$$p\,|\,q \sim \Sigma_{i=1}^{n} \sigma_i.(p_i\,|\,q) + \Sigma_{j=1}^{m} \sigma'_j.(p\,|\,q_j) + \Sigma_{i,j:Sync(\sigma_i,\sigma'_j,\sigma''_{ij})} \sigma''_{ij}.(p_i\,|\,q_j)$$

generalizing the interleaving law for CCS, described in Proposition 4.3. □

Exercise 6.18. Let $p \in \mathscr{P}_M$ be a sequential process. Prove that the following

$$\text{(iv)}\quad (\nu a)(\underline{\alpha}.p) \sim \begin{cases} \mathbf{0} & \text{if } \alpha = a \text{ or } \alpha = \overline{a} \\ \underline{\alpha}.(\nu a)p & \text{otherwise} \end{cases}$$

$$\text{(v)}\quad \underline{\alpha}.(\nu a)p \sim (\nu b)(\underline{\alpha}.(p\{b/a\})) \ \text{ if } \alpha = a \vee \alpha = \overline{a} \text{ and } b \notin fn(p) \cup bn(p),$$

hold, where we take the liberty of extending the Multi-CCS syntax to allow for the application of strong prefixing to a restricted (sequential) process term, which is not syntactically admissible. □

Interleaving bisimulation equivalence is a congruence for almost all the operators of Multi-CCS, in particular for strong prefixing.

Proposition 6.14. *Given two sequential Multi-CCS processes p and q, if* $p \sim q$*, then the following hold:*

(i) $\underline{\alpha}.p \sim \underline{\alpha}.q$ *for all* $\alpha \in \mathscr{L} \cup \overline{\mathscr{L}}$*,*

(ii) $p + r \sim q + r$ *for all sequential* $r \in \mathscr{P}$*.*

Proof. *Let R be a bisimulation such that* $(p,q) \in R$*. For case (i), It is easy to check that relation* $R = \{(\underline{\alpha}.p, \underline{\alpha}.q)\} \cup R$ *is a bisimulation. Case (ii) is left as an exercise for the reader.* □

Proposition 6.15. *Given two Multi-CCS processes p and q, if $p \sim q$, then the following hold:*
(*i*) $\quad \mu.p \sim \mu.q \quad$ *for all* $\mu \in Act$,
(*ii*) $(\nu a)p \sim (\nu a)q$ *for all* $a \in \mathscr{L}$.

Proof. *The proof is very similar to the one for Theorem 4.1, and so left as an exercise for the reader.* □

Unfortunately, $\sim$ is not a congruence for parallel composition, as the following example shows.

Example 6.9. **(No congruence for parallel composition)** Consider the CCS processes $r = a.a.\mathbf{0}$ and $t = a.\mathbf{0} \,|\, a.\mathbf{0}$. Clearly, r is bisimilar to t, $r \sim t$. However, if we consider the context $\mathscr{C}[-] = - \,|\, \underline{\overline{a}}.\underline{\overline{a}}.c.\mathbf{0}$, we get that $\mathscr{C}[r] \not\sim \mathscr{C}[t]$, because the latter can execute c, i.e., $\mathscr{C}[t] \xrightarrow{c} (\mathbf{0} \,|\, \mathbf{0}) \,|\, \mathbf{0}$, as

$$
\text{(Cong)}\ \dfrac{\text{(S-Com)}\ \dfrac{\text{(Pref)}\ \dfrac{}{a.\mathbf{0} \xrightarrow{a} \mathbf{0}} \qquad \text{(S-Com)}\ \dfrac{\text{(Pref)}\ \dfrac{}{a.\mathbf{0} \xrightarrow{a} \mathbf{0}} \qquad \text{(S-Pref)}\ \dfrac{\text{(S-Pref)}\ \dfrac{\text{(Pref)}\ \dfrac{}{c.\mathbf{0} \xrightarrow{c} \mathbf{0}}}{\underline{\overline{a}}.c.\mathbf{0} \xrightarrow{\overline{a}c} \mathbf{0}}}{\underline{\overline{a}}.\underline{\overline{a}}.c.\mathbf{0} \xrightarrow{\overline{a}\overline{a}c} \mathbf{0}}}{a.\mathbf{0} \,|\, \underline{\overline{a}}.\underline{\overline{a}}.c.\mathbf{0} \xrightarrow{\overline{a}c} \mathbf{0} \,|\, \mathbf{0}}}{(a.\mathbf{0} \,|\, a.\mathbf{0}) \,|\, \underline{\overline{a}}.\underline{\overline{a}}.c.\mathbf{0} \equiv a.\mathbf{0} \,|\, (a.\mathbf{0} \,|\, \underline{\overline{a}}.\underline{\overline{a}}.c.\mathbf{0}) \xrightarrow{c} \mathbf{0} \,|\, (\mathbf{0} \,|\, \mathbf{0}) \equiv (\mathbf{0} \,|\, \mathbf{0}) \,|\, \mathbf{0}}}{(a.\mathbf{0} \,|\, a.\mathbf{0}) \,|\, \underline{\overline{a}}.\underline{\overline{a}}.c.\mathbf{0} \xrightarrow{c} (\mathbf{0} \,|\, \mathbf{0}) \,|\, \mathbf{0}}
$$

while $\mathscr{C}[r]$ cannot. The reason for this difference is that the process $\underline{\overline{a}}.\underline{\overline{a}}.c.\mathbf{0}$ can react with a number of concurrently active components equal to the length of the trace it can perform. Hence, a congruence semantics for the operator of parallel composition needs to distinguish between r and t on the basis of their different degrees of parallelism. In other words, the interleaving semantics is to be replaced by a *truly concurrent* semantics, as illustrated in the following section. □

6.3.2 Step Semantics

Multi-CCS can be equipped with a *step semantics*, i.e., a semantics where each transition is labeled with a finite multiset of sequences that concurrent subprocesses can perform at the same time. Ordinary bisimulation over this kind of richer LTSs is known as *step bisimilarity*; this equivalence was originally introduced over Petri nets in [NT84], while [Mil85] is the first paper presenting a step semantics for a process algebra over LTSs.

The step operational semantics for Multi-CCS is given by the labeled transition system $TS_s = (\mathscr{P}_M, \mathscr{B}, \longrightarrow_s)$, where $\mathscr{P}_M$ is the set of Multi-CCS processes, $\mathscr{B} = \mathscr{M}_{fin}(\mathscr{A})$ — i.e., the set of all the finite multisets over $\mathscr{A}$ — is the set of labels (ranged over by M, possibly indexed), and $\longrightarrow_s \subseteq \mathscr{P}_M \times \mathscr{B} \times \mathscr{P}_M$ is the minimal transition relation generated by the rules listed in Table 6.8.

$$(\text{Pref}^s)\ \frac{}{\mu.p \xrightarrow{\{\mu\}}_s p} \qquad (\text{Con}^s)\ \frac{p \xrightarrow{M}_s p'}{C \xrightarrow{M}_s p'}\ C \stackrel{def}{=} p$$

$$(\text{S-Pref}^s)\ \frac{p \xrightarrow{\{\sigma\}}_s p'}{\underline{\alpha}.p \xrightarrow{\{\alpha\circ\sigma\}}_s p'} \qquad (\text{Res}^s)\ \frac{p \xrightarrow{M}_s p'}{(\nu a)p \xrightarrow{M}_s (\nu a)p'}\ a,\bar{a} \notin n(M)$$

$$(\text{Sum}_1^s)\ \frac{p \xrightarrow{\{\sigma\}}_s p'}{p+q \xrightarrow{\{\sigma\}}_s p'} \qquad (\text{Sum}_2^s)\ \frac{q \xrightarrow{\{\sigma\}}_s q'}{p+q \xrightarrow{\{\sigma\}}_s q'}$$

$$(\text{Par}_1^s)\ \frac{p \xrightarrow{M}_s p'}{p\,|\,q \xrightarrow{M}_s p'\,|\,q} \qquad (\text{Par}_2^s)\ \frac{q \xrightarrow{M}_s q'}{p\,|\,q \xrightarrow{M}_s p\,|\,q'}$$

$$(\text{S-Com}^s)\ \frac{p \xrightarrow{M_1}_s p' \qquad q \xrightarrow{M_2}_s q'}{p\,|\,q \xrightarrow{M}_s p'\,|\,q'}\ \mathit{MSync}(M_1 \oplus M_2, M)$$

Table 6.8 Step operational semantics

$$\frac{}{\mathit{MSync}(M,M)} \qquad \frac{\mathit{Sync}(\sigma_1,\sigma_2,\sigma) \quad \mathit{MSync}(M \oplus \{\sigma\}, M')}{\mathit{MSync}(M \oplus \{\sigma_1,\sigma_2\}, M')}$$

Table 6.9 Step synchronization relation

Axiom (Pref^s) states that $\mu.p$ can perform the singleton $\{\mu\}$, reaching p. Rule (S-Pref^s) assumes that the transition in the premise is sequential, i.e., composed of one single sequence. This is because $\underline{\alpha}.p$ is a sequential process, and so it cannot execute multiple sequences at the same time. Similarly, since the $+$ operator composes only sequential processes, it is assumed in the premise that the label is composed of one single sequence. Rule (Res^s) requires that M contains no occurrences of action a or $\bar{a}$ in any sequence $\sigma \in M$; we denote by $n(M)$ the set $\bigcup_{\sigma \in M} n(\sigma)$.

The highlight of this semantics is rule (S-Com^s): it allows for the generation of multisets as labels, by using an additional auxiliary relation *MSync*, defined in Table 6.9, where $\oplus$ denotes multiset union. The intuition behind the definitions of rule (S-Com^s) and *MSync* is that, whenever two parallel processes p and q perform steps M_1 and M_2, respectively, then we can put all the sequences together — yielding $M_1 \oplus M_2$ — and see if $\mathit{MSync}(M_1 \oplus M_2, \overline{M})$ holds. The resulting multiset $\overline{M}$ may be just $M_1 \oplus M_2$ (hence no synchronization takes place), according to axiom $\mathit{MSync}(M,M)$, or the multiset M' we obtain from the application of the rule: select two sequences σ_1 and σ_2 from $M_1 \oplus M_2$, synchronize them producing σ, then recursively apply *MSync* to $\sigma \oplus (M_1 \oplus M_2) \setminus \{\sigma_1, \sigma_2\}$ to obtain M'. This procedure of synchronizing sequences may go on until pairs of synchronizable sequences can be found, but may also stop at any moment due to the axiom $\mathit{MSync}(M,M)$.

Example 6.10. **(Step semantics: distinguishing parallelism from sequentiality)** Consider the CCS processes $a.\mathbf{0}\,|\,b.\mathbf{0}$ and $a.b.\mathbf{0} + b.a.\mathbf{0}$, discussed in Remark 3.11. The former is a parallel process, while the latter is a sequential process; nonetheless,

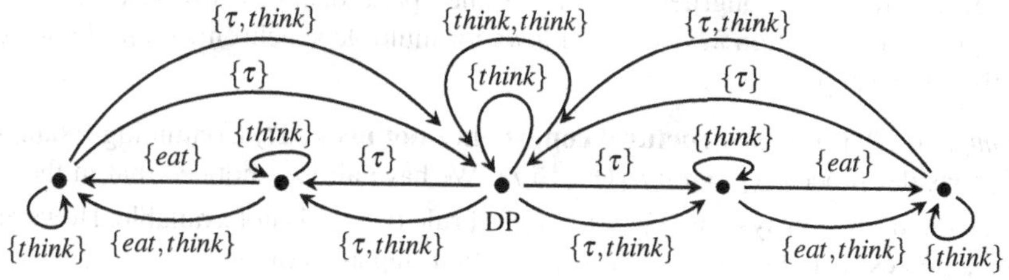

Fig. 6.8 The step labeled transition system for *DP*

they generate two isomorphic interleaving LTSs (see Figure 3.7). We argued that interleaving LTSs (i.e., LTSs with transitions labeled with a single action) are unable to distinguish *parallelism* (or *concurrency*) from *sequentiality*. Here we show that step transition systems can instead! On the one hand, $a.\mathbf{0} \mid b.\mathbf{0}$ can, besides the obvious sequential transitions $a.\mathbf{0} \mid b.\mathbf{0} \xrightarrow{\{a\}}_s \mathbf{0} \mid b.\mathbf{0}$ and $a.\mathbf{0} \mid b.\mathbf{0} \xrightarrow{\{b\}}_s a.\mathbf{0} \mid \mathbf{0}$, also do the parallel transition $a.\mathbf{0} \mid b.\mathbf{0} \xrightarrow{\{a,b\}}_s \mathbf{0} \mid \mathbf{0}$:

$$\text{(S-Com}^s) \dfrac{(\text{Pref}^s)\dfrac{}{a.\mathbf{0} \xrightarrow{\{a\}}_s \mathbf{0}} \qquad (\text{Pref}^s)\dfrac{}{b.\mathbf{0} \xrightarrow{\{b\}}_s \mathbf{0}}}{a.\mathbf{0} \mid b.\mathbf{0} \xrightarrow{\{a,b\}}_s \mathbf{0} \mid \mathbf{0}} \; MSync(\{a,b\},\{a,b\})$$

On the other hand, $a.b.\mathbf{0} + b.a.\mathbf{0}$ cannot do the same parallel step. □

Exercise 6.19. Show that the two processes $r = a.a.\mathbf{0}$ and $t = a.\mathbf{0} \mid a.\mathbf{0}$ of Example 6.9 are not step bisimilar, even if they are interleaving bisimilar. □

A nontrivial example of two step bisimilar processes is given by the two processes $(a.\mathbf{0} \mid b.\mathbf{0}) + a.b.\mathbf{0}$ and $a.\mathbf{0} \mid b.\mathbf{0}$.[4] It is not difficult to check that they are step bisimilar: on the one hand, whatever step is performed by $a.\mathbf{0} \mid b.\mathbf{0}$, the very same step can be performed also by $(a.\mathbf{0} \mid b.\mathbf{0}) + a.b.\mathbf{0}$, as it has $a.\mathbf{0} \mid b.\mathbf{0}$ as one of its two summands; on the other hand, the additional sequential steps due to summand $a.b.\mathbf{0}$ can be easily matched by $a.\mathbf{0} \mid b.\mathbf{0}$, by performing its two actions sequentially. This example shows that a parallel process can be saturated with additional summands expressing some linearizations of its step behavior, without changing its step equivalence class. Therefore, also $(a.\mathbf{0} \mid b.\mathbf{0}) + a.b.\mathbf{0} + b.a.\mathbf{0}$ is step bisimilar to $a.\mathbf{0} \mid b.\mathbf{0}$.

Example 6.11. **(Proving mutual exclusion)** Let us consider the dining philosophers system *DP* of Example 6.3. A proof that *DP* acts correctly, i.e., it never allows both philosophers to eat at the same time, can be given by inspecting its step transition system (see Figure 6.8). As a matter of fact, the step $\{eat, eat\}$ is not present. □

[4] Strictly speaking, the term $(a.\mathbf{0} \mid b.\mathbf{0}) + a.b.\mathbf{0}$ is not legal, as the sum operator is unguarded (see the discussion in Section 3.1.2); however, a completely equivalent term with guarded sum can be provided in Multi-CCS as $(\nu c)((a + \overline{c}) \mid (b + \underline{c}.a.b))$.

It is interesting to observe that the step operational rules in Table 6.8 do not make use of the structural congruence $\equiv$. The same operational effect of rule (Cong) is here ensured by relation *MSync* that allows for multiple synchronization of concurrently active subprocesses.

Example 6.12. **(Why is structural congruence not needed?)** Continuing Example 6.5, consider process $Q = \underline{a}.b.p \mid (\overline{b}.q \mid \overline{a}.r)$. We have already noticed that in the interleaving transition system $Q \not\xrightarrow{\tau} p \mid (q \mid r)$ if rule (Cong) is not available. However, such a process can do a ternary synchronization step as follows:

$$\dfrac{\dfrac{\dfrac{}{b.p \xrightarrow{\{b\}}_s p}}{\underline{a}.b.p \xrightarrow{\{ab\}}_s p} \qquad \dfrac{\dfrac{}{\overline{b}.q \xrightarrow{\{\overline{b}\}}_s q} \quad \dfrac{}{\overline{a}.r \xrightarrow{\{\overline{a}\}}_s r}}{\overline{b}.q \mid \overline{a}.r \xrightarrow{\{\overline{b},\overline{a}\}}_s q \mid r}\ MSync(\{\overline{b},\overline{a}\},\{\overline{b},\overline{a}\})}{\underline{a}.b.p \mid (\overline{b}.q \mid \overline{a}.r) \xrightarrow{\{\tau\}}_s p \mid (q \mid r)}\ MSync(\{ab,\overline{b},\overline{a}\},\{\tau\})$$

where one of the two possible proofs for $MSync(\{ab,\overline{b},\overline{a}\},\{\tau\})$ is

$$\dfrac{\dfrac{}{Sync(ab,\overline{a},b)} \qquad \dfrac{\dfrac{}{Sync(b,\overline{b},\tau)} \quad \dfrac{}{MSync(\{\tau\},\{\tau\})}}{MSync(\{b,\overline{b}\},\{\tau\})}}{MSync(\{ab,\overline{b},\overline{a}\},\{\tau\})}$$

Note that the proof of $MSync(\{ab,\overline{b},\overline{a}\},\{\tau\})$ gives a precise algorithm on how to rearrange the three sequential subprocesses of Q to obtain a process Q' in such a way that no instance of rule (Cong) is needed in deriving the interleaving ternary synchronization; first, the subprocesses originating sequences ab and $\overline{a}$ are to be contiguous: $\underline{a}.b.p \mid \overline{a}.r$ would produce sequence b. Then, we compose this system with the subprocess performing $\overline{b}$, yielding $Q' = (\underline{a}.b.p \mid \overline{a}.r) \mid \overline{b}.q$. Indeed, $Q' \xrightarrow{\tau} (p \mid r) \mid q$ and its proof makes no use of rule (Cong). Note that the proof of $MSync(\{ab,\overline{b},\overline{a}\},\{\tau\})$ is not unique: another one may start by synchronizing first ab and $\overline{b}$, and then also a and $\overline{a}$; the resulting process is $(\underline{a}.b.p \mid \overline{b}.q) \mid \overline{a}.r$, which can execute the interleaving ternary synchronization without using the operational rule (Cong), too. □

Exercise 6.20. Continuing Example 6.6, consider $p = (\underline{a}.c.\mathbf{0} \mid b.\mathbf{0}) \mid (\overline{a}.\mathbf{0} \mid \underline{\overline{b}}.\overline{c}.\mathbf{0})$. Prove that transition $p \xrightarrow{\{\tau\}}_s (\mathbf{0} \mid \mathbf{0}) \mid (\mathbf{0} \mid \mathbf{0})$ is derivable. □

6.3.3 Step Bisimilarity Implies Interleaving Bisimilarity

By *step bisimilarity*, denoted $\sim_{step}$, we mean ordinary bisimulation equivalence over the step transition system of Multi-CCS. This section is devoted to prove that step bisimilarity $\sim_{step}$ is more discriminating than interleaving bisimilarity $\sim$, at least for well-formed processes. The proof is rather technical and needs many auxiliary lemmata. First, we list some properties of relation *MSync*.

Lemma 6.5. (Additivity) *For all multisets $M_1, M_2, N \in \mathscr{B}$, we have $MSync(M_1, M_2)$ if and only if $MSync(M_1 \oplus N, M_2 \oplus N)$.*

Proof. *Take a proof tree for $MSync(M_1, M_2)$, which will end with an axiom of the form $MSync(M_2, M_2)$; then, replace such an axiom with $MSync(M_2 \oplus N, M_2 \oplus N)$ and update accordingly the proof tree; the resulting proof tree proves $MSync(M_1 \oplus N, M_2 \oplus N)$. For instance, the proof tree*

$$\dfrac{\dfrac{}{Sync(\sigma_1,\sigma_2,\sigma')} \quad \dfrac{\dfrac{}{Sync(\sigma',\sigma_3,\sigma)} \quad \dfrac{}{MSync(M_1' \oplus \{\sigma\}, M_1' \oplus \{\sigma\})}}{MSync(M_1' \oplus \{\sigma',\sigma_3\}, M_1' \oplus \{\sigma\})}}{MSync(M_1' \oplus \{\sigma_1,\sigma_2,\sigma_3\}, M_1' \oplus \{\sigma\})}$$

is transformed into the proof tree

$$\dfrac{\dfrac{}{Sync(\sigma_1,\sigma_2,\sigma')} \quad \dfrac{\dfrac{}{Sync(\sigma',\sigma_3,\sigma)} \quad \dfrac{}{MSync(M_1' \oplus \{\sigma\} \oplus N, M_1' \oplus \{\sigma\} \oplus N)}}{MSync(M_1' \oplus \{\sigma',\sigma_3\} \oplus N, M_1' \oplus \{\sigma\} \oplus N)}}{MSync(M_1' \oplus \{\sigma_1,\sigma_2,\sigma_3\} \oplus N, M_1' \oplus \{\sigma\} \oplus N)}$$

Conversely, the proof tree for $MSync(M_1 \oplus N, M_2 \oplus N)$ ends with an axiom $MSync(M_2 \oplus N, M_2 \oplus N)$; by replacing such an axiom with $MSync(M_2, M_2)$ and by updating accordingly the proof tree, we get a proof tree for $MSync(M_1, M_2)$. □

Exercise 6.21. (Additivity 2) Prove that, for all multisets $M_1, M_2, N_1, N_2 \in \mathscr{B}$, if $MSync(M_1, N_1)$ and $MSync(M_2, N_2)$, then $MSync(M_1 \oplus M_2, N_1 \oplus N_2)$.

(*Hint:* The proof tree for $MSync(M_2, N_2)$ can be adapted, as per Lemma 6.5, to a proof tree for $MSync(N_1 \oplus M_2, N_1 \oplus N_2)$; then, as the proof tree for $MSync(M_1, N_1)$ ends with axiom $MSync(N_1, N_1)$, replace such an axiom with the proof tree for $MSync(N_1 \oplus M_2, N_1 \oplus N_2)$ and update accordingly the proof tree of $MSync(M_1, N_1)$ by adding M_2 to the first argument and N_2 to the second argument of all the involved occurrences of $MSync$. The resulting proof tree proves the thesis.) □

Lemma 6.6. (Transitivity) *For all multisets $M_1, M_2, M_3 \in \mathscr{B}$, if $MSync(M_1, M_2)$ and $MSync(M_2, M_3)$, then $MSync(M_1, M_3)$.*

Proof. *The proof tree for $MSync(M_1, M_2)$ has a final axiom $MSync(M_2, M_2)$; replace that axiom with the proof tree for $MSync(M_2, M_3)$, to create a new proof tree where all the occurrence of M_2 (as second argument of MSync) are replaced with M_3 in the original proof tree: this new proof tree proves $MSync(M_1, M_3)$.* □

Lemma 6.7. (Associativity) *For all $M_1, M_2, M_3 \in \mathscr{B}$ such that there exist M', M with $MSync(M_1 \oplus M_2, M')$ and $MSync(M' \oplus M_3, M)$, there exists M'' such that $MSync(M_2 \oplus M_3, M'')$ and $MSync(M_1 \oplus M'', M)$. Also, $MSync(M_1 \oplus M_2 \oplus M_3, M)$.*

Proof. *As $MSync(M_1 \oplus M_2, M')$, by additivity Lemma 6.5 we have $MSync(M_1 \oplus M_2 \oplus M_3, M' \oplus M_3)$. By transitivity Lemma 6.6, we have $MSync(M_1 \oplus M_2 \oplus M_3, M)$. By choosing $M'' = M_2 \oplus M_3$, as $MSync(M_2 \oplus M_3, M_2 \oplus M_3)$, by additivity Lemma 6.5 we have $MSync(M_1 \oplus M_2 \oplus M_3, M_1 \oplus M_2 \oplus M_3)$, and by transitivity Lemma 6.6, we have $MSync(M_1 \oplus M_2 \oplus M_3, M)$ as well.* □

Proposition 6.16. *For all $p,q,r \in \mathscr{P}_M$, if $p\,|\,(q\,|\,r) \xrightarrow{M}_s s$, then there exists t such that $(p\,|\,q)\,|\,r \xrightarrow{M}_s t$, with $s \equiv t$, and vice versa.*

Proof. By induction on the proof of $p\,|\,(q\,|\,r) \xrightarrow{M}_s s$. We have three cases: (i) $p \xrightarrow{M}_s p_1$ and $s = p_1\,|\,(q\,|\,r)$; or (ii) $(q\,|\,r) \xrightarrow{M}_s s_1$ and $s = p\,|\,s_1$; or (iii) $p \xrightarrow{M_1}_s p_1$, $(q\,|\,r) \xrightarrow{M_2}_s s_1$, $MSync(M_1 \oplus M_2, M)$ and $s = p_1\,|\,s_1$.

In the first case, by rule (Par$^s_1$), $p\,|\,q \xrightarrow{M}_s p_1\,|\,q$, and so $(p\,|\,q)\,|\,r \xrightarrow{M}_s (p_1\,|\,q)\,|\,r$, with $p_1\,|\,(q\,|\,r) \equiv (p_1\,|\,q)\,|\,r$, as required.

In the second case, we have three subcases: (a) $q \xrightarrow{M}_s q_1$ and $s_1 = q_1\,|\,r$; or (b) $r \xrightarrow{M}_s r_1$ and $s_1 = q\,|\,r_1$; or (c) $q \xrightarrow{M_1}_s q_1$, $r \xrightarrow{M_2}_s r_1$, $MSync(M_1 \oplus M_2, M)$ and $s_1 = p_1\,|\,r_1$. In the first subcase, by rule (Par$^s_2$), $p\,|\,q \xrightarrow{M}_s p\,|\,q_1$, and so, by rule (Par$^s_1$), $(p\,|\,q)\,|\,r \xrightarrow{M}_s (p\,|\,q_1)\,|\,r$, with $p\,|\,(q_1\,|\,r) \equiv (p\,|\,q_1)\,|\,r$. The second subcase is symmetric, hence omitted. In the third subcase, by (Par$^s_2$), $p\,|\,q \xrightarrow{M_1}_s p\,|\,q_1$, and so, by (S-Com^s), $(p\,|\,q)\,|\,r \xrightarrow{M}_s (p\,|\,q_1)\,|\,r_1$, with $p\,|\,(q_1\,|\,r_1) \equiv (p\,|\,q_1)\,|\,r_1$.

In the third case, we have three subcases: (a) $q \xrightarrow{M_2}_s q_1$ and $s_1 = q_1\,|\,r$; or (b) $r \xrightarrow{M_2}_s r_1$ and $s_1 = q\,|\,r_1$; or (c) $q \xrightarrow{M'_1}_s q_1$, $r \xrightarrow{M'_2}_s r_1$, $MSync(M'_1 \oplus M'_2, M_2)$ and $s_1 = p_1\,|\,r_1$. The first two subcases are similar to the third subcase of the previous case, and so omitted. In the third subcase, by rule (S-Com^s), $p\,|\,q \xrightarrow{M_1 \oplus M'_1}_s p_1\,|\,q_1$, and so $(p\,|\,q)\,|\,r \xrightarrow{M}_s (p_1\,|\,q_1)\,|\,r_1$ — because $MSync(M_1 \oplus M'_1 \oplus M'_2, M)$ by the associativity Lemma 6.7 — with $p_1\,|\,(q_1\,|\,r_1) \equiv (p_1\,|\,q_1)\,|\,r_1$.

The symmetric cases where $(p\,|\,q)\,|\,r \xrightarrow{M}_s t$ moves first are analogous, hence omitted. □

Lemma 6.8. *For all $M, N \in \mathscr{B}$, and for all $\sigma_1, \sigma_2, \sigma \in \mathscr{A}$ such that $Sync(\sigma_1, \sigma_2, \sigma)$, if $MSync(M, N \oplus \{\sigma_1, \sigma_2\})$, then $MSync(M, N \oplus \{\sigma\})$.*

Proof. Since $Sync(\sigma_1, \sigma_2, \sigma)$, we have $MSync(\{\sigma_1, \sigma_2\}, \{\sigma\})$ by a trivial application of the rule in Table 6.9. By the additivity Lemma 6.5, we have $MSync(N \oplus \{\sigma_1, \sigma_2\}, N \oplus \{\sigma\})$. Finally, as $MSync(M, N \oplus \{\sigma_1, \sigma_2\})$, by the transitivity Lemma 6.6, we get the thesis $MSync(M, N \oplus \{\sigma\})$. □

Proposition 6.17. *If $p \xrightarrow{M}_s q$ and $M = M' \oplus \{\sigma_1, \sigma_2\}$ and $Sync(\sigma_1, \sigma_2, \sigma)$, then $p \xrightarrow{N}_s q$, with $N = M' \oplus \{\sigma\}$.*

Proof. By induction on the proof of $p \xrightarrow{M}_s q$. First note that rules for sequential processes are not applicable, as they do not produce steps with more than one element. Then, for the rule (Par$^s_1$), (Par$^s_2$), (Res^s) and (Com^s), the thesis follows by trivial induction; for instance, for the case of rule (Par$^s_1$), if $p\,|\,q \xrightarrow{M}_s p'\,|\,q$ because $p \xrightarrow{M}_s p'$ with $M = M' \oplus \{\sigma_1, \sigma_2\}$ and $Sync(\sigma_1, \sigma_2, \sigma)$, then by induction we have $p \xrightarrow{N}_s p'$, and so $p\,|\,q \xrightarrow{N}_s p'\,|\,q$. The base case of induction is rule

(S-Com^s). Assume $p \xrightarrow{M_1}_s p'$, $q \xrightarrow{M_2}_s q'$, $p \mid q \xrightarrow{M}_s p' \mid q'$, *with* $M = M' \oplus \{\sigma_1, \sigma_2\}$, $Sync(\sigma_1, \sigma_2, \sigma)$ *and* $MSync(M_1 \oplus M_2, M' \oplus \{\sigma_1, \sigma_2\})$. *Then, by Lemma 6.8, also* $MSync(M_1 \oplus M_2, M' \oplus \{\sigma\})$; *hence, by taking* $N = M' \oplus \{\sigma\}$, *rule (S-Com^s) can be applied to derive also* $p \mid q \xrightarrow{N}_s p' \mid q'$. □

The following lemma and proposition assume that (part of) the involved processes are well-formed.

Lemma 6.9. *Given two processes* p *and* q, *such that* $wf(q)$ *and* $a \notin fn(p)$, *if* $p \xrightarrow{M_1}_s p'$, $q \xrightarrow{M_2}_s q'$, $MSync(M_1 \oplus M_2, M)$ *and* $(\nu a)(p \mid q) \xrightarrow{M}_s (\nu a)(p' \mid q')$, *then there exists* N *such that* $(\nu a)q \xrightarrow{N}_s (\nu a)q'$ *and* $MSync(M_1 \oplus N, M)$, *so that transition* $p \mid (\nu a)q \xrightarrow{M}_s p' \mid (\nu a)q'$ *is derivable.*

Proof. *By induction on the cardinality of* M_2. *If* $M_2 = \{\sigma\}$, *then* $a, \overline{a} \notin n(\sigma)$, *because* $a \notin fn(p)$ *and* $a, \overline{a} \notin n(M)$ *by rule (Res^s). Therefore,* $N = M_2 = \{\sigma\}$.

In general, when the cardinality of M_2 *is greater than one, we have two cases:* (*i*) $a, \overline{a} \notin n(M_2)$; *or* (*ii*) $a \in n(M_2)$. *In the former case, take* $N = M_2$. *In the latter case, take the the sequence in* M_2, *say* γ, *with the greatest number of occurrences of* a *or* $\overline{a}$, *say* $k \geq 1$, *where* i *is the number of* a *and* $k - i$ *the number of* $\overline{a}$, *for* $0 \leq i \leq k$. *Hence, in order to let the forbidden actions disappear in* M ($a, \overline{a} \notin n(M)$ *by rule (Res^s)), it is necessary that* M_2 *contains* i *occurrences of action* $\overline{a}$ *and* $k - i$ *occurrences of action* a; *in fact, since* $wf(q)$, *all the occurrences of such* k *actions cannot be produced by a synchronization of a sequence in* q *with some action in* p; *otherwise, there would be two atomic sequences in* M_2 *not respecting the well-formedness condition of* q. *This multiset* $N = \{\gamma, i \times \overline{a}, (k - i) \times a\}$ *is a submultiset of* M_2, *and* $MSync(N, \{\overline{\gamma}\})$, *for some suitable* $\overline{\gamma}$. *Now, let* $M_2' = (M_2 \setminus N) \oplus \{\overline{\gamma}\}$. *Note that if* $q \xrightarrow{M_2}_s q'$, *then also* $q \xrightarrow{M_2'}_s q'$, *by (possibly repeated) application of Proposition 6.17. Moreover,* $MSync(M_1 \oplus M_2', M)$ *holds because the synchronization of all the sequences in* N *is strictly necessary in order to satisfy the condition* $a, \overline{a} \notin n(M)$; *moreover, if* $\overline{\gamma} = \tau$, *then such an action cannot be used for further synchronization, while if* $\overline{\gamma} = \sigma$, *then it can be used for further synchronizations, but the order of these first synchronizations (within* N*) w.r.t. the others is irrelevant by Lemma 6.2. Finally, note that the cardinality of* M_2' *is less than that of* M_2; *hence, induction can be applied: if* $p \xrightarrow{M_1}_s p'$, $q \xrightarrow{M_2'}_s q'$, $MSync(M_1 \oplus M_2', M)$ *and* $(\nu a)(p \mid q) \xrightarrow{M}_s (\nu a)(p' \mid q')$, *then there exists* N *such that* $(\nu a)q \xrightarrow{N}_s (\nu a)q'$, $MSync(M_1 \oplus N, M)$ *and* $p \mid (\nu a)q \xrightarrow{M}_s p' \mid (\nu a)q'$, *as required.* □

In the lemma above, the requirement that q is well-formed is crucial for getting the thesis, as the following example shows.

Example 6.13. Let us consider processes $p = \overline{b}.\mathbf{0}$ and $q = \underline{b}.a.\mathbf{0} \mid \underline{\overline{a}}.c.\mathbf{0}$. Clearly, q is not well-formed because $a \in ns(\underline{b}.a.\mathbf{0})$ and $\overline{a} \in ns(\underline{\overline{a}}.c.\mathbf{0})$. Note that $p \xrightarrow{M_1}_s \mathbf{0}$ with $M_1 = \{\overline{b}\}$, $q \xrightarrow{M_2}_s \mathbf{0} \mid \mathbf{0}$ with $M_2 = \{ba, \overline{a}c\}$, and also $MSync(M_1 \oplus M_2, M)$ with

$M = \{c\}$, because $Sync(\overline{b}, ba, a)$ and $Sync(a, \overline{a}c, c)$; as a consequence, we have that $(\nu a)(p\,|\,q) \xrightarrow{M}_s (\nu a)(\mathbf{0}\,|\,(\mathbf{0}\,|\,\mathbf{0}))$ is derivable. However, there is no multiset N such that $(\nu a)q \xrightarrow{N}_s (\nu a)(\mathbf{0}\,|\,\mathbf{0})$ and $MSync(M_1 \oplus N, M)$. As a matter of fact, the two sequences in M_2 cannot synchronize, and so M_2 cannot be reduced to a simpler multiset N, as done in the proof of Lemma 6.9. □

By the example above, one may be induced to think that the non-existence of a suitable multiset N is due to the fact that the synchronization relation is too restrictive; a more liberal definition of *Sync*, allowing for the synchronization of two sequences, may work: if $Sync(ba, \overline{a}c, bc)$, then the required N can be $\{bc\}$, with $(\nu a)q \xrightarrow{N}_s (\nu a)(\mathbf{0}\,|\,\mathbf{0})$. However, also with this more liberal definition of *Sync*, it is possible to find a counterexample to Lemma 6.9, as the following example shows.

Example 6.14. Let $q = (\underline{a}.b.\mathbf{0}\,|\,\underline{\overline{a}}.c.\mathbf{0})\,|\,\underline{\overline{c}}.\underline{d}.\overline{b}.\mathbf{0}$. If we assume to have a liberal synchronization relation that allows for the synchronization of two sequences, then $(\nu a, b)q$ can only do a step labeled $\{d\}$, by first synchronizing ab with $\overline{a}c$, yielding cb, which is to be synchronized with $\overline{c}d\overline{b}$, yielding d. Let $p = \underline{\overline{c}}.\underline{e}.c.\mathbf{0}$, which can only perform $\{\overline{c}ec\}$. Hence, $p\,|\,(\nu a, b)q$ can perform, as its initial steps, only $\{d\}$ and $\{\overline{c}ec\}$. On the other hand, $(\nu a, b)(p\,|\,q)$ can also perform initially the step $\{ed\}$, by first synchronizing $\overline{c}ec$ with $\overline{c}d\overline{b}$, yielding $\overline{c}ed\overline{b}$, and then synchronizing this with cb — obtained by synchronizing ab with $\overline{a}c$ – yielding ed. □

In conclusion, the validity of Lemma 6.9 strictly depends on the actual definition of the synchronization relation *Sync*: with the rules in Table 6.1, the thesis holds for well-formed processes; on the contrary, it seems that no different definition of *Sync* can work for non-well-formed processes.

Proposition 6.18. *Given two processes p and q, such that $wf(q)$ and $a \notin fn(p)$, if $(\nu a)(p\,|\,q) \xrightarrow{M}_s s$, then there exists t such that $p\,|\,(\nu a)q \xrightarrow{M}_s t$, with $s \equiv t$, and vice versa.*

Proof. *By induction on the proof of $(\nu a)(p\,|\,q) \xrightarrow{M}_s s$. We have three cases: (i) $p \xrightarrow{M}_s p'$ and $s = (\nu a)(p'\,|\,q)$; or (ii) $q \xrightarrow{M}_s q'$, and $s = (\nu a)(p\,|\,q')$; or (iii) $p \xrightarrow{M_1}_s p'$, $q \xrightarrow{M_2}_s q'$, $MSync(M_1 \oplus M_2, M)$ and $s = (\nu a)(p'\,|\,q')$. In the first case, by rule (Par$_1^s$), $p\,|\,(\nu a)q \xrightarrow{M}_s p'\,|\,(\nu a)q$, with $(\nu a)(p'\,|\,q) \equiv p'\,|\,(\nu a)q$. In the second case, by rule (Res^s), $(\nu a)q \xrightarrow{M}_s (\nu a)q'$ and then by (Par$_2^s$), $p\,|\,(\nu a)q \xrightarrow{M}_s p\,|\,(\nu a)q'$, with $(\nu a)(p\,|\,q') \equiv p\,|\,(\nu a)q'$. In the third case, by Lemma 6.9, there exists N such that $(\nu a)q \xrightarrow{N}_s (\nu a)q'$ and $MSync(M_1 \oplus N, M)$; therefore, $p\,|\,(\nu a)q \xrightarrow{M}_s p'\,|\,(\nu a)q'$, with $(\nu a)(p'\,|\,q') \equiv p'\,|\,(\nu a)q'$, as required.*

The symmetric cases where $p\,|\,(\nu a)q \xrightarrow{M}_s t$ moves first are analogous, hence left as an exercise for the reader. □

Theorem 6.3. *Let $p, q \in \mathscr{P}_M$ be well-formed processes such that $p \equiv q$. If $p \xrightarrow{M}_s p'$, then there exists q' such that $q \xrightarrow{M}_s q'$ with $p' \equiv q'$.*

Proof. By induction on the proof of $p \equiv q$. One has to show that for each axiom $p = q$ in Table 6.5, generating the structural congruence $\equiv$, we have the thesis. This is obvious for all five axioms, except for **E1** (associativity) and **E4** (scope enlargement). For these cases, we can resort to Propositions 6.16 and 6.18, respectively. Note that the assumption that the two processes are well-formed is necessary only for the case of axiom **E4**, i.e., for Proposition 6.18. □

Now we want to prove that all the step transitions labeled on a singleton are also interleaving transitions. The proof is first given for the restricted case of pool forms; then, the result is generalized to all well-formed processes.

Lemma 6.10. *For any $p \in \mathscr{P}_M$ in pool form, if $p \xrightarrow{\{\sigma\}}_s q$, then $p \xrightarrow{\sigma} q$.*

Proof. (Sketch) The proof is by induction on the proof of $p \xrightarrow{\{\sigma\}}_s q$. We proceed by case analysis.

If $p = \mu.p'$, then $p \xrightarrow{\{\mu\}}_s p'$ by (Pref^s), and also $p \xrightarrow{\mu} p'$ by (Pref).

If $p = \underline{\alpha}.p'$, then $p \xrightarrow{\{\alpha\diamond\sigma\}}_s q$ is derivable by rule (S-Pref^s) only if $p' \xrightarrow{\{\sigma\}}_s q$. As p' is sequential, hence in pool form by Lemma 6.1, induction can be applied to conclude that $p' \xrightarrow{\sigma} q$; hence, by rule (S-Pref), it follows that $p \xrightarrow{\alpha\diamond\sigma} q$.

If $p = p_1 + p_2$, then $p \xrightarrow{\{\sigma\}}_s q$ is derivable only if either $p_1 \xrightarrow{\{\sigma\}}_s q$ (rule (Sum$^s_1$)) or $p_2 \xrightarrow{\{\sigma\}}_s q$ (rule (Sum$^s_2$)). Hence, by induction (as p_1 and p_2 are sequential, hence in pool form), we have that either $p_1 \xrightarrow{\sigma} q$ or $p_2 \xrightarrow{\sigma} q$. In any case, transition $p \xrightarrow{\sigma} q$ is derivable by (Sum$_1$) or (Sum$_2$).

If $p = p_1 \mid p_2$, then both p_1 and p_2 are pool forms, by Definition 6.1. Transition $p \xrightarrow{\{\sigma\}}_s q$ is derivable only in one of the three cases: $p_1 \xrightarrow{\{\sigma\}}_s p'_1$ and $q = p'_1 \mid p_2$ (rule (Par$^s_1$)); or $p_2 \xrightarrow{\{\sigma\}}_s p'_2$ and $q = p_1 \mid p'_2$ (rule (Par$^s_2$)); or $p_1 \xrightarrow{M_1}_s p'_1$, $p_2 \xrightarrow{M_2}_s p'_2$ and $MSync(M_1 \oplus M_2, \{\sigma\})$ (rule (S-Com^s)). In the first case, by induction, we have $p_1 \xrightarrow{\sigma} p'_1$ and so, by rule (Par$_1$), also $p \xrightarrow{\sigma} q$. The second case is analogous, hence omitted. The third case is the most difficult one. As both p_1 and p_2 are pool forms, for each sequence $\sigma^k_j \in M_k$, there is a subprocess p^k_j of p_k that performs it, for $k = 1, 2$. That is, from $p_1 \xrightarrow{M_1}_s p'_1$ and $p_2 \xrightarrow{M_2}_s p'_2$ we can extract two multisets T_1 and T_2 of transitions of the form $p^k_j \xrightarrow{\{\sigma^k_j\}}_s q^k_j$ that, by induction, have their counterpart of the form $p^k_j \xrightarrow{\sigma^k_j} q^k_j$. These interleaving transitions can be rearranged suitably to build a proof tree for $p' \xrightarrow{\sigma} q'$, with $p \equiv p'$ and $q \equiv q'$, by using the actual proof of relation $MSync(M_1 \oplus M_2, \{\sigma\})$ which tells in what order the parallel subcomponents p^k_j are to be arranged by means of the structural congruence, as illustrated in Example 6.12; then, $p \xrightarrow{\sigma} q$ follows by rule (Cong).

If $p = (\nu a)p'$, transition $p \xrightarrow{\{\sigma\}}_s q$ is derivable only if, by rule (Res^s), $p' \xrightarrow{\{\sigma\}}_s q'$ is derivable, with σ not containing any occurrence of a or $\bar{a}$, and $q = (\nu a)q'$. By *induction (as p' is a pool form, too), we have $p' \xrightarrow{\sigma} q'$, and so by rule (Res), also $p \xrightarrow{\sigma} q$ is derivable.*

If $p = A$, with $A \stackrel{def}{=} r$. This case is empty, as we are assuming p in pool form (hence with prefixed constants), and $p = A$ does not satisfy this requirement. □

Proposition 6.19. *For any well-formed process $p \in \mathscr{P}_M$, if $p \xrightarrow{\{\sigma\}}_s q$, then $p \xrightarrow{\sigma} q$.*

Proof. *Given p, we can extract a pool form p' for it, by Proposition 6.3, such that $p \equiv p'$. By Theorem 6.1, we have $wf(p')$. By Theorem 6.3, if $p \xrightarrow{\{\sigma\}}_s q$, then $p' \xrightarrow{\{\sigma\}}_s q'$ with $q \equiv q'$. By Lemma 6.10, we also have $p' \xrightarrow{\sigma} q'$, and so, by rule (Cong), also $p \xrightarrow{\sigma} q$.* □

In the reverse direction, one can prove the following fact.

Proposition 6.20. *Let $p \in \mathscr{P}_M$ be a well-formed process. If $p \xrightarrow{\sigma} q$, then there exists q' such that $p \xrightarrow{\{\sigma\}}_s q'$ with $q' \equiv q$.*

Proof. *The proof is by induction on the proof of $p \xrightarrow{\sigma} q$. All the cases are trivial, except when rule (Cong) is used*

$$\frac{p \equiv p' \xrightarrow{\sigma} q' \equiv q}{p \xrightarrow{\sigma} q}$$

Note that $wf(p')$ holds, as $wf(p)$ and $p \equiv p'$, by Theorem 6.1. If $p' \xrightarrow{\sigma} q'$, by induction, we can assume that $p' \xrightarrow{\{\sigma\}}_s q''$ with $q'' \equiv q'$. By Theorem 6.3, also transition $p \xrightarrow{\{\sigma\}}_s q'''$ is derivable with $q''' \equiv q''$. So the thesis follows as, by transitivity, also $q''' \equiv q$. □

We cannot prove the stronger result $p \xrightarrow{\{\sigma\}}_s q$, because of the free use of structural congruence; e.g., $\mu.(p|(q|r)) \xrightarrow{\mu} ((p|q)|r)$ (due to (Cong)), while $\mu.(p|(q|r))$ cannot reach $((p|q)|r)$ in the step transition system.

Theorem 6.4. (Step bisimilarity implies interleaving bisimilarity) *Let $p, q \in \mathscr{P}_M$ be well-formed processes. If $p \sim_{step} q$ then $p \sim q$.*

Proof. *Let R be a step bisimulation such that $(p,q) \in R$. Then, it is easy to prove that R is an interleaving bisimulation up to $\sim$. Consider a generic pair $(p,q) \in R$ and assume that $p \xrightarrow{\sigma} p'$. By Proposition 6.20, if $p \xrightarrow{\sigma} p'$, then there exists p'' such that $p \xrightarrow{\{\sigma\}}_s p''$ with $p' \equiv p''$, and so also $p' \sim p''$ by Proposition 6.10. Since $(p,q) \in R$ and R is a step bisimulation, then also $q \xrightarrow{\{\sigma\}}_s q'$ is derivable with $(p'',q') \in R$. By Proposition 6.19, also $q \xrightarrow{\sigma} q'$ is derivable. Summing up, to move $p \xrightarrow{\sigma} p'$, q replies with $q \xrightarrow{\sigma} q'$, so that $p' \sim p'' R q' \sim q'$, as required by the definition of (interleaving) bisimulation up to $\sim$. The case when q moves first is symmetric, hence omitted.* □

Of course, the reverse implication of the theorem above is false; for instance, $a.\mathbf{0} \,|\, b.\mathbf{0} \sim a.b.\mathbf{0} + b.a.\mathbf{0}$ but Example 6.10 shows that the two are not step bisimilar as only the former can perform a step transition labeled $\{a, b\}$.

6.3.4 Properties of the Step Semantics

For step bisimilarity $\sim_{step}$ we have very similar algebraic laws as for interleaving bisimilairty $\sim$: essentially all the laws of Section 6.3.1 hold also for step bisimilarity, except the interleaving law of Remark 6.6. In particular, the following proposition shows that the structural congruence is a step bisimilarity, hence the five laws listed after Proposition 6.10 hold also for it.

Proposition 6.21. *Let $p, q \in \mathscr{P}_M$ be well-formed processes. If $p \equiv q$ then $p \sim_{step} q$.*

Proof. *It is enough to check that relation $R = \{(p, q) \mid p \equiv q\}$ is a step bisimulation. If $(p, q) \in R$ and $p \xrightarrow{M}_s p'$, then by Theorem 6.3 there exists q' such that $q \xrightarrow{M}_s q'$ with $p' \equiv q'$, hence $(p', q') \in R$. Symmetrically, if q moves first.* □

Now we prove that step bisimilarity is a congruence for the Multi-CCS operators.

Proposition 6.22. (Congruence for prefixing, parallel composition and restriction) *Let p and q be Multi-CCS processes. If $p \sim_{step} q$, then*

(i) $\mu.p \sim_{step} \mu.q$, *for all* $\mu \in Act$,
(ii) $p \,|\, r \sim_{step} q \,|\, r$, *for any process* $r \in \mathscr{P}_M$.
(iii) $(\nu a)p \sim_{step} (\nu a)q$, *for all* $a \in \mathscr{L}$.

Proof. *Assume R is a step bisimulation containing the pair (p, q). For case (i), relation $R_1 = R \cup \{(\mu.p, \mu.q)\}$ is a step bisimulation. For case (ii), relation $R_2 = \{(p' \,|\, r', q' \,|\, r') \mid r' \in \mathscr{P}\ (p', q') \in R\}$ is a step bisimulation. For case (iii), relation $R_3 = \{((\nu a)p', (\nu a)q') \mid (p', q') \in R\}$ is a step bisimulation.* □

Proposition 6.23. (Congruence for strong prefixing and choice) *Let p and q be sequential processes. If $p \sim_{step} q$, then*

(i) $\underline{\alpha}.p \sim_{step} \underline{\alpha}.q$, *for all* $\alpha \in \mathscr{L} \cup \overline{\mathscr{L}}$,
(ii) $p + r \sim_{step} q + r$, *for any sequential process* r.

Proof. *Assume R is a step bisimulation containing the pair (p, q).*

Case (i) can be proven by considering relation $R_4 = R \cup \{(\underline{\alpha}.p, \underline{\alpha}.q)\}$. Transition $\underline{\alpha}.p \xrightarrow{\{\alpha \diamond \sigma\}}_s p'$ is derivable, by rule (S-Pref), only if $p \xrightarrow{\{\sigma\}}_s p'$. As $(p, q) \in R$, also $q \xrightarrow{\{\sigma\}}_s q'$ with $(p', q') \in R$. Hence, also $\underline{\alpha}.q \xrightarrow{\{\alpha \diamond \sigma\}}_s q'$ with $(p', q') \in R_3$, as required.

Case (ii) can be proven by showing that $R_5 = \{(p + r, q + r) \mid r \in \mathscr{P}_M\} \cup R \cup \{(r, r) \mid r \in \mathscr{P}_M\}$ is a step bisimulation. □

Summing up, we have that step bisimilarity is a congruence over Multi-CCS processes. This result gives evidence that to give a satisfactory account of Multi-CCS one needs a *non-interleaving* model of concurrency, such as the step transition system. The advantages of the step semantics are essentially:

- a simpler structural operational semantics, which makes no use of the structural congruence $\equiv$, and
- a more adequate behavioral semantics, namely step bisimilarity, which is finer than interleaving bisimilarity over well-formed processes and is a congruence for all the operators of the language.

Proposition 6.22(ii) and Theorem 6.4 ensure that for any pair of well-formed Multi-CCS processes p and q, if $p \sim_{step} q$ then $p \mid r \sim q \mid r$, for any process r. One may wonder if the reverse holds: if for all r such that $wf(p \mid r)$ and $wf(q \mid r)$, we have that $p \mid r \sim q \mid r$, can we conclude that $p \sim_{step} q$? If this is the case, we can say that step equivalence is the *coarsest congruence* contained in interleaving bisimulation for Multi-CCS. The answer to this question is negative, as the following example shows.

Example 6.15. Take processes $p = \tau.\tau.\mathbf{0}$ and $q = \tau.\mathbf{0} \mid \tau.\mathbf{0}$. It is not difficult to see that $p \mid r \sim q \mid r$, for all (well-formed) processes r; however, $p \not\sim_{step} q$ as only q can perform the step $\{\tau, \tau\}$. □

The problem of finding the coarsest congruence contained in $\sim$ is open.

6.4 Case Studies

Here we briefly present three well-known coordination problems in the theory of concurrent computation, which can be solved satisfactorily in well-formed Multi-CCS (but not in CCS).

6.4.1 Concurrent Readers and Writers

According to [Pet81], there are several variants of this problem, originally introduced in [CHP71], but the basic structure is the same. Processes are of two types: reader processes and writer processes. All processes share a common file; so, each writer process must exclude all the other writers and all the readers while writing on the file, while multiple reader processes can access the shared file simultaneously. The problem is to define a control structure that does not deadlock or allow violations of the mutual exclusion criteria.

Assume we have n readers and m writers and that at most $k \leq n$ readers can read simultaneously. We can assume we have k lock resources such that a reader can read if at least one lock is available, while a writer can write if all the k locks are available, so that it prevents all the k possible concurrent reading operations. As for the dining philosophers problem, in a naïve CCS solution to this problem, a deadlock may occur when two writers are competing for the acquisition of the k locks, so that one has acquired i locks and the other one $k-i$, for some $0 < i < k$; in such a situation, both writers are stuck, waiting for the missing locks, and all the readers are not allowed to read as no lock is available. Another problem that may arise in a CCS solution to this problem is that a writer process, while trying to acquire the k locks, may be blocked in the middle because some reader process is so fast in continuously acquiring the lock, reading, releasing the lock, reacquiring the lock, reading again, and so on, that the race for the acquisition of the lock may be won always by the same (reader) process. In such a situation, less than k readers can read simultaneously, as some locks were already taken by the writer process.

A simple Multi-CCS solution to this coordination problem is forcing atomicity on the writer's acquisition of the k locks so that either all or none are taken. To make the presentation simple, assume that $n = 4, k = 3, m = 2$. Each reader process R, each lock process L, each writer W can be represented as follows:

$$R \stackrel{def}{=} lock.read.unlock.R$$
$$L \stackrel{def}{=} \overline{lock}.\overline{unlock}.L$$
$$W \stackrel{def}{=} \underline{lock}.\underline{lock}.lock.write.\underline{unlock}.\underline{unlock}.unlock.W$$

The whole system CRW is defined as

$$CRW \stackrel{def}{=} (\nu lock, unlock)(R\,|\,R\,|\,R\,|\,R\,|\,W\,|\,W\,|\,L\,|\,L\,|\,L),$$

where parentheses are omitted as $|$ is associative. Note that a writer W executes a four-way synchronization with the three instances of the lock process L in order to get permission to write:

$$CRW \xrightarrow{\tau} (\nu lock, unlock)(R\,|\,R\,|\,R\,|\,R\,|\,W'\,|\,W\,|\,L'\,|\,L'\,|\,L'),$$

where $W' = write.\underline{unlock}.\underline{unlock}.unlock.W$ and $L' = \overline{unlock}.L$.

It is easy to see that the LTS for CRW is finite-state: indeed, R, L and W are finite-state Multi-CCS processes and CRW is a regular Multi-CCS process. Note also that, in order to ensure correctness, it is not necessary to require atomicity also on the release of the locks. This choice is only done in order to have a smaller model.

Exercise 6.22. Draw the interleaving LTS for the process CRW. Check that no deadlock is present. □

Exercise 6.23. Draw also the step LTS for CRW. Check that no step $\{read, write\}$, as well as $\{write, write\}$, is present, hence showing that mutual exclusion is ensured. Argue also that the solution ensures maximal concurrency among the reader processes. □

6.4.2 Courteous Dining Philosophers

The (randomized) courteous dining philosophers algorithm, described in [LR81, RL94], can be adapted, by exploiting the multi-party synchronization capability of Multi-CCS, to obtain a symmetric, distributed (to some extent — see Remark 6.2), deterministic, divergence-free solution ensuring *strong non-starvation*: for any computation, each time a philosopher wants to eat, he will eat eventually, that is, no hungry philosopher will starve.

It is necessary to add three variables for each pair of adjacent philosophers P_i and P_{i+1}: one, called $S_{(i,i+1)}$, lets P_i to inform P_{i+1} of his desire to eat (values *on*/*off*), and vice versa, variable $S_{(i+1,i)}$ informs P_i that P_{i+1} wishes to eat (or not); then, variable $K_{(i,i+1)}$ shows which of the two has eaten last (values $i/i+1/neutral$, the last being the dummy initial value). When a philosopher P_i gets hungry, he first declares his desire to eat to his two neighbors by setting variables $S_{(i,i+1)}$ and $S_{(i,i-1)}$ to *on*; then he tries to perform a multiway synchronization with the two forks, provided that the following conditions are satisfied:

- $S_{(i+1,i)}$ is set to *off* or $K_{(i,i+1)}$ is set to $i+1$ or *neutral*; this means that the philosopher P_{i+1} is either not willing to eat or has been served last (or neither has been served yet);
- similarly, $S_{(i-1,i)}$ is set to *off* or $K_{(i-1,i)}$ is set to $i-1$ or *neutral*; this means that the philosopher P_{i-1} is either not willing to eat or has been served last (or neither has been served yet).

After eating, philosopher P_i puts down the forks and updates the variables he shares with his neighbors in the obvious way (no longer hungry and last to eat).

For simplicity's sake, here we present the (value-passing) Multi-CCS solution in the restricted case of two philosophers only. This solution is slightly simpler than the general parametric one for $n \geq 3$ (see Exercise 6.25). The whole system is

$$CDP \stackrel{def}{=} (\nu L)(P_0 \,|\, S_{(0,1)}(\mathit{off}) \,|\, K_{(0,1)}(\mathit{neutral}) \,|\, S_{(1,0)}(\mathit{off}) \,|\, P_1 \,|\, F_0 \,|\, F_1),$$

where L is the set of all the involved actions except $think_i$, $will_i$ (denoting the desire to eat of philosopher P_i) and eat_i, for $i = 0,1$. For $(i,j) = (0,1),(1,0)$, the signal variable $S_{(i,j)}(x)$ is defined as the following value-passing CCS process

$$S_{(i,j)}(x) \stackrel{def}{=} \overline{r_{(i,j)}}(x).S_{(i,j)}(x) + w_{(i,j)}(on).S_{(i,j)}(on) + w_{(i,j)}(\mathit{off}).S_{(i,j)}(\mathit{off}),$$

where action $\overline{r_{(i,j)}}(x)$ stands for the output (read operation) of the value x that $S_{(i,j)}(x)$ currently stores, while actions $w_{(i,j)}(on)$ and $w_{(i,j)}(\mathit{off})$ stand for the input (write operation) of the value *on* or *off*, respectively, to be stored. Observe that, initially in *CDP*, the stored value is *off*, meaning that philosopher P_i is not willing to eat yet. The variable $K_{(0,1)}$ is defined as

$$K_{(0,1)}(x) \stackrel{def}{=} \overline{l_{(0,1)}}(x).K_{(0,1)}(x) + v_{(0,1)}(0).K_{(0,1)}(0) + v_{(0,1)}(1).K_{(0,1)}(1),$$

where action $\overline{l_{(0,1)}}(x)$ stands for the output (read operation) of the value x that $K_{(0,1)}(x)$ currently stores, while actions $v_{(0,1)}(0)$ and $v_{(0,1)}(1)$ stand for the input (write operation) of the value 0 or 1, respectively, to be stored. Observe that, initially in *CDP*, the stored value is *neutral*, meaning that neither philosopher has eaten yet. Note that the value *neutral* cannot be assigned later on.

The two fork processes are defined as usual:

$$F_i \stackrel{def}{=} \overline{up_i}.T_1 \qquad T_i \stackrel{def}{=} \overline{dn_i}.F_i \qquad \text{for } i = 0, 1$$

Finally, the *alternating* philosopher, which alternates the activities of thinking and eating (see Example 6.3), is defined by the process P_i, for $i = 0, 1$, as

$$\begin{aligned}
P_i &\stackrel{def}{=} think_i.P_i' \\
P_i' &\stackrel{def}{=} \underline{\overline{w_{i,i+1}}(on).will_i.P_i''} \\
P_i'' &\stackrel{def}{=} \underline{up_i.up_{i+1}.Q_i} \\
Q_i &\stackrel{def}{=} r_{(i+1,i)}(off).Q_i' + l_{(i,i+1)}(neutral).Q_i' + l_{(i,i+1)}(i+1).Q_i' \\
Q_i' &\stackrel{def}{=} eat_i.P_i''' \\
P_i''' &\stackrel{def}{=} \underline{\overline{w_{(i,i+1)}}(off).\overline{v_{(i,i+1)}}(i).dn_i.dn_{i+1}}.P_i
\end{aligned}$$

where, as usual, $i+1$ is computed modulo 2. P_i first thinks and then declares his intention to eat by setting $S_{(i,i+1)}$ to *on* and, at the same time, by performing the observable action $will_i$. Now process P_i'' tries to execute a long transaction, starting with the atomic acquisition of the two forks and then ending with a safety check: Q_i checks that either $S_{(i+1,i)}$ is set to *off*, or $K_{(i,i+1)}$ is set to *neutral* or to $i+1$; if this four-way synchronization is successful, then P_i'' reaches state Q_i', i.e., the philosopher can eat now, reaching state P_i'''. Finally, P_i''', in one single atomic transaction, sets $S_{(i,i+1)}$ to *off*, also sets $K_{(i,i+1)}$ to i and finally releases the two forks, so that the cycle can be repeated.

A possible computation for *CDP* is

$$\begin{aligned}
CDP \xrightarrow{think_0} \xrightarrow{think_1}\ & (\nu L)(P_0' \,|\, S_{(0,1)}(off) \,|\, K_{(0,1)}(neutral) \,|\, S_{(1,0)}(off) \,|\, P_1' \,|\, F_0 \,|\, F_1) \\
\xrightarrow{will_0} \xrightarrow{will_1}\ & (\nu L)(P_0'' \,|\, S_{(0,1)}(on) \,|\, K_{(0,1)}(neutral) \,|\, S_{(1,0)}(on) \,|\, P_1'' \,|\, F_0 \,|\, F_1) \\
\xrightarrow{\tau}\ & (\nu L)(Q_0' \,|\, S_{(0,1)}(on) \,|\, K_{(0,1)}(neutral) \,|\, S_{(1,0)}(on) \,|\, P_1'' \,|\, T_0 \,|\, T_1) \\
\xrightarrow{eat_0}\ & (\nu L)(P_0''' \,|\, S_{(0,1)}(on) \,|\, K_{(0,1)}(neutral) \,|\, S_{(1,0)}(on) \,|\, P_1'' \,|\, T_0 \,|\, T_1) \\
\xrightarrow{\tau}\ & (\nu L)(P_0 \,|\, S_{(0,1)}(off) \,|\, K_{(0,1)}(0) \,|\, S_{(1,0)}(on) \,|\, P_1'' \,|\, F_0 \,|\, F_1) \\
\xrightarrow{think_0} \xrightarrow{will_0}\ & (\nu L)(P_0'' \,|\, S_{(0,1)}(on) \,|\, K_{(0,1)}(0) \,|\, S_{(1,0)}(on) \,|\, P_1'' \,|\, F_0 \,|\, F_1),
\end{aligned}$$

but now the reached state is such that only P_1'' can execute the fork acquisition transaction; as a matter of fact, P_0'' cannot because $S_{(1,0)}$ is set to *on* (signaling that P_1'' is willing to eat) and $K_{(0,1)}$ is set to 0 (recording that P_0'' was served last). Then, the computation can only proceed as

$$
\begin{aligned}
&(\nu L)(P_0'' \,|\, S_{(0,1)}(on) \,|\, K_{(0,1)}(0) \,|\, S_{(1,0)}(on) \,|\, P_1'' \,|\, F_0 \,|\, F_1) \\
&\quad \xrightarrow{\tau} (\nu L)(P_0'' \,|\, S_{(0,1)}(on) \,|\, K_{(0,1)}(0) \,|\, S_{(1,0)}(on) \,|\, Q_1' \,|\, T_0 \,|\, T_1) \\
&\quad \xrightarrow{eat_1} (\nu L)(P_0'' \,|\, S_{(0,1)}(on) \,|\, K_{(0,1)}(0) \,|\, S_{(1,0)}(on) \,|\, P_1''' \,|\, T_0 \,|\, T_1) \\
&\quad \xrightarrow{\tau} (\nu L)(P_0'' \,|\, S_{(0,1)}(on) \,|\, K_{(0,1)}(1) \,|\, S_{(1,0)}(\mathit{off}) \,|\, P_1 \,|\, F_0 \,|\, F_1)
\end{aligned}
$$

and the reached state will allow P_0'' to get the forks, even in case P_1 will desire to eat, because variable $K_{(0,1)}$ records that P_1'' was served last.

Exercise 6.24. Try to draw the interleaving LTS for *CDP*. Argue that *CDP* ensures strong non-starvation. □

Exercise 6.25. Provide the Multi-CCS specification of this version of the courteous dining philosophers problem for the case of n philosophers, with $n \geq 3$.

(*Hint*: Note that the component that needs major modification is the philosopher P_i, as now he has to interact with many more variables; for instance, the signal variables he has to set to *on* when declaring his willingness to eat are $S_{(i,i+1)}$ and $S_{(i,i-1)}$; similarly, when trying to acquire the two forks, he has to test variables $S_{(i+1,i)}$ and $S_{(i-1,i)}$ as well as $K_{(i,i+1)}$ and $K_{(i-1,i)}$.) □

6.4.3 Cigarette Smokers Problem

This problem, proposed by Patil in [Pat71], is a typical instance of a problem for which no deadlock-free and divergence-free solution seems to exist in CCS. Here is its description, according to [Pet81]. There are four processes: an agent A and three smokers. Each smoker continuously makes a cigarette and smokes it. But to smoke a cigarette, three ingredients are needed: tobacco, paper and matches. One of the smokers has paper, another has tobacco and the third one has matches. The agent A has an infinite supply of all three. The agent A places two of the ingredients on the table. The smokers who has the remaining ingredient can then make and smoke a cigarette, signaling the agent upon completion. The agent then puts out another two of the three ingredients and the cycle repeats. The problem is to define the code for the smoker processes to determine which of the three processes should proceed, avoiding deadlock.

A possible solution to the problem is to have an atomic transaction in which one of the smokers synchronize with the two resources offered by the smoker in that round. The three smokers can be defined as

$$
\begin{aligned}
S_{pap} &\stackrel{def}{=} \underline{tob}.mat.smoke.\overline{end}.S_{pap} \\
S_{tob} &\stackrel{def}{=} \underline{mat}.pap.smoke.\overline{end}.S_{tob} \\
S_{mat} &\stackrel{def}{=} \underline{pap}.tob.smoke.\overline{end}.S_{mat},
\end{aligned}
$$

where actions *pap* stands for paper, *tob* for tobacco and *mat* for matches; the action subscripting S signals the ingredient which that smoker already has.

The agent A decides internally (i.e., by means of an internal choice) which combination of two ingredients to put on the table; the two ingredients are to be treated as resources that are consumed and then reproduced. Agent A is defined as

$$A \stackrel{def}{=} \tau.(\overline{tob}.\mathbf{0} \,|\, \overline{mat}.\mathbf{0} \,|\, end.A) + \tau.(\overline{mat}.\mathbf{0} \,|\, \overline{pap}.\mathbf{0} \,|\, end.A) \\ + \tau.(\overline{pap}.\mathbf{0} \,|\, \overline{tob}.\mathbf{0} \,|\, end.A)$$

and the whole system is

$$Patil \stackrel{def}{=} (\nu tob, pap, mat, end)(A \,|\, S_{tob} \,|\, S_{mat} \,|\, S_{pap}),$$

where, at each round, a multi-party synchronization takes place among one of the three smokers and the two offered ingredients, so that deadlock is prevented. Unfortunately, this solution generates an infinite-state labeled transition system, because agent A is a BPP process. However, note that, by adding the equation $p \,|\, \mathbf{0} = p$ to those defining the structural congruence $\equiv$ and by identifying congruent states, the resulting labeled transition systems would be finite-state. The finite representability of *Patil* behavior is also justified by the following observation: *Patil* is a well-formed, finite-net Multi-CCS process, hence, by the Petri net semantics defined in [Gor15, GV10], the Petri net model for *Patil* is a finite P/T Petri net [DesRei98]; actually, such a net is *safe* (i.e., the number of active subprocesses is always finite), hence with a finite-state underlying (interleaving) labeled transition system.

Patil showed in [Pat71] that no sequence of P and V operations [Dij68] — which are related to the semaphore structure, described in Example 3.5 of Section 3.4.3 — can correctly solve this problem. The reason is that P and V operations offer binary synchronization only, while this problem requires a multi-party synchronization. Following the same argument by Patil, one can give evidence supporting the claim that this problem cannot be solved in CCS.

6.5 Expressiveness

As we have seen, Multi-CCS is rather expressive: so expressive that even some useful operators are actually derivable in terms of strong prefixing. In particular, we show that Multi-CCS^{-c}, where the operator + is removed, is as expressive as Multi-CCS, hence proving that the choice operator is redundant in Multi-CCS. Moreover, some operators that have been shown encodable, to some extent, into CCS in Chapter 5 are now more easily encodable. For instance, CCS^{+r}, the extension to CCS with the relabeling operator, is fully encodable into Multi-CCS. We also show that the CSP parallel operator, which is not encodable into CCS, is actually encodable into Multi-CCS. We conclude this section by presenting one problem in distributed computing that cannot be solved in Multi-CCS, hence emphasizing that Multi-CCS, albeit very expressive, cannot solve all possible problems in concurrency theory.

6.5.1 Choice

Strong prefixing can be exploited to provide an encoding of the choice operator in terms of parallel composition and restriction, so that Multi-CCS turns out to be, to some extent, redundant. The encoding is a formal compilation of a Multi-CCS term p to a term $[\![p]\!]^c$ of the Multi-CCS sublanguage not including the choice operator, called Multi-CCS^{-c}. We will show that this encoding is correct, up to $\sim$, meaning that $p \sim [\![p]\!]^c$.

The encoding of the choice operator is based on an idea originally presented in [NP96]. If we want to encode the process $s = a.p' + \overline{b}.q'$ into a process s' without the choice operator, we can split the choice subprocesses into two parallel processes $a.p'$ and $\overline{b}.q'$, both strongly prefixed by a private action c, over which a third process $\overline{c}.\mathbf{0}$ acts as a semaphore:

$$s' = (\nu c)(\underline{c}.a.p' \,|\, \underline{c}.\overline{b}.q' \,|\, \overline{c}.\mathbf{0})$$

The implementation s' reflects the intuition that the execution of an action from $s = a.p' + \overline{b}.q'$ is divided into two distinct phases, to be executed atomically: first, choose the summand (by synchronizing the strong prefix c with the semaphore $\overline{c}$), and then perform an action from the chosen summand.

If $c \notin fn(p' \,|\, q')$, the observable behavior of s and s' is exactly the same. Indeed, on the one hand, transition $s \xrightarrow{a} p'$ is matched by $s' \xrightarrow{a} (\nu c)(p' \,|\, \underline{c}.\overline{b}.q' \,|\, \mathbf{0})$, where the reached state is bisimilar to p' because $\underline{c}.\overline{b}.q'$ is blocked forever as $c \notin fn(p')$. On the other hand, transition $s \xrightarrow{\overline{b}} q'$ is matched by $s' \xrightarrow{\overline{b}} (\nu c)(\underline{c}.a.p' \,|\, q' \,|\, \mathbf{0})$, where the reached state is bisimilar to q' because $\underline{c}.a.p'$ is blocked forever as $c \notin fn(q')$. In general, the encoding of a generalized sum $\mu_1.p_1 + \mu_2.p_2 + \ldots + \mu_n.p_n$ becomes the process $(\nu c)(\underline{c}.\mu_1.p_1 \,|\, \underline{c}.\mu_2.p_2 \,|\, \cdots \,|\, \underline{c}.\mu_n.p_n \,|\, \overline{c}.\mathbf{0})$.

We can formalize this idea as follows. Let $[\![\cdot]\!]^c$ be the function from Multi-CCS processes to Multi-CCS^{-c} processes defined homomorphically with respect to most operators,

$$\begin{array}{rclrclrcl} [\![\mathbf{0}]\!]^c &=& \mathbf{0} & [\![\mu.q]\!]^c &=& \mu.[\![q]\!]^c & [\![\underline{\alpha}.p]\!]^c &=& \underline{\alpha}.[\![p]\!]^c \\ [\![q_1 \,|\, q_2]\!]^c &=& [\![q_1]\!]^c \,|\, [\![q_2]\!]^c & [\![(\nu a)q]\!]^c &=& (\nu a)[\![q]\!]^c, & & & \end{array}$$

except for constant and choice, for which it is defined as

$$\begin{array}{ll} [\![A]\!]^c = A^c & \text{where } A^c = [\![p]\!]^c \text{ if } A \stackrel{def}{=} p \\ [\![p_1 + p_2]\!]^c = (\nu a)([\![p_1]\!]^c_a \,|\, [\![p_2]\!]^c_a \,|\, \overline{a}.\mathbf{0}) & a \notin fn(p_1 + p_2) \end{array}$$

with the auxiliary encoding $[\![\cdot]\!]^c_a$ defined (only over sequential processes) as

$$\begin{array}{ll} [\![\mathbf{0}]\!]^c_a = \mathbf{0} & [\![\mu.q]\!]^c_a = \underline{a}.[\![\mu.q]\!]^c \\ [\![\underline{\alpha}.p]\!]^c_a = \underline{a}.[\![\underline{\alpha}.p]\!]^c & [\![p_1 + p_2]\!]^c_a = [\![p_1]\!]^c_a \,|\, [\![p_2]\!]^c_a \end{array}$$

E6	$p \mid \mathbf{0} = p$	
E7	$(\nu a)p = \mathbf{0}$	if $p \xrightarrow{\sigma} p'$ implies $a \in n(\sigma) \vee \overline{a} \in n(\sigma)\ \forall \sigma \forall p'$
E8	$(\nu a)(p \mid q) = (\nu a)p \mid (\nu a)q$	if $\overline{a}$ does not occur free in $p \mid q$

Table 6.10 Additional axioms generating the structural congruence $\equiv_1$

Note that the guard $\underline{a}$ is not added to $\mathbf{0}$ because it is useless: no transition is derivable from $\underline{a}.\mathbf{0}$.

Example 6.16. Consider the CCS process $p = (a.\mathbf{0} + (b.d.\mathbf{0} + c.\mathbf{0})) \mid (\overline{b}.\mathbf{0} + \overline{a}.e.\mathbf{0})$. The result of the encoding is

$$\begin{aligned}
[\![p]\!]^c &= [\![a.\mathbf{0} + (b.d.\mathbf{0} + c.\mathbf{0})]\!]^c \mid [\![\overline{b}.\mathbf{0} + \overline{a}.e.\mathbf{0}]\!]^c \\
&= (\nu f)([\![a.\mathbf{0}]\!]^c_f \mid [\![b.d.\mathbf{0} + c.\mathbf{0}]\!]^c_f \mid \overline{f}.\mathbf{0}) \mid (\nu g)([\![\overline{b}.\mathbf{0}]\!]^c_g \mid [\![\overline{a}.e.\mathbf{0}]\!]^c_g \mid \overline{g}.\mathbf{0}) \\
&= (\nu f)([\![a.\mathbf{0}]\!]^c_f \mid [\![b.d.\mathbf{0}]\!]^c_f \mid [\![c.\mathbf{0}]\!]^c_f \mid \overline{f}.\mathbf{0}) \mid (\nu g)([\![\overline{b}.\mathbf{0}]\!]^c_g \mid [\![\overline{a}.e.\mathbf{0}]\!]^c_g \mid \overline{g}.\mathbf{0}) \\
&= (\nu f)(\underline{f}.a.\mathbf{0} \mid \underline{f}.b.d.\mathbf{0} \mid \underline{f}.c.\mathbf{0} \mid \overline{f}.\mathbf{0}) \mid (\nu g)(\underline{g}.\overline{b}.\mathbf{0} \mid \underline{g}.\overline{a}.e.\mathbf{0} \mid \overline{g}.\mathbf{0})
\end{aligned}$$

Transition $p \xrightarrow{\tau} d.\mathbf{0} \mid \mathbf{0}$, due to a synchronization on channel b, is matched by transition $[\![p]\!]^c \xrightarrow{\tau} (\nu f)(\underline{f}.a.\mathbf{0} \mid d.\mathbf{0} \mid \underline{f}.c.\mathbf{0} \mid \mathbf{0}) \mid (\nu g)(\mathbf{0} \mid \underline{g}.\overline{a}.e.\mathbf{0} \mid \mathbf{0})$, where the reached state is bisimilar to $d.\mathbf{0} \mid \mathbf{0}$. □

In the proof that $p \sim [\![p]\!]^c$ we are going to provide, we make use of a generous structural congruence, denoted with $\equiv_1$, which is generated by adding to the axioms **E1–E5** of Table 6.5 also the axioms **E6–E8** of Table 6.10. Note that an instance of axiom **E7** is $(\nu a)\mathbf{0} = \mathbf{0}$, as its side condition is vacuously true when $p = \mathbf{0}$. The side condition of axiom **E8** can be made more general, but for the purposes of this section, this simple requirement is enough. Observe that axioms **E6–E8** are sound w.r.t. bisimilarity $\sim$. As a consequence, the following proposition follows trivially.

Proposition 6.24. *Let $p, q \in \mathscr{P}_M$ be Multi-CCS processes. If $p \equiv_1 q$, then $p \sim q$.*

Proof. It is enough to check that relation $R = \{(p,q) \mid p \equiv_1 q\}$ is a bisimulation. If $(p,q) \in R$ and $p \xrightarrow{\sigma} p'$, then by induction on the proof of $p \equiv_1 q$ one can prove that also $q \xrightarrow{\sigma} q'$ and $(p',q') \in R$. Symmetrically, if q moves first.

The base cases are the eight axioms generating $\equiv_1$. For the five axioms of Table 6.5, we can resort to rule (Cong): if $p \xrightarrow{\sigma} p'$ and $p \equiv q$, then, by (Cong), also $q \xrightarrow{\sigma} q'$, with $p' \equiv q'$, and so also $p' \equiv_1 q'$. For the three axioms in Table 6.10, one has to prove the thesis directly; for instance, if $p \mid \mathbf{0} \xrightarrow{\sigma} p' \mid \mathbf{0}$, then $p \xrightarrow{\sigma} p'$, with $p' \mid \mathbf{0} \equiv_1 p'$; these three cases are trivial. The inductive cases are concerned with the rules of equational deduction; these are omitted; a similar proof is reported in Lemma 6.2. □

The proof technique we adopt to prove that $p \sim [\![p]\!]^c$ is based on the idea of strong bisimulation up to, but since $\sim$ is not a congruence for parallel composition, we replace it with a sound approximation that is a congruence also for parallel composition, namely relation $\equiv_1$.

Definition 6.4. (Strong bisimulation up to $\equiv_1$) A strong bisimulation up to $\equiv_1$ is a relation $R \subseteq Q \times Q$ such that if $(q_1, q_2) \in R$ then for all $\sigma \in \mathscr{A}$

- $\forall q_1'. q_1 \xrightarrow{\sigma} q_1'$, $\exists q_2'$ such that $q_2 \xrightarrow{\sigma} q_2'$ and $q_1' \equiv_1 R \equiv_1 q_2'$
- $\forall q_2'. q_2 \xrightarrow{\sigma} q_2'$, $\exists q_1'$ such that $q_1 \xrightarrow{\sigma} q_1'$ and $q_1' \equiv_1 R \equiv_1 q_2'$. □

Proposition 6.25. *If R is a strong bisimulation up to $\equiv_1$, then $R \subseteq \sim$.*

Proof. It is easy to check that if R is a bisimulation up to $\equiv_1$, then $\equiv_1 R \equiv_1$ is a bisimulation. Hence $\equiv_1 R \equiv_1 \subseteq \sim$ by definition of $\sim$. As the identity relation $\mathscr{I} \subseteq \equiv_1$, we have that relation $R = \mathscr{I} \circ R \circ \mathscr{I} \subseteq \equiv_1 R \equiv_1$, hence $R \subseteq \sim$ by transitivity. □

Lemma 6.11. *For any sequential process p and for any $a \notin fn(p)$, if $[\![p]\!]_a^c \xrightarrow{\sigma}$, then $\sigma = a \diamond \sigma'$ for some σ'.*

Proof. By induction on the structure of p. □

Lemma 6.12. *For any process p of the form $\mu.p'$ or $\underline{\alpha}.p'$, and for any $a \notin fn(p)$, $[\![p]\!]^c \xrightarrow{\sigma} q$ if and only if $[\![p]\!]_a^c \xrightarrow{a\diamond\sigma} q$, with $a, \bar{a}$ not occurring free in q.*

Proof. If $p = \mu.p'$, then $[\![p]\!]^c = \mu.[\![p']\!]^c \xrightarrow{\mu} [\![p']\!]^c$. Similarly, $[\![p]\!]_a^c = \underline{a}.\mu.[\![p']\!]^c \xrightarrow{a\diamond\mu} [\![p']\!]^c$, and the thesis holds.

If $p = \underline{\alpha}.p'$, then $[\![p]\!]^c = \underline{\alpha}.[\![p']\!]^c \xrightarrow{\alpha\diamond\sigma} q$ provided that $[\![p']\!]^c \xrightarrow{\sigma} q$. Similarly, $[\![p]\!]_a^c = \underline{a}.(\underline{\alpha}.[\![p']\!]^c) \xrightarrow{a\diamond(\alpha\diamond\sigma)} q$, and the thesis holds. □

Lemma 6.13. *For any process p of the form $r_1 + r_2$, and for any $a \notin fn(p)$, the following hold:*

(*i*) *if $[\![p]\!]^c \xrightarrow{\sigma} q$ then there exists $\bar{q}$ such that $[\![p]\!]_a^c \xrightarrow{a\diamond\sigma} \bar{q}$ with $(\nu a)\bar{q} \equiv_1 q$, with $a, \bar{a}$ not occurring free in q;*

(*ii*) *if $[\![p]\!]_a^c \xrightarrow{a\diamond\sigma} \bar{q}$, then there exists q such that $[\![p]\!]^c \xrightarrow{\sigma} q$ with $(\nu a)\bar{q} \equiv_1 q$, with $a, \bar{a}$ not occurring free in q.*

Proof. Case (*i*). If $p = r_1 + r_2$, then $[\![p]\!]^c = (\nu a)([\![r_1]\!]_a^c \,|\, [\![r_2]\!]_a^c \,|\, \bar{a}.\mathbf{0})$. A transition $[\![p]\!]^c \xrightarrow{\sigma} q$ is possible only if either $[\![r_1]\!]_a^c \xrightarrow{a\diamond\sigma} q_1$ with $q = (\nu a)(q_1 \,|\, [\![r_2]\!]_a^c \,|\, \mathbf{0})$, or $[\![r_2]\!]_a^c \xrightarrow{a\diamond\sigma} q_2$ with $q = (\nu a)([\![r_1]\!]_a^c \,|\, q_2 \,|\, \mathbf{0})$. Note that $[\![p]\!]_a^c = [\![r_1]\!]_a^c \,|\, [\![r_2]\!]_a^c$. In the former case, $[\![p]\!]_a^c \xrightarrow{a\diamond\sigma} q_1 \,|\, [\![r_2]\!]_a^c = \bar{q}$ and $(\nu a)\bar{q} \equiv_1 q$ by axiom **E6**. Symmetrically, in the latter case when $[\![r_2]\!]_a^c$ moves.

Case (*ii*) is very similar, hence left as an exercise. □

Theorem 6.5. *For every MultiCCS process p, we have that $p \sim [\![p]\!]^c$.*

Proof. Consider relation $R = \{(p, [\![p]\!]^c) \mid p \in \mathscr{P}_M\}$. It is not difficult to prove that R is a strong bisimulation up to $\equiv_1$.

On the one hand, one has to prove that if $p \xrightarrow{\sigma} p'$, then there exists q such that $[\![p]\!]^c \xrightarrow{\sigma} q$ with $p' \, R \, [\![p']\!]^c \equiv_1 q$. This can be proved by induction on the proof of $p \xrightarrow{\sigma} p'$. We proceed by case analysis.

If $p = \mu.p_1$*, then* $p \xrightarrow{\mu} p_1$*. Also* $[\![p]\!]^c = \mu.[\![p_1]\!]^c \xrightarrow{\mu} [\![p_1]\!]^c$ *with* $p_1 \, R \, [\![p_1]\!]^c$.

If $p = \underline{\alpha}.p_1$*, then* $p \xrightarrow{\alpha \diamond \sigma} p'$ *provided that* $p_1 \xrightarrow{\sigma} p'$*. By induction, we have that* $[\![p_1]\!]^c \xrightarrow{\sigma} q$ *such that* $p' \, R \, [\![p']\!]^c \equiv_1 q$*. Also* $[\![p]\!]^c = \underline{\alpha}.[\![p_1]\!]^c \xrightarrow{\alpha \diamond \sigma} q$*, as required.*

If $p = p_1 \,|\, p_2$*, then* $p \xrightarrow{\sigma} p'$ *is possible only if* $p_1 \xrightarrow{\sigma} p'_1$ *and* $p' = p'_1 \,|\, p_2$*; or* $p_2 \xrightarrow{\sigma} p'_2$ *and* $p' = p_1 \,|\, p'_2$*; or* $p_1 \xrightarrow{\sigma_1} p'_1$*,* $p_2 \xrightarrow{\sigma_2} p'_2$*,* $Sync(\sigma_1, \sigma_2, \sigma)$ *and* $p' = p'_1 \,|\, p'_2$*. In the first case, by induction, we have that* $[\![p_1]\!]^c \xrightarrow{\sigma} q_1$ *such that* $p'_1 \, R \, [\![p'_1]\!]^c \equiv_1 q_1$*. Also,* $[\![p_1 \,|\, p_2]\!]^c = [\![p_1]\!]^c \,|\, [\![p_2]\!]^c \xrightarrow{\sigma} q_1 \,|\, [\![p_2]\!]^c$*, with* $p'_1 \,|\, p_2 \, R \, [\![p'_1 \,|\, p_2]\!]^c = [\![p'_1]\!]^c \,|\, [\![p_2]\!]^c \equiv_1 q_1 \,|\, [\![p_2]\!]^c$*, as required. The other two cases are analogous, hence omitted.*

The case $p = (\nu a)p_1$ *is very simple and left as an exercise.*

The only nontrivial case is when $p = p_1 + p_2$*: if* $p \xrightarrow{\sigma} p'$ *because* $p_1 \xrightarrow{\sigma} p'$ *(the symmetric case when* p_2 *moves is analogous, hence omitted), then by induction we can assume that* $[\![p_1]\!]^c \xrightarrow{\sigma} q_1$ *with* $q_1 \equiv_1 [\![p']\!]^c$*. Now we proceed by case analysis on the shape of the sequential process* p_1*. If* $p_1 = \mu.p'_1$ *or* $p_1 = \underline{\mu}.p'_1$*, then by Lemma 6.12, also transition* $[\![p_1]\!]^c_a \xrightarrow{a \diamond \sigma} q_1$ *is derivable. Hence,*

$[\![p]\!]^c = (\nu a)([\![p_1]\!]^c_a \,|\, [\![p_2]\!]^c_a \,|\, \bar{a}.\mathbf{0}) \xrightarrow{\sigma} (\nu a)(q_1 \,|\, [\![p_2]\!]^c_a \,|\, \mathbf{0})$*, where*

$(\nu a)(q_1 \,|\, [\![p_2]\!]^c_a \,|\, \mathbf{0}) \equiv_1 (\nu a)(q_1 \,|\, [\![p_2]\!]^c_a)$ *by axiom* **E6***,*

$(\nu a)(q_1 \,|\, [\![p_2]\!]^c_a) \equiv_1 q_1 \,|\, (\nu a)[\![p_2]\!]^c_a$ *by axiom* **E4** *as* $a \notin fn(q_1)$*, and,*

$q_1 \,|\, (\nu a)[\![p_2]\!]^c_a \equiv_1 q_1$ *by axiom* **E7***, as by Lemma 6.11,* $[\![p_2]\!]^c_a$ *is blocked by the restriction on a; summing up,* $(\nu a)(q_1 \,|\, [\![p_2]\!]^c_a \,|\, \mathbf{0}) \equiv_1 [\![p']\!]^c$ *by transitivity.*

Instead, if $p_1 = r_1 + r_2$*, then by Lemma 6.13(i), also transition* $[\![p_1]\!]^c_a \xrightarrow{a \diamond \sigma} \overline{q_1}$ *such that* $(\nu a)\overline{q_1} \equiv_1 q_1$ *is derivable. Hence,*

$[\![p]\!]^c = (\nu a)([\![p_1]\!]^c_a \,|\, [\![p_2]\!]^c_a \,|\, \bar{a}.\mathbf{0}) \xrightarrow{\sigma} (\nu a)(\overline{q_1} \,|\, [\![p_2]\!]^c_a \,|\, \mathbf{0})$*, where*

$(\nu a)(\overline{q_1} \,|\, [\![p_2]\!]^c_a \,|\, \mathbf{0}) \equiv_1 (\nu a)(\overline{q_1} \,|\, [\![p_2]\!]^c_a)$ *by axiom* **E6***,*

$(\nu a)(\overline{q_1} \,|\, [\![p_2]\!]^c_a) \equiv_1 (\nu a)\overline{q_1} \,|\, (\nu a)[\![p_2]\!]^c_a$ *by axiom* **E8***, as* $\bar{a}$ *does not occur free in* $\overline{q_1}$ *or in* $[\![p_2]\!]^c_a$*,*

$(\nu a)\overline{q_1} \,|\, (\nu a)[\![p_2]\!]^c_a \equiv_1 (\nu a)\overline{q_1} \,|\, \mathbf{0}$ *by axiom* **E7***, as by Lemma 6.11,* $[\![p_2]\!]^c_a$ *is blocked by the restriction on a; and*

$(\nu a)\overline{q_1} \,|\, \mathbf{0} \equiv_1 (\nu a)\overline{q_1}$ *by axiom* **E6***. Summing up,* $(\nu a)(\overline{q_1} \,|\, [\![p_2]\!]^c_a \,|\, \mathbf{0}) \equiv_1 [\![p']\!]^c$ *by transitivity, as* $(\nu a)\overline{q_1} \equiv_1 q_1$ *and* $q_1 \equiv_1 [\![p']\!]^c$*.*

The final case is when $p \xrightarrow{\sigma} p'$ *is due to rule (Cong), i.e., with premise:* $p \equiv q, q \xrightarrow{\sigma} q', q' \equiv p'$*. By induction, we can assume that* $[\![q]\!]^c \xrightarrow{\sigma} r$ *with* $r \equiv_1 [\![q']\!]^c$*. Moreover, it is easy to observe that if* $p \equiv q$*, then also* $[\![p]\!]^c \equiv [\![q]\!]^c$*. Hence, by rule (Cong), also transition* $[\![p]\!]^c \xrightarrow{\sigma} r$ *is derivable. Note also that* $[\![p]\!]^c \equiv [\![q]\!]^c$ *implies* $[\![p]\!]^c \equiv_1 [\![q]\!]^c$*. Summing up, to transition* $p \xrightarrow{\sigma} p'$*,* $[\![p]\!]^c$ *replies with* $[\![p]\!]^c \xrightarrow{\sigma} r$ *such that* $r \equiv_1 [\![q']\!]^c \equiv_1 [\![p']\!]^s$*, as required.*

On the other hand, one has to prove the following: if $[\![p]\!]^c \xrightarrow{\sigma} q$*, then there exists* p' *such that* $p \xrightarrow{\sigma} p'$ *with* $q \equiv_1 [\![p']\!]^c$*. This can be proved by induction on the proof of* $[\![p]\!]^c \xrightarrow{\sigma} q$*, hence by case analysis on the shape of p. Also in this case, the only non trivial case is for* $[\![p_1 + p_2]\!]^c$*, and an argument similar to the above suffices.* □

The theorem above may induce us to conclude that the choice operator is completely useless in Multi-CCS. However, for efficiency reasons, this is not the case.

As a matter of fact, a Multi-CCS process p and its encoding $[\![p]\!]^c$ may have dramatically different associated LTSs. For instance, $A \stackrel{def}{=} b.A + c.A$ is a very simple, finite-state CCS process, while its encoding $[\![A]\!]^c$ is the Multi-CCS process term $A^c \stackrel{def}{=} (\nu d)(\underline{d}.b.A^c \,|\, \underline{d}.c.A^c \,|\, \overline{d}.\mathbf{0})$, whose associated LTS has infinitely many states.

6.5.2 Relabeling

In Section 5.3 we discussed how to extend CCS with the relabeling operator and the stringent conditions under which relabeling can be encoded into CCS. Here we show that CCS, extended with a rather generous definition of relabeling, can be encoded into Multi-CCS, up to $\sim$, with little effort.

A relabeling f is a function $f : \mathscr{L} \to \mathscr{L}$ over the set $\mathscr{L}$ of inputs, which can be extended to all actions in Act as follows: $f(\overline{a}) = \overline{f(a)}$ and $f(\tau) = \tau$. Relabeling f is of *finite domain* if $f(a) \neq a$ for finitely many $a \in \mathscr{L}$ only. In such a case, f can be more explicitly represented as $[b_1/a_1, \dots, b_n/a_n]$ — where we assume that $a_i \neq a_j$ for all $i \neq j$, as well as $a_i \neq b_i$ for all $i = 1, \dots, n$ — meaning that action a_i is relabeled to a different action b_i for $i = 1, \dots, n$, and for any $c \in \mathscr{L}, c \neq a_i$ for all i, we have that $f(c) = c$. A relabeled process $p[f]$ is a process where the actions of p are renamed according to f. Its operational rule is

$$(\text{Rel}) \; \frac{p \xrightarrow{\mu} p'}{p[f] \xrightarrow{f(\mu)} p'[f]}$$

We call CCS^{+r} the extension of CCS obtained by adding the relabeling operator using finite domain relabelings only. Let $[\![\cdot]\!]^r$ be the function from CCS^{+r} processes to Multi-CCS processes, which is homomorphic w.r.t. most operators,

$$\begin{array}{llll} [\![\mathbf{0}]\!]^r = \mathbf{0} & [\![p_1 + p_2]\!]^r = [\![p_1]\!]^r + [\![p_2]\!]^r & [\![\mu.q]\!]^r = \mu.[\![q]\!]^r \\ [\![(\nu a)q]\!]^r = (\nu a)[\![q]\!]^r & [\![q_1 \,|\, q_2]\!]^r = [\![q_1]\!]^r \,|\, [\![q_2]\!]^r, & \end{array}$$

while for constants and the relabeling operator it is defined as

$$\begin{array}{lll} [\![A]\!]^r = A^r & \text{where } A^r = [\![p]\!]^r \text{ if } A \stackrel{def}{=} p \\ [\![p[f]]\!]^r = (\nu L)([\![p]\!]^r \,|\, C_f) & \text{where } C_f \stackrel{def}{=} \Sigma_{i=1}^n \underline{a_i}.\overline{b_i}.C_f + \underline{\overline{a_i}}.b_i.C_f, \end{array}$$

where, if $f = [b_1/a_1, \dots, b_n/a_n]$, then $L = \{a_1, \dots, a_n\}$. As an example, the encoding $[\![p]\!]^r$ for the process term $p = (b.c.\mathbf{0} \,|\, a.\mathbf{0})[c/b]$ is the Multi-CCS process

$$[\![p]\!]^r = (\nu b)((b.c.\mathbf{0} \,|\, a.\mathbf{0}) \,|\, C_{[c/b]}) \qquad \text{where } C_{[c/b]} \stackrel{def}{=} \underline{b}.\overline{c}.C_{[c/b]} + \underline{\overline{b}}.c.C_{[c/b]}.$$

The two processes are clearly bisimilar. For instance, $p \xrightarrow{c} (c.\mathbf{0} \,|\, a.\mathbf{0})[c/b]$ is matched by transition $(\nu b)((b.c.\mathbf{0} \,|\, a.\mathbf{0}) \,|\, C_{[c/b]}) \xrightarrow{c} (\nu b)((c.\mathbf{0} \,|\, a.\mathbf{0}) \,|\, C_{[c/b]})$, where in this latter transition a Multi-CCS synchronization takes place between the subcomponent $b.c.\mathbf{0}$ and $C_{[c/b]}$, with the effect of renaming b to c.

Exercise 6.26. Compute the encoding $[\![A[d/b]]\!]^r$ for process $A[d/b]$, where constant A is defined as $A \stackrel{def}{=} a.(b.\mathbf{0} \,|\, c.A)[b/a]$. □

Now we prove the correctness of the encoding, up to $\sim$.

Proposition 6.26. *For every process q of CCS^{+r}, $q \sim [\![q]\!]^r$.*

Proof. *Consider relation $R = \{(p, [\![p]\!]^r) \mid p \text{ is a } CCS^{+r} \text{ process}\}$. It is not difficult to prove that R is a strong bisimulation. One has to prove that $p \xrightarrow{\mu} p'$ if and only if $[\![p]\!]^r \xrightarrow{\mu} [\![p']\!]^r$. This 'iff' statement can be divided into two implications.*

The former can be proved by induction on the proof of $p \xrightarrow{\mu} p'$. The only nontrivial case is when $p = q[f]$ with $f = [b_1/a_1, \ldots, b_n/a_n]$: if $p \xrightarrow{f(\mu)} p'$ because $q \xrightarrow{\mu} q'$ by rule (Rel) (hence, $p' = q'[f]$), then by induction we can assume that $[\![q]\!]^r \xrightarrow{\mu} [\![q']\!]^r$. We have that $[\![p]\!]^r = (\nu L)([\![q]\!]^r \,|\, C_f)$ where $C_f \stackrel{def}{=} \Sigma_{i=1}^n \underline{a_i}.\overline{b_i}.C_f + \underline{\overline{a_i}}.b_i.C_f$ and $L = \{a_1, \ldots, a_n\}$. If $\mu \notin L$ and $\overline{\mu} \notin L$, then $f(\mu) = \mu$ and so to move $p \xrightarrow{\mu} p'$, $[\![p]\!]^r$ replies with $(\nu L)([\![q]\!]^r \,|\, C_f) \xrightarrow{\mu} (\nu L)([\![q']\!]^r \,|\, C_f) = [\![q'[f]]\!]^r = [\![p']\!]^r$, by application of the Multi-CCS operational rules (S-Res) and (Par$_1$). Instead, if $\mu \in L$ or $\overline{\mu} \in L$, then the relabeling applies: if $\mu = a_i$, then $f(\mu) = b_i$; instead, if $\overline{\mu} = a_i$, then $f(\mu) = \overline{b_i}$. In this case, to move $p \xrightarrow{f(\mu)} p'$, $[\![p]\!]^r$ replies with $(\nu L)([\![q]\!]^r \,|\, C_f) \xrightarrow{f(\mu)} (\nu L)([\![q']\!]^r \,|\, C_f) = [\![q'[f]]\!]^r = [\![p']\!]^r$, by application of the Multi-CCS operational rules (S-Res) and (S-Com).

The latter implication is when $[\![p]\!]^r$ moves first, and an argument similar to the above suffices. □

The relabeling operator can be added also to Multi-CCS with little effort. First of all, as now transitions are labeled with sequences in $\mathscr{A} = (\mathscr{L} \cup \overline{\mathscr{L}})^+ \cup \{\tau\}$, it is necessary to extend a relabeling function f on $\mathscr{A}$, to get $\hat{f}$ defined as follows: $\hat{f}(\mu) = f(\mu)$, $\hat{f}(\alpha\sigma) = f(\alpha)\hat{f}(\sigma)$, that is, the resulting sequence $\hat{f}(\sigma)$ is obtained by applying the relabeling f to each action occurrence in $\sigma \in \mathscr{A}$. Then, the relabeling operator for Multi-CCS can be described by the operational rule below:

$$\text{(S-Rel)} \quad \frac{p \xrightarrow{\sigma} p'}{p[f] \xrightarrow{\hat{f}(\sigma)} p'[f]}$$

We may wonder if this form of relabeling operator is encodable into Multi-CCS. Unfortunately, it seems that a correct encoding, up to $\sim$, does not exist.

6.5.3 CSP Multiway Synchronization

In this section we show how to implement into Multi-CCS the CSP parallel composition operator we have presented in Section 5.6.

$$(\text{Pref})\ \frac{}{\mu.p \xrightarrow{\mu} p} \qquad (\text{Cons})\ \frac{p \xrightarrow{\mu} p'}{C \xrightarrow{\mu} p'}\ C \stackrel{def}{=} p$$

$$(\text{Sum}_1)\ \frac{p \xrightarrow{\mu} p'}{p+q \xrightarrow{\mu} p'} \qquad (\text{Sum}_2)\ \frac{q \xrightarrow{\mu} q'}{p+q \xrightarrow{\mu} q'}$$

$$(\text{H}_1)\ \frac{p \xrightarrow{\mu} p'}{(\iota a)p \xrightarrow{\mu} (\iota a)p'}\ \mu \neq a \qquad (\text{H}_2)\ \frac{p \xrightarrow{a} p'}{(\iota a)p \xrightarrow{\tau} (\iota a)p'}$$

$$(\text{Csp}_1)\ \frac{p \xrightarrow{\mu} p'}{p \parallel_A q \xrightarrow{\mu} p' \parallel_A q}\ \mu \notin A \qquad (\text{Csp}_2)\ \frac{q \xrightarrow{\mu} q'}{p \parallel_A q \xrightarrow{\mu} p \parallel_A q'}\ \mu \notin A$$

$$(\text{Csp}_3)\ \frac{p \xrightarrow{\mu} p' \quad q \xrightarrow{\mu} q'}{p \parallel_A q \xrightarrow{\mu} p' \parallel_A q'}\ \mu \in A$$

Table 6.11 Structural Operational Semantics for TCSP

At an abstract level, in (Theoretical) CSP [Hoa85, Ros98] (also called TCSP) there is no distinction between actions (inputs) and co-actions (outputs), and so we assume that the set of all usable actions is simply $Act = \mathscr{L} \cup \{\tau\}$, ranged over by μ. The abstract syntax of the variant of TCSP we intend to study is

$$p ::= \mathbf{0} \mid \mu.p \mid p+p \mid p \parallel_A p \mid (\iota a)p \mid C,$$

where the synchronization set A is a finite subset of $\mathscr{L}$ (hence A can also be empty), and $(\iota a)p$ is the hiding operator discussed in Section 5.2. The operational semantics of TCSP is outlined in Table 6.11. Note that $p \parallel_\emptyset q$ denotes the pure asynchronous parallel composition of p and q, as no synchronization is allowed.

Example 6.17. **(Dining philosophers problem in CSP)** A symmetric, fully distributed, deadlock-free solution to the dining philosophers problem can be easily provided in CSP, by exploiting its multiway synchronization capability. The five philosophers can be described by the constants P_i as

$$P_i \stackrel{def}{=} think.P_i + up_{(i,i+1)}.eat.dn_{(i,i+1)}.P_i \quad \text{for } i = 0, \ldots, 4$$

where index $i+1$ is computed modulo 5 and the actions are indexed by a pair of numbers in the range $\{0, \ldots, 4\}$: action $up_{(i,i+1)}$ denotes the philosopher's (atomic) action of grabbing the fork of index i together with the fork of index $i+1$. A philosopher can think or can acquire the two forks; in the latter case, he can then eat, and when he has finished eating, he has to put down both forks. Process *Phils* can be defined as

$$Phils \stackrel{def}{=} P_0 \parallel_\emptyset P_1 \parallel_\emptyset P_2 \parallel_\emptyset P_3 \parallel_\emptyset P_4,$$

where no synchronization can take place among the philosophers. The five forks can be defined by means of the constants F_i as

$$F_i \stackrel{def}{=} up_{(i,i+1)}.dn_{(i,i+1)}.F_i + up_{(i-1,i)}.dn_{(i-1,i)}.F_i \quad \text{for } i = 0, \ldots, 4,$$

where indexes $i+1$ and $i-1$ are computed modulo 5. Action $up_{(i,i+1)}$ denotes availability of the fork of index i, together with the fork of index $i+1$, by philosopher of index i; analogously, action $up_{(i-1,i)}$ denotes the availability of the fork of index i, together with the fork of index $i-1$, by philosopher of index $i-1$. This means that forks F_i and F_{i+1} shares two similar summands: $up_{(i,i+1)}.dn_{(i,i+1)}.F_i$ and $up_{(i,i+1)}.dn_{(i,i+1)}.F_{i+1}$, respectively, onto which they must synchronize. As CSP parallel composition is associative when the synchronization set is fixed, we can define process *Forks* as follows:

$$Forks \stackrel{def}{=} F_0 \parallel_A F_1 \parallel_A F_2 \parallel_A F_3 \parallel_A F_4$$

where $A = \{up_{(i,i+1)}, dn_{(i,i+1)} \mid i = 0, \ldots 4\}$. Therefore, fork F_0 must synchronize either with fork F_1 (on actions $up_{(0,1)}, dn_{(0,1)}$) or with fork F_4 (on actions $up_{(4,0)}, dn_{(4,0)}$). The whole system composing the forks and the philosophers is

$$DP_{CSP} \stackrel{def}{=} (\iota A)(Phils \parallel_A Forks),$$

where the hiding operator internalizes all the ternary synchronizations occurring between each philosopher and his two forks. □

Exercise 6.27. Draw the LTS for the system DP_{CSP} when the philosophers and the forks are only two. Compare it with the LTS in Figure 6.6 for the system DP of Example 6.3. □

We complete the presentation of finitary[5] TCSP by defining the set of names $n(p)$ for a process p and syntactic substitution.

Definition 6.5. (Names and syntactic substitution) The *names* of a finitary process p, denoted $n(p)$, are defined as the set $N(p, \emptyset)$, where $N(p, I)$, with I a set of process constants, is defined as follows:

$$\begin{aligned}
N(\mathbf{0}, I) &= \emptyset \\
N(a.p, I) &= N(p, I) \cup \{a\} \\
N(\tau.p) &= N(p, I) \\
N(p+q, I) &= N(p, I) \cup N(q, I) \\
N(p \parallel_A q, I) &= N(p, I) \cup N(q, I) \cup A \\
N((\iota a)p, I) &= N(p, I) \cup \{a\}
\end{aligned}$$

[5] We assume that any TCSP process uses only a finite number of constants.

$$N(C,I) = \begin{cases} N(q, I \cup \{C\}) & \text{if } C \stackrel{def}{=} q \text{ and } C \notin I \\ \emptyset & \text{if } C \in I \end{cases}$$

Syntactic substitution was defined for CCS in Section 4.1.2. We can adapt that definition to the case of TCSP as follows, by using parametrized constants:

$$\begin{aligned}
\mathbf{0}\{b/a\} &= \mathbf{0} \\
(a.p)\{b/a\} &= b.(p\{b/a\}) \\
(\mu.p)\{b/a\} &= \mu.(p\{b/a\}) \qquad \text{if } \mu \neq a \\
(p+q)\{b/a\} &= p\{b/a\} + q\{b/a\} \\
(p \parallel_A q)\{b/a\} &= p\{b/a\} \parallel_{A'} q\{b/a\} \quad A' = \begin{cases} A & \text{if } a \notin A \\ (A \setminus \{a\}) \cup \{b\} & \text{otherwise} \end{cases} \\
((\iota c)p)\{b/a\} &= (\iota c)(p\{b/a\}) \qquad \text{if } c \neq a, b \\
((\iota a)p)\{b/a\} &= (\iota a)p \\
((\iota b)p)\{b/a\} &= \begin{cases} (\iota b)p & \text{if } a \notin n(p) \\ (\iota c)((p\{c/b\})\{b/a\}) & \text{otherwise, with } c \notin n(p) \end{cases} \\
C_\theta\{b/a\} &= \begin{cases} C_\theta & \text{if } a \notin n(C_\theta) \\ C_{\theta \circ \{b/a\}} & \text{otherwise, with } C_{\theta \circ \{b/a\}} \stackrel{def}{=} q\{b/a\} \text{ if } C_\theta \stackrel{def}{=} q \end{cases}
\end{aligned}$$

The application of an admissible substitution $\theta = \{b/a\} \circ \theta'$ to a process p can be computed as follows: $p\theta = (p\{b/a\})\theta'$, with the proviso that $p\varepsilon = p$. □

Proposition 6.27. *For any TCSP process p, for any set $A = \{a_1, \ldots, a_n\} \subseteq n(p)$ with $n \geq 1$, and any set $B = \{b_1, \ldots, b_n\}$, $B \cap n(p) = \emptyset$, let $\{B/A\}$ be the admissible substitution $\{b_1/a_1, \ldots, b_n/a_n\}$. Then, the following hold:*

- $p \xrightarrow{a_i} p'$ *if and only if* $p\{B/A\} \xrightarrow{b_i} p'\{B/A\}$,
- $p \xrightarrow{c} p'$ *if and only if* $p\{B/A\} \xrightarrow{c} p'\{B/A\}$, *when* $c \neq a_i$ *for all* $i = 1, \ldots, n$.

Proof. *By induction on the proof of the involved transitions.* □

We are now ready to present the formal encoding $[\![-]\!]$ from TCSP to Multi-CCS^{hide}, i.e., Multi-CCS enriched with the hiding operator: as we have already studied in Section 5.2 how to encode the hiding operator within CCS, we can then use one of the two encodings presented there to map Multi-CCS^{hide} into Multi-CCS.

The encoding $[\![-]\!]$ is homomorphic for all operators but parallel composition,

$$\begin{aligned}
[\![\mathbf{0}]\!] &= \mathbf{0} & [\![\mu.p]\!] &= \mu.[\![p]\!] \\
[\![p_1 + p_2]\!] &= [\![p_1]\!] + [\![p_2]\!] & [\![(\iota a)p]\!] &= (\iota a)[\![p]\!] \\
[\![A]\!] &= A' \quad \text{where } A' \stackrel{def}{=} [\![q]\!] \text{ if } A \stackrel{def}{=} q,
\end{aligned}$$

while for parallel composition it is defined as

$$\begin{aligned}
[\![p_1 \parallel_\emptyset p_2]\!] &= [\![p_1]\!] \,|\, [\![p_2]\!] \\
[\![p_1 \parallel_A p_2]\!] &= (\nu B)([\![p_1\{B/A\}]\!] \,|\, [\![p_2\{B/A\}]\!] \,|\, C_{AB}),
\end{aligned}$$

where $A = \{a_1, \dots, a_n\}$ with $n \geq 1$, $B = \{b_1, \dots, b_n\}$, $B \cap n(p_1 \parallel_A p_2) = \emptyset$, $\{B/A\}$ is the admissible substitution $\{b_1/a_1, \dots, b_n/a_n\}$ and $C_{AB} \stackrel{def}{=} \sum_{a_i \in A} \underline{b_i}.\overline{b_i}.a_i.C_{AB}$.

Note that the encoding $[\![p]\!]$ of a TCSP process p is a well-formed Multi-CCS^{hide} process;[6] as a matter of fact, only the component C_{AB} may generate sequences; the form of such sequences is $\underline{b_i \overline{b_i} a_i}$ for $i = 1, \dots, n$, and so $[\![p]\!]$ satisfies the well-formedness condition, because b_i is a new name which never appears in a sequence, and a_i is an action whose co-action $\overline{a_i}$ is never used at all.

Lemma 6.14. *For any TCSP process p, for any set $A = \{a_1, \dots, a_n\} \subseteq n(p)$ with $n \geq 1$, and any set $B = \{b_1, \dots, b_n\}$, $B \cap n(p) = \emptyset$, let $\{B/A\}$ be the admissible substitution $\{b_1/a_1, \dots, b_n/a_n\}$. Then, the following hold:*

- $[\![p]\!] \xrightarrow{a_i} [\![p']\!]$ *if and only if* $[\![p\{B/A\}]\!] \xrightarrow{b_i} [\![p'\{B/A\}]\!]$,
- $[\![p]\!] \xrightarrow{c} [\![p']\!]$ *if and only if* $[\![p\{B/A\}]\!] \xrightarrow{c} [\![p'\{B/A\}]\!]$, *when $c \neq a_i$ for all $i = 1, \dots, n$.*

Proof. *By induction on the proof of the involved transitions.* □

Proposition 6.28. *For any TCSP process p, if $p \xrightarrow{\mu} p'$, then $[\![p]\!] \xrightarrow{\mu} [\![p']\!]$.*

Proof. *By induction on the proof of $p \xrightarrow{\mu} p'$. We proceed by case analysis.*

If $p = \mu.p'$, then $p \xrightarrow{\mu} p'$. Also $[\![p]\!] = \mu.[\![p']\!] \xrightarrow{\mu} [\![p']\!]$, as required.

If $p = p_1 + p_2$, then $p \xrightarrow{\mu} p'$ is possible only if either $p_1 \xrightarrow{\mu} p'$ or $p_2 \xrightarrow{\mu} p'$. By induction, we can assume that $[\![p_1]\!] \xrightarrow{\mu} [\![p']\!]$ or $[\![p_2]\!] \xrightarrow{\mu} [\![p']\!]$. Hence, $[\![p]\!] = [\![p_1]\!] + [\![p_2]\!] \xrightarrow{\mu} [\![p']\!]$, as required.

If $p = (\iota a)q$, then $p \xrightarrow{\mu} p'$ is possible only if either $\mu \neq a$, $q \xrightarrow{\mu} q'$ and $p' = (\iota)q'$, or $\mu = \tau$, $q \xrightarrow{a} q'$ and $p' = (\iota)q'$. In the former case, by induction, we can assume $[\![q]\!] \xrightarrow{\mu} [\![q']\!]$. Hence, $[\![p]\!] = (\iota a)[\![q]\!] \xrightarrow{\mu} (\iota a)[\![q']\!] = [\![p']\!]$, as required. Similarly, for the latter case.

If $p = p_1 \parallel_A p_2$, where the synchronization set A is $\{a_1, \dots, a_n\}$, then $p \xrightarrow{\mu} p'$ is possible only if $\mu \notin A$, $p_1 \xrightarrow{\mu} p_1'$ and $p' = p_1' \parallel_A p_2$; or $\mu \notin A$, $p_2 \xrightarrow{\mu} p_2'$ and $p' = p_1 \parallel_A p_2'$; or $\mu \in A$, $p_1 \xrightarrow{\mu} p_1'$, $p_2 \xrightarrow{\mu} p_2'$ and $p' = p_1' \parallel_A p_2'$. Let $B = \{b_1, \dots, b_n\}$ such that $B \cap n(p) = \emptyset$, and let $\{B/A\}$ be the admissible substitution $\{b_1/a_1, \dots, b_n/a_n\}$. In the first case, by induction, we can assume that $[\![p_1]\!] \xrightarrow{\mu} [\![p_1']\!]$. By Lemma 6.14, then also $[\![p_1\{B/A\}]\!] \xrightarrow{\mu} [\![p_1'\{B/A\}]\!]$, because $\mu \notin A$. Hence, $[\![p]\!] =$

$$(\nu B)([\![p_1\{B/A\}]\!] \,|\, [\![p_2\{B/A\}]\!] \,|\, C_{AB}) \xrightarrow{\mu} (\nu B)([\![p_1'\{B/A\}]\!] \,|\, [\![p_2\{B/A\}]\!] \,|\, C_{AB})$$

$= [\![p']\!]$, as required. The second case is symmetric, hence omitted. In the third case, assume that $\mu = a_i \in A$. By induction, we can assume that $[\![p_1]\!] \xrightarrow{a_i} [\![p_1']\!]$

[6] The well-formedness relation can be extended to the hiding operator in the obvious way: if $wf(p)$, then $wf((\iota a)p)$. Moreover, $ns((\iota a)p) = ns(p) \setminus \{a, \overline{a}\}$.

and $[\![p_2]\!] \xrightarrow{a_i} [\![p_2']\!]$. *By Lemma 6.14, then also* $[\![p_1\{B/A\}]\!] \xrightarrow{b_i} [\![p_1'\{B/A\}]\!]$ *as well as* $[\![p_2\{B/A\}]\!] \xrightarrow{b_i} [\![p_2'\{B/A\}]\!]$. *Hence,* $[\![p]\!] =$

$(\nu B)([\![p_1\{B/A\}]\!] \,|\, [\![p_2\{B/A\}]\!] \,|\, C_{AB}) \xrightarrow{a_i} (\nu B)([\![p_1'\{B/A\}]\!] \,|\, [\![p_2'\{B/A\}]\!] \,|\, C_{AB})$

$= [\![p']\!]$, *because of a ternary synchronization among the three parallel components which has the effect of renaming* b_i *back to* a_i.

The simpler case when $p = p_1 \,\|_\emptyset\, p_2$ *is left as an exercise for the reader.*

If $p = C$, *where* $C \stackrel{def}{=} q$, *then* $C \xrightarrow{\mu} q'$ *is possible only if* $q \xrightarrow{\mu} q'$. *By induction, we can assume that* $[\![q]\!] \xrightarrow{\mu} [\![q']\!]$. *As* $[\![C]\!] = C' \stackrel{def}{=} [\![q]\!]$, *we also have* $C' \xrightarrow{\mu} [\![q']\!]$, *as required.* □

Proposition 6.29. *For any TCSP process* p, *if* $[\![p]\!] \xrightarrow{\mu} q$, *then there exists* p' *such that* $q = [\![p']\!]$ *and* $p \xrightarrow{\mu} p'$.

Proof. *By induction on the proof of* $[\![p]\!] \xrightarrow{\mu} [\![p']\!]$. *The proof is very similar to that of Proposition 6.28 and is left as an exercise for the reader.* □

Theorem 6.6. (Correctness of the implementation, up to $\sim$) *For any TCSP process* p, $p \sim [\![p]\!]$.

Proof. *It is enough to check that relation* $R = \{(p, [\![p]\!]) \mid p \text{ is a TCSP process}\}$ *is a strong bisimulation. This is an easy exercise, given the results in Propositions 6.28 and 6.29.* □

6.5.4 Last Man Standing Problem

Even if Multi-CCS is a rather powerful language, we cannot expect to be able to solve all possible problems in concurrency. The Last Man Standing (LMS) problem, introduced in [VBG09], can be solved in some process calculus if there exists a process p able to detect the presence or absence of other copies of itself without generating deadlocks or introducing divergence. In order to check whether the LMS can be solved in Multi-CCS, we would need to identify a Multi-CCS process p such that p is able to execute an action a only when there is exactly one copy of p in the current system, while p is able to perform an action b only when there are at least two copies of p in the current system. In other words, if s_i is the system where i copies of p are enabled, we should have that

$$s_1 = p \qquad s_1 \xrightarrow{a} \qquad s_1 \stackrel{b}{\not\longrightarrow}$$

while

$$s_2 = p \,|\, p \qquad s_2 \stackrel{a}{\not\longrightarrow} \qquad s_2 \xrightarrow{b}$$

$$\ldots$$

$$s_n = \underbrace{p \,|\, p \,|\, \cdots \,|\, p}_{n} \qquad s_n \stackrel{a}{\not\longrightarrow} \qquad s_n \xrightarrow{b}$$

with $a \neq b$. Rule (Par$_1$) in Table 6.3 allows us to easily prove that the LMS cannot be solved in Multi-CCS. In fact this rule states that any process p, able to execute some action a, can perform the same action in the presence of other processes as well, so that if $p \xrightarrow{a} p'$, then also $p \mid p \xrightarrow{a} p \mid p'$, which contradicts the requirement that $s_2 \not\xrightarrow{a}$. As a matter of fact, Multi-CCS is *permissive*: no parallel process can prevent the execution of an action of another process. A process calculus where the LMS problem is solvable has to possess some further features, such as the capability to express priority among its actions (see, e.g., [VBG09] and the references therein).

The LMS problem has a relationship with a well-studied problem in the theory of finite Petri nets [Pet81]: the inability to test for exactly a specific marking in an unbounded place; in our case, the LMS requires that a transition labeled a be enabled only if exactly one token is present in the place for p, while a transition labeled b is enabled only if two or more tokens are present in the place for p. Often, the variant of this problem that is commonly studied in the theory of Petri nets is the so-called *zero testing*: a transition is enabled only if no tokens are present in a certain place. However, finite Petri nets are unable to test for a zero marking in an unbounded place. This specific limitation can be overcome by extending finite Petri nets to include inhibitor arcs (see, e.g, [Pet81, BG09]).

6.5.5 Conclusion

To conclude this section about expressiveness, let us recapitulate the various approaches and techniques we adopted in this book in order to compare the expressive power of different languages.

In Chapter 3 we have compared various subcalculi of CCS on the basis of their capabilities of expressing larger and larger families of languages: so the comparison among these languages was based on trace semantics and using typical tools of the theory of formal languages. Figure 3.6 (Section 3.4) gives the overall picture of the classification, while Figure 3.11 (Section 3.4.5) describes more precisely the relationship w.r.t the Chomsky hierarchy. Among the languages described there, only finitary CCS is Turing-complete.

In Chapter 5 we have done an analogous classification for the various languages based on ACP sequential composition. In Section 5.4.5 a precise description of the syntactic and semantic relationships among these languages is given in terms of the class of languages they can represent. The overall picture is described in Figure 5.5, where the only Turing-complete formalism is PAER.

These kinds of comparisons are rooted in classical results of the theory of formal languages. However, these results are of no use when comparing two languages that are both Turing-complete.

If we want to prove that two (Turing-complete) calculi are *equally expressive*, then one possibility is to define suitable encodings in both directions, preserving the same intended behavioral semantics. This is what we have done in Chapter 5, when extending CCS with some additional operator. For instance, in Section 5.4.6 we have shown that CCS^{seq}, i.e., the calculus obtained by enriching CCS with sequential composition, is as expressive as CCS; on the one hand, CCS^{seq} is a conservative extension of CCS; on the other hand, we have shown an encoding of CCS^{seq} into CCS, up to weak bisimilarity. Similarly, in Section 5.5 we have shown that $CCS_!$ (finite CCS with replication) and $CCS_?$ (finite CCS with duplication) are equally expressive, by providing an encoding, up to $\sim$, in each direction.

A technique for proving that a Turing-complete language is *more expressive than* another Turing-complete formalism can be based on the class of processes the two languages can represent, up to some behavioral equivalence. This is the technique we have adopted in Example 3.20 (Section 3.4.6) to prove that full CCS is more expressive than finitary CCS: on the one hand, finitary CCS is a subcalculus of full CCS, and so all the LTSs representable by finitary CCS are also representable by full CCS; on the other hand, there is a full CCS process with an infinite sort, and this cannot be trace equivalent to any finitary CCS process, as any finitary CCS process has a finite sort.

A different technique has been adopted in this chapter to compare CCS and Multi-CCS: on the one hand, we have proved that Multi-CCS is a conservative extension of CCS, so that also Multi-CCS is Turing-complete; on the other hand, we have singled out a well-known problem in the theory of concurrency, namely the dining philosophers, for which a symmetric, fully distributed, deadlock-free and divergence-free solution exists in Multi-CCS, while this is not the case for CCS, as proved in [LR81, RL94].

This technique has been adopted also by others in different contexts. For instance, [LV10] proves a strict expressiveness hierarchy among CCS-like calculi equipped with n-ary synchronizations: in this setting, CCS_2 denotes ordinary CCS with binary synchronization; such a language is strictly less expressive than CCS_3 (i.e., CCS which allows also for ternary synchronizations) and, in general, CCS_n is strictly more expressive than CCS_{n-1} for any $n \geq 3$. This was proved on a parametric generalization of the dining philosophers problem, called the dining philosophers problem in the n-hypercube, where the philosophers sit at the vertices of a hypercube of dimension n, forks are at the edges, and philosophers can grab forks at their adjacent edges only. Laneve and Vitale demonstrated that the problem has symmetric, fully distributed, deadlock-free solutions in CCS_n but not in CCS_{n-1}.

Another well-known problem in distributed computing that has been used to compare the expressive power of concurrent languages is the *leader-election problem* — originally formalized by Le Lann in [L77] — which consists in requiring that the members of a network be able to elect one of their members as their leader, i.e., to reach an agreement (on the leader) in a completely decentralized way. The problem can be formulated in different manners, depending on network topology, with different difficulties in getting the solution. By using a variation of this problem, Palamidessi [Pal03] observed, in the context of the π-calculus [MPW92], the superior expressiveness of *mixed choice* (i.e., different addends in a sum may start with actions of input or output, as in CCS) with respect to *separate choice* (all the addends start with the same type of action, either all inputs or all outputs): the relative expressive power of the two constructs is stated in terms of the impossibility of a reasonable encoding[7] of mixed choice into separate choice, as only the language with mixed choice can solve the leader-election problem for a fully connected, symmetric network in a decentralized manner. By the way, note that also CCS with separate choice is Turing-complete, as the CCS modeling of the Counter Machines in Section 3.5.2 is done with separate choice only. So, Palamidessi's result may be used to show that finitary CCS (with mixed choice) is more expressive than finitary CCS with separate choice. By following the same approach, many other separation results have been obtained, each one based on the capability or impossibility of solving the leader election under appropriate conditions. We refer you to [VPP07] for an overview.

Another problem we have used in this chapter is the *last man standing problem* (LMS, for short), for which no solution exists in Multi-CCS. Nonetheless, this problem is solvable in other calculi that offer different features, such as priority among actions. In particular, [VBG09] presents a finite (i.e., with no recursion, hence not Turing-complete) language, called FAP, which can solve the LMS problem; this means there exists a problem in concurrency (i.e., the LMS problem) that a Turing complete language (i.e., Multi-CCS) cannot solve, while it can be solved by a not Turing-complete language (i.e., FAP).

These observations spur an obvious question: when is a formalism for concurrency *complete*? And with respect to what? Unfortunately, we still miss a definitive answer to this philosophical question, even if recent research is trying to unveil some aspects of this problem.

---

7 According to [Pal03], an encoding $[\![-]\!]$ is reasonable if it preserves a number of properties, of which the most important is distribution-preserving: $[\![p \,|\, q]\!] = [\![p]\!] \,|\, [\![q]\!]$.

Glossary

Symbol	Description	Where
$\mathscr{L}$	set of input actions	Definition 2.1
$\overline{\mathscr{L}}$	set of co-actions (outputs)	Definition 2.1
Act	set of labels	Definition 2.1
$\xrightarrow{\mu}$	labeled transition in an LTS	Definition 2.2
$\xrightarrow{\sigma}^*$	reflexive-transitive closure of $\xrightarrow{\mu}$	Definition 2.4
$\to^*$	reachability	Definition 2.4
$\cong$	isomorphism	Definition 2.8
$=_{tr}$	trace equivalence	Definition 2.9
$\leq_{tr}$	trace preorder	Exercise 2.16
$=_{ctr}$	completed trace equivalence	Definition 2.11
$\lesssim$	simulation preorder	Definition 2.12
$\simeq$	simulation equivalence	Definition 2.12
$\lesssim_c$	completed simulation preorder	Definition 2.13
$\simeq_c$	completed simulation equivalence	Definition 2.13
$\lesssim_{rs}$	ready simulation preorder	Definition 2.33
$\simeq_{rs}$	ready simulation equivalence	Definition 2.33
$\sim$	bisimulation equivalence	Definition 2.14
$\sim^*$	string bisimulation equivalence	Exercise 2.41
$\stackrel{\sigma}{\Longrightarrow}$	labeled weak transition in an LTS	Definition 2.16
$=_{wtr}$	weak trace equivalence	Definition 2.17
$\leq_{wtr}$	weak trace preorder	Exercise 2.45
$=_{wctr}$	weak completed trace equivalence	Definition 2.18
$\lessapprox$	weak simulation preorder	Definition 2.19
$\approxeq$	weak simulation equivalence	Definition 2.19
$\lessapprox_c$	weak completed simulation preorder	Definition 2.60
$\approxeq_c$	weak completed simulation equivalence	Definition 2.60
$\approx$	weak bisimulation equivalence	Definition 2.20
$\approx^c$	rooted weak bisimulation equivalence	Definition 2.23
$\approx_{br}$	branching bisimulation equivalence	Definition 2.24
$\approx_{br}^c$	rooted branching bisimulation equivalence	Definition 2.26

Symbol	Description	Where
$\sim_i$	i-th approximation of $\sim$	Definition 2.28
$\sim_\omega$	limit of the approximations of $\sim$	Definition 2.28
$\lesssim_i$	i-th approximation of $\lesssim$	Definition 2.93
$\lesssim_\omega$	limit of the approximations of $\lesssim$	Definition 2.93
$\mathscr{C}ons$	set of CCS process constants	Section 3.1.2
$Const(p)$	set of constants used by p	Definition 3.1
$\mathscr{P}$	set of CCS processes	Definition 3.4
$\mathscr{C}$	the CCS labeled transition system	Definition 3.5
$\mathscr{C}_p$	reachable sub-LTS from p	Remark 3.6
$\mathscr{P}_{fin}$	set of finite CCS processes	Section 3.4.1
$=_{trf}$	f-trace equivalence	Definition 5.2
$\sim_f$	f-bisimulation equivalence	Definition 5.6
$\approx_f$	weak f-bisimulation equivalence	Definition 5.7
$\approx_f^c$	rooted weak f-bisimilarity	Definition 5.8
$\cong_f$	f-isomorphism	Definition 5.9
$\mathscr{P}_{finBPA}$	set of finite BPA processes	Section 5.4.2
$\mathscr{P}_{BPA^*}$	set of BPA* processes	Section 5.4.3
$\mathscr{P}_{BPA}$	set of BPA processes	Section 5.4.4
$\mathscr{P}_{PA}$	set of PA processes	Section 5.4.5
$\mathscr{P}_{PAER}$	set of PAER processes	Section 5.4.5
$Sync$	synchronization relation for Multi-CCS	Section 6.1.3
$\mathscr{P}_M$	set of Multi-CCS processes	Section 6.2
$\equiv$	structural congruence for Multi-CCS	Section 6.2
$wf(p)$	well-formed Multi-CCS process	Section 6.2.2
$swf(p)$	strong well-formed Multi-CCS process	Section 6.2.2
$MSync$	step synchronization relation	Section 6.3.2
$\sim_{step}$	step bisimilarity	Section 6.3.3

References

[Ace94] L. Aceto, On "Axiomatising finite concurrent processes", *SIAM J. Comput.* 23(4): 852-863, 1994.

[AF+14] L. Aceto, D. de Frutos Escrig, C. Gregorio-Rodriguez, A. Ingólfsdóttir, "Axiomatising weak simulation semantics over BCCSP", *Theoretical Computer Science* 537:42-71, 2014.

[Age75] T. Agerwala, "Towards a theory for the analysis and synthesis of systems exhibiting concurrency", Dissertation, John Hopkins University, 1975.

[AH92] S. Arun-Kumar, M. Hennessy, "An efficiency preorder for processes", *Acta Informatica* 29(9):737-760, 1992.

[AILS07] L. Aceto, A. Ingólfsdóttir, K. Larsen, J. Srba, *Reactive Systems: Modelling, Specification and Verification*, Cambridge University Press, 2007.

[AIS12] L. Aceto, A. Ingólfsdóttir, J. Srba, "The algorithmics of bisimilarity", Chapter XX in [RuS12].

[B84] H.P. Barendregt, *The Lambda Calculus: Its Syntax and Semantics*, Studies in Logic and the Foundations of Mathematics 103 (Revised ed.), North Holland, Amsterdam, 1984.

[Bae05] J.C.M. Baeten, "A brief history of process algebra", *Theoretical Computer Science* 51(1/2):129-176, 2005.

[BB08] J.C.M. Baeten, M. Bravetti, "A ground-complete axiomatization of finite state processes in a generic process algebra", *Mathematical Structures in Computer Science* 18(6):1057-1089, 2008.

[BBK87] J.C.M. Baeten, J.A. Bergstra, J.W. Klop, "Conditional axioms and α/β-calculus in process algebra", in Procs. IFIP *Formal Description of Programming Concepts* III, 53-75, North-Holland, 1987.

[BBK93] J.C.M. Baeten, J.A. Bergstra, J.W. Klop, "Decidability of bisimulation equivalence for processes generating context-free languages", *Journal of the ACM* 40(3):653-682, 1993.

[BBR10] J.C.M. Baeten, T. Basten, M.A. Reniers, *Process Algebra: Equational Theories of Communicating Processes*, Cambridge Tracts in Theoretical Computer Science 50, Cambridge University Press, 2010.

[BCG07] J.C.M. Baeten, F. Corradini, C.A. Grabmayer, "A characterization of regular expressions under bisimulation", *Journal of the ACM* 54(2:6),1-28, 2007.

[BCMS01] O. Burkart, D. Caucal, F. Moller, B. Steffen, "Verification on infinite structures", Chapter 9 of [BPS01], 545-623, 2001.

[BCT08] J.C.M. Baeten, P.J. L. Cuijpers, P. J. A. van Tilburg, "A context-free process as a pushdown automaton", in Procs. CONCUR 2008, LNCS 5201, 98-113, Springer-Verlag, 2008.

[BFP01] J.A. Bergstra, W.J. Fokkink, A. Ponse, "Process algebra with recursive operations", Chapter 5 of [BPS01], 333-389, 2001.

[BG98] M. Bernardo, R. Gorrieri, "A tutorial on EMPA: A theory of concurrent processes with nondeterminism, priorities, probabilities and time", *Theoretical Computer Science* 202(1-2):1-54, 1998.

[BG09] N. Busi, R. Gorrieri, "Distributed semantics for the π-calculus based on Petri nets with inhibitor arcs", *Journal of Logic and Algebraic Programming* 78(3):138-162, 2009.

[BGZ09] N. Busi, M. Gabbrielli, G. Zavattaro, "On the expressive power of recursion, replication and iteration in process calculi", *Mathematical Structures in Computer Science* 19(6):1191-1222, 2009.

[BJ66] C. Böhm, G. Jacopini, "Flow diagrams, Turing machines and languages with only two formation rules", *Communications of the ACM* 9(5):366-371, 1966.

[BK84a] J.A. Bergstra, J.W. Klop, "Process algebra for synchronous communication", *Information and Control* 60:109-137, 1984.

[BK84b] J.A. Bergstra, J.W. Klop, "The algebra of recursively defined processes and the algebra of regular processes", in Procs. ICALP'84, LNCS 172, 82-95, Springer-Verlag, 1984.

[BK85] J.A. Bergstra, J.W. Klop, "Algebra of communicating processes with abstraction", *Theoretical Computer Science* 37(1):77-121, 1985.

[BKL83] L. Babai, W.M. Kantor, E.M. Luks, "Computational complexity and the classification of finite simple groups", in Procs. 24th Annual Symposium of Foundations of Computer Science (FOCS), 162-171, IEEE CS-Press, 1983.

[BIM95] B. Bloom, S. Istrail, A.R. Meyer, "Bisimulation can't be traced", *Journal of the ACM* 42(1):232-268, 1995.

[BLT12] J.C.M. Baeten, B. Luttik, P. van Tilburg, "Turing meets Milner", in Procs CONCUR 2012, LNCS 7454, 1-20, Springer-Verlag, 2012.

[BLV95] T. Bolognesi, J. van de Lagemaat, C. Vissers (eds.), *LOTOSphere: Software Development with LOTOS*, Springer-Verlag, 1995.

[BoBr87] T. Bolognesi, E. Brinksma, "Introduction to the ISO specification language LOTOS", *Computer Networks* 14: 25-59, 1987.

[BPS61] Y. Bar-Hillel, M. Perles, E. Shamir, "On formal properties of simple phrase-structure grammars", *Z. Phonetik. Sprachwiss. Kommunications-forsch.* 14:143-172, 1961.

[BPS01] J.A. Bergstra, A. Ponse, S.A. Smolka (eds.), *Handbook of Process Algebra*, Elsevier, 2001.

[BW90] J.C.M. Baeten, W.P. Weijland, *Process Algebra*, Cambridge Tracts in Theoretical Computer Science 18, CUP, 1990.

[Ch93] S. Christensen, "Decidability and decomposition in process algebra", Ph.D. Thesis, University of Edinburgh, 1993.

[CHL11] W. Czerwinski, P. Hofman, S. Lasota, "Decidability of branching bisimulation on normed commutative context-free processes", in Procs. CONCUR 2011, LNCS 6901, 528-542, Springer-Verlag, 2011.

[CHM93] S. Christensen, Y. Hirshfeld, F. Moller, "Bisimulation equivalence is decidable for Basic Parallel Processes", in Proc. CONCUR'93, LNCS 715, 143-157, Springe-Verlag, 1993.

[CHS95] S. Christensen, H. Hüttel, C. Stirling, "Bisimulation equivalence is decidable for all context-free processes", *Information and Computation* 12(2):143-148, 1995.

[CHP71] P. Courtois, F. Heymans, D. Parnas, "Concurrent control with readers and writers", *Communications of the ACM* 14(10):667-668, 1971.

[ClMe03] F.W. Clocksin, C.S. Mellish, *Programming in Prolog*, Springer-Verlag, 2003.

[CLN01] R. Cleaveland, G. Lüttgen, V. Natarajan, "Priority in process algebra", Chapter 12 of [BPS01], 711-765, 2001.

[CPS93] R. Cleaveland, J. Parrow, B. Steffen, "The Concurrency Workbench: A semantics-based tool for the verification of concurrent systems", *ACM Trans. Program. Lang. Syst.* 15(1):36-72, 1993.

[CRT11] S. Crafa, F. Ranzato, F. Tapparo, "Saving space in a time efficient simulation algorithm", *Fundamenta Informaticae* 108(1-2): 23-42, 2011.

[CWB] P. Stevens et al., *The Edinburgh Concurrency Workbench*. Downloadable code and manual available at `http://homepages.inf.ed.ac.uk/perdita/cwb/`

[DesRei98] J. Desel, W. Reisig, "Place/Transition Petri Nets", in [RR98], 122-173, 1998.

[Dav58] M. Davis, *Computability and Unsolvability*. McGraw-Hill, New York, 1958.

[DDM88] P. Degano, R. De Nicola, U. Montanari, "A distributed operational semantics for CCS based on C/E systems", *Acta Informatica* 26(1-2):59-91, 1988.

[DG91] P. Degano, R. Gorrieri, "Atomic refinement for process description languages", in Procs. *Mathematical Foundations of Computer Science* (MFCS'91), LNCS 520, 121-130, Springer-Verlag, 1991.

[Den07] Y. Deng, "A simple completeness proof for the axiomatisations of weak behavioural equivalences", *Bulletin of the EATCS*, 93:207-219, 2007. Full version available from `http://basics.sjtu.edu.cn/~yuxin/publications/branch.ps`

[Dij68] E.W. Dijkstra, "Cooperating sequential processes", (F. Genuys ed.) *Programming Languages*, Academic Press, 43-112, 1968.

[Dij69] E.W. Dijkstra, "Structured programming", in *Software Engineering Techniques* (J.N. Buxton and B. Randell, eds), Report on a conference sponsored by the NATO Science Committee, pages 84-88, 1969. Available from
`http://homepages.cs.ncl.ac.uk/brian.randell/NATO/nato1969.PDF`

[Dij71] E.W. Dijkstra, "Hierarchical ordering of sequential processes", *Acta Informatica* 1(2): 115-138, 1971.

[DP02] B.A. Davey, H.A. Priestley, *Introduction to Lattices and Order* (second edition), Cambridge University Press, 2002.

[DPP01] A. Dovier, C. Piazza, A. Policriti, "A fast bisimulation algorithm", in Proc. 13th Conf. on Computer-Aided Verification (CAV'01), LNCS 2102, 79-90, Springer-Verlag, 2001.

[EM94] J. Esparza, M. Nielsen, "Decidability issues for Petri nets: a survey", *Bulletin of the EATCS* 52:244-262, 1994.

[Esp98] J. Esparza, "Decidability and complexity of Petri net problems: an introduction" in [RR98], 374-428, 1998.

[ER64] C. Elgot, A. Robinson, "Random-access stored-program machines, an approach to programming languages", *Journal of the ACM* 11(4):365-399, 1964.

[FB94] R.W. Floyd, R. Beigel, *The Languages of Machines*, Computer Science Press, 1994.

[FG01] R. Focardi, R. Gorrieri, "Classification of security properties – Part I: Information flow", *Foundations of Security Analysis and Design*, FOSAD 2000, LNCS 2171, 331-396, Springer-Verlag, 2001.

[FGM02] R. Focardi, R. Gorrieri, F. Martinelli, "Classification of security properties – Part II: Network security", *Foundations of Security Analysis and Design*, FOSAD 2002, LNCS 2946, 139-185, Springer-Verlag, 2002.

[Flo62] R.W. Floyd, "Algorithm 97: Shortest Path", *Comm. of the ACM* 5 (6):345, 1962.

[Fok00] W. Fokkink, *Introduction to Process Algebra*, EATCS Texts in Theoretical Computer Science, Springer-Verlag, 2000.

[FR80] N. Francez, M. Rodeh, "A distributed abstract data type implemented by a probabilistic communication scheme", In Procs. FOCS'80, pages 373-379. IEEE Press, 1980.

[FV99] K. Fisler, M.Y. Vardi, "Bisimulation and model checking", in Procs. CHARME'99, LNCS 1703, 338-341, Springer-Verlag, 1999.

[vGl01] R.J. van Glabbeek, "The linear time - branching time spectrum I", Chapter 1 of [BPS01], 3-99, 2001.

[vGl93] R.J. van Glabbeek, "The linear time - branching time spectrum II", in Procs. CONCUR '93, LNCS 715, 66-81, Springer-Verlag, 1993.

[vGl05] R.J. van Glabbeek, "A characterisation of weak bisimulation congruence", in *Processes, Terms and Cycles: Steps on the Road to Infinity*, Essays dedicated to Jan Willem Klop, on the occasion of his 60th birthday, LNCS 3838, 26-39. Springer-Verlag, 2005.

[vGP08] R.J. van Glabbeek, B. Ploeger, "Correcting a space-efficient simulation algorithm", in Procs. Computer Aided Verification (CAV'08), LNCS 5123: 517-529, Springer-Verlag, 2008.

[vGW96] R.J. van Glabbeek, W.P. Weijland, "Branching time and abstraction in bisimulation semantics", *Journal of the ACM* 43(3):555-600, 1996.

[Gol90] U. Goltz, "CCS and Petri nets", LNCS 469, 334-357, Springer-Verlag, 1990.

[GMM90] R. Gorrieri, S. Marchetti, U. Montanari, "A^2CCS: Atomic Actions for CCS", *Theoretical Computer Science* 72(2-3):203-223, 1990.

[Gor15] R. Gorrieri, "The language of finite P/T Petri nets", submitted for publication.

[GPP03] R. Gentilini, C. Piazza, A. Policriti, "From bisimulation to simulation: Coarsest partition problems", *Journal of Automated Reasoning* 31(1):73-103, 2003.

[GR01] R. Gorrieri, A. Rensink, "Action refinement", Chapter 16 of [BPS01], 1047-1147, 2001.

[GV10] R. Gorrieri, C. Versari, "A Process calculus for expressing finite Place/Transition Petri nets", in Procs EXPRESS'10, EPTCS 41, 76-90, 2010.

[H02] H. Hermanns, *Interactive Markov Chains: The Quest for Quantified Quality*, LNCS 2428, Springer-Verlag, 2002.

[Hil96] J. Hillston, *A Compositional Approach to Performance Modelling*, BCS Distinguished Dissertation, Cambridge University Press, 1996.

[Hir93] Y. Hirshfeld, "Petri nets and the equivalence problem", in Procs. 7th Workshop on *Computer Science Logic* (CSL'93), LNCS 832,165-174, Springer-Verlag, 1993.

[HJ99] Y. Hirshfeld, M. Jerrum, "Bisimulation equivalence is decidable for normed process algebra", In Procs of 26th Int. Colloquium on Automata, Languages and Programming (ICALP'99), LNCS 1644, 412-421. Springer-Verlag, 1999.

[HJM96] Y. Hirshfeld, M. Jerrum, F. Moller, "A polynomial algorithm for deciding bisimilarity of normed context-free processes", *Theoretical Computer Science* 158(1/2):143-159, 1996.

[Hoa78] C.A.R. Hoare, "Communicating sequential processes", *Communications of the ACM* 21(8):666-677, 1978.

[Hoa85] C.A.R. Hoare, *Communicating Sequential Processes*, Prentice-Hall International Series in Computer Science, 1985.

[HMU01] J.E. Hopcroft, R. Motwani, J.D. Ullman, *Introduction to Automata Theory, Languages and Computation*, 2nd ed., Addison-Wesley, 2001.

[HR95] M. Hennessy, T. Regan, "A process algebra for timed systems", *Information and Computation* 117:221-239, 1995.

[HRS76] H.B. Hunt, D.J. Rosenkrantz, T.G. Szymanski, "On the equivalence, containment, and covering problems for the regular and context-free languages", *Journal of Computer and System Sciences* 12:222-268, 1976.

[HU79] J.E. Hopcroft, J.D. Ullman, *Introduction to Automata Theory, Languages and Computation*, Addison-Wesley, 1979.

[Jan95] P. Jančar, "Undecidability of bisimilarity for Petri nets and some related problems", *Theoretical Computer Science* 148(2):281-301, 1995.

[Jan03] P. Jančar, "Strong bisimilarity on basic parallel processes is PSPACE-complete", in Procs of the 18th Annual IEEE Symposium on Logic in Computer Science (LICS'03), 218-227, IEEE Computer Society Press, 2003.

[Jan13] P. Jančar, "Bisimilarity on basic process algebra is in 2-exptime (an explicit proof)", *Logical Methods in Computer Science* 9(1:10), 1-19, 2013.

[JE96] P. Jančar, J. Esparza, "Deciding finiteness of Petri nets up to bisimulation", In Proceedings of 23rd International Colloquium on Automata, Languages, and Programming (ICALP'96), LNCS 1099, 478-489, Springer-Verlag, 1996.

[JKM01] P. Jančar, A. Kučera, and R. Mayr, "Deciding bisimulation-like equivalences with finite-state processes", *Theoretical Computer Science* 258(1-2):409-433, 2001.

[JM95] P. Jančar, F. Moller, "Checking regular properties of Petri nets", In Proceedings of the 6th International Conference on Concurrency Theory (CONCUR'95), LNCS 962, 348-362, Springer-Verlag, 1995.

[JW+91] K. Jensen, N. Wirth, A.B. Mickel, J.F. Miner, *Pascal User Manual and Report: ISO Pascal Standard* (third edition), Springer-Verlag, 1991.

[Kel76] R. Keller, "Formal verification of parallel programs", *Comm. of the ACM* 19(7):561-572, 1976.

[KM02] A. Kučera, R. Mayr, "Weak bisimilarity between finite-state systems and BPA or normed BPP is decidable in polynomial time", *Theor. Comput. Sci.* 270(1-2): 677-700, 2002.

[Kos82] S.R. Kosaraju, "Decidability of reachability in vector addition systems", in Procs. 6th ACM STOC, 267-281, ACM Press, 1982.

[Koz97] D.C. Kozen, *Automata and Computability*, Undergraduate Texts in Computer Science, Springer-Verlag, 1997.

[L77] Le Lann, G. "Distributed systems: Towards a formal approach", in Procs. IFIP Congress,155-160, North-Holland, 1977.

[Ler11] J. Leroux, "Vector addition system reachability problem: a short self-contained proof", in Proc. 38th Symposium on Principles of Programming Languages (POPL'11), 307-316, ACM Press, 2011.

[LR81] D.J. Lehmann, M.O. Rabin, "On the advantages of free choice: A symmetric and fully distributed solution to the dining philosophers problem", In Procs. POPL'81, 133-138, ACM Press, 1981. A revised version appeared as [RL94].

[LS91] K.G. Larsen, A. Skou, "Bisimulation through probabilistic testing", *Information and Computation* 94(1):1-28, 1991.

[LV10] C. Laneve, A. Vitale, "The expressive power of synchronizations", in Procs LICS'10, IEEE-CS Press, 382-391, 2010.

[May81] E. W. Mayr, "An algorithm for the general Petri net reachability problem", in Proc. 13th Annual ACM Symp. on Theory of Comp., (STOC'81), 238-246, ACM Press, 1981.

[May84] E.W. Mayr, "An algorithm for the general Petri net reachability problem", *SIAM J. Comput.* 13:441-460, 1984.

[May00] R. Mayr, "Process rewrite systems", *Information and Comp.* 156(1-2): 264-286, 2000.

[Mil80] R. Milner, *A Calculus of Communicating Systems*, Lecture Notes in Computer Science 92, Springer-Verlag, 1980.

[Mil84] R. Milner. "A complete inference systems for a class of regular behaviors", *J. Comput. System Sci.* 28: 439-466, 1984.

[Mil85] G.J. Milne, "CIRCAL and the representation of communication, concurrency, and time", *ACM Trans. Program. Lang. Syst.* 7(2): 270-298, 1985.

[Mil89] R. Milner, *Communication and Concurrency*, Prentice-Hall, 1989.

[Mil89b] R. Milner. "A complete axiomatisation for observational congruence of finite-state behaviors", *Inf. Comput.* 81(2): 227-247, 1989.

[Mil99] R. Milner. *Communicating and Mobile Systems: The π-calculus*, Cambridge University Press, 1999.

[Min67] M.L. Minsky, *Computation: Finite and Infinite Machines*, Prentice-Hall, Upper Saddle River, NJ, USA, 1967.

[MPW92] R. Milner, J. Parrow, D. Walker, "A calculus of mobile processes", *Information and Computation* 100(1), 1-77, 1992.

[Mol90a] F. Moller, "The importance of the left merge operator in process algebras", in Procs. ICALP'90, LNCS 443, 752-764, Springer-Verlag, 1990.

[Mol90b] F. Moller, "The nonexistence of finite axiomatisations for CCS congruences", In Procs. IEEE Symp. on Logic in Comp. Scie. (LICS'90), 142-153, IEEE CS-Press, 1990.

[NP96] U. Nestmann, B.C. Pierce. "Decoding choice encodings", In Procs. of 7th International Conference on Concurrency Theory (CONCUR'96), LNCS 1119,179-194, Springer-Verlag, 1996.

[NS94] X. Nicollin, J. Sifakis, "The algebra of timed processes ATP: Theory and application", *Information and Computation* 114:131-178, 1994.

[NT84] M. Nielsen, P.S. Thiagarajan, "Degrees of non-determinism and concurrency: A Petri net view", in Procs. of the Fourth Conference on Foundations of Software Technology and Theoretical Computer Science (FSTTCS'84), LNCS 181, 89-117, Springer-Verlag, 1984.

[Old91] E.R. Olderog, *Nets, Terms and Formulas*, Cambridge Tracts in Theoretical Computer Science 23, Cambridge University Press, 1991.

[Pal03] C. Palamidessi, "Comparing the expressive power of the synchronous and the asynchronous π-calculi", *Mathematical Structures in Computer Science*, 13(5):685-719, 2003.

[Par01] J. Parrow, "An introduction to the π-calculus", Chapter 8 of [BPS01], 479-543, 2001.

[Park81] D.M.R. Park, "Concurrency and automata on infinite sequences", In Proc. 5th GI-Conference on Theoretical Computer Science, LNCS 104, 167-183, Springer-Verlag, 1981.

[Pat71] S. Patil, "Limitations and capabilities of Dijkstra's semaphore primitives for coordination among processes", Computation Structures Group Memo 57, Project MAC, MIT, 1971.

[Petri62] C.A. Petri, *Kommunikation mit Automaten*, Ph.D. Dissertation, University of Bonn, 1962.

[Pet81] J.L. Peterson, *Petri Net Theory and the Modeling of Systems*, Prentice-Hall, 1981.

[Plo04a] G.D. Plotkin "The origins of structural operational semantics", *J. Logic and Algebraic Programming* 60-61: 3-15, 2004.

[Plo04b] G.D. Plotkin "A structural approach to operational semantics", *J. Logic and Algebraic Programming* 60-61: 17-139, 2004. Revised version of the original Technical Report DAIMI FN-19, Aarhus University, 1981.

[PT87] R. Paige, R.E. Tarjan, "Three partition refinement algorithms", *SIAM Journal of Computing* 16(6):973-989, 1987.

[RL94] M. O. Rabin, D. Lehmann, "The advantages of free choice: A symmetric and fully distributed solution to the dining philosophers problem", in: A. W. Roscoe (Ed.), *A Classical Mind: Essays in Honour of C.A.R. Hoare*, Prentice Hall, 1994, Chapter 20, 333-352. An extended abstract appeared as [LR81].

[RS59] M.O. Rabin, D. Scott, "Finite automata and their decision problems", *IBM Journal of Research and Development* 3(2):114-125, 1959.

[Rei85] W. Reisig, *Petri Nets: An Introduction*, EATCS Monographs on Theoretical Computer Science, Springer-Verlag, 1985.

[Rei98] W. Reisig, *Elements of Distributed Algorithms: Modeling and Analysis with Petri Nets*, Springer-Verlag, 1998.

[Ros98] A.W. Roscoe, *The Theory and Practice of Concurrency*, Prentice-Hall, 1998.

[RSG+] P. Ryan, S. Schneider, M. Goldsmith, G. Lowe, B. Roscoe, *Modelling and Analysis of Security Protocols*, Addison-Wesley, 2000.

[RR98] W. Reisig, G. Rozenberg (eds.), *Lectures on Petri Nets I: Basic Models*, Lecture Notes in Computer Science 1491, Springer-Verlag, 1998.

[RT07] F. Ranzato, F. Tapparo, "A new efficient simulation equivalence algorithm", in Proc. 22nd Annual IEEE Symp. on Logic in Comp. Scie. (LICS'07), IEEE CS Press, 171-180, 2007.

[RuS12] J. Rutten, D. Sangiorgi (eds.) *Bisimulation and Coinduction: Advanced Topics*, Cambridge University Press, 2012.

[San12] D. Sangiorgi, *An Introduction to Bisimulation and Coinduction*, Cambridge University Press, 2012.

[Sch72] R. Schroeppel, "A two counter machine cannot calculate 2^N", Massachusetts Institute of Technology, A.I. Laboratory, Artificial Intelligence Memo 257, 1972.

[Sip06] M. Sipser, *Introduction to the Theory of Computation* (second edition), Course Technology, 2006.

[SM73] L.J. Stockmeyer, A.R. Meyer, "Word problems requiring exponential time", In Procs. 5th Annual ACM Symposium on Theory of Computing (STOC'73), 1-9, ACM Press, 1973.

[SM92] D. Sangiorgi, R. Milner, "The problem of weak bisimulation up to", In Procs. CONCUR'92, LNCS 630, 32-46, Springer-Verlag, 1992.

[SW01] D. Sangiorgi, D. Walker, *The π-calculus: A Theory of Mobile Processes*, Cambridge University Press, 2001.

[Sr02] J. Srba, "Undecidability of weak bisimilarity for PA-processes", In Procs. 6th Int. Conf. on Developments in Language Theory (DLT'02), LNCS 2450, 197-208. Springer-Verlag, 2003.

[Srba] J. Srba, "Roadmap of Infinity Results", *Bulletin of the EATCS* 78: 163-175, 2002. Continuously updated and revised version (2013) available at `http://people.cs.aau.dk/~srba/roadmap/roadmap.pdf`

[Tau89] D. Taubner, *Finite Representations of CCS and TCSP Programs by Automata and Petri Nets*, Lecture Notes in Computer Science 369, Springer-Verlag, 1989.

[Tur36] A.M. Turing, "On computable numbers, with an application to the Entscheidungsproblem". Proceedings of the London Mathematical Society, Series 2, 42, 230-265, 1936.

[VBG09] C. Versari, N. Busi, R. Gorrieri. "An expressiveness study of priority in process calculi", *Mathematical Structures in Computer Science* 19(6):1161-1189, 2009.

[VPP07] M. Vigliotti, I. Phillips and C. Palamidessi, "Tutorial on separation results in process calculi via leader election problems", *Theoretical Computer Science* 388(1-3), 267?289, 2007.

[Yi91] Wang Yi, "CCS + time = an interleaving model for real-time systems", in Procs ICALP'91, LNCS 510, 217-228, Springer-Verlag, 1991.

Index

Printed in the United States
By Bookmasters